6G: Evolution or Revolution?

Other related titles:

You may also like

- PBTE109 | Ahmadi | Digital Twins for 6G: Fundamental theory, technology and applications | 2024
- PBTE107 | Duong | Physical Layer Security for 6G Networks | 2024
- PBTE108 | Imran | The Role of 6G and Beyond on the Road to Net-Zero Carbon | 2023
- PBSE021 | Imoize | Security and Privacy Schemes for Dense 6G Wireless Communication Networks | 2023

We also publish a wide range of books on the following topics:
Computing and Networks
Control, Robotics and Sensors
Electrical Regulations
Electromagnetics and Radar
Energy Engineering
Healthcare Technologies
History and Management of Technology
IET Codes and Guidance
Materials, Circuits and Devices
Model Forms
Nanomaterials and Nanotechnologies
Optics, Photonics and Lasers
Production, Design and Manufacturing
Security
Telecommunications
Transportation

All books are available in print via https://shop.theiet.org or as eBooks via our Digital Library https://digital-library.theiet.org.

IET TELECOMMUNICATIONS SERIES 111

6G: Evolution or Revolution?

A converged view of cellular, Wi-Fi, computing and communication

Peter Smyth, Peter Willis and David Wisely

The Institution of Engineering and Technology

About the IET

This book is published by the Institution of Engineering and Technology (The IET).

We inspire, inform and influence the global engineering community to engineer a better world. As a diverse home across engineering and technology, we share knowledge that helps make better sense of the world, to accelerate innovation and solve the global challenges that matter.

The IET is a not-for-profit organisation. The surplus we make from our books is used to support activities and products for the engineering community and promote the positive role of science, engineering and technology in the world. This includes education resources and outreach, scholarships and awards, events and courses, publications, professional development and mentoring, and advocacy to governments.

To discover more about the IET please visit https://www.theiet.org/.

About IET Books

The IET publishes books across many engineering and technology disciplines. Our authors and editors offer fresh perspectives from universities and industry. Within our subject areas, we have several book series steered by editorial boards made up of leading subject experts.

We peer review each book at the proposal stage to ensure the quality and relevance of our publications.

Get involved

If you are interested in becoming an author, editor, series advisor, or peer reviewer please visit https://www.theiet.org/publishing/publishing-with-iet-books/ or contact author_support@theiet.org.

Discovering our electronic content

All of our books are available online via the IET's Digital Library. Our Digital Library is the home of technical documents, eBooks, conference publications, real-life case studies and journal articles. To find out more, please visit https://digital-library.theiet.org.

In collaboration with the United Nations and the International Publishers Association, the IET is a Signatory member of the SDG Publishers Compact. The Compact aims to accelerate progress to achieve the Sustainable Development Goals (SDGs) by 2030. Signatories aspire to develop sustainable practices and act as champions of the SDGs during the Decade of Action (2020–2030), publishing books and journals that will help inform, develop, and inspire action in that direction.

In line with our sustainable goals, our UK printing partner has FSC accreditation, which is reducing our environmental impact to the planet. We use a print-on-demand model to further reduce our carbon footprint.

Published by The Institution of Engineering and Technology, London, United Kingdom

The Institution of Engineering and Technology (the "**Publisher**") is registered as a Charity in England & Wales (no. 211014) and Scotland (no. SC038698).

First published 2024

The Institution of Engineering and Technology
Futures Place
Kings Way, Stevenage
Herts, SG1 2UA
United Kingdom

www.theiet.org

British Library Cataloguing in Publication Data
A catalogue record for this product is available from the British Library

ISBN 978-1-83953-840-7 (hardback)
ISBN 978-1-83953-841-4 (PDF)

Typeset in India by MPS Limited

Cover image: Chadchai Ra-ngubpai/ Moment via Getty Images

Contents

Foreword

'6G Evolution or Revolution?' is an important question every stakeholder is asking nowadays in the light of the huge ongoing investment in 5G and the future upcoming 5G-advanced.

This important question is thoroughly answered in the book in many chapters from different system aspects and brought together in the final chapter, where the answer is provided.

I am pleased to see such a book written by people with over 100 years of experience working in leading telecommunications research organisations and sharing their experience and expertise gained from the second generation of mobile/wireless communications, Wi-Fi and satellite communications.

Their collective experience is end-to-end of mobile systems from wireless to core to service platform and equally important economic aspects.

The book provides some commonly asked questions that are raised every time work on a new mobile generation starts. These are legitimate questions from technical and non-technical people. The answers provided are logical and honest and are based on the authors' in-depth knowledge of different generations of mobile communications and wireless systems from technology, standardisation, economy and deployment points of view.

It gives thorough answers to the important question of whether 6G is an evolution from 5G or a complete revolution, or a hybrid building on 5G evolution with some revolutionary 6G concepts. The book also provides a good reference for views of industry and ITU on the road to 6G or International Mobile Telecommunications (IMT) 2030 and the key requirements and capabilities that 6G needs to provide in meeting societal, environmental and economic challenges.

The book is important from many aspects, one of which is the fact that mobile/wireless telecom networks underpin many Critical National Infrastructures (CNI) and affects national security and a country's economy. It explores opportunities for a wider 6G role of convergence across information and communications technologies. 6G convergence provides the essential technologies for the digitisation and modernisation of many industries and, hence, a country's productivity and GDP.

The book is a good source of information for students, researchers, managers and decision-makers in industry and governments as it provides a holistic overview of the past, present, and what needs to come in future in the timescale of 2030+.

In addition to new capabilities, such as integrating sensing with communications that will make future networks more intelligent, new opportunities are unleashed for smart and individualised services that will make 6G more successful for users and mobile network operators.

Universal coverage of broadband communication is required to overcome the digital divide through the integration and convergence of space and terrestrial networks. This is probably the most economical way of addressing this 40-year-old problem since 1G. I am particularly pleased to see the 'inside to out' strategy of developing 6G technologies for terrestrial coverage. This is a fundamental shift from the past thinking of 'outside to in', where the starting point was outdoor cellular networks for indoor coverage. It is a sensible approach, as it is commonly understood and reported that more than 80% of traffic originates from indoors.

The book is unique, suggesting many novel ideas that could impact 6G. The main ones are spectrum solutions for intelligent coverage optimisation, wide area deployment spectrum sharing, new approaches to the network core and intelligent convergence between different networks, including Wi-Fi, Mobile Edge Computing (MEC), digital twins, core networks and hyperscalers.

Chapter 16, 'What 6G could be?', pulls together the whole book, exploring both evolution and what revolution could bring on top of this with the above themes.

The book also addresses revolution versus evolution from many aspects, including the economic viability of each approach.

Regius Professor Rahim Tafazolli, FREng, FIET, Fellow of WWRF
Director and founder of 5G and 6GIC
Institute for Communication Systems
The University of Surrey

Acknowledgements

This book provides an overview of the current state of 6G technology, examining both evolutionary and revolutionary scenarios. Our analysis draws on decades of experience in mobile communications, computing and telecommunications, and we have presented some unconventional ideas for the future of 6G. To refine and develop these ideas, we have consulted with experts in various fields, who have also helped review the chapters. They are, nevertheless, our ideas. We want to thank the people who have helped us – namely Bernie Mckibben, Dr. Lin Cheng, Steve Buttery, Tetsuya Nakamura, Adam Broadbent, Dr. Xuan Du, Arjun Parekh, Professor Ajith Parlikad, Izzo Wane, Dr. Francisco De Carvalho, Philip Eardley and Frank Scahill for their contributions.

We are deeply honoured to have Professor Rahim Tafazolli, the esteemed head of the 6G research centre at the University of Surrey, write the foreword for our book. Finally, we would like to thank Federico Brain, our MATLAB® expert.

We have been informed and inspired by a wide range of colleagues who, over the years, have presented at conferences, worked with us or had discussions over dinner and helped build our world views. Thanks to all of them.

Finally, to our families for putting up with us writing this book and to the publications team at the IET, led by Valerie Moliere.

About the authors

Peter Smyth has been at the forefront of developments in telecoms and computer science for over 35 years, most recently as VP of core innovation at CableLabs, USA. He previously held various senior research positions at BT, including VP of mobility and converged services and head of mobility and wireless research. His areas of interest include CBRS shared spectrum, community Wi-Fi, fibre optics, low latency DOCSIS, microwave photonics for mobile infrastructure, infra-red communications and agent-based location services. Career highlights include involvement in preparing board papers on the strategy for BT's successful acquisition of LTE spectrum and conducting successful intellectual property litigation against Google around location services and maps. He holds 19 granted patents. In 2011, he was awarded BT's most prestigious award, the Martlesham Medal for Science and Technology, which recognises pioneers in telecom technology who have made an outstanding lifetime contribution to science and engineering. He was also awarded in 2010 the BT Gold Medal for Innovation for his invention of Open Wi-Fi (Community Wi-Fi), which was launched as BT Fon in 2007. Today, this Wi-Fi network has over 5M access points in the UK and 13M globally, making it one of the world's largest Wi-Fi networks. He was elected a fellow and a chartered engineer of the Institution of Engineering and Technology in 2006. He received his MSc degree in microwaves and modern optics from the University of London (UCL), UK.

Peter Willis worked for BT Group from 1988 to 2021, during which time he researched distributed computer networks, designed, installed and operated BT's first Internet Service and contributed to BT's 21st Century Network architecture. At BT, he evaluated many new network technologies, including NFV solutions, optical burst switching, terabit ethernet and terabit routers. The evaluations went beyond lab testing to include commercial analysis and field trials. He was the lead author of the first ETSI network functions virtualisation (NFV) whitepaper. He developed NFV from a research concept in 2011 to an industry-wide technology now embedded in the 5G architecture and central to many Tier 1 network operators. Before leaving BT, he explored using network equipment to perform distributed computing operations to complement cloud computing. He holds an MEng in electronic systems engineering from the University of York, UK.

David Wisely worked at BT Group for over 25 years where he became the head of mobility research in 2009. He was the technical manager for a range of influential EU projects including BRAIN and MIND, which went on to shape 3 and 4G mobile standards. After leaving BT, he moved to Surrey University where he worked on 5G techno-economics. He published a seminal work in 2019. He holds 20 patents and is the author of two books on mobile telecoms. He has lectured at Oxford University and published over a hundred peer-reviewed articles. He holds a PhD degree in Multimedia service delivery over heterogeneous wireless networks from King's College London and another PhD degree in Techno-economic modelling of 5G networks from Surrey University, UK.

Chapter 1
Introduction

6G will definitely happen by 2030

– Nokia CEO Pekka Lundmark speaking on a panel at the World Economic Forum in Davos 2022

The industry should make 6G a no-G. There is practically nothing left that we're missing in a hypothetical new generation

– Santiago Tenorio, network architecture director at Vodafone

1.1 The backdrop to 6G

The mobile industry is at a crossroads, making now an interesting time to discuss 6G. 4G was a huge success worldwide. It was the first generation to feature a unified air interface. Moreover, the orthogonal frequency division multiplexing (OFDM) technology was efficient and adaptable in delivering mobile Internet services, especially video, which became an actual 'killer app'. Nowadays, we can get 20 GB/month for \$10 or unlimited data for \$15–\$20 (Figure 1.1). 5G is supposed to be about so much more than just existing mobile data Internet apps; otherwise, we could have stuck with long-term evolution (LTE) and Wi-Fi. Commentators said that 5G would be much faster and have much lower latency. It would be the first generation of mobile to offer quality of service (QoS) and differentiated application classes using dedicated network resources. 5G was slated to support all non-phone devices, from autonomous cars to crop dusters, smart meters and factory robots. 5G was also designed with a core that could run on a cloud platform, making it cheaper and giving more flexible in-service creation. It embraced the idea of private networks, either created from a network slice or using shared/lightly licensed spectrum such as that made available in mobile bands, notably Citizens Broadband Radio Service (CBRS) in the United States.

However, commentators agree that 5G has been costly to deploy, requiring new antenna arrays and new towers. Just as 6G research programs are ramping up and standards bodies are setting out the plan for 6G, the industry is still waiting for a significant return on its 5G investments. A good summary of the position in 2023 comes from SK Telecom [1]:

Figure 1.1 Mobile data are available at a relatively low cost. How will the investment in 6G be paid for?

> *. . . about 4 years after the commercialisation of 5G. Most of the use cases of 5G technology focus on eMBB (Enhanced Mobile Broadband) services, and several factors, such as form factor constraints, immaturity of device and service technology, low or non-existent market demand, and policy/regulatory are complexly intertwined. As a result . . . URLLC (Ultra-Reliable Low Latency Communications) and mMTC (massive Machine Type Communications) services are not activated*

Remember that 5G is still evolving, with new features being introduced into the run-up to 6G so it has time to grow. However, 5G is already shaping views of 6G with the Next Generation Mobile Networks Alliance (NGMN) issuing some guiding principles that include

> *. . . 6G must not inherently trigger a hardware refresh of 5G radio access network infrastructure. The decision to refresh 5G RAN hardware for operational reasons such as end-of-life, energy consumption or new capabilities must be an operator-driven choice, independent of supporting 6G. 6G introduction must allow certain scenarios to be realised through software-based feature upgrades of existing network elements to meet 6G requirements. . . .6G must address demonstrable customer needs across mobile, fixed and non-terrestrial networks . . . . [2]*

5G has an opportunity to deliver non-mobile broadband applications towards the end of the decade, as new mobile cores, known as standalone (SA), are deployed to enable new 5G capabilities such as network slicing, applications interfaces and support for high reliability. This two-phase approach to introducing 5G has created concerns within the industry about the scope of 6G. One purpose of this book is to assess the plans and technologies critically proposed for 6G. There is an in-depth look at the technologies behind 5G in Chapter 2 including the network slices, private networks, spectrum and actual performance to date. It sets the scene for the rest of the book, particularly as 6G evolves many of the 5G key application areas and uses cases such as enhanced mobile broadband (eMBB), machine-type communications, non-terrestrial networks (NTN) and ultra-reliable low-latency communications.

In the first part of this chapter, we describe some of the drivers that will shape 6G. There is a broad consensus on these drivers. In the second part of the introduction, we diverge from the consensus to consider different ways to think about 6G.

Our approach is to identify and analyse key questions relating to 6G. There are numerous potential paths towards achieving the 6G network and various factors that could define it. Each chapter of the book deals with one question related to 6G. We not only explore the evolutionary path of 6G, which is a continuation of the 'generation game' that involves increasing headline rates and lowering latency, but we also offer more radical and revolutionary proposals. We present the evidence behind these approaches and offer the reader an insight that equips them to critically examine the various hype and hubris that new mobile generations often create. The final, extremely detailed chapter then brings it all together, comparing two very different 6G possible futures: evolution or revolution?

1.2 Drivers for 6G

1.2.1 Technical

There is no getting away from artificial intelligence (AI) these days, from worries that it will cause mass redundancy and take over the world to predictions of a significant increase in productivity and living standards. Essentially, AI has three components: data, algorithms and computing. It is important to delineate two ways AI and 6G could interact. First, AI could play a significant role in optimising network performance, such as higher spectral efficiencies and lower power consumption. A typical example might be advanced multiple input multiple output beam steering and radio resource scheduling. Base stations have access to very significant amounts of data on the radio environment and the time-variant traffic patterns over changing conditions. If the native ability to use AI to optimise the network is embedded in 6G, it will boost network performance. Second, 6G could facilitate wider AI use. AI needs large amounts of data, and 6G could be the mechanism that delivers it to the algorithms and computing elements securely and privately. 6G could also incorporate AI computing services within its network. 5G is deploying mobile edge computing (MEC) servers at the network's edge, and the core runs virtualised services. These could evolve to enable 6G to offer general AI-based services. Linked to this is the technological trend for networks to minimise the amount of custom hardware and shift everything to general-purpose computer resources (be that cloud or not). However, some physical constraints will not be overcome by centralising 6G, such as some time-critical services. Chapters 4 and 11 look at the trade-offs between latency and the number of locations for these local computing resources. The other constraint to virtualisation is the radio front end. 5G was a first step towards virtualisation and removal of specialist hardware. 6G could go much further, offering greater flexibility in optimising and upgrading functionality and creating a more open network ecosystem.

Another key technical issue is the availability of spectrum for mobile use. The general rule for wide-area mobile networks is that the lower the frequency, the better the coverage, especially indoors where the most users are. However, the trend has been for each generation to use higher frequency bands as larger channel bandwidths are required for the higher data speeds demanded by successive generations. These higher frequencies make offering full indoor coverage from a macro network increasingly challenging. This trend is known as the spectrum staircase. For example, 3G was the first to use 2.1 GHz, 4G used 2.6 GHz and 5G used 3.5 GHz.

Mobile network operators (MNOs) initially believed that using spectrum above 3 GHz in wider area networks would be economically challenging. Although 5G currently operates on 3.5 GHz, this has only been technically possible through large element arrays and advanced beam steering technology that has proved expensive to deploy. The search for 6G spectrum in the 6–24 GHz range or above 100 GHz for even higher speeds means that it will be challenging to densify a wide area network using this spectrum. This spectrum is ideal for small cell applications that require ultra-high speeds for indoor and campus-type scenarios or fixed wireless access. We refer to this as the spectrum dichotomy, and this book explores this theme extensively.

However, new semiconductor material systems with phased arrays may enable us to proceed to even higher frequencies in the sub-terahertz region (100–300 GHz), challenging the commonly held belief that operating at higher frequencies is problematic. This is because the link budgets improve with frequency, and these systems allow a system-on-chip (SoC) solution for all base station electronics, potentially lowering costs. Nonetheless, these systems offer no realistic penetration of solid objects, so obstructions become the limiting factor in cell sizes. We explore the millimetre waves (mmWaves) and higher frequencies in Chapter 8.

Moreover, only 15% or so of the most valuable spectrum under 3 GHz, ideal for wide-area coverage, is actually used on average at a given location and time. It is not available for telecommunications mainly because it is dedicated to other users such as broadcast services such as TV, defence or point-to-point links and is often only used at a minimal number of locations. Spectrum reform and new schemes to share this valuable spectrum could liberate an additional 1900 MHz for 6G mobile services, transforming the deployment scenarios for 6G and its economics. In Chapter 7, we show how to achieve this technically. This new proposed sharing system could be applied to the entire electromagnetic spectrum.

1.2.2 Societal

There are societal trends that will impact 6G. Over the next 20 years, several countries' populations are predicted to age significantly as people live longer than previous generations. 6G could support these ageing populations with health monitoring, support for home living and greatly improve hospital productivity. Examples of possible 6G services include virtual wards, robot carers and greater human–machine interaction. One proposed 6G feature is digital twinning – a detailed real-time model of anything from a house, factory, person or location. These digital twins could facilitate high-resolution virtual travel, virtual groups and simulated experiences in virtual reality. A person's digital twin might integrate information from body-worn sensors, mobile terminals and sensing from base stations or Wi-Fi access points. 6G privacy and security functions and AI resources could promote better health outcomes while improving healthcare provider efficiency. This emphasises the importance of personal data in 6G, which we explore in Chapter 12.

More widely, 6G could continue the post-pandemic trend of more flexible working by facilitating more efficient cooperation, better training and access to AI services and resources.

1.2.3 Economic

It is becoming obvious that 6G will need a significant investment. According to Thales [3], MNOs worldwide are currently spending around £175B/year on 5G (although figures from different sources vary considerably). There is certainly enough money to potentially pay for 6G – currently, mobile technologies generate about 5% of global GDP ($5 trillion) [4].

Chapter 5 explores how 6G could enhance Industry 4.0 and shows how 2% of the improvement in productivity from Industry 4.0 alone could pay for a 6G deployment. The challenge for MNOs is to capture enough of this value chain to justify a major investment in 6G at a time when their investment in 5G still needs to show a significant return.

5G has necessitated major upfront investment in the radio network, which has delivered on the anticipated increase in speed and capacity for mobile broadband, at least for users in urban areas. This has not, however, yielded more than a slight uptick in user revenue. Other use cases that entail private networks, network slicing and low latency require the introduction of a new mobile core. These cores are starting to be delivered in 2024, and the expectation is that they will offer new revenue streams from a diverse range of non-mobile broadband sources such as

Industry 4.0, IoT services and private networks. Some see the success of these new revenue streams as a crucial precursor to investment in 6G.

There is usually conflict among the value chain players involved in delivering mobile services. This is unlikely to be any different in 6G. What 6G offers mobile operators, however, is control over network functions that are difficult to duplicate on a best-efforts IP-based network. These might be functions such as time-sensitive computing [multi-access edge computing or mobile edge computing (MEC)], privacy, radio interface sensing and facilitating network convergence. This is another theme that is explored throughout the book and in particular in Chapter 9, where we explore the opportunities for collaboration with hyperscalers.

1.2.4 Security

Protecting data, allowing specific access for authorised use, complying with local and national restrictions (e.g. GDPR) and orchestrating its use in AI are key aspects of the future digitally connected society. Only the network can provide the fundamental level of security and trustworthiness. 6G AI network elements and advanced sensing have access to information that could greatly improve many areas of security and trustworthiness in a way that over-the-top security could not. Chapter 12 takes an in-depth look at privacy and trustworthiness and describes how these might be delivered in 6G.

1.2.5 Climate change

Communications networks consume significant amounts of energy, with ICT estimated at 2% of global carbon emissions, and mobile networks account for about 10% of that figure [5]. It is, however, important to look at the entire carbon budget of the services being provided. 6G has a major theme of helping to achieve the UN targets for net zero, but as is detailed in Chapter 6, there are no systematic measures of 5G energy use today. Such measures must capture the service substitution gains that 6G offers, such as video conferencing versus travel. Interestingly, it has been proposed that 6G will consider a type of energy audit of new services to ensure there is no unwarranted deployment; however, the enforcement measures are unclear.

A future mobile network will consist of radio base stations, backhaul and computing resources. AI optimisation of the network, with lower power consumption as one of the optimising factors, could save energy by powering off base stations at quiet times, anticipating traffic patterns and using highly focused beams that offer high signal-to-noise ratios for higher data rates but contain less total energy. Another energy saving on the radio side is to locate the base stations physically closer to the users, which is discussed in Chapter 9, avoiding energy wasted on radio propagation losses. SK telecom [1] estimates that the RAN is responsible for 73% of a mobile network energy consumption and so should be the focus of energy savings.

Backhaul and computing optimisation will likely involve a trade-off between larger numbers of local nodes, offering lower latency, and a smaller number of more centrally located nodes, which are more efficient due to their scale but will increase latency.

1.2.6 Inclusion

It is estimated that 90% of the world's population has access to 3G or 4G, which leaves a large area of land and most of the world's seas and oceans with no coverage. One of the primary goals of 6G is to offer a truly global service, with satellites providing a fill-in for terrestrial base stations. It was a goal in 5G, 4G and 3G before that, highlighting how difficult and expensive it is to provide high levels of coverage. According to Oughten, in a techno-economic analysis of 5G terrestrial roll-out in Holland [6], covering the first 30% of the population accounted for only 5% of the cost. The following 40% constituted only 20% of the costs, and the final rural 30% of the population was responsible for 75% of the total rollout cost. The expense of improving coverage has, traditionally, also increased directly with the carrier frequency used because of the shorter-range propagation characteristics of higher frequencies. Getting anything like '100 per cent' coverage in 6G is not economically feasible without using NTNs (a term that includes drones and high-altitude balloons/aircraft). The most notable NTN development of late is the launch, or planned launch, of large constellations of low Earth orbit satellites offering much greater capacity than previous GEO (geostationary) satellites that orbit at a much higher altitude. Satellites might provide global coverage for narrowband services, such as voice, messaging and tracking. Still, they will not work indoors without external antennas outside houses, factories, buses, etc. In addition, even these constellations have very limited capacity compared to either mobile or fixed networks and currently cost at least an order of magnitude more per bit delivered, making them unsuitable for providing 'eMBB from the sky'.

Chapter 10 explores NTNs in detail, including estimating the capacity and costs of the new satellite constellations and considering alternative solutions for better broadband coverage, such as balloons or aircraft. The chapter also considers the types of 6G applications that might better fit with the limited capacity of NTNs, such as narrowband IoT applications and sensing by remotely piloted drones.

1.2.7 Industrial and machine communications

It is fairly easy to imagine that if machines, meters, vehicles, robots, bodies and just about everything else were able to connect to the Internet and if the communications system that connected them could also provide low latency for control, AI and accurate mapping information, then huge leaps in productivity could be achieved. This level of integration might finally enable driverless vehicles and lights-out factories. The advantages of driverless vehicles have been clear for a long time: they could be safer, cause less pollution and reduce transport costs. Similarly, a futuristic factory could move control of its processes, such as machines, robots or cyborgs, from local computers to generalised remote AI-driven processes running in a 6G network. The whole factory could benefit from real-time sensing of the entire operation from the radio environment at various frequencies, some of which might provide non-line-of-sight mapping. Chapter 5 analyses the communications needs of Industry 4.0, and Chapter 14 describes joint sensing and communications for 6G. According to Mckinsey, Industry 4.0 has the potential to provide [7]:

- Inventory cost reduction of 15–20%
- Labour productivity increase of 15–30%
- Machine downtime reduction of 30–50%
- Throughout increase of 10–30%
- Forecasting accuracy increase of 85%.

To date, mobile communications (connections and data volume) have been dominated by personal users. The second half of the decade will see 5G introduce a range of new capabilities for IoT and machine-type communications. There is also a clear need for lower power and smaller form-factor 6G connection mechanisms – including those that can harvest power from the RF field – to enable a much wider range of devices.

However, the transition to a fully digital society will require much more than new standards – in fact, there are already many competing short-range and wide-area connection standards for IoT and machine-type devices. 6G will need to solve some of the integration, coverage and service creation issues that have held back the expected growth of these machine-type services. One 6G revolutionary vision includes a 'network of networks' concept that brings together mobile, fixed, NTN, many wireless IoT protocols, Edge Cloud and hyperscalers' clouds into a common service framework. This unification of networks into a single service platform may facilitate many new non-mobile broadband applications. The concept of network of networks is analysed in Chapter 9.

1.3 6G futures – difficult questions

1.3.1 Why do we need 6G at all? The generation game

Why are there mobile generations? (Figure 1.2). The real driver for mobile generations has been the changing air interface to facilitate higher capacity and speed in combination with the introduction of new spectrum bands. That has meant introducing new base stations and handsets to access the higher speed or lower latency on offer. It is interesting to compare this to Wi-Fi generations, which have always been backwards compatible, although this has come at the expense of absolute performance.

Successive mobile generations have, however, offered progressively lower efficiency gains in terms of bits transmitted per Hertz of spectrum used, and it has been suggested that the OFDM 5G air interface be reused in 6G. Now that so much of the network is just software running on general-purpose chips, including many elements on the handset, the upgrade is more like a major firmware upgrade. This would be analogous to new operating system releases that can be installed on most modern phones or laptops, as suggested by the NGMN in the quote at the start of the chapter. New capabilities, such as seamless integration with NTNs, sensing and a 'network of networks', might then be the differentiators from 5G with a stronger emphasis on machine-type communications and Industry 4.0.

Figure 1.2 First generation mobile phone – does 6G require a new air interface and new spectrum?

There is also a narrative that even generations finally deliver the visions of the odd mobile generations: 4G delivers the mobile broadband promise of 3G, and 6G delivers the 5G promise of digitising society and industry. A 6G derived and evolved from 5G by the same players and standards bodies is the 'business as usual' or 'evolutionary' approach. Many people are questioning this approach, and in every chapter, we can find a proposal for some aspect of 6G that differs from this view. Chapter 6 reviews the global definition of 6G with a critical analysis of white papers and presentations from various vendors, operators, chip suppliers, standards bodies and industry organisations.

1.3.2 Can't we use Wi-Fi instead of 6G?

Wi-Fi 7 has now been certified with routers available and leading smartphones, including Wi-Fi 7 interfaces. Work on Wi-Fi 8, offering ultra-high reliability, is now progressing towards launch around 2028. Wi-Fi also now operates in (up to) 1 GHz of a new spectrum from about 6 to 7 GHz, and Wi-Fi 7 can achieve maximum data rates of 1–5 Gbit/s in real-world scenarios. Wi-Fi 8 will introduce interference management between access points – with an ultra-reliable low latency service offering. Chapter 3 looks at Wi-Fi in depth and provides some real-world performance figures for Wi-Fi generations 6, 7 and 8.

None of the above, however, answers the question of why Wi-Fi cannot be used instead of 6G indoors, in shopping malls and in many industrial settings. This is part of a wider question of whether services will continue to migrate to being delivered over a best-effort IP network. Even in Wi-Fi 8, the shared nature of the spectrum will rule out any absolute quality of service (QoS) guarantees. If 6G services just require more speed and lower latency, then Wi-Fi could deliver them indoors and across campuses and malls. A wide-area macro 6G network would still be needed, but this might only be required to offer IP delivery without supporting voice, security or handover. These functions could then be provided over-the-top, as is the case with Wi-Fi today, and would reduce 6G to simply a best-efforts IP delivery service. But 6G is undoubtedly much more than this. If 6G services involve sensing, require time-sensitive computing or AI or security support from the network, then a best-efforts IP delivery network alone will not be enough. It seems more likely that Wi-Fi will be incorporated within 6G as part of a 'network of networks' that will provide these advanced functions that are unified across a range of terrestrial and non-terrestrial networks. Chapter 9 is an in-depth look at the convergence aspects of 6G, including the delivery of 6G services over Wi-Fi.

1.3.3 Mobile networks are just capacity and coverage – is that true of 6G?

If you improve the capacity of a mobile network, then lots of good things happen: the headline rates generally go up, the latencies come down and the reliability improves. More capacity means the network is less stressed, which leads to less interference, fewer re-transmissions and ability to send data with higher signal-to-noise margins – all of which contribute to lowering the latency and improving reliability. Maldonado *et al.* [8] analysed the trade-off between capacity and latency/reliability for several networks. They found that both Wi-Fi and 5G can be configured to offer ultra-low latency (1 ms) and high reliability (99.999%) in a local network. However, this trade-off came at the cost of up to a 90% reduction in capacity.

There is a straightforward formula for the capacity of a radio network:

$$\text{Capacity} = \text{Spectrum} \times \text{spectral Efficiency} \times \text{Density of base stations}$$

Every generation has been allocated more spectrum, using a more efficient air interface and densifying the network. 6G could be a continuation of this trend. There are chapters on using low bands spectrum (7 GHz and below) using spectrum more effectively (Chapter 7) and on further expansion into cm and mmWave frequencies (Chapter 8). There is certainly a case to be made that the current lower frequency spectrum needs to be allocated and licensed for efficient use. Making better use of it would help 6G because, traditionally, the cost and power consumption of delivering data have increased as a higher frequency spectrum has been utilised. Chapter 7 includes a radical idea for making much more efficient use of the sub-3 GHz spectrum. However, for ultra-high rates, say, above 10 Gbit/s, there is no alternative but to move to cm and mmWave (millimetre Wave) frequencies. There are considerable barriers to deploying these frequencies across wide areas, meaning that most usage may be confined to indoor, small cells in city centres and campus-like scenarios.

In contrast, spectral efficiency is getting harder to improve with every generation. LTE and, especially, 5G are running close to the Shannon limit in many situations. One way efficiency can easily be improved is to locate the base stations closer to the predominantly indoor users. The savings in terms of the radio budget are very substantial when compared to using a macro base station to serve indoor users. That extra radio budget can allow a higher-order modulation scheme, improve efficiency or reduce power consumption. Base stations located indoors could serve nearby users outside the building. This is described in the convergence chapter and is part of a more radical possible 6G future. Semantic communications leveraging AI is another alternative to improving the efficiency of communications (Chapter 13).

1.3.4 Isn't 6G just some target key performance indicators?

A different way to look at 6G is to define it in numbers – called key performance indicators (KPIs). Table 1.1 looks at the KPIs put forward in the early days of standards development by the NGMN and International Mobile Telecommunications (IMT) for 5G and compares them with the actual 5G performance in 2023. The table also includes 6G KPIs from Tong [9] and those we have derived as being needed for the applications and use cases put forward by the International Telecommunication Union (ITU).

The original visions of 6G, developed in 2020–2022, were often based on KPIs. These included specifications such as 1 Tbit/s headline rates and sub-1 ms latency. The KPIs proposed for 5G were mostly not met, as Chapter 2 explains in detail, and it is telling that more recent visions and ideas on 6G have mostly abandoned KPIs in favour of key value indicators, although it is unclear how they are measured. Every extra requirement, whether latency reductions or bandwidth increases, will increase the overall cost of 6G deployment, often only for outliner use cases with limited revenue potential. Chapter 4 presents an in-depth analysis of the network performance required to deliver many proposed 6G services and clearly shows that terabit data rates and sub-milli-second latencies are not needed for any but the most demanding applications. The revolutionary approach to 6G, which we describe in the various chapters and the extensive conclusion, offers a lower-cost, higher functionality alternative to the more evolutionary approach of simply multiplying the KPIs of the previous generation by a factor of 10 or so. The last column of Table 1.1 shows the 6G KPIs derived in Chapter 16 to meet the use cases described in a seminal report issued by the ITU in November 2022 entitled 'Future Technology Trends International

Table 1.1 5G and 6G KPIs: 5G failed to meet its KPIs – does it make sense to specify them for 6G?

	NGMN 5G	**Nokia 5G**	**5G 2024 Actual**	**6G [9] 2020**	**Derived from IMT 2030 report**
User-experienced data rate	50 Mbit/s everywhere	100 Mbit/s minimum	0.1–1 Gbit/s in urban areas only	10–100 Gbit/s	10–100 Gbit/s
Peak data rate	1 Gbit/s	10 Gbit/s	0.1–1 Gbit/s	1 Tbit/s	100 Gbit/s
Connection density	150,000/km^2	10–100 × number of devices	Similar to LTE 5000/km^2	10,000,000/km^2	1,000,000/km^2
End-to-end latency	1–10 ms	1 ms	10–50 ms	0.1 ms	1–10 ms
Traffic volume density	1–15 Tbit/s/km^2	10,000 × more traffic than LTE	100–500 Mbit/s/km^2	1000 times 5G	
Mobility	500 km/h	More than 500 km/h	500 km/h	1000 km/h [10]	
Energy efficiency	2000 times better than LTE	Much lower J/bit	More efficient per bit (90%) but network consuming more power than LTE	100 times better than 5G (per bit)	Very high energy efficiency (the same 5G KPI target)
Reliability	99.999%		Similar to LTE 97–99.5%	99.99999%	Same as 5G frame error rate 10^{-9}

Mobile Telecommunications (IMT) Towards 2030 and Beyond'. Our analysis shows that the early KPIs suggested for 6G are unnecessary in delivering virtually all the 6G services now being considered.

1.3.5 What actually are the 6G applications?

Having looked for 6G from the bottom up, it is also possible to look from the top down: What applications will it enable? Most 'ordinary' people (including investors) want to know what 6G will deliver? If the applications are not compelling, then why bother?

Every generation has had a multitude of 'visions', white papers, R&D and have proposed a series of use cases. Many of these never actually succeed or make it to market; instead, successful applications are often those not considered during the standardisation process, and so some caution is needed. As an example, SMS was never designed to be a public service; it was introduced for engineers to exchange short messages about the network. 3G mobile operators were going to set up a walled garden of content such as TV sports and movies. When the fixed Internet took off in the late 1990s, that model failed, and there was a clear need for mobile Internet, which 3G was not optimised to deliver. Only with the advent of LTE did the mobile Internet experience approach that of the fixed network.

5G certainly has the potential to support applications like AR and VR (augmented and virtual reality) in a way that LTE cannot. 5G also provides private networks for factories and campus-like sites, some of which use lightly licensed spectrum and local network cores to support lower latencies. 5G will be a commercial failure if all it does is massively increase network capacity for existing, best-effort, mid-latency traffic. 6G will need to build on the 5G momentum away from being dominated by mobile broadband and look to new services that require additional elements of 6G, such as sensing, privacy or local processing.

Chapter 14 looks at joint sensing and communications and details how mmWaves could unlock new applications in this area. Powered by 6G sensing and communication innovations could transform monitoring, control and testing. There is also a primary emphasis in 6G on Industry 4.0 applications. It is envisaged that combining 6G elements, such as sensing, low latency, ultra-reliability and enhanced security, will enable significant gains in productivity and innovation. Chapter 5 describes not only Industry 4.0 but also analyses how these new 6G elements will bring about new industrial applications, such as digital twins that are online, real-time models of the entire plant/process or function and can be optimised and controlled with 6G provided AI.

The Internet of Bio is the concept of a local network monitoring and connecting via a range of protocols and frequencies to a hub that connects to the 6G network. You can read about the use of the smartphone as a hub in Chapter 15.

6G could also transport you into the Metaverse – possibly with full surround, all-sense tactile support. The Metaverse is a kind of 3D virtual world with various

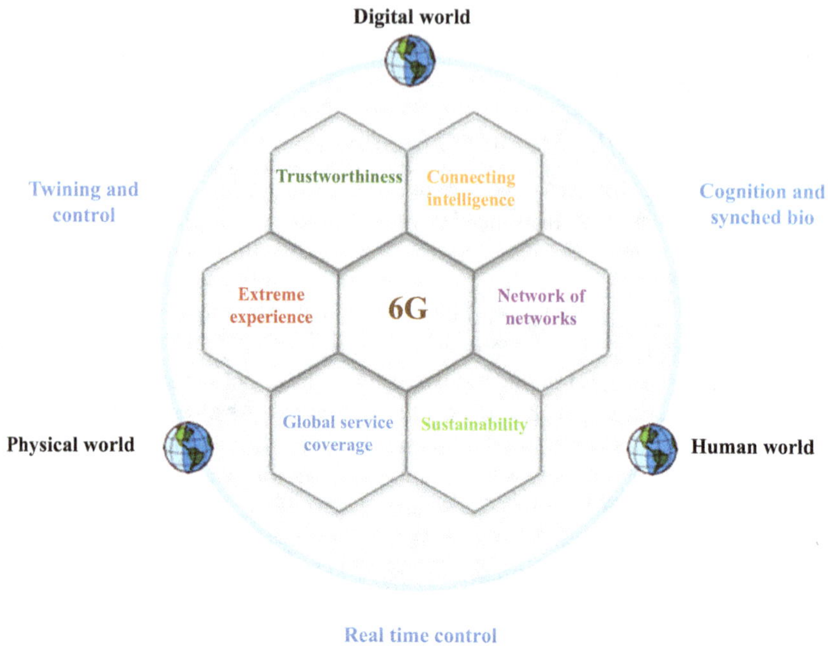

Figure 1.3 6G vision from the EU project Hexa-X. Can 6G deliver all these goals?

avatars representing users. That would require high data rates and low latency processing and probably be delivered in the home over Wi-Fi 8.

The addition of non-terrestrial networks will enable more IoT applications. Satellites will also increase the coverage for narrowband applications such as voice, messaging, disaster alerts and tracking. NTNs could also enable ultra-reliable connections by providing backup links for terrestrial networks in critical situations. Drones, operating singularly or in swarms, also have the potential for monitoring, sensing and rapid delivery.

You can read about current 6G visions in Chapter 6 (Figure 1.3), with an assessment of future applications and their real network requirements in Chapter 4. Chapter 15 is a futuristic look at the 6G iPhone and potential new device form factors and their capabilities. While our final Chapter 16 provides an in-depth analysis of what 6G could deliver.

1.3.6 Will 6G finally merge fixed and mobile comms?

People have discussed convergence in telecoms for over two decades, specifically fixed-mobile convergence (FMC). Still, there has been little to no progress apart from mergers and acquisitions between fixed and mobile operators. Will this change in 6G?

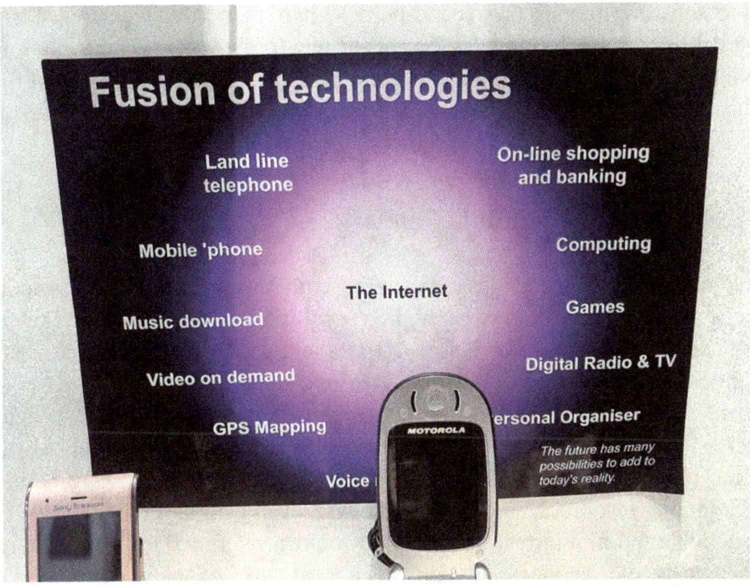

Figure 1.4 The BT Fusion phone (2009) – an early attempt at fixed-mobile convergence. Will 6G deliver a 'network of networks'?

One of the first FMC products was the BT (British Telecommunications Plc.) Fusion phone, which used a combination of Wi-Fi, focusing on use indoors and in hot spots from its broadband network, with mobile coverage via its partnership with Vodafone in an Mobile Virtual Network Operator (MVNO) arrangement (Figure 1.4). BT Fusion was not a success, but 6G could offer a new approach with a 'network of networks' that mixes fixed and mobile telecom networks with cloud AI providers. The challenge is that the consolidation of networks increases the risk of security breaches, requiring robust measures to protect sensitive data, as discussed in Chapter 12 on Privacy Enhancing Technology. 6G convergence could encompass Wi-Fi in Trusted or Non-trusted forms as a true partner, enabling 6G services inside buildings for domestic and industrial use cases. In industry and shared spaces, there is the opportunity to encompass private networks and neutral hosted infrastructure. Neutral hosting supports multiple operators and could offer significant radio efficiency and energy use gains. New 6G convergence standards and protocols could enable more widespread adoption of neutral hosting, as described in Chapter 9.

Convergence would also make sense with the rise of players with both fixed and mobile plays – who want to offer users a unified service. Some fixed-network players are looking to 5G, using shared or lightly licensed spectrum such as the US CBRS spectrum, mobile band 48 and Wi-Fi to augment their MVNO deals. The network topology of mobile and fixed networks is certainly converging – with only the very minimum of essential hardware at the base sites and as much of the functionality moved to software running on general-purpose hardware in data centres. Chapter 4 discusses the new possibilities this brings for new providers and innovation within the value chain. If the underlying architecture is similar, the cost/overhead of running specialised radio resource management software, such as providing ultra-reliable packet delivery, may be negligible.

1.4 Structure of the book

If an evolutionary 6G is 'business as usual' – with new spectrum, higher data rates and a new air interface – then what might a revolutionary 6G look like? It could try to reuse the existing spectrum below 3GHz more efficiently to reduce costs and development time. It might not even have a new air interface. Small 6G base stations may be deployed indoors to reduce power consumption and provide an alternative to a macro network in urban areas. The RAN and core could be powered by AI and reduced to open and fully virtualised, cloud-native software running on basic computing resources. The core might even disappear as a recognisable entity. Wi-Fi, cellular and non-terrestrial networks could converge to the point where they offer seamless 6G services as part of a 'network of networks' New, non-communication applications, might develop, such as scanning your food for allergens. Finally, the ecosystem and value chain might feature new or existing players with greater or reduced roles. You can read about all these ideas and more in the rest of the book, directly comparing the evolutionary and revolutionary paths in the concluding Chapter 16 (Figure 1.5).

The book has three sections. The first, Chapters 1–5, deals with the lessons from history and the drivers that are moving cellular to 6G and Wi-Fi to version 8. The second part, Chapters 6–13, examines the key components of 6G and includes more radical proposals related to each topic. Finally, Chapters 14–16 present a vision of what 6G might become. Each chapter of the book tackles what we consider to be a major question in 6G. Each can be read as a standalone chapter, and the order is not critical. Readers familiar with all aspects of 5G, including its current progress and future upgrades, could skip Chapter 2. It does, however, set the scene for the discussion of 6G, and if 5G take-up is slow and returns poor, then 6G may be delayed, and it may be less ambitious in that it only delivers the original 5G vision.

Each of Chapters 4 to 15 also provides a more radical idea, something different from the general trend of new generations being largely evolutions of the previous generations. One example is the idea that, rather than using higher and higher frequency spectrum, we could simply use the spectrum we have below 3 GHz more effectively. There are clearly barriers to doing this, but the benefits for 6G would be enormous. All these radical ideas come together in Chapter 16, where we compare the evolutionary and revolutionary futures for 6G – of course, they are not mutually exclusive, and the revolutionary scenarios build on those of the evolutionary; the real future probably lies somewhere in the middle.

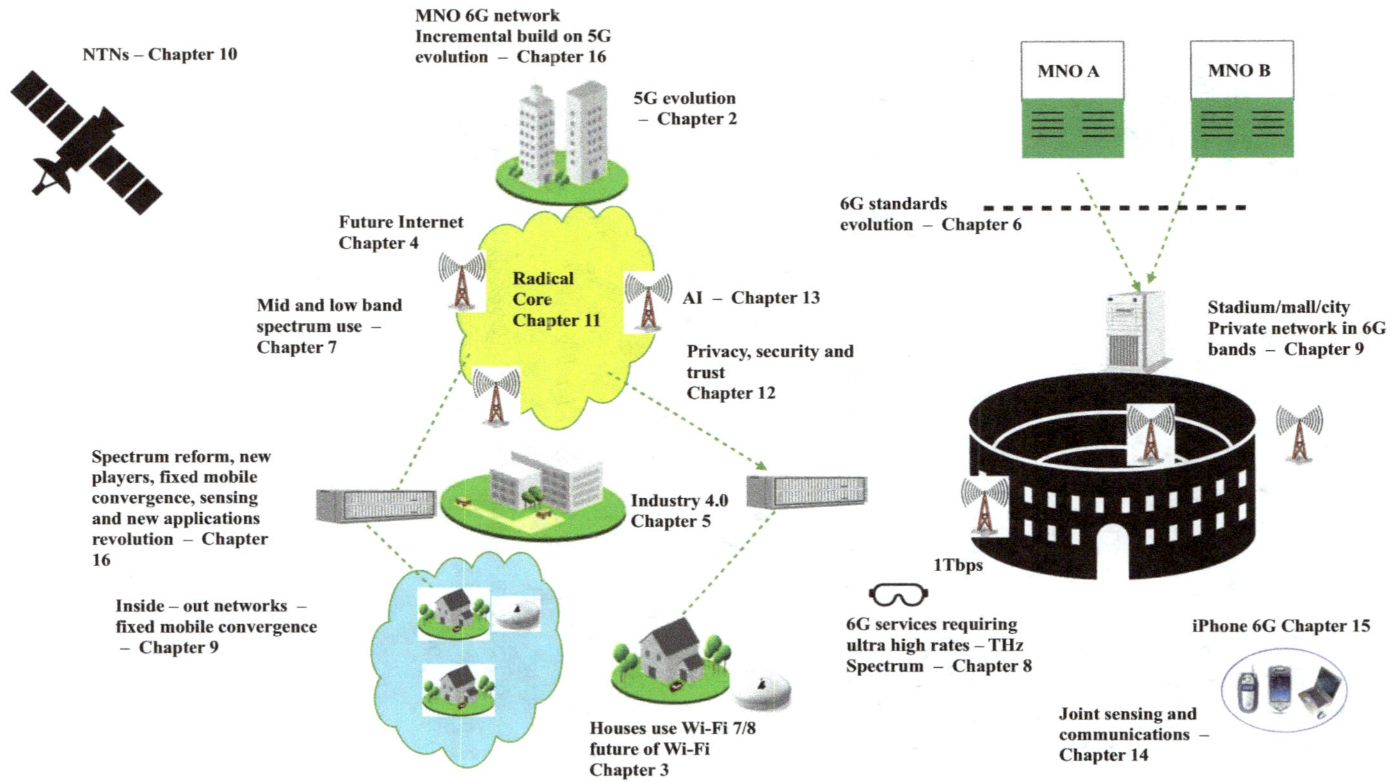

Figure 1.5 The structure of the book

References

[1] *SK White Paper* [Online]. [Cited Jan 2024] Available from https://www.sktelecom.com/img/pds/press/SKT_5G%20White%20Paper_V1.0_Eng.pdf.

[2] *NGMN 6G Whitepaper* [Online]. [Cited Jan 2024] Available from https://www.ngmn.org/wp-content/uploads/NGMN_6G_Position_Statement.pdf.

[3] *Thales Statistics on 5G* [Online]. [Cited Jan 2024] Available from https://www.thalesgroup.com/en/worldwide/digital-identity-and-security/magazine/5g-progress-report-launches-subscribers-devices#:~:text=5G:the commercial impact, money into 5G network infrastructure?

[4] 'The mobile economy 2020' Published by the GSMA 2020 [Online]. [Cited Jan 2024] Available from: https://www.gsma.com/mobileeconomy.

[5] News Story 'ICT sector helping to tackle climate change', Dec. 2016 [Online]. [Cited Jan 2024] Available from: https://unfccc.int/news/ict-sector-helping-to-tackle-climate-change, accessed Jan 2024.

[6] Oughton, E. J., Frias, Z., van der Gaast, S., and van der Berg, R. 'Assessing the capacity, coverage and cost of 5G infrastructure strategies: Analysis of the Netherlands'. *Telematics and Informatics*. 2019; 37: 50–69, https://doi.org/10.1016/j.tele.2019.01.003.

[7] Mckinsey & Co Research, 2023 [Online]. [Cited Jan 2024] Available from: https://www.mckinsey.com/capabilities/operations/our-insights/capturing-the-true-value-of-industry-four-point-zero#.

[8] Maldonado, R., Karstensen, A., Pocovi, G., *et al.*, 'Comparing Wi-Fi 6 and 5G downlink performance for industrial IoT', *IEEE Access*, 2021; 9: 86928–86937, doi:10.1109/ACCESS.2021.3085896.

[9] Tong, W. and Peiying Zhu, P. (eds), *6G: The Next Horizon: From Connected People and Things to Connected Intelligence*, Cambridge University Press, Cambridge, 2021

[10] 5G vs 6G Spectrum Comparison, 2022. [Online]. [Cited Jan 2024] Available from: https://electronics360.globalspec.com/article/18760/5g-vs-6g-spectrum-comparison.

Part I

Lessons from history and drivers for 5G transition to 6G and Wi-Fi 6 transition to 8

Chapter 2
5G next steps to 5G advanced

There is a potential to overhype and under-deliver on the 5G promise. If anyone tells you they know the details of what 5G will deliver, walk the other way.

–FCC Chairperson Tom Wheeler in 2019

2.1 Introduction

There are compelling reasons to consider 6G through the lens of 5G. First, many of the visions and use cases of 5G overlap with those proposed for 6G. Second, as explained below, 5G introduced several critical radio and network innovations that are only just being rolled out (Q1 2024). It is fair to say that their commercial and technical success still remains uncertain. Not everyone universally supported the development and standardisation of 5G in its current form. Some commentators said that many of the proposed 5G use cases were niche applications, and that most could be satisfied by combining 4G cellular Wi-Fi with some regulatory changes. Nevertheless, 5G is being deployed faster than any preceding generation, but that does not yet make 5G a commercial success. This chapter also examines the techno-economics of 5G, highlighting how commercially challenging some areas of early 6G concepts might be. Mobile networks, much as iPhones, computers, cars, and other tech stuff, continually evolve. There are several enhancements to 5G, some of which go under the banner '5G-advanced' – 6G will have to coexist with 5G for decades, and investment will only be made in 6G if benefits (in the broadest sense) outweigh the cost. This chapter concludes with a look at these 5G enhancements in the pipeline and an analysis of what light 5G can shed on the various aspects that make up 6G.

2.2 5G development and launch

Why was 5G developed, standardised and deployed in the first place? What need was it designed to meet? What was the 'Killer app'? To answer these questions and fully appreciate the genesis of 5G, we need to delve slightly deeper into the past of earlier generations.

The development of 3G is often thought to be about data and the mobile Internet. It was never designed for that. 3G was conceived in the late 1990s, before the Internet revolution. It was envisaged that services such as two-way video calling and direct viewing of premium video content from the mobile operator would be offered. As such, it had several architectural features that later proved problematic for Internet traffic. The air interface technology was also less efficient when measured against 4G and 5G. In some markets worldwide, however, 3G was a great success at first, primarily when there was a shortage of voice capacity. In the United Kingdom, however, 3G struggled in the early years. Two-way video calling never took off for the mobile operators, and there was a continual search within the industry for the grail-like 'killer app'. In the end, the killer app did appear, but it was mobile Internet access. The operator's dream of selling IP messaging and connectivity services based on an IMS Internet Multimedia Subsystem (a multimedia service creation platform) also failed commercially when much cheaper, over-the-top (OTT) offerings from Microsoft and Yahoo appeared. Operators also tried to sell premium content (such as music) but were commercially unsuccessful when faced with global content providers (such as Apple) with existing customer relationships.

3G then became a 'bit-transport' mechanism, with networks offering data bundles and the price of data volumes as differentiators. With the advent of app stores for Apple and Android OS and video traffic rising rapidly, new capacity at a much lower cost per bit was required from 4G.

Long-term evolution (LTE) emerged as the de facto 4G standard, and the terms are often used interchangeably (although strictly speaking, LTE is the technology developed to meet the standard). This was a significant step forward in standardisation as multiple air interfaces met the 3G standard.

The key innovations in 4G – aimed at improving capacity, reducing latency, and lowering costs – were:

- All IP networks – no circuit-switched element
- Voice carried over IP.
- A new, more efficient air interface – offering high headline rates (100 Mbit/s) and excellent scheduling flexibility.
- Reduction in network nodes and separation of control and data function.
- A significant increase in the available spectrum.
- Provision for the Internet of Things (IoT) – low bandwidth and low power device support.

All of this made LTE perfect for delivering Internet data. Voice proved slightly more problematic because most early implementations fell back to 3G for voice calls, and IP voice solutions took several years to perfect. The first trial system operated in 2007; the first deployment was in 2009, and since then, it has grown to deliver nearly all of the mobile traffic in the world until recently. It was further enhanced by LTE-Advanced (2013), which increased efficiency and headline rates to 1 Gbit/s and later LTE-Advanced Pro (with headline rates of 1.2 Gbit/s).

2G and 3G networks still exist, partly to provide backwards compatibility but also because they tend to inhabit lower frequency bands with better coverage. Fourteen years after LTE's launch, its coverage is still far from complete or universal. This is an essential point for 6G and is taken up later in the chapter in the techno-economic analysis of 5G.

5G was researched, trialled and standardised during 2012–18, with deployments starting in 2019. The claims for what 5G will offer users by politicians, industry commentators, and suppliers are wide-ranging. If we are to believe what we have been told, then 5G will enable [1,2]:

- Almost limitless download capacities – 10,000× that of today's network,
- Headline speed up to 20 Gbit/s,
- Driverless cars,
- Remote robotic surgery,
- Drones that can deliver blood supplies in an emergency,
- Smart fridges and smart rubbish collection that will save the average UK household £450/year,
- Augmented and virtual reality,
- Coverage everywhere in the UK,
- 99.99% reliability.

5G was based on three visions or general use cases that we will now look at. They do not always align with the hype quoted above, which are media and political interpretations of what 5G will be and the benefits it will bring. This is relevant to 6G because it illustrates how these things get disconnected from reality and because one of the complete sets of 6G visions is fivefold, three of which are extensions of the 5G visions detailed below. Figure 2.1 shows an early 5G vision (2014).

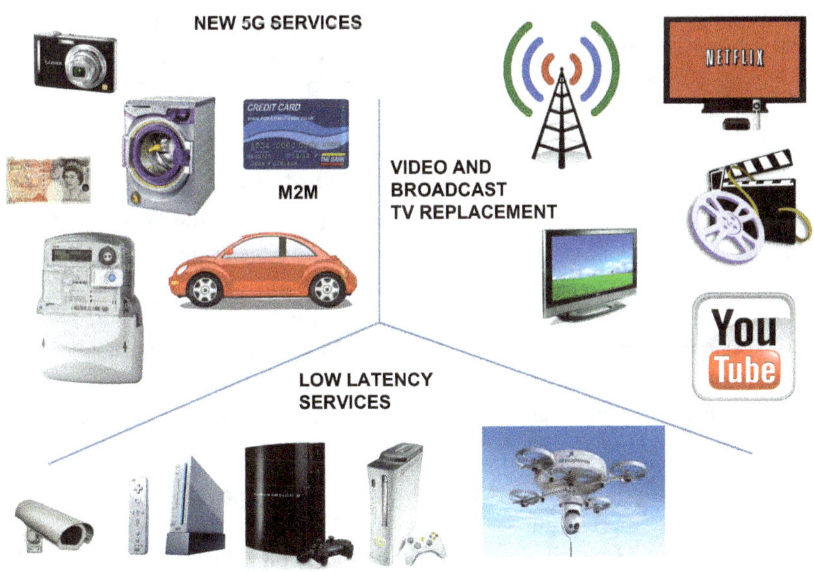

Figure 2.1 Early vision for 5G services

2.3 5G visions (use cases)

Is a vision the same as a use case? Either way, it is probably best to be cautious about them. For past generations, they have ended up as niche uses, and other use cases, totally unimagined at the time, have been the key. Here is a list of use cases (dreamt up or launched by the industry) that never really came to fruition or flopped for the mobile operators at least:

- Video calling;
- Location-based services;
- Femtocells;
- Picture messaging;
- eHealth;
- Payment from a mobile account;
- Walled Garden premium content;
- Wireless application protocol (WAP) – to access news, web chat, instant messaging;
- Gaming service.

Video calling and location services were a great success as over-the-top services, but the original plans and offerings from mobile operators were not. You could be forgiven for believing that mobile broadband and voice are the only genuinely successful services/use cases and visions. (SMS as an accident; it was designed for network engineers to communicate.)

The following three sections look at the main application areas for 5G, formulated during the gestation of 5G, from around 2012 until its launch in 2020 [3–7]. After describing the technology that powers 5G, we will look at real-world deployments, applications, and future directions and enhancements to 5G. This is an exciting story as 6G is going through the same cycle, and you can read about some of the cases and applications proposed for 6G in the following chapters.

2.3.1 5G – massive machine-type communications (mMTC)

This area is extensive – typically with applications covering transport, energy, health, factory automation, robotics, intelligent wearables, video surveillance and home automation. All the non-human things that might want to connect. The term IoT (usually pronounced 'I', 'O', and 'T') is essentially the same thing. The key is that this vision covers various devices with wildly varying requirements: From a vending machine that sends one small message daily ('I've not been stolen') to a driverless car controlled by an external network. Some of these requirements cover many orders of magnitude:

- *Power consumption.* Many sensors will not have a power source and rely on either energy harvesting or a long-life battery. The power consumption involved in connecting, transmitting and disconnecting to current cellular networks is too high for these devices.

- *Cost.* Many envisaged applications, such as pallet-tracking and tiny sensors, are inexpensive. The cost of 3G or 4G radios alone is too much to support these applications. More straightforward, cheaper radios are required.
- *Bit rate.* There is an enormous variation of potential bit rates. A vending machine might log on once daily to report that it is still working and does not need a service (a few bits/day), or a security camera may need to send HD video at 30 Mbit/s.
- *Number of devices.* Current cellular networks are not easily scalable to hundreds of thousands of devices per km^2. It is easy to imagine very dense sensor networks frequently trying to establish a radio connection to send small amounts of data. The random access channel (RACH – used to start a data session) of existing systems would not be able to cope with such a signalling storm.
- *Security.* SIM cards are unsuitable for many IoT applications. The cost and form factor alone would preclude several applications. In addition, the security overhead for the vending machine mentioned above, which sends 1 bit (i.e. 'everything is OK'), would be unsustainable.
- *Latency.* The typical IoT application requiring very low latency is autonomous vehicles. Trains and cars are imagined to be interchanging information that activates braking and steering systems in response to jaywalkers.

Excluding 5G, there are quite a lot of solutions for IoT being employed to meet some of these needs, and many applications fall within the parameters of these networks. LTE introduced two specific technologies to support IoT: known (with very little imagination!) as LTE-NB (narrow band 10–100 kbit/s but with high latency) and LTE-M (machines 1–5 Mbit/s, lower latency). Both offer lower capabilities than full LTE, but that lowers device cost, complexity and power consumption. These are now widely deployed around the world and used extensively. However, vast numbers of 2G and 3G connected devices remain, which makes it difficult to switch these networks off or re-use their spectrum. In addition, many non-cellular solutions, such as WEIGHTLESS, LoRA and Sigfox, are used for IoT solutions. Wi-Fi, Zigbee, Bluetooth and active RFID technology are also used for shorter-range networks.

A combination of support for limited, low-power devices as well as much improved coverage [with non-terrestrial networks (NTNs), such as satellites] and vehicle-to-vehicle or network communications (V2X) might unlock new verticals. These would include autonomous vehicles, agriculture, smart grids and mapping. To support mapping applications, the position resolution of the network is being greatly enhanced in 5G, with 10 cm resolution indoors and outdoors. In addition, the network will offer timing and synchronisation services, which means that devices will not need separate global positioning service (GPS) receivers. An example application is the programme to replace GSM-R (the GSM variant used on many of the world's railways for signalling from trains to control centres, including voice) with a 5G advanced solution tailored to the existing GSM-R bands. More widely, 5G will offer much greater flexibility in frequency band use and sharing.

Another very significant part of 5G is non-terrestrial networks. These encompass satellites, high-altitude platforms (HAPs – balloons and aircraft), and uncrewed aerial vehicles (UAV – drones). One significant development in the last few years has been the launch (or planned launch) of low Earth orbit (LEO) satellite constellations. These can offer global coverage, low latency, and higher data rates. By integrating them within 5G, it would be possible to offer truly global connectivity, at least out of doors. HAPs are being actively trialled to extend 5G coverage to areas that are not economical to provide terrestrial coverage. 5G will also allow remote control of drones and locate mini-base stations on the drones for temporary capacity increases and emergency use.

2.3.2 5G – enhanced mobile broadband (eMBB)

In this context, enhanced means 'faster' and more data. It is not quite that simple. A slight diversion into two key concepts is required to understand what enhancing mobile broadband means. Capacity, coverage and cost are the primary metrics of a mobile network. The secondary metrics include headline rate, latency, energy efficiency and reliability. Capacity has traditionally been measured in Gbit/s/km^2 in the downlink with a typical mixture of users, applications and uplink/downlink split. It depends on the spectrum available, how efficiently it is used and how many base stations there are per km^2. Each technology tends to have a maximum theoretical bit rate – for LTE, it was initially 100 Mbit/s but was extended to 1.2 Gbit/s in later upgrades. This headline rate is not that useful in real networks. Far more significant is the typical user download speed during the busiest hour(s). This is often much lower than the headline rate and reflects much better real-life conditions. It is also directly connected to the capacity, like the size of the cake available. In theory, a few people could get a large slice of the cake (radio resources), but as more people appear, the slices get smaller. The second key metric is coverage. In data terms, we could set a minimum data rate of 1 Mbit/s and go to representative locations around the network, such as indoors, outdoors, close to the base stations and then the cell edge. Inevitably, there would be places where the data rate dropped below 1 Mbit/s. That coverage % would be highest when the network was quiet and reduced to a minimum at the busy hour. If we decided we needed 10 Mbit/s, the coverage would still be lower. Sometimes, it is possible to boost coverage at the expense of capacity, for example, by allocating more resources (power, bandwidth) to those users with a significant radio loss between themselves and the base stations. These users can only achieve modest data rates because of the high propagation losses, plus possible interference near the cell edge. However, that lowers the overall capacity as these resources would have enabled users with better radio signals (e.g. those close to the base station) to have much higher data rates. We will return to the concepts of capacity, coverage and cost later in the section on 5G

Table 2.1 Global mobile data traffic growth [8] in the 5G standardisation era. 1 EByte is a billion bytes

Year	Global data (EB/month)
2017	10
2018	25
2019	40
2020	55

techno-economics. Still, the point is that 5G eMBB aims to increase capacity, coverage, headline rates and average user rates.

The demand for mobile data continued to increase in the years running up to the launch of 5G (Table 2.1) [8] – typically doubling every 18 months. If a mobile generation lasts ten years, then that implies at least a hundred-fold increase just for existing services. Much of this data growth was driven by video [5]. During the gestation of 5G, there were many compelling reasons to believe that even more video traffic would need to be supported on 5G mobile:

- *End of broadcasting:* Following the switch from analogue to digital TV and the auctioning of the cleared 800 MHz spectrum, together with increased delivery of TV from the non-broadcast spectrum (satellite, IP multicast, Iplayer, etc.), discussions have started about removing terrestrial broadcasts and switching to IP-based delivery. 5G (including satellites or HAPs) are often considered to reach the final 5% of users for whom it is deemed uneconomic to serve with fixed Internet and represent a barrier to an Internet-only TV broadcast.
- *Higher definition TV:* 8K was considered the new standard for TV broadcasts. Although not needed for current mobile screens, the advent of VR and new headsets have increased bandwidth requirements.
- *End of real-time delivery:* Users increasingly want to watch TV on-demand rather than live TV, meaning many separate deliveries of the same content and more delivery over mobile networks.

The data growth rate has now (2024) levelled off at around 20% increase per annum [9]. This is a saturation effect in that only so much video can be watched daily. It also reflects the move to smaller screens (laptops, tablet and phones) and that 8K TVs have not caught on.

2.3.3 5G – ultra-reliable low latency communications (URLLC)

One of the major themes of 5G was much better support for XR – virtual, augmented and extended reality, especially in a highly mobile environment.

Augmented reality (AR), glasses that interpret what you are seeing, and visual overlays require that the camera output is sent from the device for AI processing. A visual overlay is returned with a very low delay. Support for these is provided by making the network (especially the uplink) much more efficient for the types of traffic these applications generate – bursty, small, time-sensitive packets. Further changes are being made in 5G to better support energy efficiency in a broader range of XR devices, including those that require low power consumption. With a strict delay budget, the 5G network has been designed to offload much of the computational work onto network elements (such as computing). The applications for XR are seen as gaming and a large number of industrial and diagnostic uses. Later, in the genesis of 5G, the emphasis switched towards Industry 4.0, where the main applications are low-latency. Chapter 5 describes Industry 4.0 applications in detail.

One area that 5G targeted was autonomous vehicles. New interfaces have been added between cars and from vehicle to network. Some of these links must have low latency and be very reliable. Overall, the 5G visions supported a much more comprehensive range of terminals and non-eMBB applications across various verticals, more efficiently and with lower power.

2.4 5G key performance indicators

So much for the visions. How did these translate into the complex (numerical) requirements for 5G? During the genesis of 5G, many organisations, from vendors to industry organisations to operators, issued white papers detailing their vision for 5G. Analysing about 20 white papers [4–19] led to a distillation of a standard set of 5G requirements representative of all the organisations involved (Table 2.2). The final column shows the improvement required over LTE Rel. 12 (do not worry if Release 12 does not mean anything; we'll get to releases in the section on 5G Advanced). Global standards are extensively covered in Chapter 6, and the IMT 2030 report on 6G requirements is analysed in great detail in Chapter 16.

Despite differences in views amongst the various organisations involved, one of the most surprising things about this table is the commonality of requirements. 5G was, by and large, understood to mean a massive increase in capacity, data rates, connections and traffic densities – coupled with ultra-low latency, much-reduced energy consumption and improved reliability compared to LTE. Later in the chapter, we will look at 5G five years after launch. How do these use cases and visions stack up with actual usage? How does the actual performance of 5G compare to these KPIs from its gestation period? It is interesting to note that 6G development started with sets of KPIs (see the introduction for one example set) but has now moved away from these – this is, in part, a reaction to the story of the 5G roll-out.

Table 2.2 Summary of 5G requirements – colours indicate the change from LTE advanced on a traffic light basis. The next-generation mobile networks (NGMN) Alliance is an operator body. The 5G Infrastructure public private partnership (5G PPP) is a joint initiative between the European Commission and the European ICT industry. METIS was an EU Horizon project on 5G. The ITU (International Telecoms Union) represents global regulators and issued a report on 5G called IMT 2020, and ARIB (Association of Radio Industries and Businesses) is an Asia PAC association.

	South Korea 5G Forum	**ARIB 2020**	**IMT 2020**	**METIS**	**5GPP**	**Nokia**	**NGMN**	**Improvement on LTE Rel 12**
User-experienced data rate	1 Gbit/s downlink; 0.5 Gbit/s uplink		0.1–1 Gbit/s	1–10 Gbit/s	10–100 times more than 4G	100 Mbit/s min	50 Mbit/s everywhere	10× average 100× on edge
Peak data rate	More than 50 Gbit/s downlink; over 25 Gbit/s uplink	More than 10 Gbit/s	10–100 Gbit/s	No data	No data	10 Gbit/s	1 Gbit/s	10× average 100× on edge
Connection density	1000 times more connected devices	10,000 per cell	1 million connections per km^2	300,000 per access node	10–100 times more than 4G	10–100 × number of devices	150,000/km^2	×100
End-to-end latency	1 ms user plane	1 ms	About 1 ms	About 1 ms	1 ms	1 ms	1–10 ms	×10
Traffic volume density			10–100 Tbit/s per km^2	1000 times mobile data volumes	More than 1000 times higher cell throughput than 4G	10,000 × more traffic	1–15 Tbit/s per km^2	×1000
Mobility	More than 350 km per hour	500 km per hour	More than 500 km per hour	500 km per hour	1000 times higher wireless area capacity than in 2010		500 km/h	×1.5
Capacity	10 bit/s per hertz downlink; 5 bit/s per hertz uplink	More than 1000 times greater per km^2	5–15 times more efficient than 4G	36 TB per month per user	More than 10 times greater than 4G	10,000 × more traffic		×10 (With Wi-Fi and microcells)
Energy efficiency	100 times more efficient than 4G		100 times more efficient than 4G	About 10% of current energy consumption	80% reduction in the radio access network	Much lower J/bit	1/2000 that of LTE!	×100
Reliability				99.999% within time budget			99.999%	×10

2.5 5G key innovations

In this extended section, we will look at the key aspects of 5G that underpin the delivery of these requirements and differentiate 5G from LTE. The basic structure of the section is to move up the protocol stack from the spectrum, air interface, and radio network to the core network. If you are familiar with 3GPP and the standard releases, this section relates to releases 15 and 16. If this makes no sense, we will cover the releases in depth in the section on 5G Advanced.

2.5.1 Spectrum

LTE introduced 2.6 GHz bands, extensively used worldwide in urban areas but far less in non-built areas because of the lower coverage. 5G has accessed a relatively small amount of new sub 1 GHz spectrum – made available in some countries by the move from the analogue to the digital TV broadcast. However, much more new spectrum is being utilised at 3.5 GHz worldwide and 25 GHz in the United States. Further bands up to 60 GHz are being considered. Rather confusingly, the bands are given the following terminology:

- FR1 – Sub-450 MHz to 6 GHz (changing to 7.125 GHz) to (5, 10, 15, 20 or 100 MHz spectrum chunks),
- FR2 – 24.25–52.6 GHz (50, 100, 200 and 400 MHz chunks),
- FR3 – 10–20 GHz (later added and more difficult to exploit).

Table 2.3 outlines central new spectrum allocations for 5G in different geographic areas. It is essential to realise that there are 49 other bands designated for 5G, which cover both the new spectrum and many existing mobile allocations. An example is band n25, which comprises 1850–1915 MHz in the uplink and 1930–1995 MHz in the downlink. This standard LTE band is now being re-farmed for 5G, along with many other bands in the existing mobile spectrum by mobile operators. 5G can also use the 900 MHz, 2.1 GHz and 2.6 GHz bands previously used for 2G, 3G and 4G, respectively, and some operators are dynamically sharing spectrum between 4G and 5G in these bands. A complete list of the 5G bands is available in the standards document 3GPP TS 38.104. It is also important to note that many bands overlap, meaning they are only partially allocated in some countries. For example, only part of brand n78 is available in the EU and the UK. The bands are, in any case, often split up and allocated piecemeal to multiple operators. As of Q1 2024, the only use of bands above 10 GHz for 5G was in the USA. There is more about 5G at higher frequencies later in this chapter in the section on 5G progress. In the vision stage of 5G, it was anticipated that these high-frequency systems would be a significant component of 5G – delivering very high data rates and low latencies.

There are a lot of potential 5G candidate spectrum bands that are being considered for licensed/unlicensed or even shared use. In the next chapter, there is an extensive discussion on the 5.9–7.1 GHz band and its possible use by mobile technologies. A higher frequency spectrum is considered for both 5G and 6G in Chapter 9 on microwaves.

Table 2.3 New 5G mobile spectrum bands

	China	Japan	UK and EU	USA
Low FR1 (MHz) FDD	703–748 (up) 758–793 (down) Part band n28	Re-use of existing spectrum 700/800	703–748 (up) 758–803 (down) Band n28	824–849 (up) 869–894 (down) Band n5 663–698 (up) 617–652 (down) Band n71
Mid FR1 (MHz) TDD	2300–2400 Band n40 2496–2690 Band n41	Re-use of 2300		2496–2690 Band n41 – reuse of 4G band
High FR1 (MHz) TDD	3300–3800 (Band n78) 4800–5000 (Part band n79)	3600–4100 (Bands n77 and n78) 4500–4900 (Band n79)	3400–3800 Most of the band n78	3450–3550 3700–3980 Part band n77
FR2 (GHz) TDD	24.75–27.5 37–43.5 Studies	27.0–28.2 28.2–29.1 29.1–29.5 All n257	24.75–27.5 Band n258	37–40 (band 260) 27.5–28.35 (band 261)
Unlicensed/ shared spectrum (GHz) TDD		5.9–6.4 57–66	5.9–6.4 57–66	3.55–3.7 – Band n48 (CBRS) share 4.94–4.99 5.9–7.1 57–71

Spectrum is also an excellent place to start when looking at a mobile generation. One excellent reason is the (simplistic but very insightful) capacity equation from the introduction:

Network capacity (Mbit s^{-1} km^{-2}) = spectrum (Hz) $\times$ spectral efficiency (bit s^{-1} Hz^{-1}) $\times$ density of base stations (km^{-2})

If you want a massive increase in capacity (Table 2.2), then you can play with these three options and pretty much nothing else – that is the reality of the laws of physics.

We will look at spectral efficiency in the section on multiple antennas in more detail, but for now, suffice it to say that each successive generation has found it more challenging to increase this value, and the gain expected in 5G when compared to LTE-A, is in the range 1–3 (Table 2.4). Operators often complain that real deployments never match lab-measured and theoretical spectral efficiencies. A UK operator stated that a mixed 4G/5G network achieved about 1.5–4 bit/s/Hz.

To increase the capacity by thousands, you could densify the network by adding many smaller-scale base stations. A typical cellular network has large macro

Table 2.4 Typical spectral efficacy for different generations

Technology	Spectral efficiency (bit/s/Hz)
2G	0.17–0.33
3G (HSPA 2 × 2) 0.7–2.1 GHz	1.3
4G (2 × 2 MIMO) 0.7–2.1 GHz	1.5
4G (4 × 4 MIMO) 0.7–2.1 GHz	2.0
4G 8 × 4 (multi-user MIMO) 2.3–2.6 GHz	2.4
5G massive MIMO 3.5 GHz	5.8
Real-life mixed 4G/5G network (UK)	1.5–4

base stations (Figure 2.2) that transmit high power and offer comprehensive area coverage (1–10 km). They are often divided into three or six sectors. The illustration shows three pairs of antennae (white oblongs). Each one of the white oblongs contains an eight-by-two array of actual antennae that comprise two elements that transmit in opposite polarisations. The shape of the beam from each is tailored in a sideways lobe to cover a third of the surrounding area and angled in a vertical direction for best building coverage. The next tier of base-stations are microcells (often used in urban areas with a much shorter range (100–500 m) – usually sited on building roofs in urban areas and arranged with lobes of coverage aimed at nearby buildings.

Is the answer to a 1000× increase in capacity lots more base stations? No, because new base stations are expensive (we will quantify this later). Putting up new macro and micro cells means finding new sites, negotiating leases, getting planning permission, overcoming local opposition and commissioning a fibre backhaul connection. And that is all before you buy the new radio kit. It is much cheaper to install it on existing sites, but that does not cause any improvement in the densification factor.

So, that takes us back to the spectrum and partially explains why each generation seeks as much spectrum as possible. Not all spectrum is created equal; at least not in cellular systems. In general, the lower the frequency of the spectrum, the greater the coverage. So, in first building out a cellular system, you want a sub-2.5 GHz spectrum to keep the number of base stations to a minimum. As traffic grows and the need switches from coverage to capacity, the cheapest solution is to try and use more low-frequency spectrum on the same base stations (unless, of course, there is a shortage that has caused the price to be exorbitant, Chapter 7 looks at the whole question of spectrum in detail). Only when there is no more low-band spectrum would it make much economic sense to use a higher frequency spectrum.

2.5.2 5G air interface – the new radio (NR)

Every new mobile generation has had a new air interface (see Chapter 9 for a discussion about whether 6G should have one). Mainly, this has been driven by a desire for greater spectral efficiency, but it has also reflected the development of

Figure 2.2 Typical macro cell

electronic design, manufacturing, and processor capability. Thus, when the 3G air interface was selected, it was known that orthogonal frequency division multiple access (OFDMA) was a more efficient solution than the wideband code division multiple access (W-CDMA). However it was impossible to engineer a viable solution with existing electronic technology then.

There are three essential things to consider when assessing a new air interface. First, how it shares the radio resource between the base station and the mobiles – this is the multiple access scheme. Second, how it avoids transmitting and receiving simultaneously (using either time or frequency separation) and, third, how it mitigates radio propagation problems at typical mobile frequencies. This section will look at how 5G compares with previous generations. This is a slightly prolonged explanation, and those familiar with radio basics could skip to the end of the section. Still, these radio fundamentals do crop up again in the context of many ideas for 6G radio.

To serve different users, GSM used a time-division multiple access scheme (TDMA) – whereby each user was allocated a time slot within a frame structure and obtained all the resources for that short time. A good analogy was a teacher in charge of a raucous class. In TDMA, the teacher speaks to each pupil in turn in the downlink and listens to each in turn in the uplink (with the rest being quiet). 3G introduced a Code division scheme. The analogy now is that the class is more international, and each speaks only their language (Mandarin, Spanish, Scottish, etc.). Now, there is a teacher for each language, and they talk simultaneously, and everyone listens. However, the French student can understand the French version because all other versions fade into background noise. Each teacher can hear only their own language when they all reply together.

LTE introduced a new multiple-access scheme called OFDMA. In our analogy, this is the equivalent of splitting the large classroom into lots of little classrooms – all isolated. Only information for the (smaller number of) pupils is given out in these little classrooms (on a TDMA basis). There is much more about OFDMA in the section on radio propagation issues – but essentially, 5G NR (New Radio) uses a very similar OFDMA scheme to LTE.

The next essential thing to consider is separating uplink and downlink signals, known as the duplex scheme. If you try to transmit and receive simultaneously at the same frequency, all that happens is that the receiver is overwhelmed by the transmitter's leakage or back reflections/scattering. If you want to use the whole spectrum, the simple answer is to transmit or receive, but not both simultaneously. This is called TDD – time division duplex – precisely what happened in the classroom example – either the teachers or the pupils are talking but not both simultaneously! Another scheme called FDD (frequency divisions duplex) allows simultaneous transmission and reception on different frequencies that must be sufficiently far apart so that filters can provide a reasonable degree of isolation.

So, GSM can be classified as TDMA/FDD and 3G as CDMA/FDD. LTE was both OFDMA/TDD and OFDMA/FDD. In some bands, it uses FDD (usually the lower frequency bands) and in some TDD (mostly higher frequency bands 2 GHz+). 5G NR is typically OFDMA/FDD in bands up to 2.6 GHz and OFDMA/TDD in higher bands.

The LTE air interface was a significant improvement over 3G. What do we mean by a major upgrade? The most important metric for air interfaces is their spectral efficiency – for a given amount of spectrum, how much data can be transmitted? But you need to be careful with these numbers and any quoted ones. What it should mean is that averaged over many base stations and users in a typical scenario (maybe mixed rural and urban) and with typical traffic patterns (Internet use, video download) and an agreed acceptable quality of service (buffering, lost packets, delays, etc.) that the (downlink) spectral efficiency is the average throughput (bit/s) divided by the total spectrum used (Hz). It is standard to see values quoted for isolated cells (which are much higher as there is no interference) or in an unspecified scenario. It is not easy to calculate and hard to get accurate comparisons, but Table 2.4 gives a fair guide. The uplink efficiency is always lower – it is much harder to coordinate and synchronise the uplink transmissions of the mobiles back to the base station, which lowers efficiency in all the schemes. LTE uplink efficiency is significantly lower.

2.5.3 Radio propagation and the 5G NR details

Radio propagation is quite complicated – many different effects are in play – some directly related and some tangentially to others. In this section, we focus on these propagation issues and then look at how 5G NR negates these more efficiently than any previous air interface.

The best place to start is to imagine a very simple transmitter and an even simpler aerial (Figure 2.3). This is just transmitting a pure sine wave at 800 MHz (a standard cellular frequency). Suppose we have a simple handheld receiver that indicates the received power (there is no interference – in this test, no other nearby transmissions in or near 800 MHz are happening). In that case, we can map the power as a function of the receiver position. This reveals three different effects at three very different scales. The power drops off exponentially on a large scale (with the power measured over 100 m cells). In a rural environment, this might be close to the expected square law (twice the distance, a quarter the power) of free space. In urban environments, however, the exponential fall-off is more like distance to the power 3.5 – much more severe. At a smaller scale (typically 10 m), there is a noticeable fall-off in power due to moving inside a building (with each wall between the receiver and transmitter reducing the power by 2–10× – depending a bit on the wall material) or with a building or trees or vehicles between the transmitter and receiver. This is called shadowing and is superimposed on top of the general exponential loss. Finally, there is a finer grain variation in signal strength, typically on a scale of 20 cm. Move the receiver by 10–20 cm, and the signal can vary dramatically. This variation is caused by all the different ray paths from the transmitter to the receiver – variously scattered, diffracted and reflected around the environment – sometimes arriving in phase and constructively interfering and sometimes out of phase and destructively interfering.

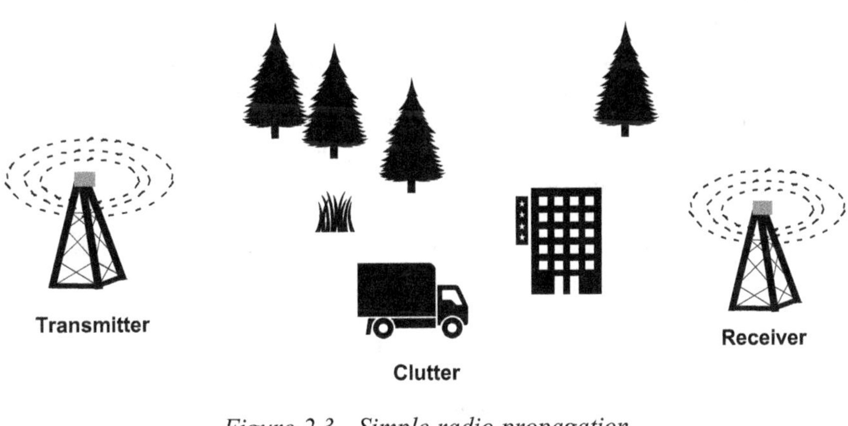

Figure 2.3 Simple radio propagation

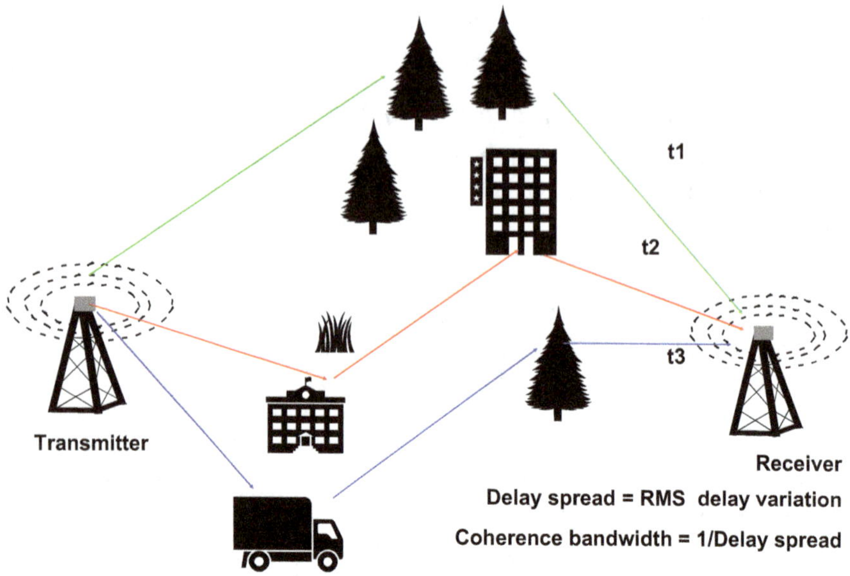

Figure 2.4 Multipath fading

A pure 800 MHz sine wave is not transmitting any information – if you start modulating it, the frequency spreads to a region centred on 800 MHz. If we swept the transmitter's frequency from 790 to 810 MHz, we would find that, at any given location, the received power would vary substantially on a scale of about 100–200 kHz.

The cause of this frequency selective fading is that the different ray paths (Figure 2.4) have a spread of flight times (typically called the delay spread td); as the frequency changes, these move from constructive to destructive interference over a frequency range known as the coherence bandwidth (which is about 1/td). Typically, td ranges from 0.25 μs in rural areas to 2 μs in dense urban areas at cellular frequencies). For completeness, there is further degradation if the receiver moves – the Doppler effect causes a frequency shift that must be accommodated in a radio system. Figure 2.5 provides a handy taxonomy of fading types.

How does the 5G NR overcome all of these issues? We will tackle the fading types in reverse order – because exponential and shadowing loss are mitigated by using multiple antennas in the next section! We will start by looking at frequency selective fading and how 5G turns this into an advantage.

What 5G does instead is to chop the spectrum available – typically 10–100 MHz – up into a whole load of small channels – called tones – each of which is narrower than (or at least comparable with) the coherence bandwidth. These are the different frequencies in OFDM we met earlier. Choosing tones with an excellent radio path (low attenuation) for a particular user can significantly increase the data rate – even though the average propagation to the users might be pretty poor. This matching of tones to individual users overcomes frequency selective fading.

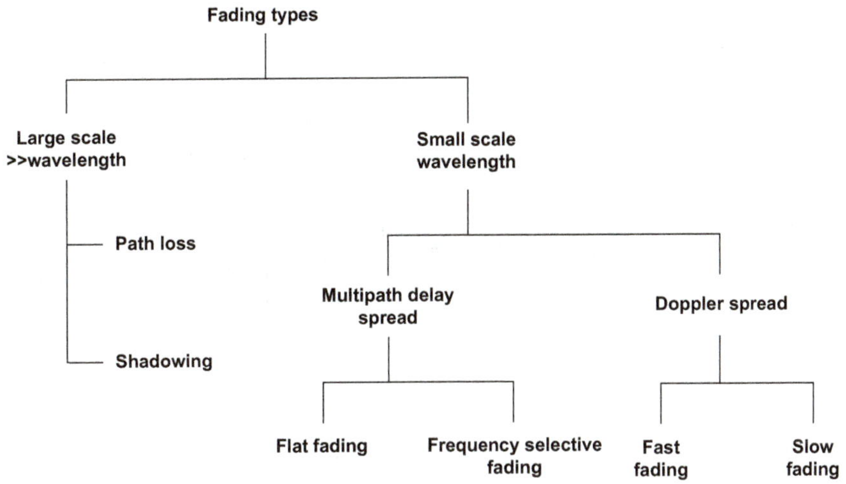

Figure 2.5 Fading types

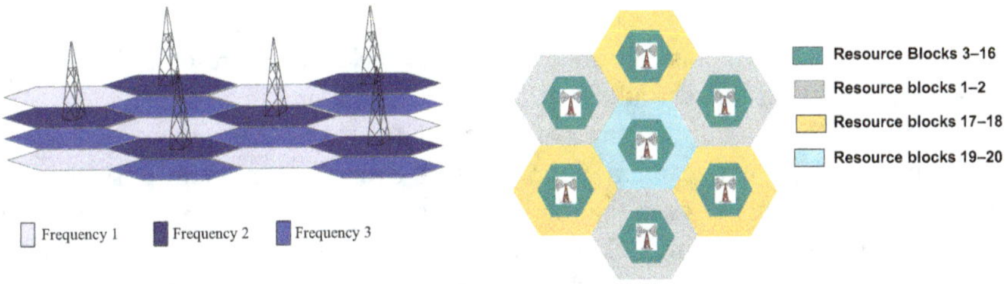

Figure 2.6 OFDM tones and fractional re-use

Another advantage of OFDM (there is no A when discussing the technology) is the ability to re-use the whole frequency block between neighbouring base stations. In GSM, neighbouring cells were allocated a third of the spectrum each, which enabled neighbouring cells to avoid interference. With OFDM, it is possible to allocate 80% of the tones to an area around each base station (i.e. serving users in this region) and then use the remaining 20% on a three-pattern repeat for users on the cell edge. This is called fractional re-use (Figure 2.6). In addition to fractional reuse, more than one base station can transmit the same tones to the same mobile – usually at the edge of both cells – to boost the data rate in a Coordinated Multipoint (ComPt) scheme. Neighbouring cells can also avoid transmitting certain tones causing interference in neighbouring cells – by time sharing them – and this scheme goes under the name eICIC (enhanced inter-cell interference coordination). These advantages mean the frequency re-use with OFDM approaches 3× that of GSM – one of the reasons for the overall spectacular spectral efficiency gain (Table 2.4) – along with the fact that GSM uses a wide 200 kHz channel that requires equalisation.

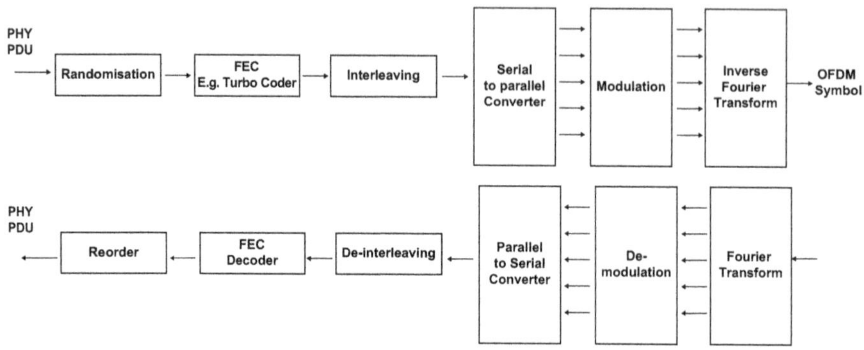

Figure 2.7 OFDM transmitter and receiver block diagram

OFDM has proved possible to implement in hardware for 4G (following advances in chip capabilities) – using inverse fast Fourier transform (iFFT) and FFT chips that are available and widely used for a variety of other applications (Figure 2.7). In 5G, the subcarriers can have a flexible spacing of 15, 30, 60, 120 or 480 kHz with a maximum of 3330 subcarriers. The optimum spacing varies greatly depending on the carrier frequency (with the higher spacings used at mmWave frequencies). Twelve subcarriers for the duration of 14 symbols constitute a resource block, and it is resource blocks that are the radio resources that are allocated for data transmission in 5G.

What is the maximum data rate of the 5G NR? Like a car's top speed or the resolution of a shiny new camera, people obsess over the maximum possible data rate – when the average or even the minimum matters much more in actual use. We need further diversion into modulation and channel capacity to get to this. This is important because it impacts 6G and leads nicely into the next section. So the problem or question might be: what is the maximum bitrate a radio channel can support? The answer is found in an equation first proposed by Claude Shannon in 1949:

$$\begin{aligned}\text{Channel capacity } C \text{ (bits/s)} &= \text{Channel band width } B \text{ (Hz)}\\ &\quad \times \log_2(1+ \text{ Signal power } (S)/\text{Noise power}(N))\end{aligned}$$

It is intuitively obvious that the capacity should be proportional to channel bandwidth and that the better the signal-to-noise ratio is, the more information can be transmitted on that channel. In a concrete example, I could have 10 MHz of spectrum and transmit 1 s and 0 s – turning the radio power from nothing to max (1 or 0) every 100 ns – so I get 10 Mbit/s. If, however, the signal-to-noise ratio is high, it is straightforward to tell the 1 s from the 0 s, and maybe I could use 0, 1/3, 2/3 and full power and still quickly tell the difference. If I take 2 bits and code them as 00, 01, 10 and 11 onto these values, I can get 20 Mbit/s out of my 10 MHz. This is a modulation scheme, and the power values are symbols. The symbol rate is 10 Mbit/s. In real systems, both phase and amplitude are expected to be used to create symbols. In 5G, a maximum of 8 bits/symbol is possible with 256QAM modulation (quadrature amplitude modulation), which means 256 different combinations of

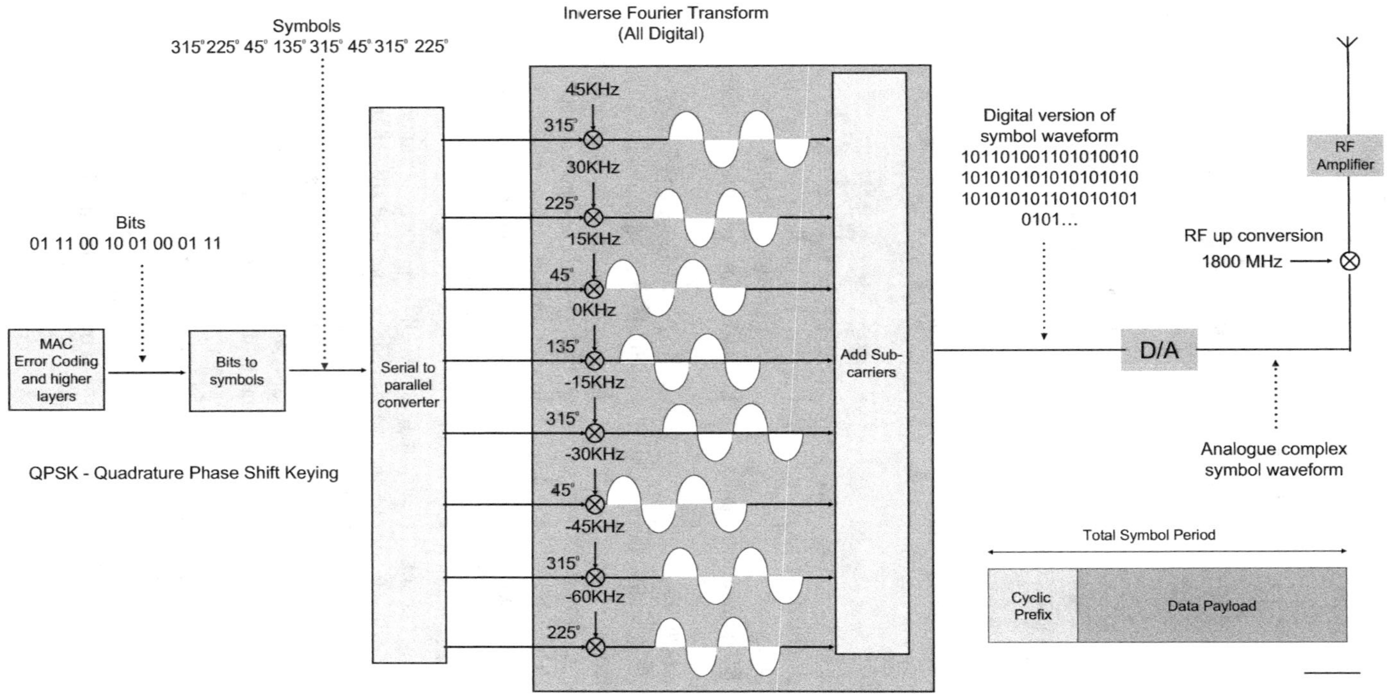

Figure 2.8 OFDM transmitter in detail forming symbols for transmission – note the iFFT block is all digital – the waves shown only exist as digital representations

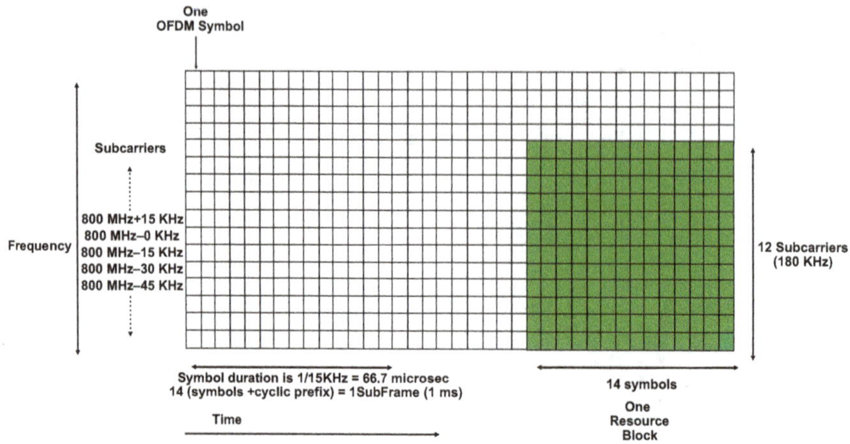

Figure 2.9 Resource blocks in 5G New Radio – shown for the example of 15 kHz sub-carrier spacing

amplitude and phase are being used and must be separated to decode the data correctly. Higher modulation rates and larger spectrum chunks have been the main drivers for increasing headline rates across the generations. Figure 2.8 shows OFDM with QPSK (quadrature phase shift keying) – forming a symbol from the tones (subcarriers)

5G is limited to 1 symbol per Hz, down to the orthogonal part of OFDM. Suppose you modulate (with frequency or amplitude) the subcarriers at the same frequency as the subcarrier spacing. In that case, the subcarriers remain orthogonal – you can recover each one at the receiver as if they were independent channels. Use a higher or lower symbol rate; they no longer remain orthogonal. There is an exquisite and straightforward mathematical way of showing this, but it involves knowledge of Fourier analysis. (Remember, there are lots of bits/symbols, so the efficiency of bits/Hz is much higher than 1.)

Coming back to one of the advantages of OFDM, each resource block of 12 subcarriers can be evaluated for all the users in each period and allocated so that the users get the RBs with the best radio conditions and then can use the highest modulation possible. As they move or radio conditions change, the whole resource scheme can be reallocated, and different blocks can be allocated. Even if a terminal goes into a building, say (and all blocks have a poor signal-to-noise ratio), the control algorithm can allocate lots of blocks (at other users' expense) if this is a high-priority connection. Figure 2.9 shows the formation of resource blocks, for example, sub-carrier spacing of 15 kHz – which is the lowest option in 5G and the same as LTE.

The highest data rate that 5G can support is 8 bits/symbol, and since there is 1 symbol per Hz of bandwidth, and the maximum bandwidth chunk available in 5G is 100 MHz, you would think the max headline rate is 800 Mbit/s (actually, a bit lower due to overheads, error coding non-data control traffic, etc., and retransmissions) – more like 600 Mbit/s. This is a lie because 'everyone knows' that 5G supports rates up to 20 Gbit/s. How that happens is down to two technologies known as MIMO ('My Mo') and carrier aggregation (CA). MIMO is a significant innovation in 5G and merits its section. CA will feature in the section following that on the radio network.

2.5.4 Multiple input multiple output

The multiple refers to transmitters and receivers. This is an important topic – one of the critical innovations in 5G is massive MIMO. Undoubtedly, this will be further developed in 6G – although the higher frequencies will be more about multiple narrow beams. It is (also) quite a complicated topic. It can be formulated mathematically (using matrices), but that does not give insight into what happens between the transmitter and the receiver. Cox [20] describes the matrix approach. However, that requires degree-level matrix algebra for those interested. For a more intuitive approach, imagine two transmitters – T1 and T2 – transmitting to two receivers, R1 and R2. Imagine T1 and T2 are coincident (in the same place), as are R1 and R2. This has no advantage since both receivers get the same signal from T1 and T2 with the same attenuation and delay. Now, move T1 and T2 far apart – typically several wavelengths, if you recall from the previous section. Now, the path from T1 to the two receivers is uncorrelated and utterly different from that of T2. In 5G, each resource block of subcarriers (just a chunk of the spectrum) is narrowband, so either path could have all the different rays adding up constructively or destructively. The best strategy for the two receivers might be to pick the transmitter with the best channel. Alternatively, as is more usual, both channels are neither good nor bad, and the receiver can combine the two signals in several ways to improve the overall channel. This is called diversity gain and has been used in mobile generations since 2G – often using antennas with crossed polarisation – each sending the same signal. The two polarisations have different propagation characteristics.

The next stage is to move the transmitters half a wavelength apart. Now, the beams from both transmitter antennas – spaced 1/2 wavelength apart – interfere. Suppose there is a suitable phase difference between them. In that case, they constructively interfere in a forward direction, as shown in Figure 2.10. If there is a line of sight to the receivers, then the power received is increased – meaning the signal-to-interference and noise ratio (SINR) is increased, and so a higher order modulation (e.g. 256QAM) could be used, or the coverage distance increased. This is called

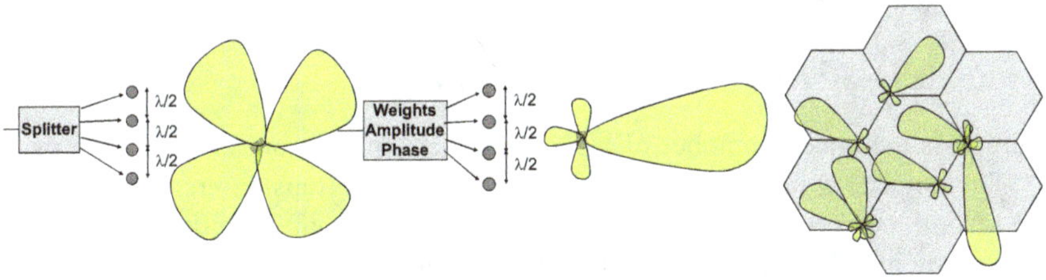

Figure 2.10 Basics of beam-forming

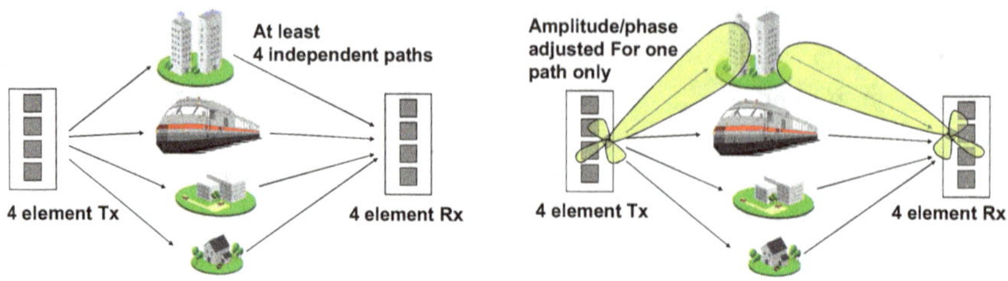

Figure 2.11 Beam-forming creates low loss from transmitter to receiver

beam-forming gain. There does not need to be a line of sight between transmitter and receiver. Often, there is a dominant path (possibly a reflection of a nearby building) and pointing the beam in that direction increases the received signal.

The same trick can be played at the receivers – place them far apart, and there is diversity gain but separated by 1/2 wavelength, and the receiver beam width can be adjusted in the same way as the transmitter – effectively producing extra gain. Figure 2.11 shows the case of four transmitters and four receivers – the buildings add radio clutter in this example. If you transmit a data stream from all four transmitters but with different amplitudes and phases so that the lobe of the subsequent beam points at only one of the paths and has less energy towards the other (it is unusual that the paths can be isolated in this way, but it shows the principle). Doing the same thing at the receiver points the joint lobe towards the same path. That creates a strong signal along that path and a very high loss along the other route.

You can also adjust the phase shifts and amplitudes to use the second path at the transmitters and receivers. The clever part is to use both paths simultaneously for the same or different data streams.

Sending the same data on all four paths can improve SINR, but sending different streams makes it possible to increase the data rate significantly.

The data for stream one are sent to the four transmitters with appropriate phase and amplitudes to go on path one and stream 2 (at the same time) with different amplitudes and phases for the second channel and so on. Because everything is linear, the output is the sum of both. At the receiver, I get two received signals, and by applying different amplitudes and phases to them, I can recover the two transmitted data streams. It is as if, by magic, the channel's capacity is quadrupled; this is called array gain. Shannon's law for MIMO systems must be re-written asfollows:

$$C\ \left(\text{bits Hz}^{-1}\right) = W\,B\,(\text{Hz})\log_2(1 + \text{SINR})$$

where W is the rank or number of layers in the channel. In our example, it was 4.

The obvious extension is to use 8, 16, 32, 64, etc., transmitters and receivers. The scheme being used (and it can change under different conditions, as we will see) is written 4T4R (meaning four transmitters are being joined with four receivers to create up to four layers, each of which can, in theory, carry an independent stream of traffic) (Figure 2.12).

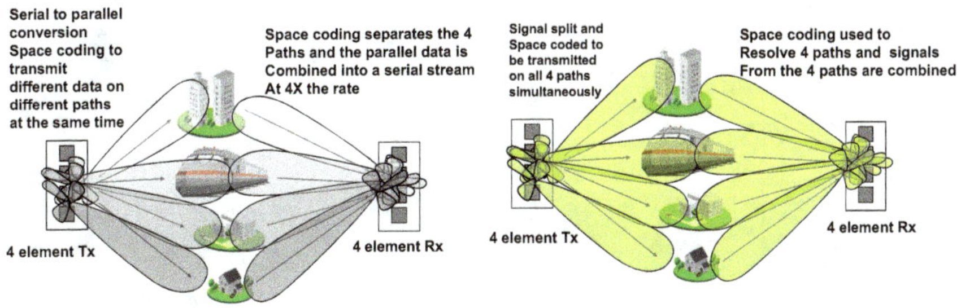

Figure 2.12 Four-layer MIMO

It sounds great, but there are quite a few practical issues involved. First, the simplistic example of Figure 2.11 never really exists. Often, a transmitter and receiver pair settle on the best combination of amplitude and phase at the transmitter (pre-coding) and receivers (post-coding) for maximum throughput that does not correspond to identifiable paths or clear ray directions. Somehow, the complete jumble of paths – with different rays and attenuations throws up weights that translate into odd-looking beams that are often hard to relate to the physical environment.

Second, to work effectively, the system needs channel information. In a 2Tx2Rx example, it is possible to write:

$$R1 = H11\ T1 + H21\ T2$$

$$R2 = H12\ T1 + H22\ T2$$

where the (Complex – i.e. including phase) loss from T1 to R1 is H11 (and so forth).

You see now where the matrices come in handy! But, if you have got first-grade maths, you know that these can be rearranged to find T1 and T2 – but that means you need to know H. In 5G, there are channel measurement and reporting mechanisms. The uplink and downlink are taken as reciprocal in TDD, but that cannot be assumed in FDD, and both must be measured. Channels can also change very quickly if the mobiles are moving.

The third thing is that layers rely on diverse paths between the different transmitters and receivers – if the path is the same (say, in a line-of-sight situation), it does not work. Try solving the above equations with H11=H22 and H21=H12 (answer you cannot – it signals that the paths are the same and there is no diversity and, consequently, only one layer can be formed). Figure 2.13 shows the formulation of MIMO coding for 4T2R.

Diversity, beam-forming, null-forming, and layer gain are confusing because each is a different facet of multiple Tx/Rx. Hopefully, an example will clarify how these are used in practice. Consider a macro cell in an urban area on a frequency of 800 MHz – the low band ideal for macrocells as it has the lowest propagation loss. Terminals close by have an excellent SINR – so even using the highest modulation

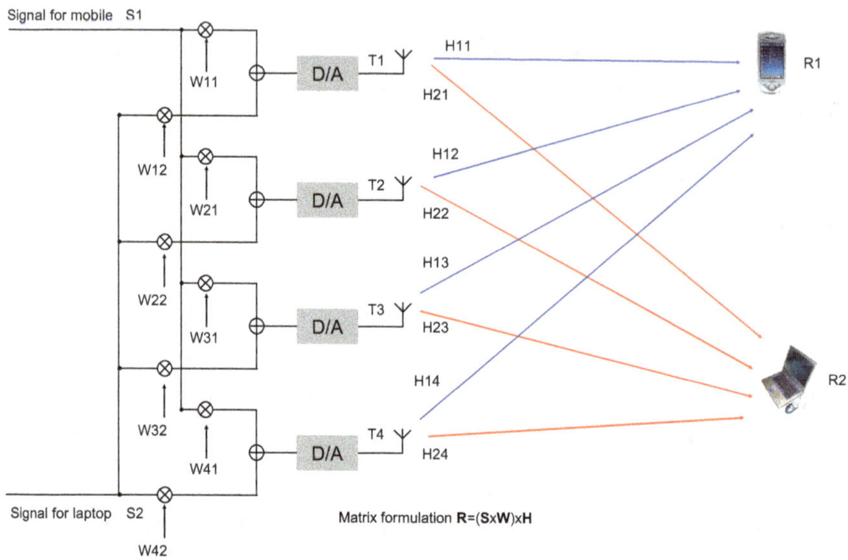

Figure 2.13 MIMO coding for a 4T2R arrangement

scheme (256QAM), the SINR is way above that for 10% errors or whatever is the optimum for the system. If they are not in the line of sight, these terminals often have multiple paths to the base station and using 4T4R can get very close to 4× capacity and maximum speed on the connection. Typically, the maximum number of layers is limited to 4 and increasing to 8T4R or 16T4R does not offer any gain (this is the downlink, and the small mobile terminal only has 4R and 4T – as on an iPhone 14, say). If there are four layers, each layer only has 1/4 of the power, and the RF amplifier output sets the maximum power. Mobiles far from the base station don't have enough signal for 4T4R and might drop back to 2T2R – two layers. At the cell edge, the propagation loss is substantial, and the interference from neighbouring base stations is much higher (very low SINR). Here, the best option is to use all transmitter elements to beam from a narrow lobe towards the mobile – or at least along the dominant path. This is one layer; here, polarisation is often used for diversity. Some trade-offs are needed in the transmit array – the optimum spacing for MIMO layers is one wavelength, and for beam-forming, it is 1/2 wavelength. So often, they are spaced at 0.67 as a compromise.

The restrictions on handset size and complexity/power consumption limit the Tx/Rx numbers. Base stations have unlimited space and power and can support much larger arrays. As we have seen, however, a 256Tx array would not be that useful for communicating with a single mobile. Remember, the base station is only transmitted to a single mobile at a given time and frequency range (resource block). Instead of using all the elements for one user, which might have hit diminishing returns at 8T, multi-user MIMO targets multiple users simultaneously. Figure 2.14 shows how different beams can target two terminals. The first beam can be

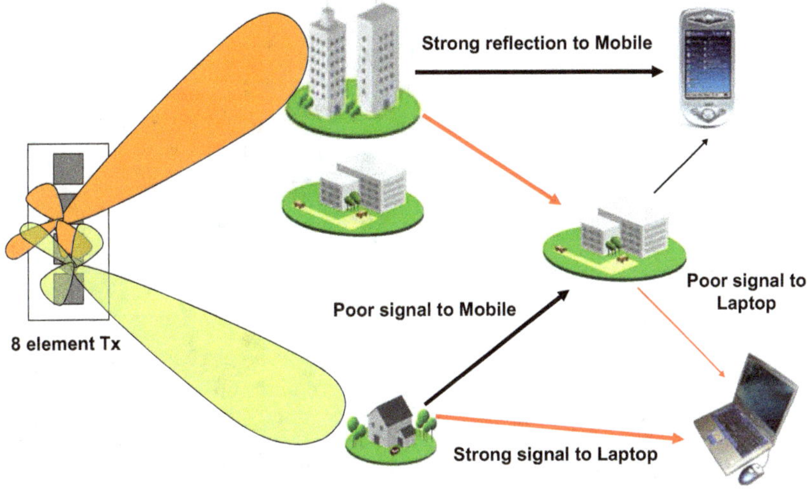

Figure 2.14 Multi-user MIMO in action

generated by a set of weights (phase and amplitude) at the 8Tx and applied to the data stream for user 1. The second beam uses a second set of weights and is applied to the stream for user 2. Because the beams have good isolation, there is little overlap (seen as interference), and now the same resource block can be used twice – effectively doubling capacity.

Moreover, if a 32Tx is used, both mobiles can have 4T4R and four layers, so the sector's overall capacity has increased (theoretically) by a maximum of 8×. It is very similar to dividing up a sector into multiple sub-sectors. It is beneficial towards cell edges where a transmitting cell can put a null of the beam (destructive interference) at a mobile communicating with a different cell – much-reducing interference.

Just a final point – all these techniques rely on digital coding – the coding takes place before the D/A converter – creating multiple streams that all need a separate D/A (and A/D at the Rx). These devices use a lot of power, and this (as well as the cost) increases with the frequency. It is possible to simply steer a beam with a simple analogue phase shift (much cheaper and lower power), and this is often combined with digital coding in a hybrid system (Figure 2.15).

This leads nicely to massive MIMO. There is no universally accepted definition of massive MIMO. Still, many consider it applies when the number of transmitters at the base station dramatically exceeds the number of mobiles, which is also quite large. Typically, in 5G, this might be 64Tx transmitting to something like 16 mobiles. It turns out to be quite challenging to adjust the phase and amplitude of the array to create 16 beams that also put the remaining mobiles into a null. Instead, one approach is taken to measure the propagation from the mobile to each transmitter element – 64 values – and then compute the optimum phase and amplitude to maximise the signal at that mobile. The same procedure happens with all the other mobiles, and the transmitter sends these different codings simultaneously. If the

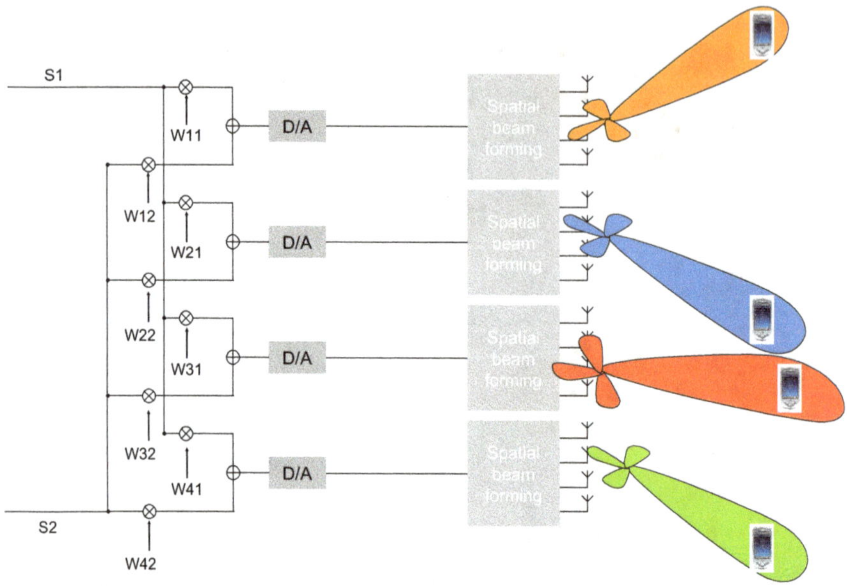

Figure 2.15 Hybrid MIMO – with limited coding and simple (phase-only) beam-steering

propagation is favourable, the strong signals at each mobile are enough to overcome any interference from transmitting to other mobiles. Even if one or two mobiles have a poor SINR, they can use a lower modulation (data rate), but the overall capacity is significantly boosted – up to 16×. Massive MIMO combines beam-forming gain, null-forming and multiple layers – Figure 2.16 shows how these can combine.

Massive MIMO is mainly being used at mid-band 5G (3–5 GHz). This frequency range is being introduced in urban sites for capacity enhancement. We will look later at actual deployment strategies, but it is evident that adding this frequency to existing sites is the cheapest way to deploy it. The problem is that 3–5 GHz has much worse coverage than sub-2 GHz frequencies – it does not propagate as far and does not penetrate buildings well. In an urban environment, there are many buildings and rarely a line of sight – most signals bounce around a lot from base station to mobile – creating many different paths that are often well isolated. Massive MIMO then compensates for the poor propagation of 3–5 GHz. So, as a concrete example, the Ericsson Air 6488 has 64Tx and operates on frequencies from 2.5 to 3.7 GHz. These units offer complete 64 digital coding, hybrid digital coding, and analogue beam steering. The computations for 64 digital coding are significant and will be featured again when we look at the radio access network (RAN). Massive MMIO can somewhat mitigate exponential propagation loss and shadowing – returning to how 5G NR is tackling these fundamental radio issues.

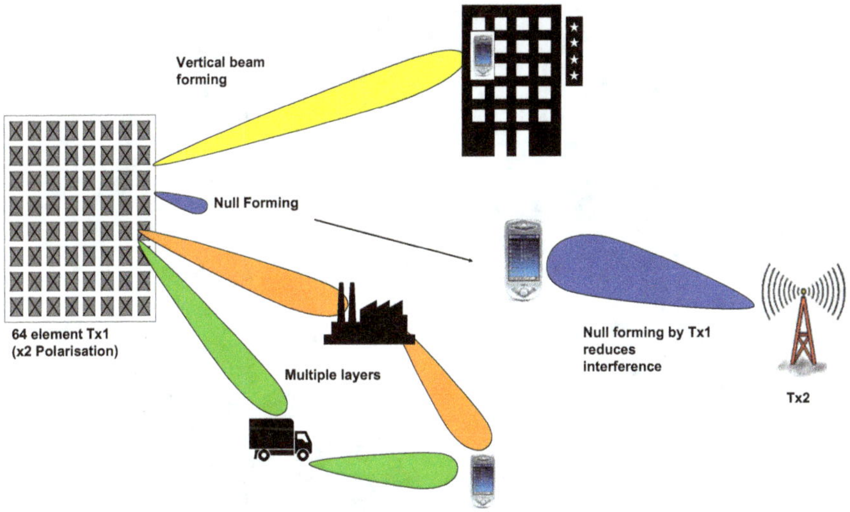

Figure 2.16 Massive MIMO

In the high band – mmWave frequencies – the path from base station to mobile is often dominated by a few dominant ray paths. It is possible to spatially select these using analogue beam steering – D/A's are expensive and power-hungry at these frequencies [21]. Figure 2.17 shows a typical urban 5G base-station operating at 3.5 GHz and including 64–256 element arrays..

2.5.5 5G RAN

Having covered the 5G spectrum, radio and massive MIMO, it is time to put the base stations together and look at the part of the 5G mobile network connected to and controlling the base stations. This section will analyse a stand-alone (SA) 5G RAN. This is the RAN a mobile operator would deploy if they introduced 5G independently of LTE or a new deployment. In practice, this is less common in the early implementations, and the 5G base stations are often attached to existing LTE networks (called non-SA) – we will look at this in the section on '5G deployments and progress' below.

Figures 2.18 and 2.19 show the evolution of the RAN over the generations. In 2G, the radio base station was known as the base transceiver station (BTS) and was connected to a base station controller (BSC) via an interface known as the A-bis. The 3G base station is called the NodeB, a bit of an in-joke by the people who developed the standard. The 3G NodeB has less functionality than the BTS, and more capability was placed in the radio network controller (RNC). In addition, mobility management was moved from the core to the RAN, necessitating a new inter-RNC interface called the Iur. This approach was not considered entirely successful. LTE did away with the network controller and put all of the RAN functionality in the base station, which is now (not very imaginatively) called the eNodeB. To make this work, each eNodeB must be connected to all its neighbours to manage mobility and interference. This was done using an interface called X2.

Figure 2.17 Example 64Tx64Rx 5G active antenna – new microcell site

5G (SA) is, in fact, quite similar to LTE (Figure 2.20). The eNodeB has evolved into a gNodeB, which is connected to all neighbouring cells via an Xn interface. What exactly is the gNodeB doing? First, managing access to the network allows mobiles to discover its frequency and synchronisation and to request a connection. It also manages the radio resources under its control, allocating resource blocks, managing MIMO layers, and any beam-forming/null-forming. In addition, the gNodeB has to take the IP packets moving across the RAN. If you are browsing a page, watching a movie or making a voice call, the requests and data are all assembled and encapsulated in IP packets. The RAN is responsible for placing the IP packets within a further packet-based protocol – including encryption with IPSec – transmitted over the air. This protocol also provides forward error correction, meaning that extra bits provide redundancy against errors and backward error correction in that missing or corrupted packets are retransmitted. The 5G RAN has been designed to reduce the delay due to backward error correction as

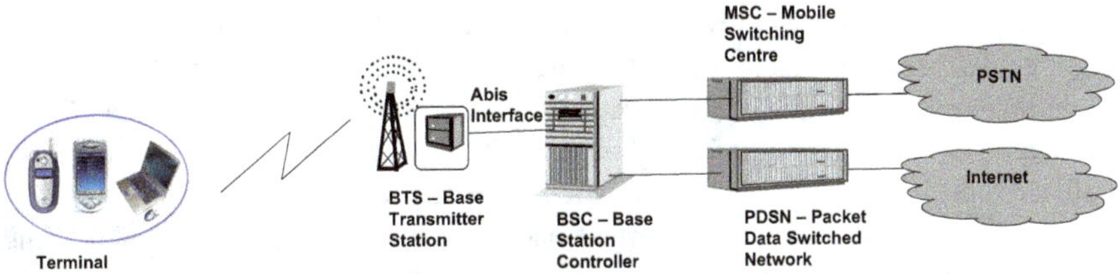

Figure 2.18 2G RAN

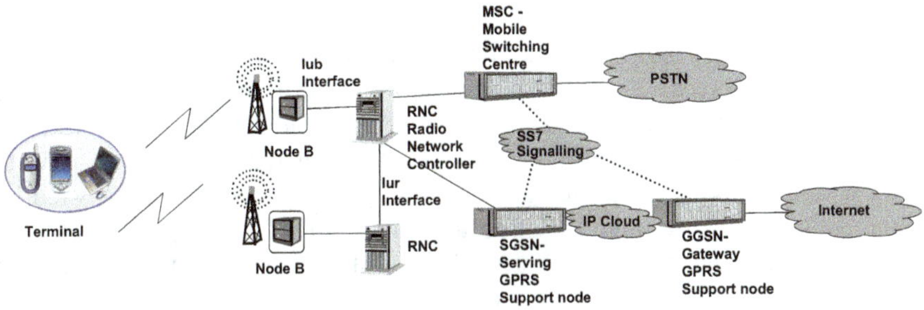

Figure 2.19 3G RAN

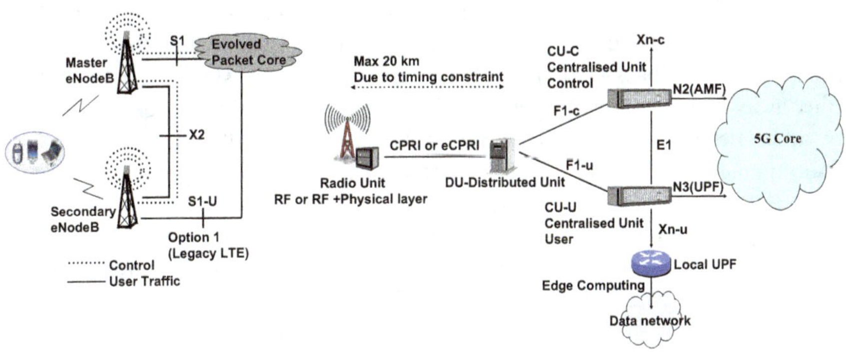

Figure 2.20 LTE RAN (left) and 5G SA RAN (right)

part of the drive to reduce latency. The gNodeB also coordinates its radio resources with neighbouring cells. This allows seamless mobility for users and neighbouring cells to utilise various interference management techniques. These include modifying the frequencies and power to alter the coverage area to provide less overlap with neighbours. Another method is forming a beam to create nulls for mobiles at the cell edge connected to a next-door gNodeB. Two or more gNodeBs can transmit the same resource block simultaneously to a terminal in the cell edge between them – significantly increasing signal power (SINR). This is called ComPt, which we have already met alongside eICIC. It has been described as 'distributed MIMO'.

To understand the next level of detail of the 5G (SA) RAN, it is necessary to look at actual base stations and how they have evolved. In the days of 2G, a typical site might look like Figure 2.2. Because so much functionality is in the BTS, a small cabin would be required to house it. This would include a powerful RF amplifier that fed a passive antenna, often with two inputs for different polarisations. By the time of 3G, the equipment had shrunk in size, and the NodeB functionality had been split into a remote radio unit (called an RRU) that handled most of the lower layer functions – including up-conversion to the carrier frequency and the RF amplification. The remaining functions were purely digital and now contained in a small base-band unit (BBU). Figure 2.21 shows the evolution of the base station.

The BBU only handles signals after down-conversion and A/D and D/A – allowing a digital feed to the RRU by a fibre. Typical sites had three passive antennas (Figure 2.2) – each with a vertical coverage of about 60° and a horizontal coverage of 120° – achieved using an array of antenna elements. However, all were driven with the same signal. Each aspect could transmit two polarisations fed by separate radio chains from the RRU for diversity or 2T2R MIMO.

In 4G, the architecture and interfaces allow the BBU to be separate from the radio site. This was facilitated by replacing microwave links, which were low-cost but had limited data capacity, with fibre connections, which were low-cost but had limited data capacity. The BBUs were centralised in a building (sometimes known as a BBU hotel). This was called C-RAN (there is some confusion over this term – and it is often said to be Cloud RAN – but C-RAN originally referred to Centralised RAN as described). The hotel could not be too far from the radio sites because the BBUs were responsible for the backward error correction and retransmissions over the air. The further away they were, the longer the delay was in retransmitting packets, which, in turn, added delay to time-sensitive services such as voice. Further developments have led to running the BBUs for a cluster of base stations on general-purpose computing hardware using virtualisation techniques, which will be discussed later.

The 5G RAN takes all this to the next level by splitting up the gNodeB into several sub-components with defined interfaces between them (Figure 2.20). The RU takes a digital connection from the distributed unit (DU) that carries all the user data and the control information for users and the network. This is known as the front haul and is carried using a protocol called common public radio interface (CPRI) or eCPRI (enhanced to be taken over ethernet, often offered by fibre connection providers and provides compression). Theoretically, the DU could send a digitised version of the baseband signal – leaving the RU to just up-convert and amplify. In practice, for massive MIMO sites, the RU often includes the logic for computing the phase and amplitude of the elements. The front haul distance is limited to 20 km or so because the DU is responsible for retransmissions – leading to a physical layout as shown in the figure with DUs clustered in locations at about one per town and a few per city (not unlike old telephone exchange distribution). The motive for this is to reduce costs. The less equipment at the network's edge, the better – that reduces rental, maintenance and power costs. Moving elements of the RAN to a data centre and running on general computing resources centrally allows

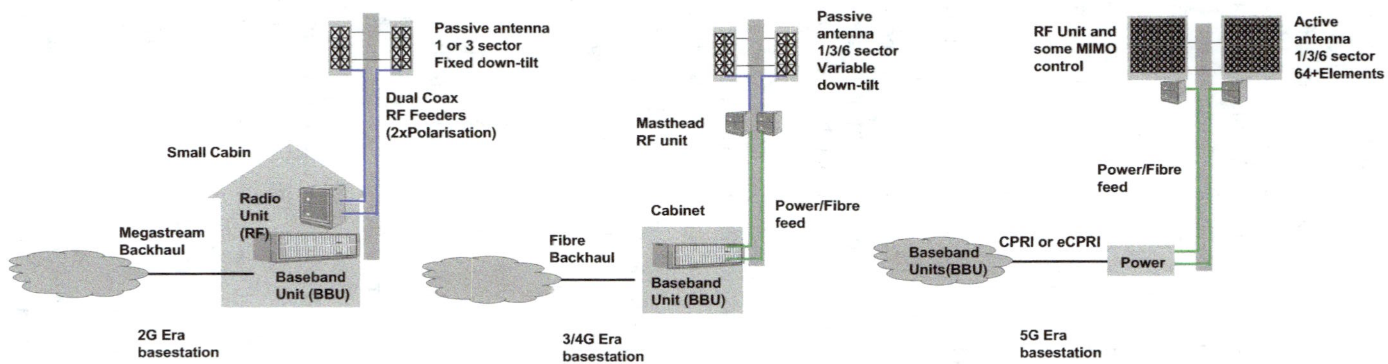

Figure 2.21 Evolution of the base station site (showing only the DU of the gNodeB)

sharing and scale cost savings. The centralised unit (CU) covers mobility and security management, radio resource management and running the packet data convergence protocol (PDCP) that takes user and control traffic (usually in IP packets), applies header compression, re-orders packets and applies encryption. The connection between the DU and the CU is called mid-haul and can be more like 100 km because of less strict time constraints. One CU could connect to multiple DUs. The CU connects back to the 5G core using the N2 and N3 interfaces for control and user plane functions (UPFs), respectively (and the CU is capable of user and control separation, as shown). As of Q1 2024, many operators had not taken advantage of these splits, and the DU was still co-located with the base site.

Another hot topic for 5G is Open RAN (Sutton [22] provides a good summary). The rationale for Open RAN is that going back to the Abis interface in 2G and Iur interface in 3G, they were proprietary – meaning that the standards did not fully specify them and, consequentially, all RAN elements had to be purchased from the same vendor. In LTE, the X2 interface was standardised by 3GPP. Still, many vendors added proprietary extensions that made interworking eNodeBs from different vendors difficult – requiring the core network to cover these and limiting functionality. These extensions often provided additional functionality that was not envisaged at the time the standards were developed. An excellent example of 5G is the ability to dynamically share a single spectrum band between 4G and 5G, depending on demand. Ericsson has a technology called Ericsson Spectrum Sharing that achieves just this. It has subsequently been realised this would be an excellent feature to add to 5G, and it is now being standardised. Much the same thing has happened in 5G, with vendors adding extensions to the Xn interface.

It is essential to distinguish between Open RAN, an initiative within the industry to disaggregate base station hardware and software, and OpenRAN, a project group that defines and builds base stations on general-purpose hardware. O-RAN refers to the O-RAN alliance – a group of mobile operators developing specifications and testing interworking. The O-RAN alliance is working with the Linux Foundation to create open-source software for the RAN. Open RAN is looking at ways to split apart the functions of the radio chain – Figure 2.22 shows the radio chain from the antenna to the CU and how Open RAN is considering new reference points.

So, in summary, mobile operators would like to buy a minimum amount of radio equipment and connect it to data centres with dark fibre or switched ethernet and run open-source software on the O-cloud (this is the underlying computer infrastructure) to provide all the functions of the gNodeBs. This could lower costs and improve the interworking between radios from different vendors. A final advantage is claimed to be the innovation of network control and management. In the Open RAN architecture (Figure 2.23), two new elements are introduced – the non-real-time RAN Intelligent Controller (nRT-RIC) and the real-time RAN intelligent controller (RT-RIC). The RT-RIC provides control of the algorithms and settings of the DU and CU in the timeframe 10 ms to 1 s. The nRT-RIC includes input into the RT-RIC in a slower time frame. These elements can run different RAN control algorithms like XApps and rApps.

An example would be a small, innovative software company that developed efficient machine learning algorithm apps that could learn traffic patterns and offer

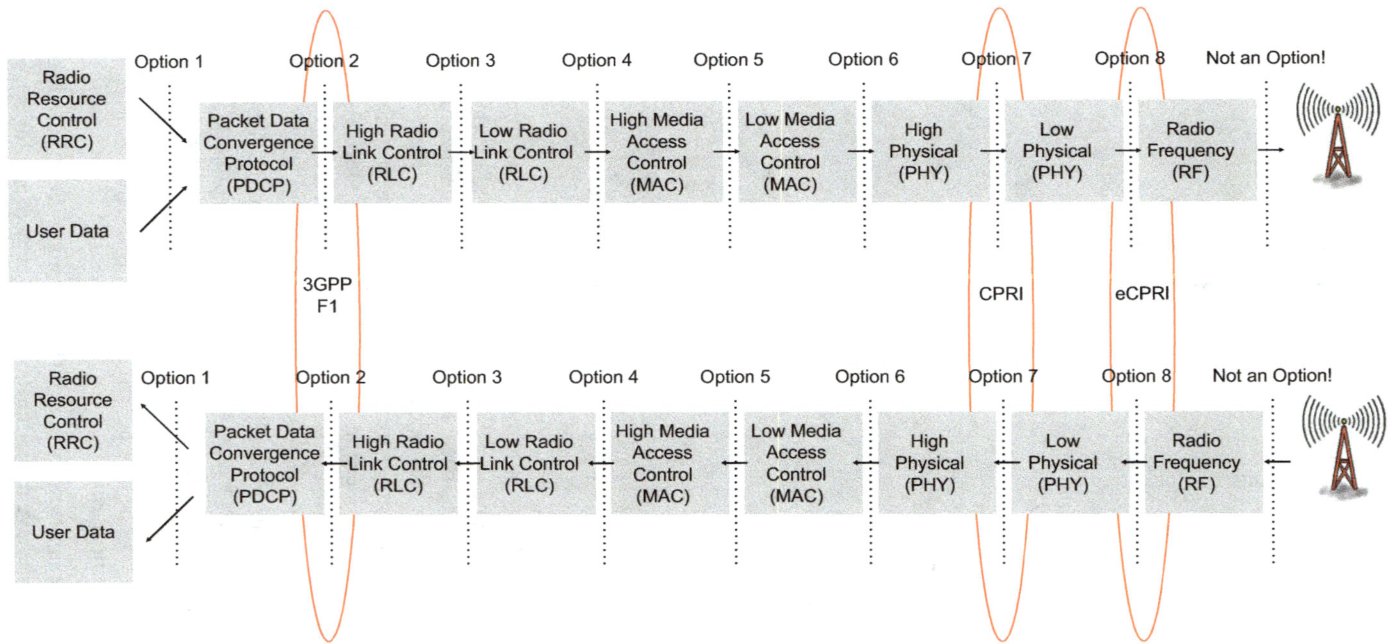

Figure 2.22 RAN split options as discussed by Open RAN

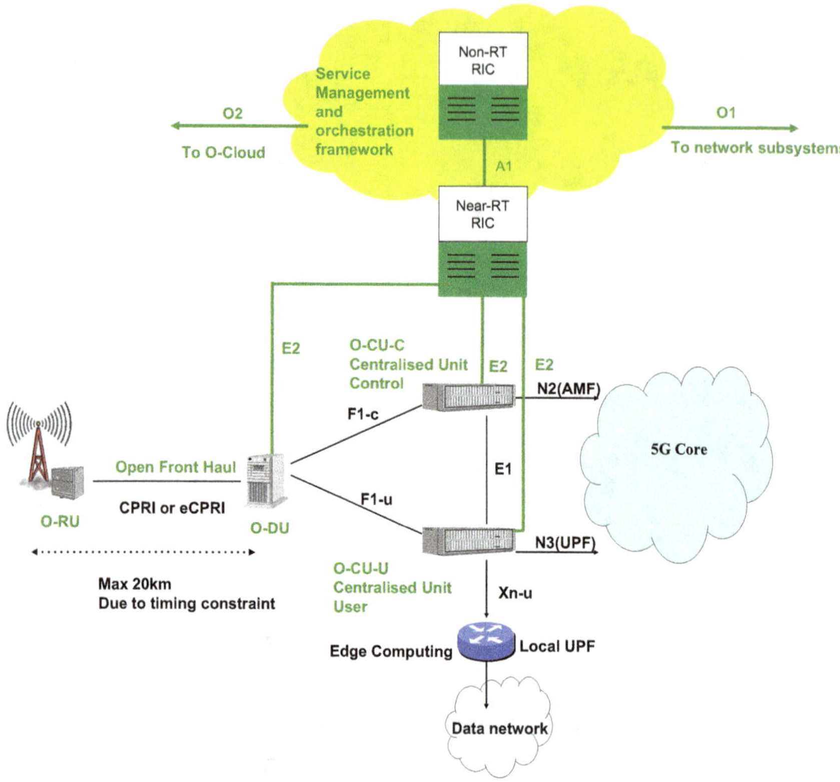

Figure 2.23 Open RAN architecture

high network throughput with dynamic cell resource coordination. If they performed better than existing apps, they could be swapped overnight. The idea is to stimulate innovation and speed up the introduction of new control algorithms instead of waiting for the next release from a traditional vendor.

It turns out that making hardware and software from different vendors work together is quite tricky – despite the specifications in Open RAN. Open RAN is taking about 6%–10% of the RAN market [23], but practically all these sales are made by existing mobile vendors supplying the whole RAN. NEC, Fujitsu and Mavenir are new entrants to the RAN market, but all pitch themselves as ‘end-to-end’ suppliers. Open RAN features heavily in many 6G projects but has yet to make much of an impact on 5G. Vodafone has a trial [24], but it is not clear whether this is a multi-vendor. They have said 30% of their network will be OpenRAN by 2030.

2.5.6 5G core network

Having seen the complexity of the RAN, you might be excused for thinking the core network did not have many functions. It is responsible for allowing users onto the network (while rejecting stolen or blocked accounts/equipment). The core needs to tell the network elements what services the user has subscribed to (LTE/5G? Unlimited or restricted data? Max speed? Roaming?). The core also coordinates mobility management and has a role in handovers – diverting the user traffic to the new cells. It has to provide VoIP (Voice over IP) services and connect to the PSTN (public switch telephone network – aka fixed phones and call centres). Finally, it transports the user traffic, in Ethernet frames or IP packets, to a suitable interface point. This might be a server on the Internet for a video from Netflix. A voice call to a fixed phone might have to go to a voice gateway for conversion to a circuit voice connection. If it is a sensor reading from a factory robot, it might need to go, with very low latency, to a controller within the factory. Figure 2.24 shows the 5G core.

Each different connection from the terminal to an external data network (DN – e.g. VoIP/Internet/low latency industrial network) is associated with a session management function (SMF), and each SMF function can control multiple UPFs – in reality the UPFs are the encapsulation and routing of the user traffic from NG-RAN to the DN. Before a terminal can send any data, it has to interact with the Access and mobility management function (AMF) – this, in turn, consults the unified data management function (UDM) – which is an extensive database of user information [authentication is handled by the authentication server function (AUSF)]. The policy control function (PCF) is responsible for discovering the parameters that should apply to each connection [maximum data rate, quality of service (QoS), maximum delay, etc.] and communicating these to the SMF and AMF. The remaining entities – AF (application function) and NSSF (network slice selection function) will be explained below.

Essentially, the core is responsible for getting the user data from the terminal to the correct DN connection with the appropriate QoS while managing the user's movement and collecting any charging information. It also handles the case of roaming. At a superficial level, it looks similar to an LTE core – which provides all the same functions. However, the 5G core has been modernised to introduce many of the concepts of cloud computing and virtual machines, and to explain this, a slight digression is needed.

In the 1990s, all the elements of a mobile network were physically separate boxes, and the interface between them depended on protocols specially created for mobile networks (in truth, they often looked like a computer rack). These units had very close integration between hardware and software. Upgrading for more capacity usually required trading them in for a bigger unit, and they were often not expandable. Similarly, new software and features came only very rarely from the vendors and only then at a cost. There was no way to add new functionality without waiting for the vendor to develop it and offer it as a product upgrade. Most computer systems of the time were very similar – dedicated software built from the ground up to run on a specific machine or operating system – with elements like databases running on stand-alone machines and being accessed by remote protocols.

The first stage in moving away from a traditional telecoms architecture is to create a software-defined network (SDN; Figure 2.25). Essentially, this means

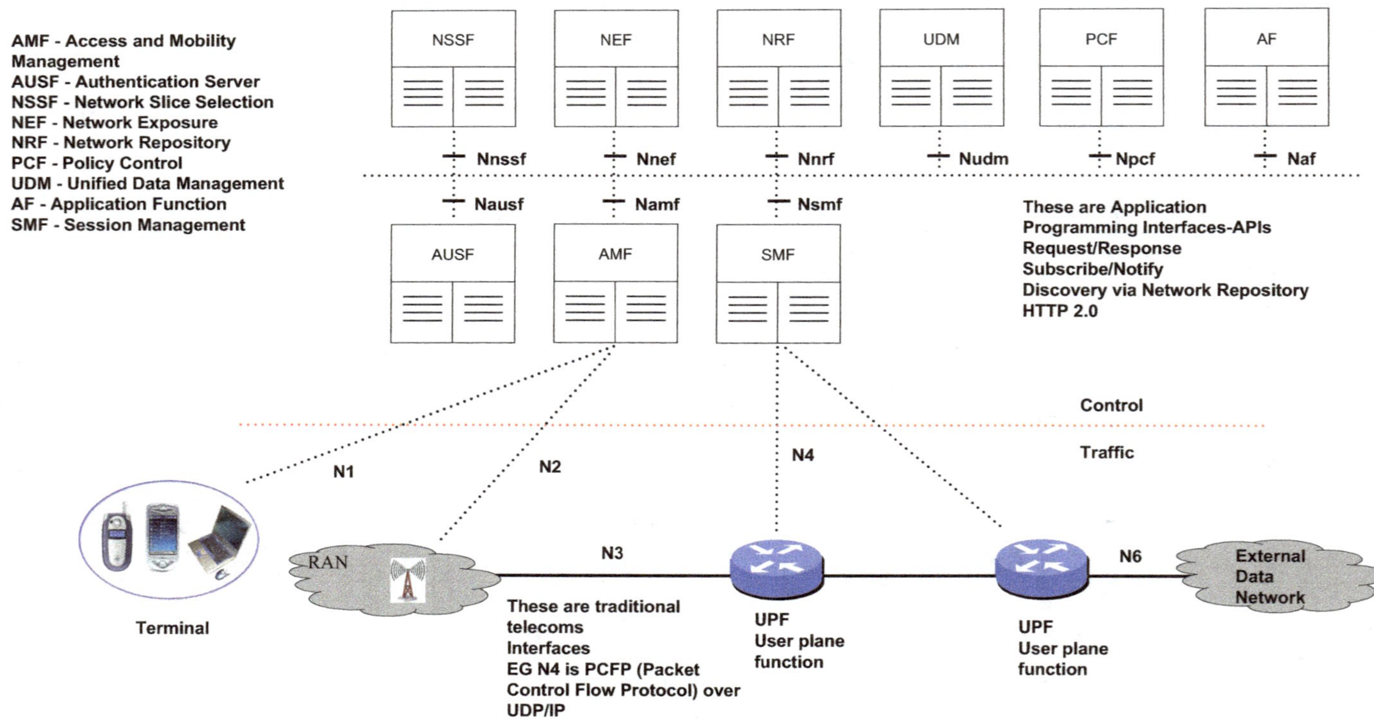

Figure 2.24 5G core

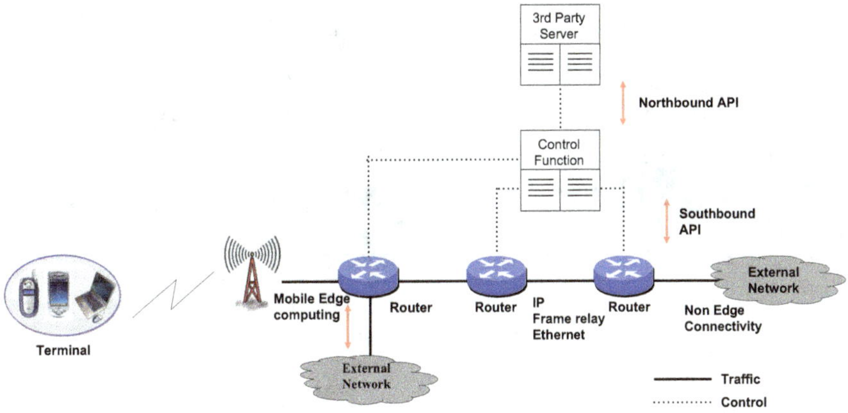

Figure 2.25 5G network as an SDN

separating the control elements from the traffic elements. This began in LTE and progressed as LTE-A was developed. The essential advantage is that the control elements can be separately dimensioned and updated. Because they are far less hardware-intensive than the user elements, they are easier to run on virtual platforms. It also means the underlying transport networks (such as IP/Frame Relay/Ethernet) no longer need to be dedicated to mobile data only.

Fast forward to 2024, and many large-scale software systems are being written to run in the cloud on platforms such as Amazon Web Services (AWS). It is possible to create databases, call centre management control, invoicing, credit checking, ordering and almost everything else needed to automate any business using cloud services' software framework. These functions run on general-purpose hardware and can be expanded and upgraded easily as the company grows. In practice, cloud services are data centres – with servers based on Intel X86 chips, memory and software capable of creating virtual machines from these elements combined with massive connectivity.

The significant advantages of running services from the cloud are that the hardware is cheap (mass-produced and used for everything), the capacity can easily be increased and scaled, and it is built on a platform that allows interworking with other services built in the same way. Virtualised machines (VMs) are nothing new – Java virtual machines (JVM) have allowed Java programs to run on various hardware and operating environments since 1994. It has become more critical for telecoms as the hardware has continued to get more powerful (Moore's Law) and now can perform many of the functions that used to require dedicated hardware. It is critical to distinguish between virtual machines on a single server and those running in a data centre using shared processing, memory and storage – cloud computing (Figure 2.26). Cloud computing dates back to its origins in time-shared on 1960s mainframes. However, the rise of offerings such as AWS, coupled with the growth of hardware speeds and optical high-speed connectivity, has fuelled the

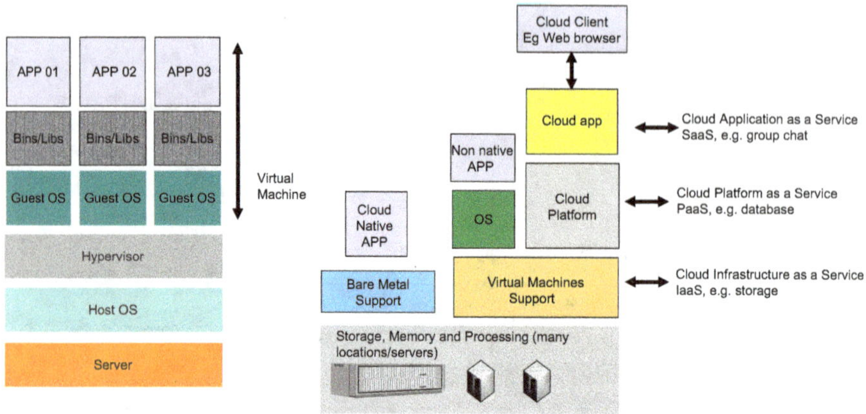

Figure 2.26 Virtualisation and cloud computing

cloud revolution. Some vendors are offering 'cloud native' 5G cores that are designed from the ground up for cloud implementation.

The 5G core has been standardised to assist it in running on disturbed general-purpose platforms (although the standard itself is not fully cloud-native). Instead, some of the traditional interfaces between elements have been specified as APIs (Application Programming Interfaces), able to issue function calls such as:

Nudm_SubsrciptionDataManagement_Get

To any programmer, it just looks like a function call to get data about a user from the UDM. All the calls work on the Request/Response or Subscribe/Notify model, and processes, discover or register other processes via the registry function. HTTP 2.0 (over UDP/IP) conveys these messages. These are much closer to Internet protocols and architectures than previous generations. The advantages of moving to this architecture are reduced interfaces, a much simpler and scalable design, hardware and software independence, and excellent compatibility with third-party services and interfaces (these all use the API model). The interfaces N1/N2/N3/N4 and N6, however, use traditional custom telecoms-only signalling protocols, which limits the overall flexibility. All major vendors now offer 5G cores that run on cloud services – a leading UK operator has a core that runs over 2200 servers distributed in its cloud network at diverse locations.

2.5.7 Network slicing

Another prominent feature of 5G is network slicing. This is a mechanism to partition the network resources in many different ways. In the first example, a mobile network operator might create subscription classes (Gold, Silver and Bronze) with different guarantees for minimum bandwidth, average bandwidth, and latency. Another simple way of slicing the network would be eMBB, URLLC and mMTC – illustrated in Figure 2.27. Slices can also be targeted at different consumer market

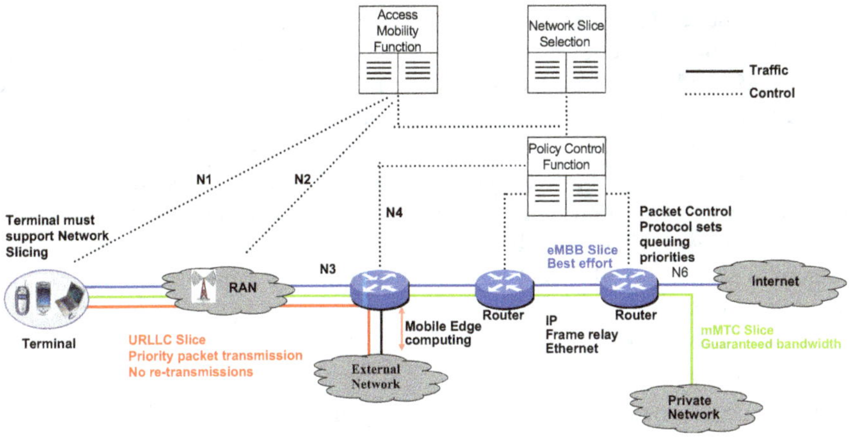

Figure 2.27 Network slicing

sectors – one leading candidate is cloud gamers. Many of these multi-person games require low latency for high performance. A gaming slice would offer significantly lower latency and, possibly, minimum and average throughput. It could be tied to specific games (perhaps activated by buying the gold version of a game, with the producers paying the MNOs for the slice), or it could be a user subscription or even an in-game or app purchase.

Another example might be creating a slice suitable for a particular industry (often called a vertical). A concrete instance of this on a real network is outside broadcast. If CNN or the BBC want to cover the book launch of '6G Evolution or Revolution', then, traditionally, that would involve lots of cameras feeding back to mobile trucks, often with long fixed cables. The trucks then relay the feeds to the studio via satellite or microwave links. A 5G network can create a network slice for camera feed upload – this might have a guaranteed uplink of at least 25 Mbit/s per camera with a stable latency of 20 ms – getting towards that offered by fibre. This service would then be provided to any broadcasters – subject to network coverage more or less on demand. This might open up opportunities for new broadcasters (Tiers 2 and 3) and make integrating remote cameras and drone feeds easier. (This is a real trial example used on 5G today).

Another example of a slice might be a factory (let us say it is full of robots). If the factory is a stand-alone plant, it might be better to have a 5G private network (explained below). Suppose the factory is on campus or on the ground floor of a sizeable multi-tenanted building. In that case, 5G might be an ideal solution for connecting all the robots to controllers, feeding back video, and allowing operators to take remote control. Without a slice, the factory runs very well at night. However, the moment the gamers gather for their gaming sessions in the cafe next door, the whole thing grinds to a halt because latency has increased and the bandwidth has been reduced (or, more realistically, network use just rises, and everyone's quality of service is degraded equally). The MNO can create a slice for the factory that might again guarantee latency and minimum bandwidth.

5G introduces many new features to support slicing – The NSSF (network slice serving function) – ascertains the slices the mobile has subscribed to and ensures that the AMF and SMF allocated to the mobile can support these slices. There are four standardised slice categories: eMBB, URLLC, and mMTC – the standard 5G use cases joined by V2X (Vehicle to something). MNOs then have 24 bit slices (16 million) that they can create. Phones like the iPhone now support URSP (User Equipment Route Selection Policy), allowing the OS or the apps to request a specific QoS. Network slicing can be very dynamic in that a single terminal can run different apps, or even different sessions within an app, on different slices, and these can be created dynamically.

2.5.8 5G private networks

Your own private mobile network? With 5G, you can have one – but why would you want one? Mostly, the applications for private networks are in industrial settings such as a factory or a port, but they can also be used in hospitals, airports, or university campuses. There are several different ways a private network can be created, as well as various spectrum options. The simplest is for an existing MNO to create a network slice that becomes a virtual private network. The MNO may add additional base stations in or close to the premises. A very different example is a factory that might build a mini-5G network with base stations operating in a licence-exempt or shared spectrum. A 5G core would be dedicated to the network, and the factory would issue SIM cards. Alternatively, the factory might buy or lease base stations for operation in the shared or licence-exempt spectrum and have this network managed by an MNO or remote cloud core supplier. Figure 2.28 shows how a private network could be deployed in a factory setting.

So, what makes a private network so attractive? One of the early applications is in several shipping ports. A port usually covers a wide area and often has poor to middling existing mobile coverage. The geographic separation makes using unlicensed spectrum attractive as there are no nearby users to share it or having an MNO use licensed spectrum as, again, there are few other users to share it. In a port, security is critical, and the ability to control access to the network by controlling SIM issuance is vital. In addition, reliability might have to be greater than that of existing networks. A private port network can include redundant links and hot-swappable servers and components. Finally, there might be automated trucks, cranes, etc., that need to be controlled with very low latency (possibly 5–10 ms or so), and public networks can't provide that. By containing network infrastructure elements, on-site delays can be reduced significantly. So, in summary, the benefits of a private network are improved security, reduced costs, lower latency, better coverage, lower overall costs and more excellent reliability than can be potentially provided by a public network.

Several elements in 5G mainly support private networks. The vast number of bands that 5G supports means that terminals covering these bands are widely available. Also, some shared/unlicensed/lightly licensed frequency bands are close to mid-band 5G including Germany (3.7 GHz) and Citizens Broadband Radio Service (CBRS) in the United States (3.5 GHz). Although CBRS is lightly licensed,

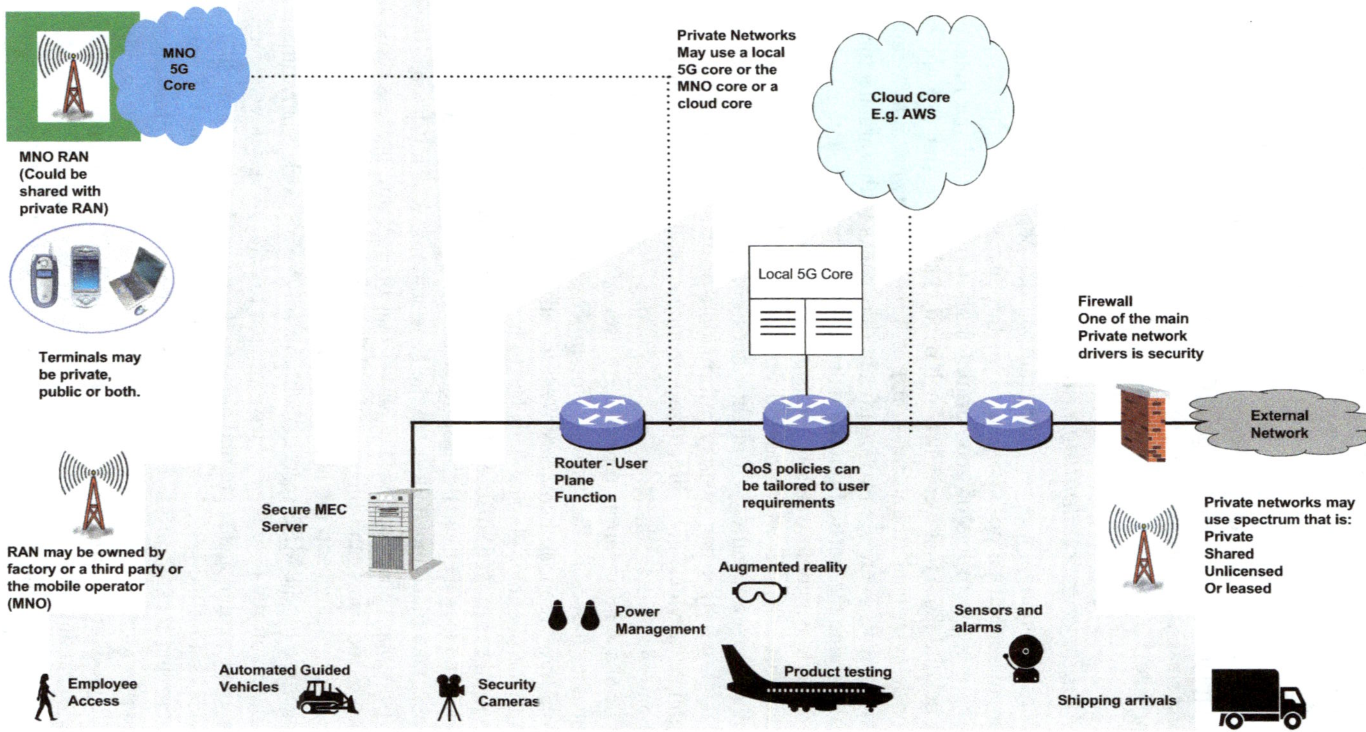

Figure 2.28 Private network in an industrial setting

it is available for non-MNOs to create private networks. Many other countries have also allocated a spectrum of 5–6.4 GHz for use in private 5G networks. 5G also includes several ways to integrate non-licensed spectrum, including New Radio Unlicensed (NR-U), Licensed-Assisted Access (LAA) and Licensed Shared Access (LSA) – all of which seek to include the spectrum access within a 5G RAN architecture – with or without licensed spectrum as a second carrier. These schemes will crop up later in the Wi-Fi chapter (Chapter 3). Eswaran and Honnavalli [25] give a good survey article on 5G private networks with many references and states that the private 5G makes will reach $14B by 2028.

2.5.9 Mobile edge computing/multi-access edge computing

Mobile edge computing was renamed multi-access edge computing in 2016 to make it more general purpose. If you recall the discussion on the 5G RAN and Open RAN, there were lots of good arguments for moving all the functions and processes beyond those that were ultra-high bandwidth/processing (RF and massive MIMO) or delay-critical (retransmissions) of a mobile network into a centralised cloud environment. There are two significant drawbacks to this architecture – first, when you examine the traffic from the data centre to the base station, a lot of it is duplicated, such as people watching the same video content, costing extra money for the transport network, which could have been avoided by caching the content closer to the users. Secondly, you might want to offer an AR service whereby the users wear glasses that can overlay information on the viewed image, translate signs, highlight services, give directions, recognise faces, etc. The glasses connect to the phone via Bluetooth. However, the phone does not have the computing power or database information to provide the service, so it sends the feed from the camera on the glasses to the network for processing. This might include facial recognition, translation, and content insertion (adverts) before the overlay video is sent back to the phone or glasses. This has to happen in about 5–10 ms; otherwise, the user's head has moved sufficiently that the overlay is out of sync with the actual view, which induces nausea. Because light travels at 0.75 ns/ft in glass, the processing must take place much closer to the base stations, which is about 20–50 km at most. To support these services, a server can offload and process some of the data closer to the users, and this is called MEC (pronounced 'Meck'). There is a tension between delays and the number of MEC sites, explored in Chapter 4.

Another use for the MEC server is within a private network. Many time-critical applications might run over the 5G network, and to achieve very low latency, these will have to be created and supported close to the network edge. It might also be a vital service so that the MEC server can be ultra-reliable (e.g. on hot standby) or super-secure and located inside a deep firewall, and the secret information never leaves the private network. The big players in cloud services (Amazon, Microsoft, etc) are also looking to offer Edge computing via MEC servers in 5G. Verizon and AWS are expanding their MEC services to major US cities using Verizon's 5G ultra-wideband network. This is targeted at self-driving cars, autonomous industrial equipment and low-latency applications. Mach [26] gives a good survey on MEC.

2.6 Dissenting views

Not everyone was sold on the 5G dream. Several leading commentators suggested 5G would be an expensive white elephant. Chief among these was Webb, whose book – *The 5G Myth* [27] expounded the argument that mobile traffic growth would not increase indefinitely, that 4G radio technology was reaching the limits set by the laws of physics (meaning at the Shannon limit for a particular MIMO layer) and that the demand for the new services that 5G would create would not generate anywhere near enough revenue to offset the cost of a total 5G roll out. He set out an alternative future that used Wi-Fi to provide mobile services indoors and in-demand areas:

> *5G – the next generation in mobile telephony – is heralded as a huge advance in global connectivity. But the vision is flawed. It is flawed because users will not value the promised data rates and will not need the higher-capacity forecast. It is flawed because technological advances are insufficient to realise the vision and because the mobile operators are insufficiently profitable to afford it [27].*

Webb was essentially talking about the consumer aspects of 5G. The decade's second half will determine whether industrial and IoT applications generate significant revenue for 5G operators and presage a less consumer-focused 6G. This is further discussed in the section below on 5G progress. The 5G question matters because these arguments already appear in the context of 6G, and it matters because if 5G is only a qualified success and generates only modest returns, then that will influence the direction 6G will take.

We have spent the last few pages talking about all the great new technology in 5G – massive MIMO, virtualised core, low latency, network slicing, new radio, etc. But do you need to upgrade from LTE? Table 2.5 provides a comparison.

If you examine the table closely, you will see that the critical things that 5G can offer and that LTE cannot are low latency, very high data rates, and satellite integration. 5G is also far more cloud/virtual machine-friendly. Do these differences matter? The latency certainly could – LTE has too high an intrinsic latency to support applications like AR, remote drone control and robotic factory control. Very high data rates, meaning over 1 Gbit/s, since that is where LTE tops out, might be needed for holographic video or for feeding multiple broadcast cameras, say. The extent to which 5G is more cloud-friendly is a slightly moot point in that LTE cloud cores have long been available, as well as Open RAN software for eNodeBs. The core typically represents 10% or less of overall network costs, so virtualising the core is not really about cost saving but more about flexibility, scalability and upgradability.

In the early days of 5G – comparable to the current phase of 6G – there were differing views about 5G:

- '5G is an end-to-end ecosystem to enable a fully mobile and connected society. It empowers value creation towards customers and partners, through existing

and emerging use cases, delivered with consistent experience, and enabled by sustainable business models.' *NGMN 5G Vision* [28]
- '5G is on its way, and rather than being another 'next generation', it will be a better integration of old and new technologies. Integrating different systems will enable more stringent requirements in some areas to be met and relaxed needs in others, focusing on keeping overall costs and energy dissipation low. The combination of evolution and revolution, wide and local area, big and small cells and different carrier frequencies will enable a fully scalable service experience on demand, where people and machines will enjoy a virtual zero latency gigabit experience when and where it matters'. – Nokia [29]
- 'View 1 – The hyper-connected vision: In this view of 5G, mobile operators would create a blend of pre-existing technologies covering 2G, 3G, 4G, Wi-fi and others to allow higher coverage and availability …' 'View 2 – next-generation radio access technology: This is more the traditional 'generation-defining' view, with specific targets for data rates and latency being identified such that new radio interfaces can be assessed against such a criterion …' – GMSA [30]

View 2 is, slightly disconcertingly, not unlike many current visions for 6G. If you read chapter 16, remember this quote about 6G.

Figure 2.29 is a version of a diagram first produced by the GSMA [30]. The question about 5G can be re-phrased into 'How valuable are the services that lie outside of the LTE-A envelope?' Are most of these services niche applications (remote surgery), or could they be provided over Wi-Fi since they mostly happen

Table 2.5 5G compared to LTE

Category	LTE-A	5G
Max speed	1.2 Gbit/s	20 Gbit/s
Typical latency	40–50 ms	5–10 ms
Spectral efficiency	2.4 bit/s/Hz (MU MIMO –2.6 GHz)	5.8 bit/s/Hz (massive MIMO 3.5 GHz)
Virtualised core	Yes	Yes – but there is much more cloud support in the standard
C-RAN	Yes	Yes
Private network	Yes	Yes – more frequencies supported
MIMO	Eight layers	Massive MIMO 64Tx+
Network slicing	QoS support, e.g. Emergency Services Network (UK)	Yes
Air interface	OFDM	OFDM – more flexible subcarrier spacing
mmWave	No	Supported
Satellite	No integration	In Release 17
IoT support	NB-IoT and M-LTE	RedCAP

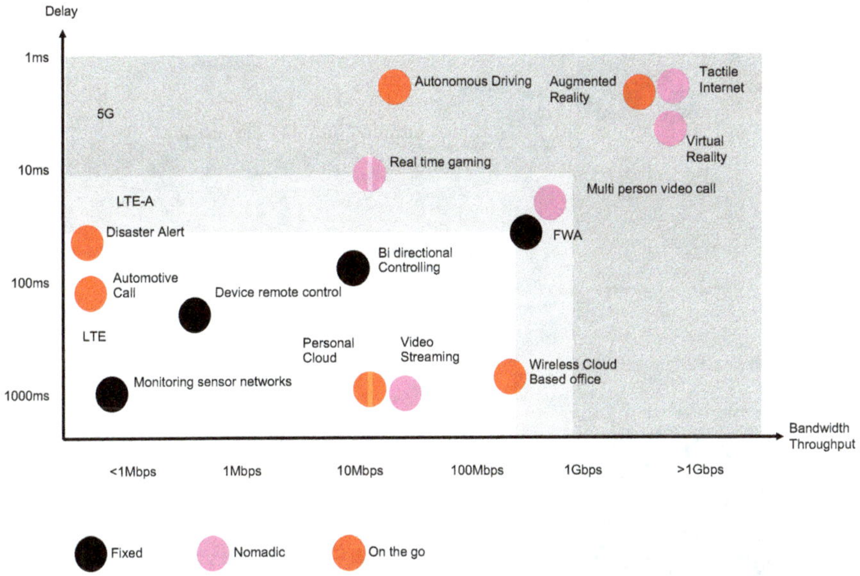

Figure 2.29 Applications that only 5G can support (after GSMA)

indoors (gaming, tactile Internet)? Autonomous vehicles have proved much more challenging to automate fully than first imagined and, as one of the authors famously said: 'Cars that can't drive themselves without real-time network input are not truly autonomous and could be subject to network jamming.'

A 'modular' 5G architecture was proposed to be developed at a much lower cost [31]. The concept was to build a baseline macro network in the sub-6 GHz spectrum. Spectrum usage and re-use is critical here:

- Existing LTE spectrum will continue to be used by operators for LTE and evolved LTE
- Existing 2/G and 3G spectrum will be re-farmed to a new OFDM-based air interface that is backward-compatible with LTE or simply LTE (Rel 12 + any 5G features that can be added)
- A small amount of 2G and 3G spectrum will be used for a national network that supports the legacy devices of all operators. This network is controlled by a non-MNO body.
- New sub-6 GHz is used to offer three overlay networks:
  - The lowest frequency (most excellent coverage – probably 400 or 700 MHz) spectrum available is used to create an IoT/M2M network. This would require approximately 2000 base sites and a national network (i.e. one only). It would run an M2M radio protocol (such as WEIGHTLESS [32]) and support low to mid-data rate applications. This could be a regulated – open-access – network – in much the same way that Openreach provides access to the fixed copper network.

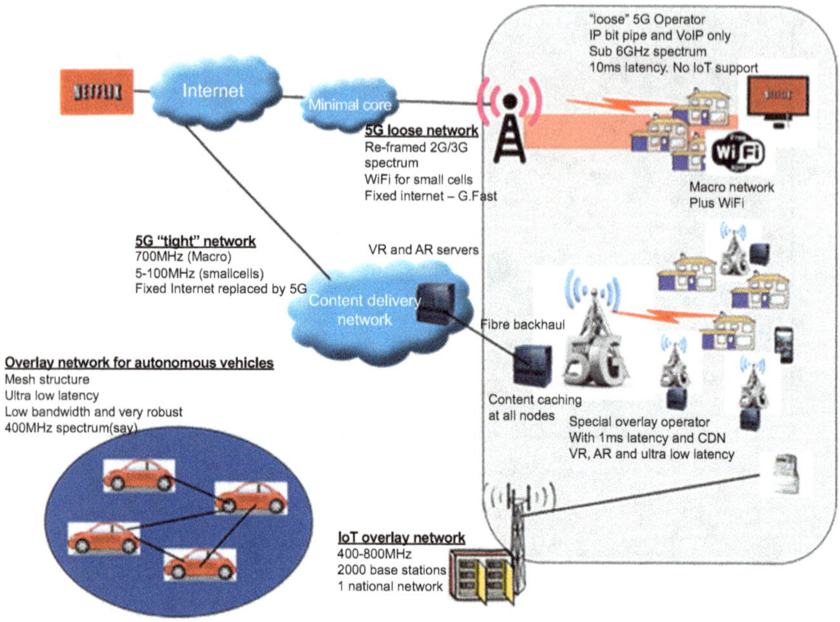

Figure 2.30 Modular 5G architecture proposed in 2015 by Wisely et al.

- ○ Some new spectrum would be used for a new TDD-based air interface with 1 ms latency. This would be deployed in urban areas with an existing fibre network as a network for premium services. It would offer full mobility (for autonomous vehicles) and nomadic capability (for tactile Internet and VR – which are not likely to be mobile). There is unlikely enough commercial pull for multiple networks in any given location.
- ○ Another significant, contiguous block of spectrum, at the highest available frequency, would be used for a 500 Mbit/s service. Following an analysis by the EU project METIS, it might be possible to offer 300–500 Mbit/s at 4–6 GHz. This spectrum has a short range, and multiple operators can be licensed in the same spectrum. It would mainly be deployed indoors or in urban areas where fibre backhaul is available.

Figure 2.30 shows possible operators and the general architecture.
There are many advantages to this modular approach:

- Each operator does not need to build a high-cost 5G network that meets all requirements.
- The niche requirements of 5G (ultra-low latency, very high headline rate and IoT) are all provided by overlay networks whose economics and commercials can be assessed separately. Moreover, the cost of these networks does not affect the general IP bit-pipe delivery success of 4G.
- Services can be introduced gradually in different locations without offering a significant upgrade for 4G in one step.

- Costs are more transparent.
- There is no more confusion and overlap with fixed Internet (+ Wi-Fi)
- There is enormous scope for value-chain innovation – particularly network subsidies for specific functions.

The modular approach never got off the ground, and all significant operators quickly moved to introduce 5G, as defined in 3GPP standards, in some form or another. The key argument for the modular approach was that it provided the same benefits as a widespread 5G roll-out for a much lower overall cost. It is appropriate, therefore, to look at the costs of 5G.

2.7 5G Techno-economics

There is much less research into the techno-economic side of cellular than any other performance aspect. It is also noticeable that the requirements for 5G didn't say anything about making it lower cost. Mobile operators don't publish full costs; independent research can be hard to find. In this section, we will look at some techno-economic modelling of 1 km^2 of central London based around Marlybone [33], which sheds some light on 5G costs. This study examined how much capacity and coverage could be achieved for minimum rates of 64 Mbit/s (to mirror the 5G requirement of 50 Mbit/s everywhere) and 100 Mbit/s. 5G was considered at low band 700 MHz, mid-band 3.5 GHz and mmWave 25 GHz, and WLAN; the technologies modelled are detailed in Table 2.6.

The authors created radio propagation models for all four technologies and used statistical models for shadowing and fading. For each cell in the model, a signal-to-noise interference ratio (SINR) for both uplink and downlink on all technologies could be computed with a pre-set distribution of base stations in typical locations (macrocells on roofs and wireless LANs on lamp posts). The SINR could then be translated into throughput (with a given load), and the overall capacity was determined as a function of the base station density for each technology (all of which were independent due to their different frequencies).

Table 2.6 Technologies modelled by Wisely et al.

Technology	Frequency/bandwidth	Base station density	MIMO
5G FDD	700 MHz (2 × 10 MHz)	1–64 macrocell/km^2	2 × 4
5G TDD	3.5 GHz (100 MHz)	1–256 microcell/km^2	4 × 4
5G TDD	24.5–27 GHz (1 GHz)	1–128 hotspots/km^2	8 × 8
WLAN 802.11ac	5.8 GHz 3 × 160 MHz 1 × 480 MHz	1–1664 access points /km^2	4 × 4

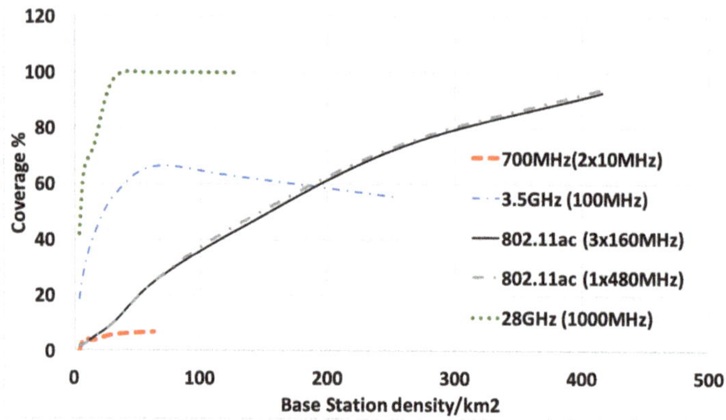

Figure 2.31 Outdoor coverage at 64 Mbit/s for different technologies

Table 2.7 Outdoor capacity gains (excluding mmWave – see Figure 2.32)

Technology	Cell density	Capacity
LTE (2017) (2 × 30 MHz)	16 cells/km^2	1 Gbit/s/km^2
700 MHz 5G FDD Umbrella Macro (2 × 10 MHz)	32 cells/km^2	0.83 Gbit/s/km^2
3.5 GHz microcells	256/km^2	30.6 Gbit/s/km^2
802.11ac WLAN 3 × 160 MHz	416/km^2	20.9 Gbit/s/km^2

Figure 2.31 shows the outdoor coverage as a function of base station density for minimum download speed 64 Mbit/s – with a 9:1 download/upload split. WLANs and mmWave can easily cover the area at intermediate cell densities. 3.5 GHz can provide 65%–70% coverage (at 64 Mbit/s+) at a density of 50 base stations/km^2. This is surprisingly similar to actual deployment densities in London. The coverage is better than this because the deployed base stations often use 64T4R(downlink), and the model was based on 8T4R. The 700 MHz spectrum is too small (2 × 10 MHz) to reliably support 64 Mbit/s. Coverage was, however, almost 95% at 30 Mbit/s. In practice, operators have re-farmed 2G/3G/LTE spectrum as well as shared spectrum bands between LTE and 5G dynamically (one of the advantages of both being OFDM) to create more like 50+ MHz of 5G at 2.1 GHz or below.

Table 2.7 and Figure 2.32 show the capacity that these technologies can add (outdoors) and compare these with the pre-5G (circa 2017) LTE network for a typical operator in London (as one of 4 in total). 3.5 GHz technologies offer capacity gains of ×30 greater than the pre-5G LTE macrocell network. While this is a considerable increase, it is much closer to the generational trend (×10) than to

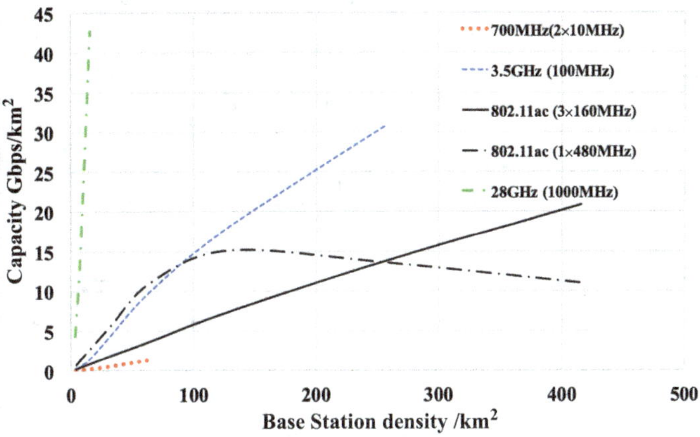

Figure 2.32 Outdoor capacity gains

the 5G requirements of ×1000-×10,000 capacity increases. Only mmWave technology offers almost an ×1000+ increase in capacity in this modelling.

Next, the authors estimated the costs of the RAN part of the modelled networks. This included Spectrum costs, site rental, equipment costs, maintenance and upgrades, and backhaul connection and transmission costs. However, expenses related to the core, advertising, or handset subsidy were excluded.

They concluded that a typical UK MNO with 2×30 MHz of 800 MHz spectrum using LTE Release 10 with 16 macro base stations per km^2 (three sectors) has a capacity of around 1 Gbit/s/km^2 calculated on the 1 km^2 modelled and averaged (weighted) over indoors and outdoors. If such a network was built from scratch (instead of upgrading), the annualised cost was calculated as £0.53 M/yr/km^2. Further, they concluded that if an MNO deployed a very dense 5G 3.5 GHz network, with a base-station density of 256/km^2 each, then the total capacity added could be 30 Gbit/s/km^2 (weighted average across all users with a 7:1 indoor: outdoor ratio). The annualised cost would be £1.8 M/yr/km^2 – a 30× capacity increase for an ×3 rise in price over LTE. With 802.11ac WLAN deployments, a further 110 Gbit/s/km^2 could be added internally and externally to 2080 access points/km^2. However, this would again have to be shared between 4 operators, meaning a maximum of 27 Gbit/s/km^2 per operator, at an annualised cost of around £0.5 M/yr/km^2 per operator for 2000 access points/km^2 in total and costs shared.

Both approaches would increase the available capacity by a factor of ×30 to ×50 over the 2017 LTE cellular capacity. The only technology modelled that can offer ×1000 or more capacity increases is mmWave with 1.5 Tbit/s/km^2, requiring 256 base stations/km^2 with 1 GHz of the spectrum. However, this does not address indoor users without repeaters or in-building distribution, which we will discuss in later chapters. Table 2.8 and Figure 2.33 summarise a typical operator's likely capacity gains and costs (of 4).

Table 2.8 Capacity gains and costs for different technology options – for a typical MNO (one of four)

Technology	Cell density (km^2)	Capacity ($Gbit/s/km^2$)	Cost ($£M/yr/km^2$)
LTE	16	1	0.53
5G 700 MHz	32	0.6	0.6
5G 3.5 GHz	256	30.6	1.8
WLAN (Shared)	2000	20.9	1
5G mmWave	128	740	0.64

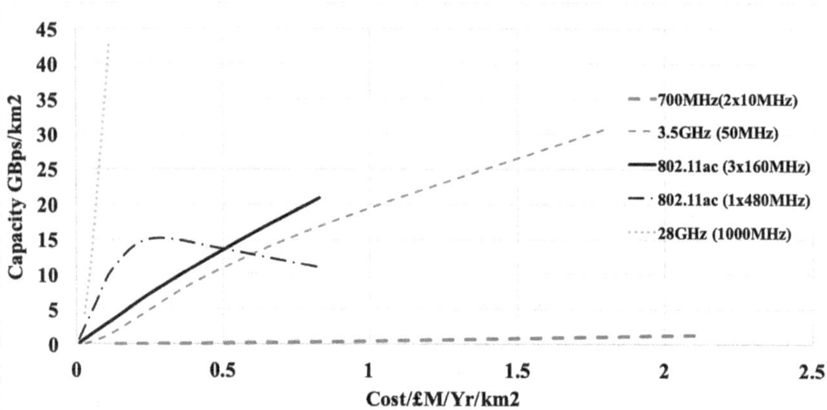

Figure 2.33 Costs of outdoor capacity in London model

Marylebone is part of Westminster and has a resident population density of 11,500/km^2, which doubles to (approximately) 20,000/km^2 on a working day. This equates to 5000 users per operator. Other UK cities have similar user densities in the centre: for example, Southampton has an average population density of 5000/km^2 and double that in the centre. Taking just the 802.11ac and 3.5 GHz networks (because mmWave can only serve outdoor users and most are located indoors), then the cost of providing eMBB is £2.3M/yr/km^2 or £460/yr/user (assuming 5000 users per operator). This figure only covers RAN costs, and the total cost would be higher due to other factors such as core network, handset subsidies, advertising, etc. At a density of 5000 users/km^2, a capacity of 56 Gbit/s/km^2 (adding the 802.11ac and 3.5 GHz capacity) translates (with appropriate assumptions about busy hours and overbooking) to 800 Gbytes/month/user – or a continuous average rate of 2.5 Mbit/s non-stop 24 hours a day to all users. It is difficult to conceive that users will be willing to pay significantly more than they do today for the eMBB service of 5G, meaning that there is likely to be a significant revenue gap that needs to be filled by new services (such as ultra-low latency, IoT services, network slicing and private networks).

Oughten and Frias [34], using a whole country model, concluded that each MNO in the UK would need to invest £2B/yr in capital costs alone from 2020 to 2028 to achieve 90% population coverage (5G with 50 Mbit/s in urban and 30 Mbit/s in rural areas). £8B/year or £64B in total, equivalent to over £1000 per mobile user in the UK. That is the capital costs without ongoing site rental, maintenance, and backhaul transmission costs. It is clear from these sorts of numbers that eMBB might have been better served with LTE alone. 5G economics rests on new services, and the next section looks at progress beyond eMBB.

2.8 5G Progress report

Before we look at progress towards 5G deployments, we need to look at the non-stand-alone (NSA) core. This comes about because the fastest and cheapest way to launch 5G is to parent it on a 4G core network. Figure 2.34 shows what is known in the standards as option 3 – a simple way to connect a gNodeB to an existing LTE network. In this case, the eNodeB acts as a controller node in an arrangement known as dual connectivity (DC). DC and another vital technology for 5G early deployments called Carrier Aggregation (CA) were introduced into LTE to increase headline rates, capacity and coverage. Figure 2.35 Shows both technologies on a pure LTE network.

Looking at the top master cell in Figure 2.35, only this cell is operating CA: typically, a lower frequency (say 800 MHz) is used for the primary cell, and a higher frequency (say 2.6 GHz) is used for a secondary cell. They come from the same base station, although likely through different antennas. In CA, the primary cell carries user data and signalling, whereas the secondary cell carries only user data and can be uplink/downlink or both. The obvious advantage is that the mobile has a reliable connection to the base station at a lower frequency with good coverage and indoor penetration. Suppose the mobile has a strong signal at a higher frequency, which could be used for downlink traffic, something like a video that might tolerate the odd variation in throughput. Both data channels are combined seamlessly at the mobile and base stations and seem like a single, higher-capacity connection. CA has allowed LTE to combine up to 5×20 MHz chunks of spectrum (the limit in the standard) to offer a maximum of 100 MHz of spectrum and headline rates of up to 1 Gbit/s (now 1.2 Gbit/s in LTE-A pro).

DC is a further development in that a secondary eNodeB is added (right part of the figure). Somewhat confusingly, this also has primary and (optional) secondary cells (secondary, secondary cells!). The core network can send user data to both cells (S1 and S1-U interfaces). However, all control messages to the terminal go via the master primary cell, and the master cell controls the DC operation via the X2 interface. These interfaces have a latency of several milliseconds – too long for coordinated transmission from the two base stations, so they transmit independently. They can't offer the same data rates and coverage as CA and are only employed when the base stations are not co-located. For example, several 2.6 GHz micro base stations may have been deployed across a city, some with existing macro base stations, but (because

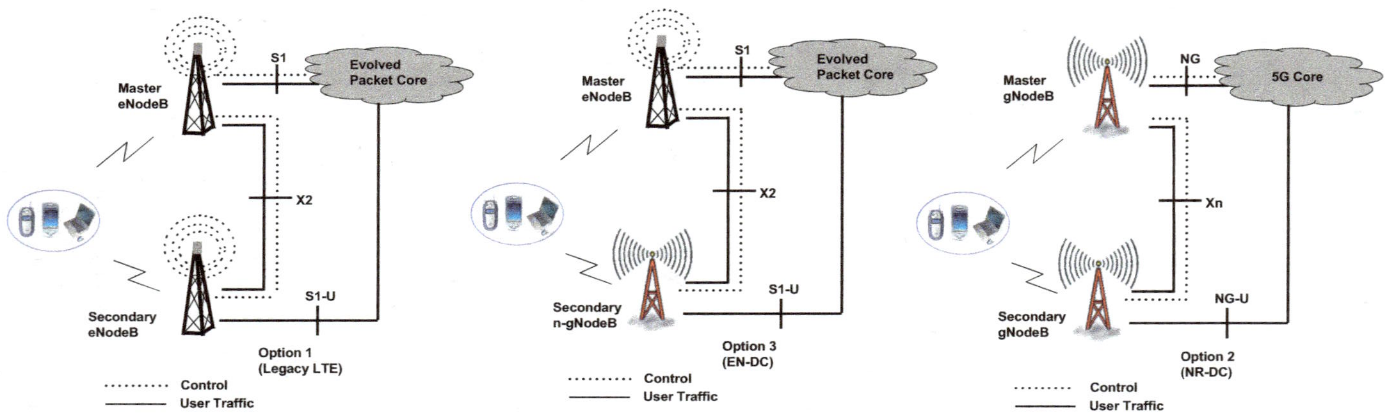

Figure 2.34 5G architecture progression to stand-alone (LTE, non-SA, SA)

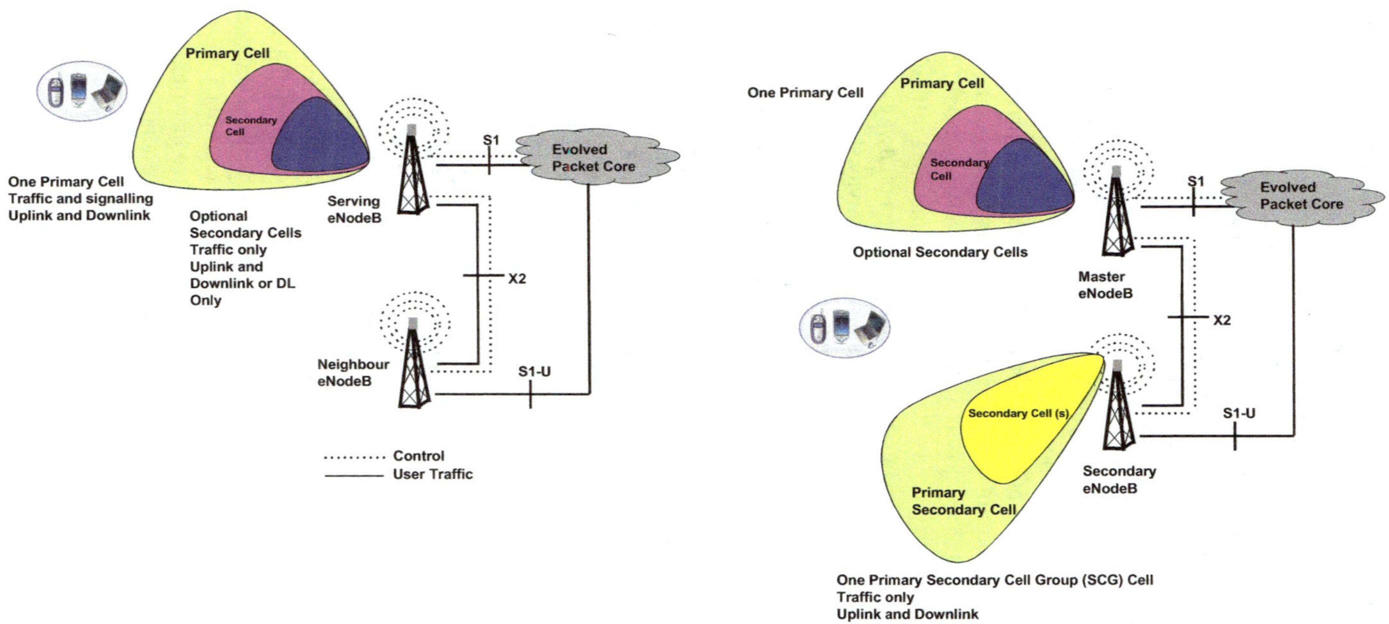

Figure 2.35 Carrier aggregation (CA) (left) and dual connectivity (DC) (right)

of the poor coverage) others are on new sites. Coming back to option 3 (Figure 2.34) for adding 5G to this scheme, a gNodeB is added as a secondary NodeB in a DC scheme and communicates with the LTE core and master LTE eNodeB via the S1-U and X2 interfaces, respectively. It is called an en-gNodeB to show that it is backwards compatible with these interfaces. We can see how 5G is being rolled out in cities like London. Operators have purchased or re-farmed as much low-frequency spectrum as possible (for good coverage) and added microcells in the 3–4 GHz range. These are part of a CA scheme if they are co-located and in a 5G DC scheme if they are separated. These nodes are then added to the existing LTE core using DC.

Around the world, many spectrum bands have been used for 5G. Regulators have generally auctioned or allocated low-band spectrum at 700–800 MHz. Most spectrum comes from mid-bands, especially 3.3–3.5 GHz. Some regulators have set aside spectrum in priority 5G bands for local users to create private networks. The US has adopted a spectrum-sharing framework in the 3.5 GHz range, the CBRS. A problem that seems to have arisen with this spectrum is the lack of device support for these many frequency bands. Reportedly, it is sometimes difficult to determine the supported frequency range, and specifications are not always a good guide.

All bands over 3 GHz use TDD in anticipation of heavily skewed downlink traffic. Suppose the same band is being used on neighbouring base stations for TDD. In that case, synchronisation issues mean that it must either be used for downlink eMBB or uplink or ultra-low latency. However, these uses can't be mixed and matched in overlapping cells on the same frequency due to synchronisation issues. These applications must use another band. The GSMA [35] pushes for much greater harmonisation and more spectrum/re-farming at 600 MHz, 2.1 GHz, 2.6 GHz, 4.8 GHz, 6 GHz and 40 GHz. They are pressing for 2 GHz of the midband spectrum per market by 2030.

5G rollout speed has been relatively fast. The build-out of 5G continues, with around 280 networks launched worldwide. Global 5G population coverage is forecast to reach 45% at the end of 2023 [36]. Given that the first deployment of 5G was in mid-2019, that is a fantastic rate of growth, much faster than any previous generation. 5G has been on the iPhone since the 12th iteration. Nearly every major operator in North America, Asia Pac and Europe has launched 5G, and users are regularly averaging downlink rates of roughly 90–230 Mbit/s (based on the currently available data from RootMetrics, Point Topic, Open signal and other validated tests, across the UK's networks [37]).

That makes 5G sound like a great success, but there is also a catch. Non-SA core does not support low latency; it is about the same as LTE (40–50 ms). It also does not support network slicing or 5G private networks. It is not virtualised and does not offer 5G interfaces to third-party applications. Of course, you can offer LTE private networks, control QoS on LTE and buy virtualised LTE cores that run on cloud platforms. But these have generally not been widely used. One example might be the new emergency services network (ESN) in the UK that replaces the old Airwave private network, effectively building a private network for emergency services on LTE. At the time of writing, the project was seven years behind schedule and way over budget.

Figure 2.34 also shows the end-game architecture for 5G (option 2) that includes only gNodeBs and 5G interfaces. Many other options involving various LTE and 5G elements can be used during migration to option 2. Many operators were considering or trailing SA when writing, but only relatively few had launched SA (39 Q4 2022 [38] – compared to 200 NSA and 700 LTE deployments). A further update suggested only 15% of 5G networks were SA at the end of 2023 [39].

There are examples of 5G private network launches or trials [40] – typically at airports (Vienna, Brussels), ports (Hamburg, Rotterdam) or factories (Siemens, BASF). Some seem to be examples of private networks created by network slicing. There are many predictions of growth in private networks: the US private LTE and 5G network market is expected to grow at a compound annual growth rate of 24.1% from 2022 to 2030 to reach USD 13.65 billion by 2030 [41], with the most significant user being manufacturing. The US market is perhaps the most promising private network market with the mid-band spectrum available. It is important to note that many players (beyond the traditional mobile operators) are moving into the 5G private network market. AWS offers a '5G private network in a box' service. Much of this expected revenue might not flow into the mobile operators.

A typical example is the US tractor and agricultural manufacturer John Deere, who previously used Wi-Fi, Ethernet and ZigBee. They have now decided to equip around 20 facilities with a private network. They are purchasing spectrum licenses and deploying an onsite core at each site. In the United Kingdom, the power grid outfit UK Power Networks uses a private network created on a network slice by Vodafone. It appears to be a test network at the time of writing.

According to Heavy Reading [42], only 2 out of 80 surveyed operators had a network-slicing product in the fall of 2022. Most were anticipating private network extension into a wide area, secure enterprise VPNs, speed boost, events and gaming as major applications for slicing. Most operators were planning ten or fewer slices. The biggest challenge was the need for an SA core network.

It is essential to say something about 5 G's progress towards energy efficiency and improved reliability. There is no doubt that 5G is more efficient than LTE. However, the rise in traffic has more than offset this, leading to an overall increase in energy consumption (although some of this may be offset by reduced travel, etc., due to the better connectivity). Ericsson recently reported [43]. 'A high-performing mid-band 5G deployment requires 150 times more compute power than 4G'. Reliability has very little to do with the mobile generation – all the generations can be engineered to be more reliable – at a price. 5G operators are now using satellite connections at base stations to improve reliability, which will become easier as satellites are better supported in the standards and offer greater capacity in the next few years.

The years 2024–30 are critical for 5G. The eMBB scenario has already been met in that capacity has increased 10–50-fold in urban areas from a combination of new spectrum, spectrum re-farming, massive MIMO and densification of base stations (and mmWave in the United States). Headline rates are approaching 1 Gbit/s, and average speeds of 100–300 Mbit/s are typical in urban areas. But all of that could have been achieved with LTE. For the eMBB use case, the operators now have a massive uplift in capacity available and more than sufficient for compound data volume growth rates of 20%–30% per annum. However, the average Revenue Per User (ARPU 'Arr-Poo') has hardly changed. Statistica [44] shows that UK

ARPU, along with leading supplier Vodafone, has not increased in the past four years, which coincides with the launch and build-out of 5G. With many users on unlimited data packages with no supplementary charge for 5G, there has been minimal revenue boost to pay for 5G. According to SDXCentral [39], the compound annual growth of subscriptions (2010–20) was 4%, and the compound annual growth of ARPU was −3% across global operators. With many leading mobile operators suffering falling profits in the last few years, there is a great expectation that 5G SA will reduce costs and significantly increase revenues via slicing and private networks. If this revenue is disappointing, then it will be a significant setback to the progress of 6G. In addition, many of the applications and use cases slated for 6G build on those of 5G (faster, lower latency, higher density, etc.) and are not viable if the 5G use cases do not materialise.

5G in late 2023 indeed falls well short of the original visions (Table 2.9), but it offers a significant upgrade in capacity, headline rate and latency compared to LTE. It also uses lower energy per bit. The figures for latency in Table 2.9 represent the current status in real-world networks. There is plenty of scope for operators to reduce these figures. ZTE has demonstrated 5G at 2.6 GHz with a 1 ms latency [45]. The question is whether operators see a market for lower latency services. It has been suggested that 5 ms could be achieved in typical urban deployment without investment. Measurements in Munich on a live 5G network [46] suggest current latency figures are about 7 ms in both up and down links (16 ms round trip) on a stand-alone 5G urban network with low traffic, and rising to 7 ms downlink, 13 ms in the uplink and a 22 ms round trip time in heavy traffic. LTE managed a round trip time of 22 ms (low traffic) and 36 ms (high traffic).

Table 2.9 Comparison of 5G (2023) with the original vision

	NGMN 5G	**NOKIA 5G**	**5G 2023 Actual**
User-experienced data rate	50 Mbit/s everywhere	100 Mbit/s	50–500 Mbit/s in urban areas only
Peak data rate	1 Gbit/s	10 Gbit/s	0.1–1 Gbit/s
Connection density	150,000/km^2	10–100× number of devices	Similar to LTE 5000/km^2
End-to-end latency	1–10 ms	1 ms	7–40 ms
Traffic volume density	1–15 Tbit/s/km^2	10,000× more traffic	100–500 Mbit/s/km^2
Mobility	500 kph	More than 500 kph	500 kph
Energy efficiency	1/2000 that of LTE	Much lower J/bit	More efficient (90%), but the network is consuming more power
Reliability	99.999%		Similar to LTE 97–99.5%

Maldonado *et al.* [47] have modelled the performance of 5G NR and Wi-Fi 6 when used in an Industrial setting to offer low latency. They placed between one and twelve base stations (or access points) on the ceiling in a typical factory setting (120 m × 50 m with a height of 5 m). In the 5G NR example, they first considered 2×20 MHz of FDD spectrum, setting the radio packet reliability at 99.999% (i.e. one in one hundred thousand radio packets would be lost) and determined the maximum latency as the load was increased. All the traffic was a single QoS class, with a packet size of 50 Bytes and a Poisson arrival pattern, with no pre-emption. The Indoor-Factory sparse-clutter (InF-SH) propagation model from 3GPP was used to determine the link budget. The authors adjusted the 5G NR parameters for the lowest possible latency. This included using a short TTI (Transmission Time Interval) of 4 OFDM symbols (a mini-slot), 0.143 ms with the standard 30 kHz subcarrier spacing. This provided the best case: 0.5 ms for a single transmission and 1 ms after a single re-transmission.

It should be noted that using mini-slots in some current 5G deployments will not be possible due to licence constraints related to adjacent frequency bands. Typically, a 5G network, for the eMBB use case, runs with a BLER (block error rate) of 10%, which tends to provide the optimum trade-off between a high rate modulation and coding scheme and the resulting re-transmissions. For low latency traffic, the optimum figure is 1% at low loads, but it increases as the traffic rises. This figure was variable in the model.

Table 2.10 shows the result of the modelling for 5G NR FDD and TDD as well as Wi-Fi 6 and 5G NR-U (5G NR operating in unlicensed Wi-Fi spectrum) – with two gNBs or 2 APs. The result shows how the spectral efficiency falls rapidly as the maximum delay is constrained – giving some idea of the possible trade-offs that will be needed in 5G to support URLLC.

All the technologies could support 1 ms latency, but their efficiency was markedly reduced in all cases. Interestingly, the efficiency at 5 ms is relatively close to that at 100 ms. Given that 5G networks are currently operating at 7 ms latency (Munich stand-alone 5G core network, lightly loaded), it suggests that 5–10 ms is a reasonable target for 5G, but extending to 1 ms will represent a technical challenge. It is also an economic one in that the cost of URLLC traffic will be much higher, bit for bit, than eMBB traffic. At the time of writing, there were no reports of URLLC being launched as a commercial service and only a few trials (such as by ZTE [48] in Thailand).

Table 2.10 Spectral efficiency with different maximum delays – from the modelling in Maldonado [47]

Delay (ms)	5G NR FDD (bit/s/Hz)	5G NR TDD (bit/s/Hz)	5G NR U (bit/s/Hz)	Wi-Fi 6 (bit/s/Hz)
1	0.72	0.5	0.31	0.16
4	2.4	3.7	3.7	1.125
10	3.5	3.7	3.7	2.25
100	3.6	4	4	2.5

2.8.1 mmWave progress

It is also essential to look at the progress of higher-frequency networks. The only area in the world where they have been deployed appears to be the United States, where some operators have built outdoor networks in dense urban areas. These networks offer very high download speeds with Ookla (all operators combined Q4 2023) [49] quoting the following median download speeds in the United States:

2.5 GHz (n41) – 400 Mbit/s,
3.5 GHz (n48) – 200 Mbit/s,
3.7 GHz (n77) – 250 Mbit/s,
25 GHz (n258) – 1 Gbit/s,
28 GHz (n261) – 1.6 Gbit/s,
39 GHz (n260) – 1.7 Gbit/s.

It seems that there was initial enthusiasm for wide area mmWave in the United States, but that confidence has waned with the realisation that 25–39 GHz signals do not penetrate indoors and have a minimal range. Ookla found that in 2021, 5G users in the United States were connected to an mmWave network less than 1% of the time.

It seems more likely to be used in the short term for fixed wireless access with Light Reading reporting [50]:

> *And other mmWave uses appear to be on the horizon. For example, US cellular phones use technology for some fixed wireless access (FWA) services. And executives from T-Mobile – one of the world's leading FWA providers – have hinted that the operator might use mmWave connections to bolster its mid-band FWA offerings.*

There has been little interest in mmWave 5G outside of the US, with South Korea's regulator stripping two operators of their mmWave licences after failing to realise a network [51]. The GSMA reports that only 9% of current 5G devices support mmWave, increasing the selling price of a Google Pixel 6 from $599 in the United States without mmWave to $699 with the technology [50]. It may take new services and indoor base stations for mmWave to become a significant part of 5G/6G. There is a further discussion on this in chapter 9.

2.9 5G advanced

The actual 3G/4G/5G standards have been developed by a body called 3GPP [52]. 3GPP is an umbrella organisation for seven major and many minor regional and national standards bodies. 3GPP started with GSM and developed a unified 4G and 5G system. It is the critical player for 6G standards. Just to note that although they liaise with the IETF (responsible for Internet standards) and Wi-Fi standards bodies, they are entirely separate, attended by different ecosystems of vendors, chipmakers, operators, etc. 3GPP standards are released much more frequently than each generation (Figure 2.36) – with some features being updated and new services (as well as fixes for issues discovered after introducing previous standards).

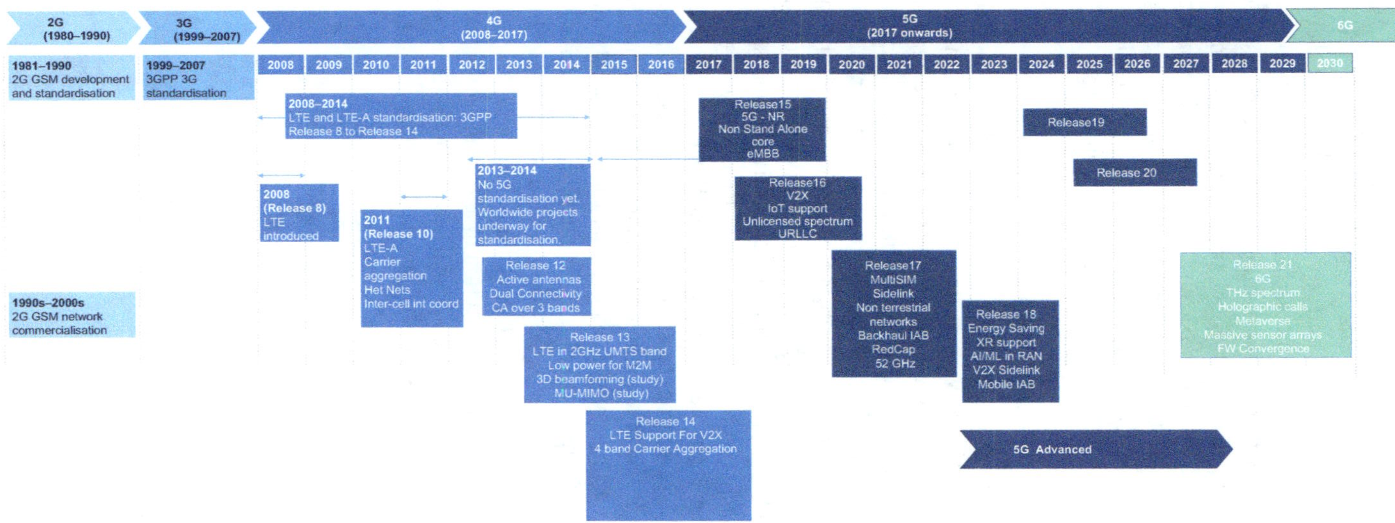

Figure 2.36 Standards progress to 6G

These releases cover both hardware and software and are not always backwards compatible. Releases 15 and 16 covered the essentials of 5G. Releases 17 and 18 are enhancements to 5G, with 18/19 and 20 dubbed 5G Advanced. The release numbers were supposed to mirror calendar years, and there was a release 99, but delays have meant that alignment has long since been lost.

The section below will examine the upgrades these releases will introduce. These are significant in the context of 6G, partly because they may fulfil many of the use cases suggested for 6G or because they may have minimal commercial appeal, thereby setting back the timescale for 6G. 5G Americas [53] give a detailed breakdown of what to expect in 5G-A, and Lin [54] describes Release 18. At the time of writing (early 2024), 3GPP had begun work on Release 19 [55].

2.9.1 5G Advanced technologies

5GA will offer a new use case (in addition to eMBB, ULLC and mIoT) known as Vehicle-to-Everything (V2X). V2X is the flow of information from vehicles to other vehicles, pedestrians, servers in the network and roadside units (such as speed cameras and charging points). Release 16 brings in four categories of applications for V2X. Extended sensors allow the exchange of information gathered from video cameras. This might enable early warning of queues ahead and signal braking. Advanced driving allows nearby vehicles to share driving intentions and coordinate movements, especially in semi or full-automated driving modes. Vehicle platooning gives a lead vehicle control of those behind. Remote driving allows control of the vehicles by either a remote driver or V2X application. Figure 2.37 shows the connection to a V2X server from a car. There are many possible applications, from supporting different levels of autonomous vehicles to charging and enforcing local traffic policies and integration into smart cities. Careful inspection of Figure 2.37 will show the new PC5 interface. This is an example of a sidelink – from one 5G terminal directly to another (as opposed to a base station). Sidelinks have been available in LTE, but the PC5 sidelink has new features, such as much better Doppler shift tolerances, to allow connection to and from moving vehicles. Further enhancements to 5G sidelink are planned: sidelink CA and sidelink over unlicensed spectrum may support new industrial and safety scenarios.

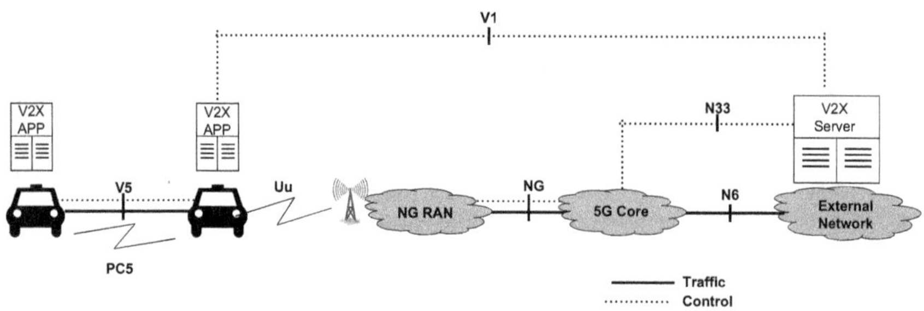

Figure 2.37 V2X services

Another upgrade to 5G basic is an enhancement to location services. Location services allow an application to find out the location of a terminal. They were introduced into LTE Release 9 for emergency services and lawful intercept using satellite-based systems such as GPS. When the GPS signal was unavailable, air interface measurements (such as base station triangulation) could estimate a less accurate location. In release 15, 5G uses satellites and base station measurements to achieve a horizontal accuracy of 5m for 80% of mobiles with a vertical accuracy of 5 m – all with a latency of 30 ms. Release 16 – using gNodeBs – increases the accuracy to 3 m indoors. No less than six different air interface measurements (mostly time of flight and beam-forming angle) are used to achieve this. Release 18/19 will get the location accuracy down to 10 cm and also offer a timing/synchronisation service that will be at least as good as that provided by GPS – this will be especially applicable to use cases such as smart grids – where the precise timing of supply and demand is critical.

5G advanced is also addressing the issue of ever-increasing capacity. The spectrum below 10 GHz is increasingly crowded. To continue adding capacity and increasing headline rates, 5GA will expand mmWave frequencies. Outdoors, in dense urban environments, mmWaves can offer surprisingly good coverage, although they have a high cell density. Because mmWave does not penetrate walls, it is necessary to have base stations in every office or enclosed space that needs coverage. The problem is that equipping each of these mini base stations with a fibre optic backhaul is extremely expensive. A feature called 'Integrated Access and Backhaul' (IAB) allows 5G NR to be used as a backhaul (Figure 2.38). The IAB nodes are initially intended to be fixed – so there might be one on each building floor with external and internal antennas. In later releases, 5G advanced will support moving nodes such as trains, trams, buses and rickshaws.

It also intends to include support for higher mmWave frequencies in 5G Advanced, such as 52.6 to 71 GHz – offering large spectrum blocks for higher

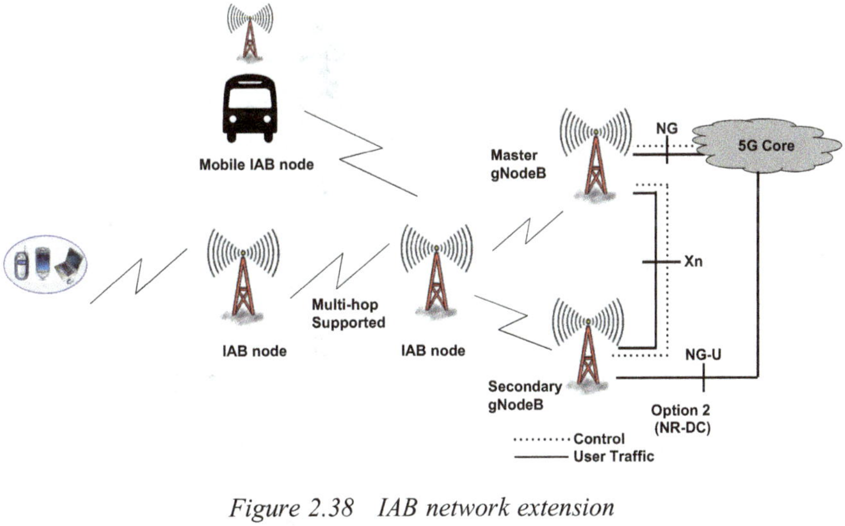

Figure 2.38 IAB network extension

capacities and headline rates. There is also better support for dynamic spectrum sharing (DSS) between 5G and LTE. The two air interfaces are sufficiently alike (OFDM) to allow efficient resource sharing dynamically. Currently, this is available from some vendors as a proprietary solution.

Releases 18 and 19 aim to enhance uplink capacity, allowing terminals to transmit simultaneously to two base stations. This primarily supports better XR applications, including a significant upgrade for AR, which requires uploading video for processing by edge computing and returning results with very low latency. It is understood that future applications might need a higher performance in the uplink. Cameras will likely stream video to the network for computation (e.g. object or face recognition by MEC computing resources). Metaverse/XR applications generate significant uplink traffic with strict reliability and latency requirements.

Another central area addressed in 5GA is NTNs, which refers to satellites, HAPs (usually balloons or aircraft) and UAVs (usually drones). What has changed recently is the emergence of LEO (low Earth Orbit) constellations. These are potentially much cheaper and orbit lower (under 1243 miles) than previous communications satellites; consequently, they have much-reduced latency. The flip side is that more satellites are needed to provide coverage, and the tracking, alignment, and handover are more complicated. Chapter 10 covers NTNs in greater depth.

Figure 2.39 shows how 5G NR could be seen by the mobile transparently through a satellite link, although extensions are being introduced to cope with the

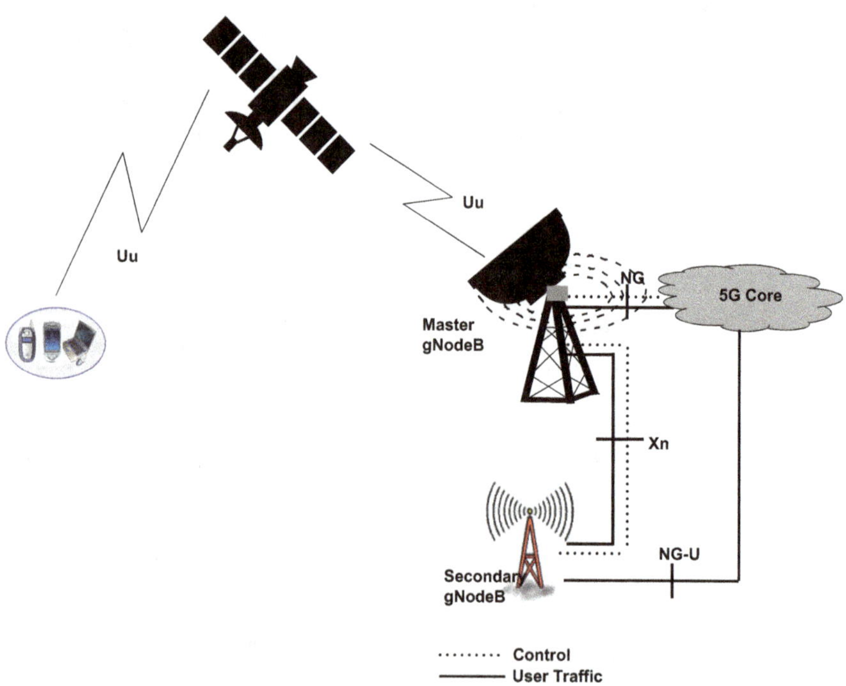

Figure 2.39 Satellite connection to 5G

extended delay. Further enhancements are planned to allow VoIP over (LEO only) satellites. One major UK operator aims for 80% population coverage by 5G by 2028, with the option of using satellites to fill in the remainder. Satellites could offer proper global connectivity, at least out of doors, and increase reliability for links with terrestrial connections. HAPs are being considered and trialled for 5G coverage in rural areas, and Chapter 10 describes several systems involving light aircraft trials. The connection of drone swarms and their remote piloting is being standardised in 5GA, allowing applications such as 3D mapping, delivery, and temporary capacity increases from mini-base stains attached to the drones.

The original specifications of 5G NR did not support the mMTC – specifically, long battery life and enhanced coverage. Three changes have been made to 5G to help cover this situation. Firstly, the ability for (currently LTE) devices using NB-IoT (narrowband IoT) and MTC (machine type communications) to access the 5G core (slices, private networks, low latency, etc.). Secondly, NB-IoT is being developed as a standard that can be deployed in 5G networks, and, lastly, a new 'flavour' of the 5G radio – called RedCap (reduced capability) – has been developed. Table 2.11 handily summarises the differences between these similar technologies. RedCap devices are lower-cost, lower-complexity battery-powered devices such as wearables and industrial sensors. They are sold as 'mid-tier' devices between NB-IoT and fully 5G devices such as iPhones. Further enhancements to allow devices to enter a 'deep sleep' state – whereby they only need to wake up very infrequently to check if they are being paged are also planned in 5GA, as well as remote recharging of batteries, possibly involving drones emitting an RF field that is used by an ultra-low power device to replenish a small battery.

Table 2.11 Comparison of different 4G and 5G support for lower capability devices (note the eMBB rules are for simple devices with limited transmitters) FD = full Duplex, HD = Half Duplex.

	R15/16 NR eMBB	**R17 NR RedCap**	**R18 RedCap**	**LTE-M (Cat M1)**	**NB-IoT (Cat NB1)**
UE BW	100 MHz (FR1) 200 MHz (FR2)	100 MHz (FR1) 200 MHz (FR2)	20 MHz or 5 MHz (FR1 only)	1.4 MHz	180 kHz
Duplex	FD-FDD TDD	FD/HD-FDD TDD	FD/HD-FDD TDD	FD/HD-FDD TDD	HD-FDD TDD
UE-antenna	1T2R(FDD) TDD1T4R (TDD)	1T1R/1T/2R	1T1R/1T2R	1T1R	1T1R
Max modulation	256QAM (DL) TDD64QAM (UL)	64QAM TDD(256QAM optional)	64QAM TDD (256QAM optional)	16QAM	QPSK
Peak rate DL	2.3 Gbit/s	220 Mbit/s	10 Mbit/s	588 kbit/s	26 kbit/s
Peak rate UL	468 Mbit/s	120 Mbit/s	10 Mbit/s	1119 kbit/s	66 kbit/s

5G advanced aims to get more out of the air interface. One of the major hopes is that using AI (Artificial Intelligence) over the radio interface and across the RAN dramatically increases overall performance. In the 1990s, operators bought RANs from a single vendor that had spent many years developing optimisation software to control the RAN. This is always a complex optimisation problem because of the rapidly changing mobile conditions, mobility and QoS requirements (e.g. not to drop voice calls), all while balancing capacity in different cells (that interfere with neighbouring cells) and getting the UL/DL split to match the traffic. The idea behind using AI is that the network and terminals collect large amounts of data, which is the basis of ML algorithms used in AI. Release 18 introduces new interfaces for data collection, although it leaves the actual algorithms as vendor-proprietary. An example of where AI might be helpful is a railway station: the algorithm might hone the MIMO to focus on the platform and concourse. It might be primed to cope with a sudden rush of handovers from terminals leaving a train (anticipating the need for extra signalling resources) or, perhaps, power down much of the network at less busy times. Another radio improvement investigated in 5GA is to look (again) at transmitting simultaneously in the uplink and downlink (no more TDD or FDD). This remains an enormous challenge, even at the base station where better Tx/Rx isolation is possible. Nvidia has published a paper that provides a helpful guide to how the NR air interface could be enhanced with AI [56]. There is a lot more about AI and its role in 6G in Chapter 13.

Finally, 5G advanced might support advanced multiSIM (MUSIM). Non-advanced MUSIM is effectively two completely different SIMs that can be switched but are never active simultaneously. Advanced MUSIM would allow two or more connections simultaneously over the same radio. This could be because of different policies or charging reasons (say, to avoid carrying around a corporate and a private mobile).

2.10 Summary and conclusions

On one level, 5G has been a great success with a worldwide rollout in many frequency bands. The ability to connect to an LTE core and (with proprietary support) dynamically share spectrum with LTE has resulted in the fastest rollout of any generation. Median download speeds are approaching 100 Mbit/s, and peak rates are 1 Gbit/s. Capacity has also been massively boosted. In urban areas, it is reasonable to judge 5G as successful in delivering the eMBB use case with '50 Mbit/s everywhere' in regions of 5G coverage. Rural service depends on further network roll- out and possible satellite connections outside urban areas. Part of the original vision of 5G was also to support industry with low latency, URLLC, and network slicing with both public and private spectrum. To implement all these innovations, operators must migrate from LTE cores to a 5G stand-alone (SA) architecture. A significant expense is a fully-fledged SA 5G network with multiple frequency bands, base station upgrades, and new sites. As of Q1 2024, this transition is still happening, and it is too early to judge demand and revenue from non-eMBB services. The later part of the decade could significantly increase non-human 5G services and revenues. 5G is also being improved with satellite connectivity additions, 5G radio backhaul and side-link extensions. New use cases for vehicles, XR, drones, satellites and moving base stations are being supported in R17/18. In addition, extra developments in machine-type communications continue in R19 onwards.

References

[1] UK Standard News Article. [Online]. [Cited Jan 2024]. Available from: https://www.standard.co.uk/tech/5g-future-potential-o2-mark-evans-a3791556.html.

[2] Samuthira Pandi, V. and Lakshmi Priya, J. 'A Survey on Mobile 5G Technology', *published in the 2017 IEEE International Conference on Power, Control, Signals and Instrumentation Engineering (ICPCSI)*. Chennai, India, 2017.

[3] Cisco Visual Networking Index: Forecast and Methodology, 2014–2019. [Online]. [Cited Jan 2024]. Available from: http://www.cisco.com/c/en/us/solutions/collateral/service-provider/ip-ngn-ip-next-generation-network/white_paper_c11-481360.pdf.

[4] Droste, H., Zimmermann, G., Stamatelatos, M., *et al.* 'METIS 5G Architecture: A Summary of METIS Work on 5G Architectures', *Vehicular Technology Conference (VTC Spring)*, 2015 IEEE 81st Pages: 1–5, DOI:10.1109/VTCSpring.2015.7146131.

[5] South Korea Telecom White Paper, 201. [Online]. [Cited Jan 2024]. Available from: https://www.sktelecom.com/img/pds/press/SKT_5GWhite Paper_V1.0_Eng.pdf.

[6] Next Generation Mobile Networks forum White paper. [Online]. [Cited Jan 2024]. Available from: https://www.ngmn.org/work-programme/5g-white-paper.html.

[7] GSM Association Intelligence. 'Analysis: Understanding 5G: Perspectives on Future Technological Advancements in Mobile', Dec 2014. [Online]. [Cited Jan 2024]. Available from: https://data.gsmaintelligence.com/research/research/research-2014/understanding-5g-perspectives-on-future-technological-advancements-in-mobile.

[8] Ericsson Mobility Report 2022. [Online]. [Cited Jan 2024]. Available from: https://www.ericsson.com/en/reports-and-papers/mobility-report/reports/november-2023?gad_source=1&gclid=Cj0KCQiA2eKtBhDcARIsAEGTG41i_nKIMkvSNQZ7JZeLVnGc8E4wY2ej2a8-jPeD9Sv2U6zEUBMPkVYaAu58EALw_wcB&gclsrc=aw.ds.

[9] Ericsson Report. [Online]. [Cited Jan 2024]. Available from: https://www.ericsson.com/en/reports-and-papers/mobility-report/dataforecasts/mobile-data-traffic-growth-rate#:~:text=Yearlymobile data traffic growth, in the latter forecast years.

[10] Analysis Mason Research Report, 2014. [Online]. [Cited Jan 2024]. Available from: https://www.analysysmason.com/globalassets/x_migrated-media/media/analysys_mason_5g_spectrum_oct20142.pdf.

[11] Dannewitza, C., Kutscherb, D., Ohlmanc, B., Farrelld, S., Ahlgrene, B., and Karla, H., 'Network of Information (NetInf) – An Information-Centric Networking Architecture', *Computer Communications*. 2013;36(7):721–35.

[12] Huawei Innovation Research Programme, 2015. [Online]. [Cited Jan 2024]. Available from: https://www-file.huawei.com/~/media/CORPORATE/PDF/Downloads/Inaugural-Issue-5G-Research-and-Innovation.pdf.

[13] Fraunhofer Fokus, Towards the 5G Environment: Making Research Steps with Meaningful Results Version 1.0, Aug 2014. [Online]. Available from: www.Open5GCore.net.

[14] 4G Americas, 'Recommendations on 5G Requirements and Solutions'. [Cited Jan 2024]. [Online]. Available from: http://www.4gamericas.org/files/2714/1471/2645/4G_Americas_Recommendations_on_5G_Requirements_and_ Solutions_10_14_2014-FINALx.pdf.

[15] Nokia Networks, '5G Use Cases and Requirements', 2014. [Online]. Available from: https://www.ramonmillan.com/documentos/bibliografia/5GUseCases_Nokia.pdf.

[16] Nakamura, T., Benjebbour, A., Kishiyama, Y., Suyama, S., and Imai, T. '5G Radio Access: Requirements, Concept and Experimental Trials', *IEICE Transactions on Communications*. 2015;E98.B:1397–406. 10.1587/transcom.E98.B.1397.

[17] ARIB 2020 and Beyond Ad Hoc Group White Paper. 'Mobile Communications Systems for 2020 and Beyond'. Version 1.0.0. October 8th. [Online]. [Cited Jan 2024]. Available from: 2014,. https://www.arib.or.jp/english/image/committee/adwics/02-00_2020bah/20bah-wp-100.pdf.

[18] Gupta, A. and Jha, R.K. 'A Survey of 5G Network: Architecture and Emerging Technologies', *IEEE Access*. 2015;3:1206–32.

[19] Qualcomm Presentation, '5G Vision', 2015. [Online]. [Cited Jan 2024]. Available from: https://www.qualcomm.com/content/dam/qcomm-martech/dm-assets/documents/whitepaper_5g_vision_for_the_next_generation_of_connectivity.pdf.

[20] Cox C., *An Introduction to 5G*. New York: Wiley; 2021.

[21] Massive MIMO Primer. [Online]. [Cited Jan 2024]. Available from: https://www.ericsson.com/4947d3/assets/local/ran/doc/03142022-massive-mimo-handbook-extended-1st-edition-e-book.pdf?utm_campaign=2203-BNEW-WEC-GEN-Massive-MIMO-Extended-Version&utm_source=WebContent&utm_medium=email.

[22] Sutton, A., 'Open Radio Access Network'. [Online]. [Cited Jan 2024]. Available from: https://www.engagingwithcommunications.com/publications/ITP_Papers/5G_network_architecture/ITP_Journal_Vol12_Part1_Pages9-15_5G_Network_Architecture.pdf.

[23] Light Reading, 'Standalone 5G Remains a Disappointment'. [Online]. [Cited Jan 2024]. Available from: https://www.lightreading.com/the-core/standalone-5g-progress-remains-a-disappointment/d/d-id/782748.

[24] News Item, 'Vodafone deploys OpenRAN in Urban Locations in European First'. [Online]. [Cited Jan 2024]. Available from: https://www.vodafone.co.uk/news-centre/press-release/openran-deployed-in-urban-locations-in-european-first/.

[25] Eswaran, S. and Honnavalli, P., '5G Private Networks: A Survey on Enabling Technologies, Deployment Models, Use Cases and Research Directions', *Telecommunication Systems*. 2023;82:3–26. [Online]. [Cited Jan 2024]. Available from: https://doi.org/10.1007/s11235-022-00978-z, Accessed Jan 2024.

[26] Mach, P. and Becvar, Z., 'Mobile Edge Computing: A Survey on Architecture and Computation Offloading', March 2017. [Online]. [Cited Jan 2024]. Available from: https://arxiv.org/pdf/1702.05309.pdf.

[27] Webb, W., 'The 5G Myth', Jan 2017, CreateSpace Independent Publishing Platform. ISBN-13:978-1540465818.

[28] NGMN. [Online]. [Cited Jan 2024]. Available from: https://www.ngmn.org/about-us.html.

[29] Nokia 5G Whitepaper. [Online]. [Cited Jan 2024]. Available from: https://docplayer.net/21520832-Nokia-networks-futureworks-looking-ahead-to-5g-building-a-virtual-zero-latency-gigabit-experience-white-paper-looking-ahead-to-5g.html.

[30] GSMA Views of 5G. [Online]. [Cited Jan 2024]. Available from: https://www.slideshare.net/AngelicaNaguit/2014-gsma-understanding-5-g.

[31] Wisely, D., 'Capacity and Costs for 5G Networks in Dense Urban Areas', PhD Thesis University of Surrey 2014. [Online]. [Cited Jan 2024]. Available from: https://openresearch.surrey.ac.uk/esploro/outputs/doctoral/Capacity-and-costs-for-5g-networks-in-dense-urban-areas/99512629102346.

[32] 'WEIGHTLESS – A Standard for M2M Communications'. [Online]. [Cited Jan 2024]. Available from: http://www.weightless.org/.

[33] Wisely, D., Wang, N., and Tafazolli, R., 'Capacity and Costs for 5G Networks in Dense Urban Areas', 30 October 2018. [Online]. [Cited Jan 2024]. Available from: https://doi.org/10.1049/iet-com.2018.5505.

[34] Oughton, E. J. and Frias, Z., 'The Cost, Coverage and Rollout Implications of 5G Infrastructure in Britain'. *Telecommunications Policy*. 2017;42(8): 636–652.

[35] 5G Spectrum GSMA Public Policy Position, June 2022. [Online]. [Cited Jan 2024]. Available from: https://www.gsma.com/spectrum/wp-content/uploads/2022/06/5G-Spectrum-Positions.pdf.

[36] GSMA Intelligence. [Online]. [Cited Jan 2024]. Available from: https://www.gsmaintelligence.com/data/.

[37] UK 5G Speed from Rootmetrics. [Online]. [Cited Jan 2024]. Available from: https://www.ispreview.co.uk/index.php/2023/01/rootmetrics-benchmark-the-top-uk-mobile-networks-for-h2-2022.html, Accessed Jan. 2024.

[38] Lightreading 5G Core. [Online]. [Cited Jan 2024]. Available from: https://www.lightreading.com/the-core/standalone-5g-progress-remains-a-disappointment/d/d-id/782748.

[39] 5G Success or Failure? [Online]. [Cited Jan 2024]. Available from: https://www.sdxcentral.com/articles/opinion-editorial/five-years-in-is-5g-a-success-or-failure/2023/11/.

[40] 5G Observatory. [Online]. Available from: https://5gobservatory.eu/5g-private-networks/.
[41] US LTE and 5G Private Networks. [Online]. [Cited Jan 2024]. Available from: https://www.grandviewresearch.com/industry-analysis/us-private-lte-5g-network-market.
[42] Brown, G., '5G Network Operator Slicing Survey', 2022, Heavy Reading. [Online]. [Cited Jan 2024]. Available from: https://mrkt-lp.flywheelsites.com/heavy-reading-5g-network-slicing/?gf_protect_submission=1.
[43] Ericsson News. [Online]. [Cited Jan 2024]. Available from: https://www.ericsson.com/en/ran/cloud.
[44] Statistica. [Online]. [Cited Jan 2024]. Available from: https://www.statista.com/statistics/685135/vodafone-mobile-contract-arpu-in-the-uk/.
[45] News Item. [Online]. [Cited Jan 2024]. Available from: https://www.thaienquirer.com/47065/ais-zte-launch-1st-trial-of-urllc-showcasing-low-latency-of-1-ms-on-the-2-6-ghz-5g-network/.
[46] 5G Latency in Munich. [Online]. [Cited Jan 2024]. Available from: https://www.rohde-schwarz.com/uk/solutions/test-and-measurement/mobile-network-testing/stories-insights/5g-nr-and-lte-latency-analysis-in-a-public-network_255963.html.
[47] Maldonado, R., Karstensen, A., Pocovi, G. *et al.*, 'Comparing Wi-Fi 6 and 5G downlink performance for industrial IoT', *IEEE Access*, 2021;9:86928–37, doi:10.1109/ACCESS.2021.3085896.
[48] URLLC trial. [Online]. [Cited Jan 2024]. Available from: https://www.thefastmode.com/technology-solutions/29670-ais-zte-first-to-trial-5g-urllc-in-thailand.
[49] US 5G Speeds by Band. [Online]. Available from: https://img.lightreading.com/2023/02/783211/426.png.
[50] Lightreading. [Online]. Available from: https://www.lightreading.com/5g/mmWaveave-5g-in-2023-fast-but-still-expensive-and-hard-to-find.
[51] Lightreading. [Online]. [Cited Jan 2024]. [Cited Jan 2024]. Available from: https://www.lightreading.com/5g/korean-govt-falls-for-5g-fallacy-over-28ghz.
[52] 3GPP. [Online]. Available from: https://www.3gpp.org/.
[53] 5G Americas. 'Becoming 5G Advanced: The 3GPP 2025 Roadmap. [Online]. Available from: https://www.5gamericas.org/becoming-5g-advanced-the-3gpp-roadmap/.
[54] Lin, X. 'An overview of 5G advanced evolution 3GPP release 18', *IEEE Communications Standards Magazine*. 2022;6(3):77–83.
[55] Release 19. [Online]. [Cited Jan 2024]. Available from: https://www.nokia.com/blog/5g-advanced-shifts-to-the-next-gear-with-release-19/#.
[56] AI Overview. [Online]. [Cited Jan 2024]. Available from: https://arxiv.org/pdf/2308.05315.pdf.

Chapter 3
Future Wi-Fi

It is a good thing to have two ways of looking at a subject and to admit that there are two ways of looking at it.

– James Clerk Maxwell

3.1 Introduction

Having a chapter on Wi-Fi in a book predominantly about 6G might seem slightly incongruous. However, there are good reasons to examine the latest Wi-Fi developments and standards. Wi-Fi has always carried several times more traffic than cellular networks, doing the heavy lifting of wireless connectivity indoors where macrocellular networks are less efficient. Recent Wi-Fi developments have made much more spectrum available, promise higher maximum data rates and have the potential to offer much greater control over latency and reliability than past generations. With 5G introducing private networks enabled with advanced features such as network slicing, high reliability and low-latency support, the two technologies will offer potential solutions in the Industry 4.0 setting. In the domestic setting, 6G may deploy higher frequencies via home gateways to provide 6G services that require ultra-high data rates or processing within the network. With support from third-party network elements, many of these 6G services could also be offered within the home using Wi-Fi. Wi-Fi-cellular convergence and fixed-mobile convergence (FMC) have long been discussed. Standards exist that allow full integration of Wi-Fi networks within a cellular core but have never been widely deployed. All of these questions and tensions will form part of the development of 6G and present a strong motivation for examining recent developments in Wi-Fi and the potential implications of these. This is an exciting time for Wi-Fi, with Wi-Fi 7 certification completed January 2024, and quite a few smartphones with Wi-Fi 7 capable chipsets are appearing, such as the GooglePixel 8 Pro. Intel has also developed a BE200 Wi-Fin7 card, and the expectation is that the iPhone 16 Pro will also support Wi-Fi 7.

The chapter begins with a brief history of Wi-Fi and the origins of the critical differences with cellular networks. It then describes the key innovations in the current Wi-Fi 6 generation and its extension into the 6–7 GHz band in 6E. The following section then evaluates Wi-Fi 7. This introduces significant new features supporting higher throughput, lower latency and improved reliability. Wi-Fi 8 is focused on further improving reliability by reducing interference, enhancing coordination and offering better latency and packet delivery bounds. This is due to complete standardisation in 2028, and Wi-Fi 8 devices will be commonplace around the time 6G is scheduled to appear (2030). The chapter then looks at Wi-Fi using mmWave frequencies, describing the emerging standards and potential use cases. The penultimate section describes ways Wi-Fi and cellular can be coupled. Finally, a brief look at the future of Wi-Fi with AI and sensing is likely incorporated into future standards.

3.2 Wireless local area network (WLAN) – how it works and critical shortcomings

The first WLAN was a wireless incarnation of the original Ethernet standard, where many of its strengths and weaknesses originate. The name Ethernet comes from Robert Metcalfe, who in 1974 called it Ethernet after the Luminiferous Ether, which was the once-proposed medium that electromagnetic waves propagated through until special relativity removed the need for it [1].

This section contrasts WLAN's operation with traditional cellular networks, mainly in terms of how radio resources are managed and the spectrum used. Note that the term Wi-Fi was only later adopted for marketing purposes, as explained in the following sections. WLANs can be created with a range of different technologies, only some of which are branded Wi-Fi.

3.2.1 Origins

In the early 1970s, the University of Hawaii set up a radio packet data network using a protocol called ALOHA. This was very simple: if a station had some data to transmit, it just transmitted it – regardless of what anybody else was doing. If the network was lightly used and traffic levels low, there was rarely a collision (two stations transmitting simultaneously on the same frequency). However, as traffic levels increased, the throughput fell. Metcalfle's insight for fixed Ethernet was to use carrier sense with collision detection. However, collision detection is not possible for radio networks. The first WLAN Protocol used Carrier Sense Multiple Access (CSMA). To avoid several stations sensing the carrier as free (perhaps at the end of a previous transmission) and all transmitting at once, they ran a random wait algorithm that usually resulted in one station transmitting first and by the time the station with the next shortest wait was ready to transmit the signal from the first station had arrived and been detected. This was called collision avoidance (CA); the original WLAN ran this CSMA/CA protocol. Figure 3.1 shows how it improved on ALOHA. The receipt of each packet was confirmed with an explicit ACK (Acknowledgement) – such that if the transmitting station did not receive an ACK, it was assumed a collision had taken place, and the frame was re-transmitted. Collisions can occasionally occur because stations chose the same random number for the countdown or because more distant stations are causing interference.

Commercial WLANs were kick-started by the FCC, allowing the 2.4 GHz band to be used for unlicensed use. This was known as the Industrial Scientific and

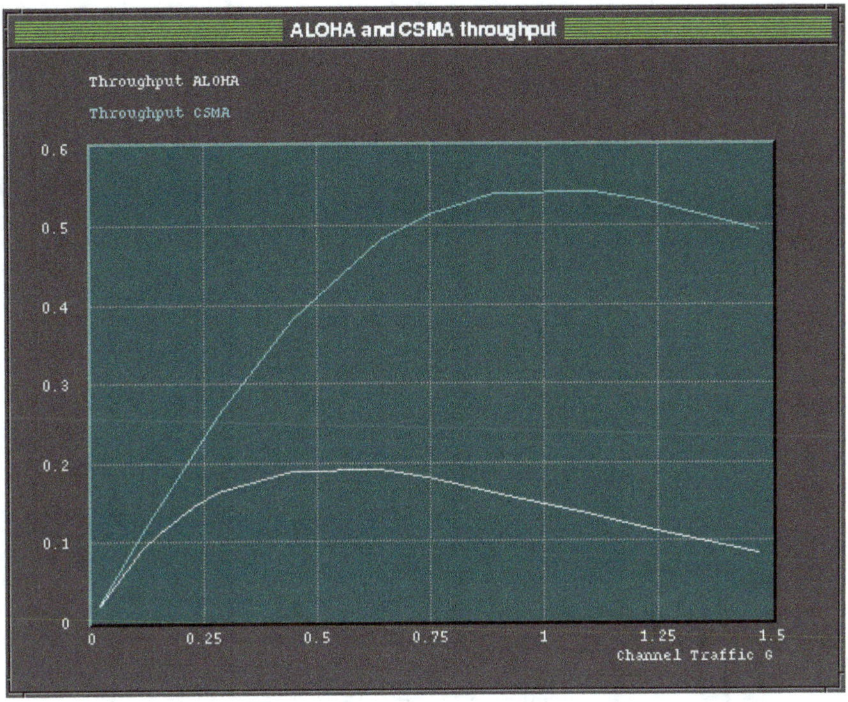

Figure 3.1 Throughput of ALOHA and CSMA/CA – for eight stations and with increasing traffic load

Medical band and could be used by any radio protocol subject to transmit power restrictions. Work at Nokia in the mid-1990s led to 802.11 (1997 with 1–2 Mbit/s), 802.11b (1999 with 1–11 Mpbs) and 802.11a (1999 with 1–54 Mbit/s). The IEEE is producing these standards [2]. The IEEE is the Institute of Electrical and Electronics Engineers and is separate from 3GPP, attended by a completely different eco-system of vendors and operators. It covers only the protocol stack's first two layers (PHY and MAC) rather than the total system design of 3GPP. Early 802.11 systems suffered from compatibility issues, which led to the formation of the Wi-Fi Alliance (Wi-Fi is not short for Wireless Fidelity – it is just a logo and meant that the system had been tested and could sport the familiar logo). The alliance is a global non-profit organisation of about 550 members [3]. The term Wi-Fi now means the whole family of 802.11 standards from the IEEE.802.11. Standards continue to evolve – reaching much higher data rates, operating in new frequencies (notably 5 GHz), and developing wide area standards (such as WiMAX).

3.2.2 Generations

Each IEEE working group has a designation 802.11xx, from which the various standard names have come (Table 3.1). There are many more standards than the main

Table 3.1 Wi-Fi generations. There are many more 802.11 standards, and some are described later in the chapter – these are the primary consumer standards and the ones adopted by the Wi-Fi Alliance.

Technology	**New designation**	**Frequency band**	**Chanel bandwidth**	**Multiplexing method**	**RF technology**	**MIMO**	**Data rates (absolute Max at PHY later)**
802.11a (1997)	Wi-Fi 2	5 GHz	20	CSMA/CA	OFDM		6–54 Mbit/s
802.11b (1997)	Wi-Fi 1	2.4 GHz	20	CSMA/CA	HR-DSSS		1–11 Mbit/s
802.11g (2003)	Wi-Fi 3	2.4 GHz	20	CSMA/CA	OFDM		6–54 Mbit/s
802.11n (2009)	Wi-Fi 4	2.4 GHz 5 GHz	20,40	CSMA/CA	OFDM	Four layer	72–600 Mbit/s
802.11ac (2013)	Wi-Fi 5	5 GHz only	20,40,80, 80 + 80, 160	CSMA/CA	OFDM	Eight-layer max DL multi-user MIMO	433 Mbit/s–6.933 Gbit/s
802.11ax (2019)	Wi-Fi 6	2.4 GHz 5 GHz	20,40,80, 80 + 80, 160	OFDMA	OFDM	Eight-layer max DL and UL multi-user MIMO	0.574–9.608 Gbit/s
802.11ax (2020)	Wi-Fi 6E	6 GHz	20,40,80,160	OFDMA	OFDM	Eight layer max DL and UL Multi-user MIMO	0.574–9.608 Gbit/s
802.11be (2024)	Wi-Fi 7	2.4 GHz 5 GHz 6 GHz	2,04,08,01,60,320	OFDMA	OFDM	Sixteen layer max DL and UL Multi-user MIMO	1.376–46.120 Gbit/s

consumer ones listed in the table (several are described later in the chapter). However, only these primary standards have been given the designations Wi-Fi 5/6/7/8, etc.

802.11 ac has been (retrospectively) called Wi-Fi-5 by the Wi-Fi Alliance (and by extension, 802.11n is now Wi-Fi 4). Wi-Fi 6 is a significant change to the fundamental operation of WLANs – going back to the original principles of operation and a ground-up re-design to improve the CSMA/CA protocol. One of the main drivers of the development of Wi-Fi 6 has been the low throughput of 802.11ac (Wi-Fi 5) with multiple users and in areas with many competing WLAN systems. One factor involved in this is that many Internet applications transmit small packets. As an example, good-quality voice uses 70-byte packets. This compares to an FTP file transfer, which uses the default maximum Ethernet frame size of 1500 bytes. Each 70-byte IP packet must be transferred individually over WLAN, which incurs a significant overhead per packet. The same overhead applies to larger packets, but that is a much smaller percentage of the payload. The overall efficiency will be much lower if many applications use smaller packets.

There are mechanisms to aggregate frames in Wi-Fi 4 and 5, and a technique called aggregated MAC protocol data unit is extensively used and essential to delivering high throughput. However, there are still considerable inefficiencies, and the level of aggregation is limited for voice traffic as it adds delay in forming the aggregated packets. According to a recent book, 75–80% of 802.11 frames are under 256 bytes, 60% of 802.11 traffic is control frames and 15% are management frames [4].

Another problem with Wi-Fi 5 is the need for interference management. Any WLAN not under your control can choose any channel (and there are only three 160 MHz channels in the 5 GHz band). Access Points under your control are not coordinated at a frame level and offer no interference avoidance. There is no beam forming or null forming as there is in 5G. Although modern access points can scan the radio environment for the optimum channel, the net result of this interference is that as more WLANs operate in a densely populated area, such as an office block, the efficiency falls dramatically.

3.2.3 Comparison with cellular

It is helpful to compare WLANs and cellular systems (Table 3.2). The table highlights particular issues with Wi-Fi that have driven the innovations in Wi-Fi 6, 6E and Wi-Fi 7 and the developments towards Wi-Fi8. The use of shared spectrum is much more effective if you have more spectrum to choose from and you can use parts of it more effectively. To achieve higher rates, it is generally necessary to utilise wider channels. Better control of interference and base station coordination is needed to provide lower latency and higher reliability calls for multiple links. All of these are either already introduced or being considered for future Wi-Fi. The following three sections look at these innovations across Wi-Fi 6, 6E and Wi-Fi 7 and what is being studied for inclusion in Wi-Fi 8.

Wi-Fi generates many more standards, which are issued more frequently than cellular standards. Some of these are more commercially successful than others.

Table 3.2 Comparison between cellular and Wi-Fi systems

	Wi-Fi	**Cellular**
Spectrum	Shared spectrum	Dedicated spectrum
Transmit power	Low (regulated)	High
Range	Short range	Long range
Network type	Self install	Planned and built by MNOs
Main use	Indoor/campus/factory	Outdoor and indoor (but indoor performance reduced)
Interference management	Limited	Many interference avoidance techniques
Radio sensitivity	Medium sensitivity	Noise limited
Handover	Not generally	Yes
Security	Medium	High
Backwards compatibility	Yes	No (each generation features a new air interface)
QoS Support	In standards but not widely used	Yes

Examples of this are the mmWave Wi-Fi standards described later in the chapter. The more recent move to having a more straightforward numbering of generations follows the cellular approach. Wi-Fi generations are also backwards compatible, which is not true of cellular.

3.3 Wi-Fi 6 – innovations

In this section, we describe the main changes in Wi-Fi 6, its actual performance in real-life situations and typical installation scenarios.

3.3.1 Major innovations in Wi-Fi 6

The first significant change from Wi-Fi 5 is using OFDMA (orthogonal frequency division multiple access). OFDM is the chopping up a spectrum band into a series of smaller sub-channels, called tones. Each tone can support a different modulation, and these are assembled (by an inverse Fourier transform) into a symbol. If the symbol rate is the same as the sub-carrier spacing, then the sub-carriers remain orthogonal. 802.11g and 802.11n used OFDM but only between any two stations simultaneously. So, the base station would use the whole bandwidth to transmit to Station A (including any MIMO layers it could form) and then use the same bandwidth for Station B. Each transmission used all the good or bad sub-channels for that radio path. OFDMA, however, transmits to multiple stations at the same time. In Wi-Fi 6, the base station transmits in the downlink to multiple stations

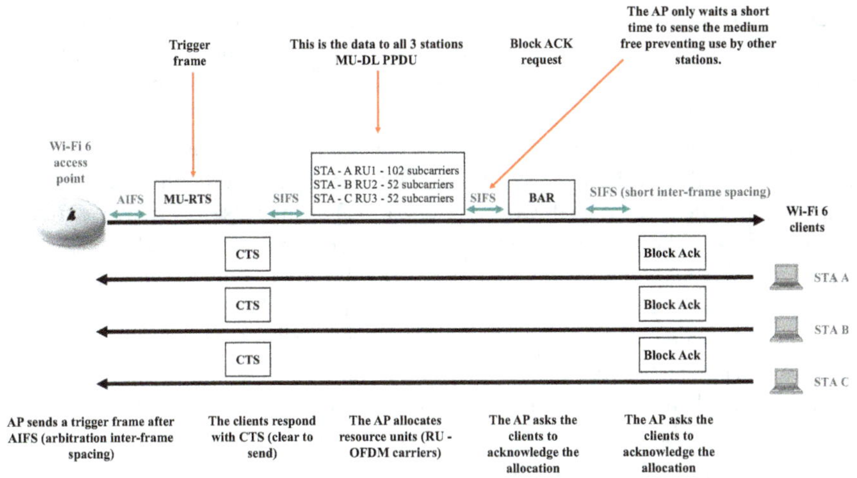

Figure 3.2 Wi-Fi 6 downlink transmission cycle

using OFDMA, with each station having its sub-carriers allocated. The main advantage is that small amounts of data sent to several stations are sent within a single frame, significantly reducing overhead costs and latency. Wi-Fi 6 also recognises the base station as managing radio resources. In Wi-Fi 5, all stations, including the base station, must compete for TXoPs (transmission opportunities) on an equal footing. The overall downlink cycle of Wi-Fi 6 is shown in Figure 3.2, which details how the AP divides the resource units (RUs – the OFDM carriers). The cycle is used to allocate RUs to the clients, and shorter spacing allows the AP to do this without being interrupted by other users — these must wait a longer time to sense the medium free. By this time, the Wi-Fi 6 AP has started the allocation cycle.

Wi-Fi 6 also introduced MU-MIMO (multi-user MIMO – the principles of multi-user MIMO are described in Chapter 2). This allows the access point to transmit multiple streams to multiple users simultaneously (and on the same OFDMA sub-carriers). Figure 3.3 shows an AP transmitting two layers to STA-A, 1 to STA-B and 1 to STA-C. All of these streams use the same set of sub-carrier tones. In practice, most laptops and phones only support two-layer MIMO. This is a handy way to increase capacity and flexibility. In each downlink transmission, the AP can change the different streams and stations receiving them and the sub-carrier

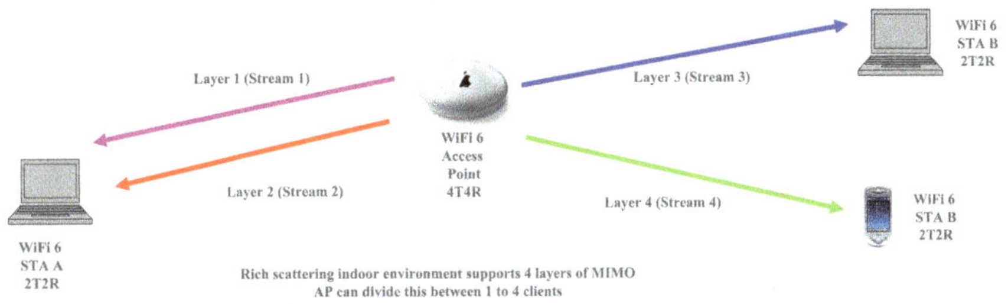

Figure 3.3 MU-MIMO in Wi-Fi 6

allocations. How it does that is proprietary, and other vendors will have different algorithms to share the resources and make the best use of them. AI/ML (artificial intelligence/machine learning) will significantly improve performance in this area.

Downlink MU-MIMO was (optionally) included in Wi-Fi 5 but seldom used. Sheth *et al.* [5] report only about 10% of current equipment supports this. In Wi-Fi 6, the maximum MU-MIMO is 8T8R, and a maximum of eight different users are supported (all of which would be down to one stream). That increases capacity eightfold – although this would only happen in a rich scattering environment. OFDMA also helps with latency as multiple clients receive data from the AP within a single transmission and do not wait for their turn for a downlink cycle. Uplink MU-MIMO is also supported, with up to eight receivers at the AP used to decode multiple streams. As with MIMO in cellular systems, there is a need to know the radio channel conditions between each client and the AP. After the clients' measurements, this has to be reported back to the AP, with slightly added overhead. It also makes MU-MIMO unsuitable for moving terminals at any speed above walking pace. MU-MIMO will only be applicable where the radio environment is diverse enough to support it.

Wi-Fi 6 also addresses a fundamental issue with the CSMA/CA mechanism and the limited number of channels at 2.4 and 5 GHz. A station or AP must sense that the medium can utilise a transmission opportunity. The standards mandate that Wi-Fi receivers must be able to decode the preamble of packets at a received level of −82 dBm or better. If a preamble is detected, then the channel is considered busy. There is also an energy detection threshold of −62 dBm for non-WLAN systems, such as TV signal repeaters (TV senders).

Figure 3.4 shows the issue of the preamble detect range being much larger than the data transmission range for Wi-Fi 5. This means that distant WLANs can prevent transmissions that would, in all likelihood, have not interfered.

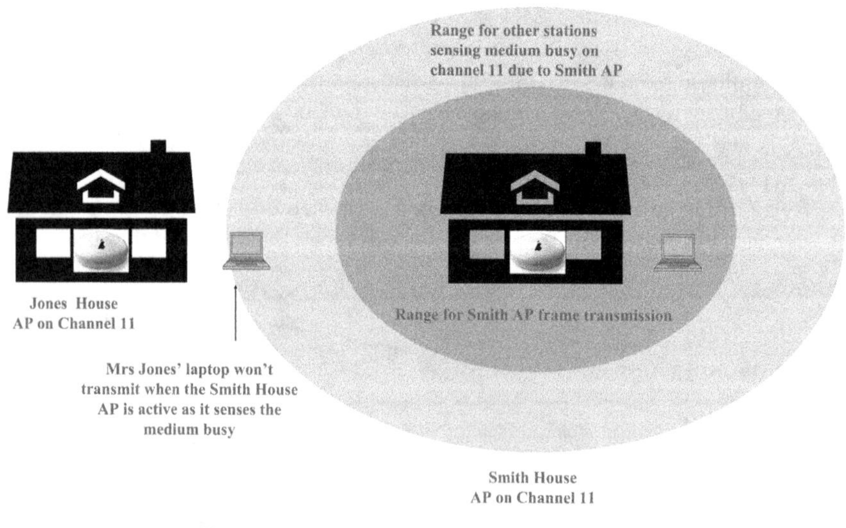

Figure 3.4 Different 802.11 ranges cause co-channel interference issues

The obvious solution is to have neighbouring APs operating on different channels. However, in the 2.4 GHz band, only three independent channels (70 MHz) exist. In the 5 GHz band (480 MHz), there can be 24 × 20 MHz channels or, more usually, in enterprise deployments, 12 × 40 MHz. However, the extended range of the preamble threshold still causes problems with the free-channel assessment. In Figure 3.4, the householder, Mr. Smith, has many neighbours. Whatever channel he uses, there are access points and Wi-Fi equipment (which might be printers, storage devices, smart TVs, gaming consoles, cameras, etc.) that use all available channels, causing his system to sense the channel as busy. This is even though nearly all these transmissions from neighbours would not be strong enough to disrupt transmissions between his devices. To combat this, Wi-Fi 6 now includes a 'colour' field (there are only 64 colours in this Wi-Fi 6 world) that allows nodes to differentiate between Own-WLAN (Smith) and Other-WLAN (Jones *et al.*). That on its own is not helpful, but Wi-Fi 6 also allows a variable energy or preamble detect threshold for other-colour WLANs (Figure 3.5). Changing the threshold is a bit of a double-edged sword. Keep it too low, and there is no benefit; raise it too high, and the number of packet errors due to interference rises rapidly. It will be helpful in domestic settings and enterprises – although current enterprise systems often allow the network managers to adjust the energy threshold of the APs. Another benefit of this BSS 'colouring' is that packets not for a node because they are a different 'colour' can be identified early in the decoding process, saving energy.

There are a few other, less significant improvements in Wi-Fi 6 (see Sheth *et al.* [5] for details). Target wake time (TWT) is the only one of any significance. It is an agreement between the AP and clients in power save mode to allow the AP to coordinate communications with the clients when they are awake. This will enable devices to sleep for long hours or more compared to Wi-Fi 5 devices. It is aimed at low-power IoT devices and offers very significant power savings.

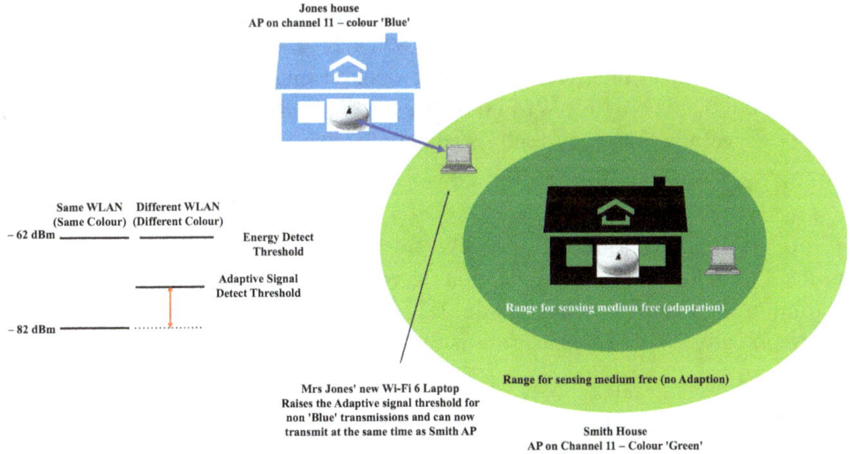

Figure 3.5 Variable energy detects threshold in Wi-Fi 6

At the time of writing (2024), Wi-Fi 6 was on the iPhone from version 11, with two layers of MIMO supported. In the UK, most ISPs have been sending new customers a Wi-Fi 6 hub since early 2022. According to the Wi-Fi Alliance, 3.5 billion Wi-Fi 6 devices were shipped in 2022 – 80% of all shipments [6].

One excellent data source on Wi-Fi 6 performance is the WBA Wi-Fi 6 trials report [7]. The WBA conducted a series of controlled trials across a variety of key application areas:

An aerospace manufacturing plant near Birmingham (England). A Wi-Fi 6 network, including IoT connections, was installed in the trials. Previous attempts with 802.11n at the plant had failed. It was reported, 'During the trial, speeds of 700 Mbit/s using 80 MHz channels were achieved, and low-latency applications, like video calling and video streaming, performed well with results below 6ms' [7].

A large shopping mall in South Korea with 272 shops, 89 restaurants, a movie theatre and an aquarium. The WBA reported an average IP/TCP throughput of 2.3 Gbit/s at each access point and 800 Mbit/s on a smartphone. When OFDMA was used in the trial, latency was also reduced from 21 to 4 ms.

Single-family and multi-dwelling units (houses and flats). For the single house, three APs were used. The trial was near Denver Co., so the house was large by European and Asian Standards. The report gives TCP up and downlink average throughputs at 700–800 Mbit/s. Latency was typically 8–11 ms.

3.4 Wi-Fi 6E – more spectrum

Of course, you should treat the results of trials by the WBA with caution. How much these networks would cost to deploy and run in real situations is still being determined. What is clear, though, is that to reach 1 Gbit/s and 5 ms is not realistic with Wi-Fi 6 in demanding conditions. Two significant issues hold back Wi-Fi 6 performance – legacy nodes and a lack of channels. The backwards compatibility of Wi-Fi 6 means that older nodes can disrupt Wi-Fi 6 AP scheduling, reducing efficiency and increasing latency. Legacy APs will not have a 'colour', and there will still only be 3 × 20 MHz channels at 2.4 GHz and 6 × 80 MHz at 5 GHz. This will still cause significant co-channel interference in dense deployments. What is needed is some new spectrum that is only for Wi-Fi 6 devices and, ideally, a much more substantial amount of spectrum than is currently available for WLANs. This is what 6E brings.

3.4.1 Spectrum

Wi-Fi 6E is the most critical Wi-Fi development to date. The starting gun for 6E was the FCC (Federal Communication Agency – the US body responsible for spectrum management), which made 1.2 GHz of new spectrum available for WLANs in early 2020. The spectrum is in four bands:

- U-NII-5 5.925 GHz – 6.425 GHz,
- U-NII-6 6.425 GHz – 6.525 GHz,
- U-NII-7 6.525 GHz – 6.875 GHz,
- U-NII-8 6.875 GHz – 7.125 GHz.

That can be divided into 59 × 20 MHz channels, 14 × 80 MHz or even 7 × 160 MHz.

The best thing about this new spectrum is that no legacy devices are operating or will be certified for it. Wi-Fi alliance-certified APs and devices will be given the 6E branding.

There are three significant caveats on this new spectrum. First, only a limited number of countries are licensing the whole band (notably the United States, Canada, Brazil, Ethiopia, Columbia, Costa Rica and South Korea), with many others only considering the U-NII-5 band (5.925 – 6.425 GHz) (EU, UK, Russia, Mexico, Hong Kong and Japan) – some are noted as 'considering' the whole band (including the UK). Check out the latest position at the Wi-Fi Alliance [8].

The second caveat is that WLANs are not the primary users of this spectrum. In the United States, much of the 6E spectrum is used for point-to-point microwave links. Indoors, there are no significant restrictions as the 6 GHz signal has an even shorter range than 5 GHz and is significantly absorbed by walls, doors and ceilings. Table 3.3 shows slight variations in the maximum power that can be used indoors across the different subbands in the United States. Note – limits are different between the United States and the UK/EU. The UK limits are 23/30 dBm for 5 GHz (C and B bands), 24 dBm for 6 GHz lower band, and 20 dBm for 2.4 GHz. The upper 6 GHz band is currently used by various services in the UK, including Fixed Links, Fixed Satellite Services, certain Short-Range Devices, Earth Exploration Satellite Services and Radio Astronomy.

Note that the regulatory limit is EIRP, not Tx Power, so attaching high-gain antennas to an AP can cause you to break the regulatory limit. This also means that transmitter beam-forming can have a limited effect since you must reduce the transmitted power per antenna to account for the beam-forming gain.

In the United States, the U-NII-5 band has a maximum allowed EIRP of 4W, but only with a technology called AFC (automated frequency coordination), which we will explain below. AFC is only used in the United States. In addition, indoor APs will not be allowed to be weatherproofed (to prevent people from mounting them outdoors) and will have integral antennas (to prevent higher-gain ones from being attached).

Outdoors, in the United States, no operation is allowed in U-NII-6 and U-NII-8 (satellite operators are using them). In bands 5 and 7, it is possible to have an outdoor AP with a maximum EIRP of 36 dBm, but that must use AFC because there

Table 3.3 Indoor EIRP (equivalent isotropic radiated power) limits for Wi-Fi 6E devices (USA)

Device class	Band	EIRP
Low-power (indoor) AP	All bands	30 dBm (1 W)
Standard power (indoor) AP with AFC	U-NII-5	36 dBm(4 W)
Client connected to low-power AP	All bands	26 dBm (0.4 W)
Client connected to high-power AP	U-NII-5	30 dBm (1 W)

are primary users of this spectrum who do not want interference from Wi-Fi. AFC is an extensive database of these (fixed) primary installations and standard (outdoor and Indoor). APs need to communicate their location (3D) to the database – probably using GPS – and algorithms running on the database server decide which (if any) channels and at what maximum EIRP can be used at that location. In Feb 2024, the FCC approved seven players to offer AFC services: Qualcomm, Broadcom, Comsearch, Federated Wireless, Sony, Wi-Fi Alliance and the WBA.

The third major issue is that the WRC 23 (World Radio Conference) designated the upper part of the band (6425–7125 MHz) for mobile use in the fall of 2023. That means 5G devices and cellular radio could use this band. That would be subject to local licence conditions and devices becoming available. The regulatory debate about using the upper 6 GHz band continues as of 2024. There are several options that regulators may take:

- Keep it as low-power shared spectrum for Wi-Fi.
- Use a sharing mechanism that allows both cellular and mobile.
- License higher power cellular use.

Ofcom (UK) has proposed a hybrid sharing approach [9]. This could be based on

> *Indoor-outdoor split. Wi-Fi routers tend to be indoors, carrying broadband traffic within a localised indoor area, whereas mobile transmitters are mostly outdoors, providing wider coverage. So, we are exploring the possibility of enabling Wi-Fi indoors and licensed mobile use outdoors.*
>
> *Geographical sharing. Most data traffic carried across mobile networks is concentrated in a relatively small proportion of sites. It might be possible to enable licensed mobile use in specific high-traffic locations while allowing Wi-Fi use elsewhere. It might also be possible to prioritise Wi-Fi use in particular areas of high demand while allowing mobile use in other areas.*

There is some expectation that AFC will be used to allow sharing in the United States since this is already working to enable Wi-Fi and primary spectrum users to share.

One issue with 5G use of the upper band is that only 100 MHz (7025–7125 MHz) is available globally. Nevertheless, the upper band (where available) could be extremely useful for 5G, especially for private networks and small cells inside shopping malls, factories and houses. These applications suit the short range of the spectrum and allow the relatively scarce 3–4 GHz spectrum to be used entirely for wider area networks. Mobile operators are keen to utilise the spectrum. Vodafone reported a trial using a 200 MHz channel in this spectrum:

> *6GHz equipment was installed on an existing 5G site in Madrid covering Vodafone's campus and the surrounding area, including several indoor locations. Massive MIMO antenna technology was used. This technology is already being deployed for 5G networks today to 'beamform' signals in the direction of individual users, maximising their signal quality and minimising*

network interference. A 200 MHz channel (the anticipated amount of spectrum made available per mobile operator in each European country) was used, which is approximately double the bandwidth used for 5G services today, enabling higher speeds and capacity for evolved 5G networks. Competing needs for the band's use by mobile and Wi-Fi industries have resulted in some proposals for shared use of the band between these technologies in the same service area. However, this would only be possible if significant restrictions were imposed on mobile base station power levels, sacrificing the performance benefits for customers as demonstrated in this trial [10].

3.4.2 6E performance and latency

Commentators are excited about 6E for Industry 4.0 (robots, factory automation), campuses and sports stadiums. Early large-scale installations include 16,000 6E APs across two sites by the University of Michigan and the Chase sports arena in San Francisco. Orange has announced a home router with 6E in early 2023.

Maldonado *et al.* [11] modelled the performance of Wi-Fi 6 with 5G in an industrial IoT setting. In various modelling scenarios, they found that it was possible to hold the 99.999% upper packet latency below 10 ms for loads representing up to 2.5 bit/s/Hz—a significant improvement on Wi-Fi 5.

Table 3.4 shows how the spectral efficiency of Wi-Fi 6 compares with 5G as the latency maximum is reduced. The efficiency gap between Wi-Fi 6 and 5G NR using the same spectrum was essentially because gNodeBs cooperate and coordinate to reduce interference and protect low-latency traffic. These same features are being introduced into the next generations of Wi-Fi 7 and 8.

3.5 Wi-Fi 7

Wi-Fi 7 brings further significant improvements to Wi-Fi performance. Wi-Fi 7 certification was launched in January 2024, and quite a few smartphones with Wi-Fi 7 capable chipsets are being delivered in 2024, including the GooglePixel 8 Pro. Intel has developed the BE200 Wi-Fi 7 card, and the expectation is that the iPhone 16 Pro will be Wi-Fi 7. This section starts with a look at how new functionality has been divided between Wi-Fi 7 and 8, describes the significant changes in Wi-Fi 7 and presents evidence of how this improves performance in real-life deployments.

Table 3.4 Spectral efficiency with different maximum latencies – from the modelling by Maldonado [11]

Latency (ms)	5G NR FDD Bit/s/Hz	5G NR TDD Bit/s/Hz	Wi-Fi 6 Bit/s/Hz
1	0.72	0.5	0.16
4	2.4	3.7	1.125
10	3.5	3.7	2.25
100	3.6	4	2.5

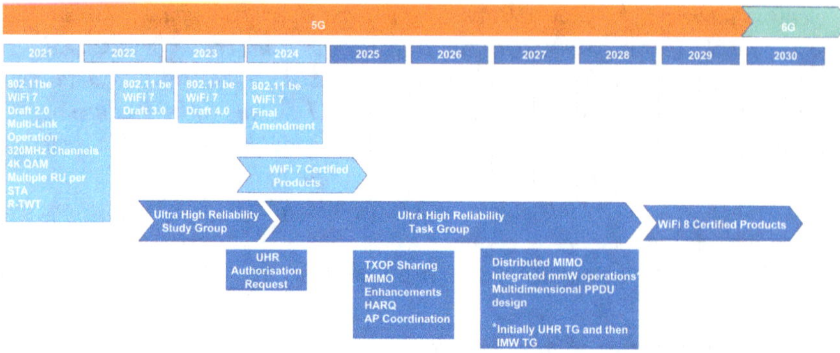

Figure 3.6 Standards timeline for Wi-Fi 7 and 8 (as of 2024)

3.5.1 Standards

There is a lack of clarity over the timing and features of Wi-Fi 7 and 8. This arises from the IEEE shortening the five-year development cycle to four years precisely as the pandemic hit. Figure 3.6 shows the most recent information (2024) on the status of the standards. Many of the improvements studied for Wi-Fi 7 have yet to be included in the 802.11be (Wi-Fi 7) release in 2024. Instead, they have become study items for Wi-Fi 8. This section will examine the key innovations that Wi-Fi 7 brings and assess how they will impact throughput, latency and reliability. The following section on Wi-Fi 8 will continue with the proposed study items for Wi-Fi 8. The first Wi-Fi 8 devices will likely ship in 2028–29 – meaning that 6G will co-exist with a mixture of 6E/7/8 devices in the early years of the 2030s. Many current details, modelling and assessment of Wi-Fi 7 can be found in Refs. [12–17]. The primary focus in developing Wi-Fi 7 was to increase headline speeds and reduce latency. That part of the Wi-Fi 7 standards was termed EHT – Extremely High Throughput. With more clarification over the timelines, people now talk of Wi-Fi 8 – Ultra High Reliability – clearly showing the different emphasis.

3.5.2 Innovations

Two changes are introduced into Wi-Fi 7 to bolster the data rate: 320 MHz wide channels and two new coding and modulation schemes (including 4096QAM). These two features offer a 2.4× increase in (theoretical) maximum rates. In practice, 320 MHz channels will only be available in the 6 GHz band, and 4096QAM will require such a high SNR that a line-of-sight connection (or close to it) will be needed.

However, the most significant change in Wi-Fi 7 is the ability to allow multi-link operation (MLO). These links can be channels spread across the 2.4, 5 and 6 GHz bands or channels within a single band. An example use might allow traffic with different QoS classes to be sent on different bands. Alternatively, MLO could be used for very high throughput with concatenated channels. Other configurations use various bands for up and downlinks (FDD) and for identical transmissions on

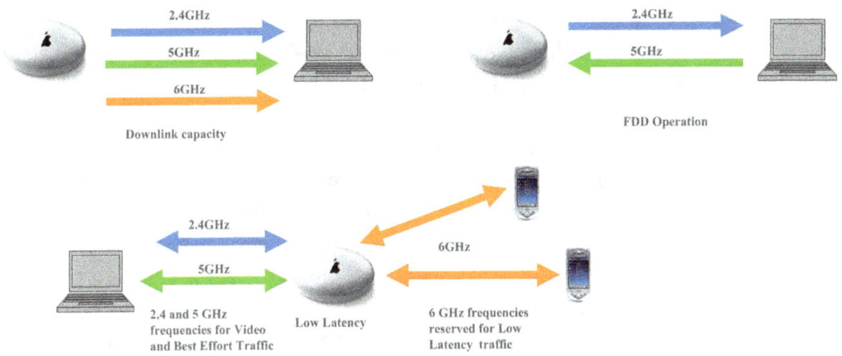

Figure 3.7 Multi-channel operations. (The top left shows channels in different bands concatenated for high throughput. The top right shows different bands used for uplink and downlink. The bottom shows different bands used for different QoS classes.)

different bands for ultra-reliability (Figure 3.7). This is very similar to carrier aggregation used in cellular systems (such as LTE or 5G).

MLO also helps reduce latency, as waiting for a particular channel to become accessible to transmit a time-sensitive packet is unnecessary. The following section examines predictions, modelling and measurements of Wi-Fi 7 performance.

3.5.3 Wi-Fi 7 performance

Bellalta *et al.* [18] have modelled the throughput and latency of packets in a typical cell with various loading factors and interference from neighbouring cells. They find that adding a second, third and fourth channel for five stations increases throughput by $\times 2.4$, $\times 3.7$ and $\times 5$, respectively, over Wi-Fi 6 for a constant 95th percentile latency of 5 ms. As the interfering traffic increases, the advantage grows to $\times 5$, $\times 8$ and $\times 11$, respectively. These results suggest Wi-Fi 7 will offer a significant upgrade throughput (and headline rates) over Wi-Fi 6.

Initially, Wi-Fi 7 was meant to include features that would also improve latency (especially the long tail of re-transmitted packets that cause latency to be uncertain and unbounded). These features will appear in Wi-Fi 8, leaving the question of whether Wi-Fi7, using MLO, could reduce latency. In any given scenario, more capacity will reduce queuing delays and re-transmissions. There is also much more agility in working around interference from overlapping WLANs. Wi-Fi 7 will also significantly impact latency with Multi-Link as this allows the device to use the first band to become free instead of waiting for a free channel.

In a corporate setting, assuming that the premises are large enough to exclude other WLANs, there is complete control over channel allocations and delay-sensitive traffic can be allocated to reserved channels. Active delay monitoring and admission control can prevent new streams from degrading existing traffic. WLANs are at the mercy of interference outside this scenario, such as a shared office block or a domestic setting. That is the nature of a shared spectrum and the fundamental difference from cellular systems. We will see in the next section that Wi-Fi 8 will attempt to coordinate access points outside central control to mitigate this to some extent.

In a typical scenario, with interfering WLANs, will Wi-Fi7 reduce latency? Caracas *et al.* [19,20] have looked at the performance of Wi-Fi 7 in several settings. In an isolated AP, transmitting in the downlink to several stations, it was found that, for a median latency of 1 ms, using only a single link, a maximum traffic of 0.5 Gbit/s could be supported. With two radios (i.e. two links), this rose to 1 Gbit/s, and for four independent radios (and hence four links), it was 2 Gbit/s. There is always a trade-off between latency and throughput, but this result demonstrates that multi-link can increase the throughput for a given latency average. In many applications (such as Industry 4.0 and XR), the 99% (or more) percentile latency counts. With a load of 0.5 Gbit/s, the modelling showed 99-percentile latencies of 1.6 ms with a single link, 1.7 ms with two radio multi-link and 0.5 ms with four links. When the traffic increased to 1 Gbit/s, the single link operation caused unbounded latency while the 2 and 4 radio multi-links could keep the 99 percentile to 1.7 and 0.7 ms, respectively. The researchers went on to investigate a crowded scenario with four BSSs (four APs and numerous stations) – all in overlapping coverage and with only 4 × 80 MHz channels available between them. They compared a static assignment of one channel per AP to two channels per AP and a fully flexible system in which all APs could dynamically switch between all four channels and simultaneously use as many as possible. Under low loads, the more radios used, the lower the latency – as in the isolated AP case. However, they found that using multiple radios could also increase the 99 percentile latency – due to blocking effects in which some APs are 'frozen' out of transmission opportunities for long periods. Increasing the available channels to five significantly improved performance for the multi-link scenario, and the authors conclude that intelligent strategies for channel selection and coordination may be needed to get the optimum performance from MLO in crowded environments.

Further modelling of how Wi-Fi 7 will reduce delays is provided by Naik *et al.* [21] and Bellalta *et al.* [18]. Both these groups of researchers created models of contested WLANs. They studied how the average and 95/99 percentiles increased with general loading for configurations of single and multiple link APs and stations. From all these studies, it is possible to draw some early conclusions as to what Wi-Fi 7 will offer, namely:

- Latencies can be very low – sub-1 ms – in lightly loaded conditions.
- Average and 90–99 percentile latency increase as loads increase – latency can be guaranteed if the total load is controlled.
- Extra radios and multi-links increase the total load that can be carried for a given latency specification. With 4/5 radios, Bellata suggests a 5×–11× improvement. Carrascosa 3×–8× and Lopez-Perez *et al.* [12] 3.2×–4.6×.
- The throughput can increase further if time-sensitive traffic is prioritised over best effort.
- Ultimately, all traffic becomes delayed at a high enough load, and the 90–99 percentile becomes unbounded or very high.

Early reports of real-life measurements on the throughput possible with Wi-Fi 7 are starting to be made. 1–5 Gbit/s is reported to be achievable with Wi-Fi 7 at the IP layer:

> *Using the iperf3 networking benchmark set to simulate 20 users, a Mac mini M2 with a 10Gbit/s Ethernet interface and Analiti's Speed Test app on the OnePlus 11 5G phone, I recorded peak download speeds of 3.85Gbit/s at 5 feet. That's more than triple the throughput of Netgear's Nighthawk RAXE500's 1.15Gbit/s over the 6GHz band using Wi-Fi 6E. This slowed to 2.55Gbit/s at 15 feet, though, showing the limited range of the 6GHz band [22].*

Although a long way from the maximum theoretical limit of 46 Gbit/s, several real-world constraints limit Wi-Fi 7 performance. These include the inability of many devices to use adjacent channels (such as 2 × 320 MHz) due to insufficient filtering between them. Second, most terminals will be limited to 2T2R MIMO with some premium devices capable of 4T4R (8T8R is the maximum in the standard). Finally, the highest rates are only possible with an SINR of 40 dB+, which requires a short, direct path between the AP and the terminal – as in the example quoted above at 5 ft. The 46 Gbit/s figure is also measured at the PHY layer, and the throughput at the TCP/IP layer is much lower. The consensus is that Wi-Fi 7 will offer 1–4 Gbit/s (at the TCP layer) in typical domestic settings, allowing 1–2 Gbit/s to be distributed around typical flats and houses with repeaters.

3.6 Wi-Fi 8 – ultra-high reliability (UHR)

Wi-Fi 7 will offer both higher headline rates (×6.5) and a capacity boost (×5) over Wi-Fi 6 (under the right conditions). What it will not do is, however, provide reliable packet delivery or guaranteed packet delays in crowded conditions. Wi-Fi 8 looks set to address this.

The IEEE has identified the following critical use cases for indoor connectivity in unlicensed bands for 2030 + [23,24]:

- Immersive Comms: AR/VR migrating to Holographic presence.
- Digital twins for manufacturing: a digital model of a complex environment/process.
- e-Health for all: remote monitoring and surgery.
- Cooperative, mobile robots.

Chapter 4 examines these and many other use cases for 6G, carefully analysing the critical reliability, bandwidth and delay requirements. Slightly anticipating that discussion, however, these applications are said to require:

- Reliability: 99.9% to 99.999999%,
- Delay: 1–10 ms,
- Bandwidth 1–100 Gbit/s.

In this section, we will look at the key technologies being studied for Wi-Fi 8 and report some early modelling suggesting that these targets can be met. In the following section, we look at another aspect of Wi-Fi 8 – new spectrum – evaluating how a move to mmWave frequencies could further boost data rates. The current situation with Wi-Fi 8 standardisation is that a Technical Group (TG) on UHR has been formed, assuming the standard will be released in 2028 [25]. There may also be a TG on AI/ML to explore the possibility of AI/ML features in Wi-Fi 8 and beyond [26]. In addition, an IEEE 802.18 mmWave Ad Hoc Group is exploring opportunities in the 45 and 60 GHz bands, which will be discussed in the next section. The rest of this section looks at the technology that will underpin UHR.

3.6.1 Major innovations

Latency and reliability are very much different sides of the same coin. As we saw in discussing Wi-Fi 7 performance, what is needed is a joint specification – 99.9% of packets with a latency under 10 ms and an average latency of 5 ms or less (say). An isolated AP (with no other users of the shared spectrum within range) can assign packets to different QoS classes and perform admission control to prevent delays if more traffic is offered in a given QoS class. A typical example would be multiple voice clients, all attempting a VoIP call simultaneously. The AP would block new call attempts when the latency for existing calls rises to a pre-set quality threshold. Likewise, the system administrator can install an entire WLAN system from a single manufacturer in an isolated office block. These also allow the creation of QoS classes, sophisticated admission control based on almost any criteria (e.g. packet headers) and support layer two handovers. With Wi-Fi 5 and 6, these corporate networks are not particularly efficient – often allocating channels on a per AP basis – in the manner of the fixed frequency allocations of GSM. Wi-Fi 8 will offer a very significant performance improvement for these corporate networks. For WLANs suffering from interference, the gains will be even more pronounced.

To deliver UHR, Wi-Fi 8 APs will cooperate to reduce interference and coordinate their transmissions to improve performance radically. The first element of this will be distributed MLO operation over a group of APs forming a virtual cell. This is intended to enhance significantly handover and make roaming between APs much smoother – with reduced packet loss and reduced considerably latencies (Figure 3.8).

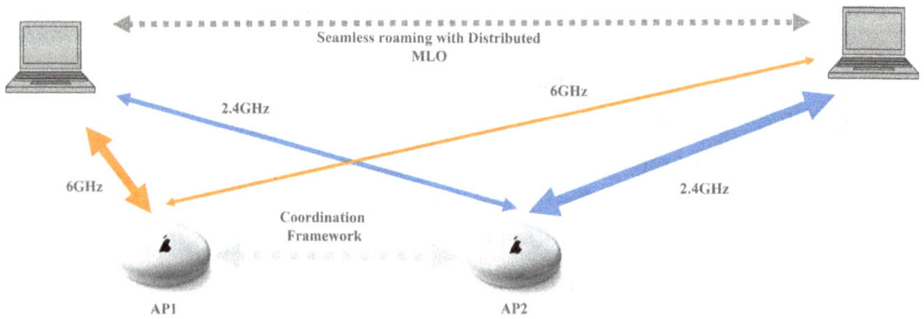

Figure 3.8 Distributed MLO

The second study item consists of PHY and MAC layer enhancements. These include HARQ (Hybrid Automatic Repeat Request) and increasing the number of spatial streams to 16 (from 8). HARQ can combine corrupted packets to recreate the original transmission rather than require multiple attempts until a perfect packet is transmitted, thereby reducing latencies.

A more complicated mechanism being proposed to give priority to URLLC (Ultra Reliable Low Latency Communication) packets is pre-emption. In Figure 3.9, the Wi-Fi 8 AP can aggregate data packets (PDDUs) in time and frequency. The two that are best-effort traffic are destined for the laptop. During transmission, a time-sensitive packet arrives for the voice application from the network. The extra flexibility in the Wi-Fi 8 MAC allows one of the best-effort packets to be abandoned and replaced immediately by the high-priority packet. The other best-effort packet is transmitted unaffected, and the lost packet is later re-transmitted.

Another feature studied is AP coordination, which allows Wi-Fi APs to adjust how they transmit to (and from) clients, sharing frequency bands and sub-carriers and causing interference. The APs can decide to share the sub-carriers by transmitting them at different times or by sharing them. They can also adjust their power levels to optimise the overall network performance. Again, this is very similar to fractional frequency re-use and eICIC interference mitigation technologies that both LTE and 5G use (as described in Chapter 2). In the proposal, access points also coordinate their beam formation (Figure 3.10), effectively boosting the signal-

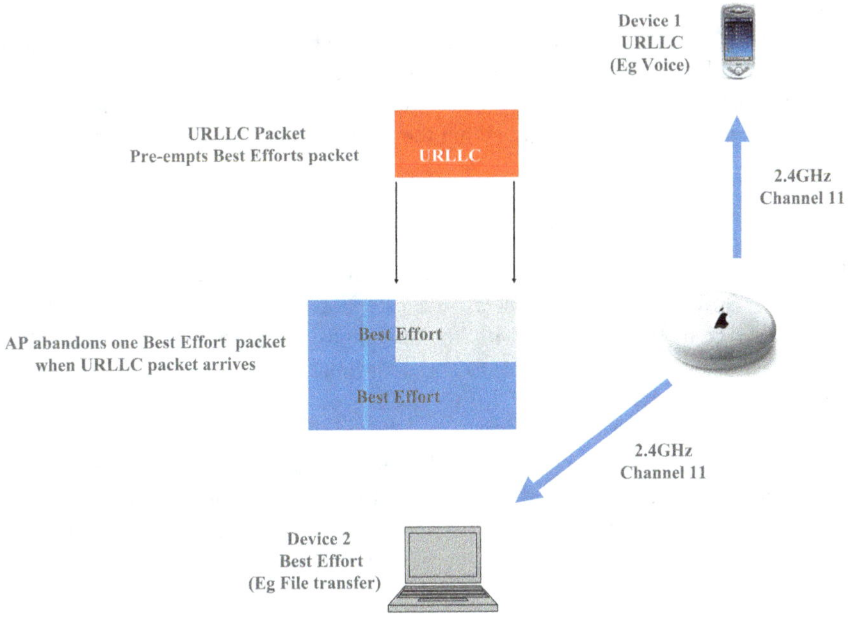

Figure 3.9 Pre-emption for URLLC packets

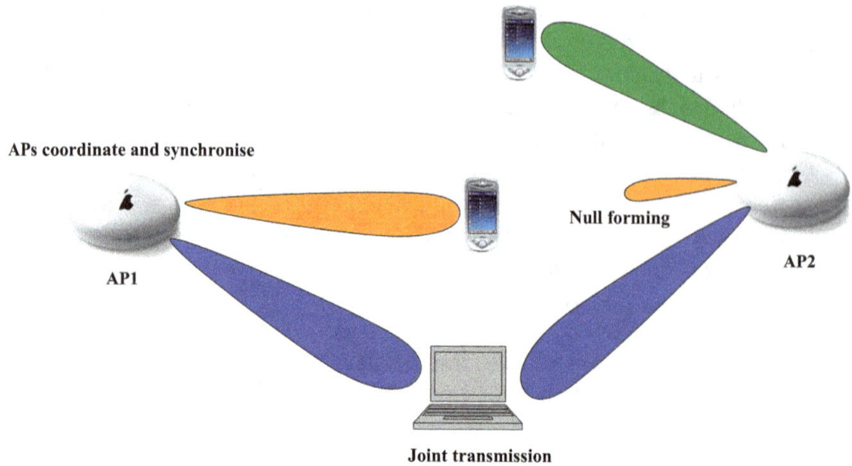

Figure 3.10 Coordinated beam-forming

to-noise ratio of clients near the edge of cells and reducing interference with null formation from interfering nodes. The final shared AP technology being studied in Wi-Fi 8 is joint transmission. Here, two or more linked APs transmit on the same sub-carriers simultaneously with the same data for a single client. Again, this is similar to CoMPT (Coordinated Multi-Point) and coordinated MIMO in LTE and 5G. Both of these cellular technologies are based on cooperation between neighbouring base stations.

AP coordination does require that the APs be configured as a group. The group's control will need software that coordinates all these resources, monitors all the radio channels between the APs and the clients, and sets the priorities for different types of traffic – all the things that the gNodeB and X2 interfaces accomplish for 5G.

3.6.2 Wi-Fi 8 performance

There is limited modelling of how these technologies reduce latency and improve reliability. Still, Galati-Giordano *et al.* [25] have modelled two APs using 2 × 160 MHz in the 6E band with four antennas and a single client with two antennas. The authors modelled MLO and coordinated beam forming (CBF). They compared stand-alone MLO and MLO + CBF performance with the following results (Table 3.5).

These results indicate that substantial latency reductions will be possible (for a given required reliability), but that strong nulling is needed to realise these gains fully. Latencies are worse at 10 dB nulling as the interference level is still high, forcing down the modulation and coding scheme compared to the other options.

Another group [27] has also modelled the performance of CBF – using 2 APs with eight antennas and 24 users distributed over a sizeable indoor enterprise. They

Table 3.5 Packet latencies (modelled) with CBF [25]

Scheme	Median packet latency (ms)	99 percentile latency (ms)	99.9999 percentile latency (ms)
Stand-alone MLO	16	64	109
MLO + CBF (10 dB nulling)	94	102	124
MLO + CBF (20 dB nulling)	7	34	56
MLO + CBF (30 dB nulling)	1	4	14

compared Wi-Fi 6 to Wi-Fi 7 with CBF. They found the 99-percentile latency reducing from 79 to 12 ms and the 99.99-percentile from 208 to 22 ms. Again, this suggests a significant performance gain for CBF in Wi-Fi 8. As envisaged by Industry 4.0, this could offer very controlled latency performance in an industrial setting.

3.7 Wi-Fi at mmWave frequencies

The IEEE also issued a WLAN standard that operates in the 60 GHz band. This is called the V-band and was initially used by the military (in the United States). In 1994, the FCC allowed unlicensed operation from 57 to 64 GHz, extending this to 71 GHz. The band was only used once the IEEE published the 802.11ad standard in 2009. It was dubbed WiGiG by the Wi-Fi Alliance in 2013, but it was not until 2016 that chip sets were available for interoperability testing. The initial use cases of WiGiG were transmitting HD TV signals around a house, cable-less docking for laptops, Ethernet bridges and connecting VR headsets. These have taken off slowly [28], partly because lower frequency Wi-Fi (5 and now 6/6E) offers a much greater indoor range and similar data rates. The V-band has been more successful in short-range, outdoor applications such as backhauling video cameras and providing fixed-wireless access (FWA). A typical 100–300 m range is possible [29].

3.7.1 60 GHz

In recognition of this limited uptake and further interest in higher bands, the IEEE has completed the standardisation of 802.11ay – an upgraded protocol for use in the 60 GHz band that supports a much higher data rate. The standard was only finally released in 2021 (following the pandemic). Where 802.11ad uses a maximum bandwidth of 2.16 GHz, the new standard allows four channels to be bonded together and adds up to four spatial streams – giving a 16× improvement of the maximum rate to 176 Gbit/s. Higher rates may be possible if higher modulations are used. 802.11ad/ay uses the same MAC as the lower frequency variants but with a modified PHY to allow channel sounding, beam forming and multi-user MIMO at these higher frequencies. Sun *et al.* [30] give a full description.

The key strategic questions, however, are whether 802.11ay will compete with or complement Wi-Fi 7/8 and whether it will be more commercially successful than

802.11ad. Wi-Fi 7 has a significant advantage over 802.11ad because lower-frequency devices can penetrate walls and ceilings and avoid beam blockages. Power consumption has also increased at these higher frequencies (Chapter 9). Chipsets for 802.11ay are starting to appear [31], aimed at indoor and outdoor applications.

Wi-Fi 8 could incorporate mmWave bands in an expanded Multi-Link Operation set [25]. There is currently a debate as to whether to adopt a PHY layer similar to the one in Wi-Fi 7 or to re-use those of 802.11ad/ay for mmWave use in Wi-Fi 8. Using an OFDM-based PHY layer for mmWave would avoid the need for very wide channels (these are 8.64 MHz wide in 802.11ay) and the requirements for power-hungry and costly amplifiers that go with it, which is thought to be a significant disadvantage of the 802.11ay standard. The trade-off would be a lower maximum throughput.

One mode of MLO operation (Figure 3.11) being considered in Wi-Fi 8 for a mixed mmWave and lower frequency set involves using the lower frequency bands for control and only implementing 60 GHz operation in favourable conditions and for data only [25]. For example, if the video was being transmitted, then this arrangement allows the control information and keyframes to be transmitted over the more reliable lower frequencies and the detail frames, which provide much higher resolution, over the mmWave link. That way, if the mmWave beam is temporarily interrupted, then at least a minimum resolution continues, and the control algorithm can re-arrange the overall radio resources to restore the original data rate as soon as possible and much sooner than if a mmWave-only link is used.

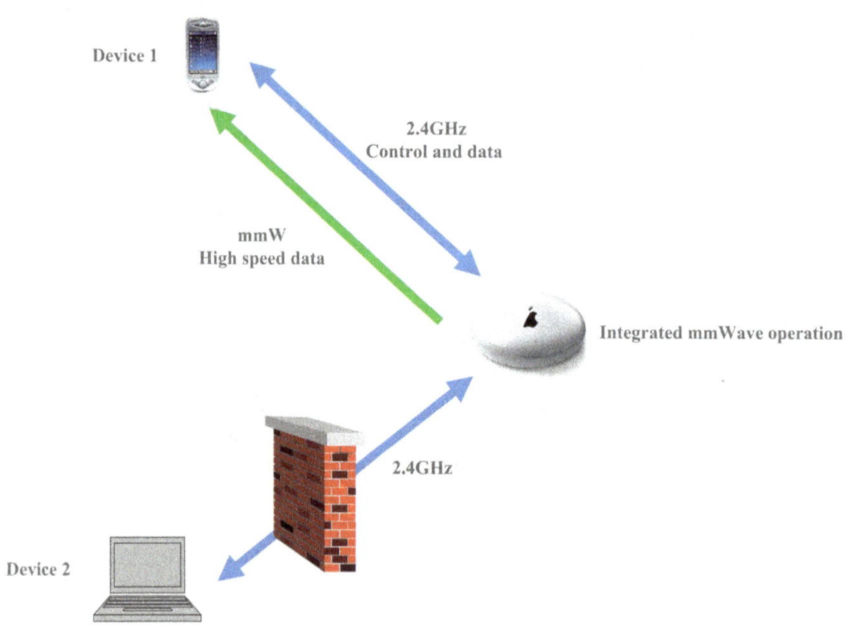

Figure 3.11 Possible integration of mm-wave bands in MLO operation of Wi-Fi 8

3.8 Wi-Fi – cellular interaction

There have been several attempts to create 'Wi-Fi' cities and towns (e.g. Felixstowe UK, Figure 3.12). In general, these have not been successful, and many have now turned off their networks as LTE has become the default wide area mobile Internet access (e.g. Norwich in the UK). There are examples of campus area Wi-Fi networks, such as the University of Michigan, with 16,000 APs – download speeds of 500–600 Mbit/s have been reported. Wi-Fi systems are also required for legacy devices. Many things, like printers and cameras, only connect with Wi-Fi – so even locations with 5G private networks generally have a Wi-Fi network.

Wi-Fi is, therefore, likely to co-exist with cellular systems in the 6G era. It is, therefore, essential to examine how they might interact. This has traditionally been called 'coupling', the following sections look at increasingly close coupling and what each might offer. Chapter 9 provides more detail about using Wi-Fi as a 'trusted' and 'untrusted' network and considers Wi-Fi within a converged fixed-mobile framework.

3.8.1 No coupling

With no coupling, there is no association or coordination between the cellular network and the Wi-Fi provided by a third party. An excellent example of this

Figure 3.12 Felixstowe free Wi-Fi

might be an iPhone user with an MNO subscription simply using the connection manager on iOS to access free Wi-Fi at the local grocery store or diner. User preferences can always be set to make the connection when this SSID (Service Set IDentifier) is encountered. Many free Wi-Fi systems connect automatically once the user has initially registered. A second example is home Wi-Fi – most users typically set their home SSID to connect automatically and enter a WPA 2 key (wireless protected access). Some broadband providers (such as BT and, recently, Virgin Media in the United Kingdom) have also adopted technology that allows registered users to connect to any compatible access points. BT has introduced an app that offers automatic connection to hot spots and is also rolling out 802.1X – a more robust security protocol – offering three-way security (avoiding bogus access points) and more straightforward, automatic authentication once set up.

The no-coupling option has a variety of authentication and security mechanisms (based on certificates, passwords, MAC addresses and so on). Security on the wireless link may or may not be present – depending on the third-party Wi-Fi set-up (meaning that secure applications, such as banking or corporate VPN, require over-the-top end-to-end security). Today, most devices select either Wi-Fi or cellular, but bonding is now possible with some of the latest mobiles.

The no-coupling option offers no handover support. Users could install an MIP (Mobile IP) client, but this is a little used. The OS connection manager generally makes the network switch: many applications pause and then continue (browsers, downloads, iPlayer, etc.) and some break (typically real-time apps such as VoIP or secure links such as VPNs that cannot tolerate a change of IP address). Similarly, there is no QoS support. Users often associate with Wi-Fi access points that fail to offer connectivity (e.g. due to IP address pool exhaustion) or are congested.

3.8.2 Loose coupling (untrusted Wi-Fi)

In this case, users establish a tunnel to a gateway (in LTE, the enhanced packet data gateway – ePDG) from an 'untrusted' Wi-Fi – one not associated with the mobile operator. Untrusted Wi-Fi does not provide guaranteed carrier-grade encryption over the air or support three-way authentication. Until recently, loose coupling has not been extensively used, but several operators (e.g. EE in the United Kingdom) have now utilised it to provide a Wi-Fi voice service [32]. Some MNOs take text or voice minutes from your monthly bundle this way if you are not on an unlimited package.

After accessing the Wi-Fi application, the terminal uses IPsec to create a secure tunnel to the gateway. Authentication is made to the mobile network using, for example, the SIM card over extensible authentication protocol (EAP). All data associated with that application is then tunnelled into the mobile network. In the case of a voice-over Wi-Fi application, this allows full integration with Voice-over cellular services.

The advantage of loose coupling is that it can support mobile services (packet inspection, lawful intercept, integration with cellular services) or allow automatic authentication and billing using the user's mobile account. Typically, traffic that does not need to pass through the mobile core is offloaded to the Internet. Network discovery and selection depend mainly on user intervention (HotSpot 2.0 is changing this with more automated selection). Even for the Wi-Fi voice service, there is still no QoS support when using Wi-Fi. The call can fail when the Wi-Fi becomes congested, or users wander into a poor or no-coverage area.

A cable service provider (CSP) offering fixed broadband and mobile networks could offer both networks to the end terminal using multipath TCP/IP technologies. Multipath TCP (MPTCP) enables the simultaneous use of several IP addresses/interfaces by modifying TCP to present a regular TCP interface to applications while spreading data across several networks.

For fixed broadband operators with an MVNO, using an MPTCP platform [33,34] gives them better control over their customers' experience, whether on Wi-Fi or cellular (Figure 3.13). MVNOs can set granular business rules to differentiate their product in the marketplace, whether on price, QoE or both. Rules for aggressive Wi-Fi offloading can significantly lower payments to the host MNO. Users can also move gracefully between Wi-Fi and cellular without a connection

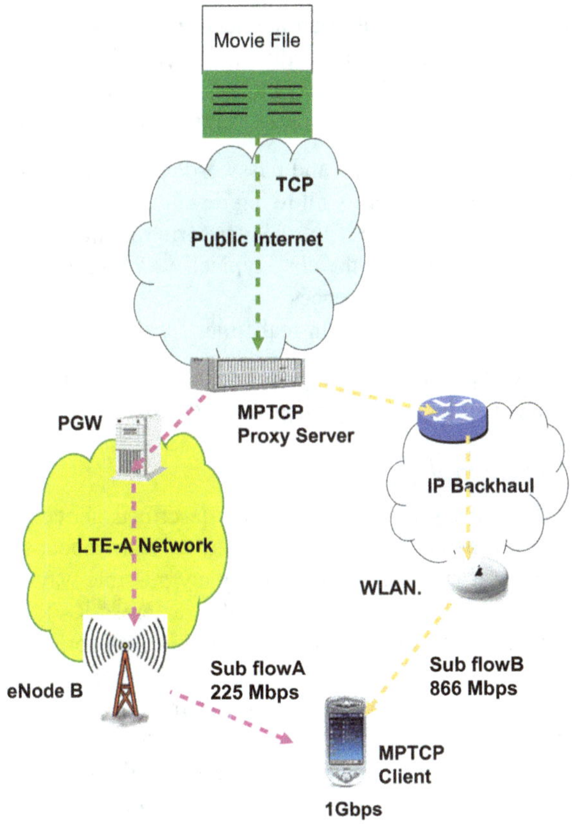

Figure 3.13 MPTCP (after [33]). Note the data rates are just illustrative.

dropping. Alternatively, MVNOs can charge for a premium product by guaranteeing the fastest speeds through Wi-Fi and cellular connectivity link aggregation. There are similar advantages for CSPs who own both networks or mobile operators with wholesale access to a fixed broadband network.

Another example of an untrusted service is EE's 'Wi-Fi Speed Boost' in the UK. This allows mobiles to connect to 150,000 APs (owned by EE partner BT) and on the London Underground (owned by a third party) and combines both Wi-Fi and cellular data (with control traffic remaining on cellular) to improve speeds or coverage using MTCP. It is automatically available on iPhones and requires a small number of settings on Android devices.

3.8.3 Tight coupling (trusted Wi-Fi)

Tight coupling always takes place over 'trusted' Wi-Fi. The MNO does not necessarily own this but does support carrier-grade authentication over the air and three-way authentication (so that the network, access point and user are all mutually authenticated and bogus access points are detected). 802.1X offers these facilities and can encapsulate several authentication credentials (SIM, certificate, name/password.) over EAP. However, the most critical step forward for tight coupling is HotSpot 2.0 (2014). This technical specification (by the Wi-Fi Alliance) combines several industry standards. Access Points supporting HotSpot 2 offer:

- EAP authentication – SIM/name and password/certificate authentication.
- Access network query protocol – allowing devices to interrogate access points before attaching (to determine if the AP supports roaming to their provider).
- Carrier-grade encryption over the air – typically, WPA2/3 and AES.
- User and carrier network preferences.
- Account set-up and provisioning in real-time.

Chapter 9 explains the standards that support trusted Wi-Fi in more detail. It is fair to say that it is not widely used and that many current devices do not support it.

3.8.4 Very tight coupling

The coupling and seamless mobility solutions described above suffer from one serious drawback: the device and network need to learn about the state of the WLAN and whether it is congested. There are many possible ways to achieve very tight coupling. Galinina *et al.* [35] describe a cross-RAT (Radio Access Technology) coordination function. Essentially, in this arrangement, the Wi-Fi falls under the control of the cellular network – with two types of control. The first is to assign (semi-statically) users to Wi-Fi or cellular, and the second is a MAC-layer joint scheduling scheme. The latter was a complicated scheduling system that considered the time-importance of packet delivery and the bandwidth required. It was found that a simple RAT-assignment scheme significantly outperformed the uncoordinated use of both radios. With a more complicated and dynamic assignment algorithm, considering QoS requirements, further gains were made (and the highest gains of all were from joint scheduling – with three times as many users reaching their QoS threshold in this scenario).

A version of very tight coupling is LTE-U – LTE in unlicensed spectrum, initially proposed by Qualcomm to work with releases 10/11/12. Designed for quick launch of LTE in certain countries in the 5 GHz (shared) Wi-Fi band, it was not good at sharing the spectrum with other Wi-Fi systems (no listen before talk). By 2019, it was reportedly only in use in three countries. Licensed Assisted Access (LAA) was added in Release 14 with listen before the talk but was for downlink traffic only. eLAA debuted in release 14 for uplink and downlink, and Release 15 (5G) further enhanced LAA (efLAA). In practice, however, LAA is yet to gain much traction.

A slightly different version is called LWA (LTE WLAN Aggregation). In this version, the 5 GHz spectrum uses the standard 802.11 protocols but also carries some cellular traffic. This has the great advantage that no hardware modifications are needed in the handset and that standard WLAN chipsets can be used. Chunghwa Telecom launched a commercial LWA LTE/Wi-Fi network in 2017 in Taiwan [36] – the first use of the technology. It is unclear if the small cells it employed are still in use. However, no reports or plans exist for significant use of LWA with 5G.

It is also possible for 5G terminals to access the 5G core solely using Wi-Fi – this was standardised for untrusted networks in Release 15 and trusted networks (as well as wireline access) in Release 16. A new network function – the Non-3GPP Interworking Function (N3IWF) – operates a secure, encrypted connection to the mobile via a new NWu interface. This is further described in Chapter 9.

3.8.5 Cellular Wi-Fi coupling futures

All the standards needed for Wi-Fi-cellular convergence exist. However, many still need to be implemented in current handsets, and whether that will change significantly in 6G remains to be seen. MNOs could offer 6G services over Wi-Fi, and broadband suppliers could offer 6G over Wi-Fi with an MVNO deal for wider coverage. Third parties, such as handset manufacturers or hyperscalers, may also see an opportunity to provide 6G. All of these factors make for some engaging future scenarios. Chapter 9 provides much more detail on FMC standards, and chapter 16 explores some of these scenarios in much greater depth.

3.9 Beyond Wi-Fi 8 – the future

This section looks at three areas where active Wi-Fi research exists and how these might impact 6G.

3.9.1 Artificial intelligence and machine learning

WLANs are excellent candidates for applying AI/ML. It is essential to distinguish between WLANs collecting information for wider AI/ML, such as passenger movements, and AI/ML used to manage and the WLAN settings – such as channel bandwidth and beam forming. Only this last use is considered here. WLANs are ideally suited to the use of AI/ML because there are many possible combinations of channel and bandwidth, modulation and coding, spatial beam forming, contention window and many other parameters. In addition, a considerable amount of data is available for training algorithms, including channel state information, NACKs, SNR, and preambles. Traditional algorithms need help with the non-linearity of the problem space and typically, use relatively simple decision rules.

Hundreds of papers have been published on using AI/ML to improve WLAN performance, and Sun *et al.* [30] is an excellent, if lengthy, field survey. All aspects of WLANs have been tried with different algorithms, and they can be roughly divided as

- Channel access – choosing the optimum channels to reduce collisions and improve scalability and user fairness
- Link configuration – rate adaption, SGI (Short Guard Interval) adaptation to improve throughput and reduce latency
- Frame aggregation – to achieve a better trade-off between aggregation and packet loss (latency)
- PHY features – Interference estimation is used to enhance the selection of modulation/coding and improve MIMO performance
- Beam-forming – including initial beam alignment for mmWave links
- Multi-user comms – optimum combinations of spatial filtering, OFDMA and beam forming to serve multiple users with improved fairness
- Multi-link operation – how to dynamically choose the combination of channels/bands (e.g. there are 59 × 20 MHz channels in the 6E band)
- Connectivity management – optimising the parameters of a group of overlapping WLANs

These have generally been studied in isolation and (mainly) only through modelling. Nevertheless, the potential for increased throughput, lower latency or greater fairness resulting from adopting AI/ML algorithms has been firmly established. Manufacturers of APs will likely use some combination of the most effective of these algorithms to improve the performance of a single AP – even when in a highly congested area. Moreover, there will also be cooperative and distributed algorithms that can be configured across a range of APs from a single vendor to optimise something like a campus or office WLAN. What looks unlikely is that a standard will emerge to allow APs from different vendors or competing, interfering WLANs to optimise jointly. This is a problematic issue: would you be happy to throttle back your domestic WLAN performance so that your neighbours could significantly improve theirs? Would you be happy if this happened without your knowledge? Either way, Wi-Fi 7 and 8 will start to include AI/ML, significantly boosting performance. The modelling summarised by Sun *et al.* [30] could justify a claim for 100–200% possible improvement.

3.9.2 Sensing

Using the WLAN physical layer (PHY) to sense environments and people is familiar. The authors of a 2013 paper [37] describe beam-forming to determine the number of people in a room behind a wall and to target a single person with a beam array to detect gestures (such as raising an arm or moving). Applications proposed for WLAN sensing include intruder detection (burglar alarm), eHealth (monitoring the sick and old), human communication (e.g. using gestures) and building management. One of the main issues has been that standard WLAN chips do not offer direct access to the channel readings, and many experimental systems use software-defined radio (SDR) solutions. There are a minimal number of Wi-Fi sensing systems on the market at present. However, the Wireless Broadband Alliance (WBA) published a seminal white paper on Wi-Fi sensing in 2019 [38] that sets out the building blocks that need to be added to 802.11 standards to enable a complete sensing eco-system to develop. They have promised to work with the IEEE to include these components in future Wi-Fi standards. This would fit the concept that 6G combines communications and sensing. Chapter 14 covers sensing in the context of 6G in detail.

3.9.3 IoT

Since their inception, standard WLANs have been used to connect IoT devices to the Internet. There have been two IEEE standards for IoT connection. 802.11af used Cognitive Radio to access TVWS (TV White Space – see Chapter 2), but it was not a commercial success. 802.11ah has had better success operating in varying sub-1 GHz bands around the world (e.g. 863–868 MHz in the EU) and using 802.11g technology downsampled give up to 26 × 100 kbit/s channels and a range of about a mile. Commercially, devices are branded as HaLow ('Halo'). HaLow is effectively a low-power wide area network and, as such, competes with several other technologies such as LoRa, SigFox, NB-IoT (see Chapter 2) and WiSUN. 802.11ah potentially offers higher data rates, traded off for a shorter range than alternatives.

So far, few chipsets have been available for HaLow, but Methods2Business, Morse Micro, and Newracom have products available or under development. For example, the Newracom NRC7292 supports operation in 750–950 MHz band (1, 2 and 4 MHz channels) and offers data rates from 15 kbit/s to 15 Mbit/s. The standards were published in 2017, so it is still being determined whether it will have a widespread impact in the coming years. Readers interested in 802.11ah should see Tain's excellent survey of current research [39].

3.10 Summary and conclusions

From humble beginnings in 1997 with 802.11 at 1 Mbit/s to 2024 and 802.11be – now dubbed Wi-Fi 7 – at 2–5 Gbit/s represents a 5,000X increase in maximum speed in 27 years. This is not that far off doubling every two years: reasonably close to following

Moore's law for electronics. WLANs have been very successful in cafes and houses as DSL/fibre Internet tails, the fact that the standards were limited to PHY and MAC layers dovetailing with the Ethernet/IP world. Wi-Fi did less well in large, dense deployments. The limited amount of spectrum and its shared nature meant that any QoS or latency bounds were very difficult to deliver in practice. Wi-Fi 6E starts to manage interference better and, most importantly, can exploit up to 1.2 GHz of new spectrum. However, the upper part of the 6–7 GHz band looks likely to be either shared with cellular or allocated directly to mobile in at least some jurisdictions, limiting the impact of 6E.

Wi-Fi 7 EHT looks set to offer increased capacity and headline rates and reduce latency through Multi-Link Operation. Early reports and modelling suggest realistic rates of 1–5 Gbit/s with modern phones and laptops and a significant reduction in latency in many situations.

Wi-Fi 8 is just beginning standardisation but will ship in 2028/29, so its device will exist alongside 6G. The main focus is on UHR, with new AP coordination strategies and CBF allowing much better upper limits on packet re-transmission and latency. In addition, Wi-Fi 8 is looking to integrate mmWave WLANs, which could push data rates up further.

Wi-Fi will co-exist with 6G because it is needed for legacy devices and is better suited to serving indoor traffic than external macro-cellular networks. There have been many standards and initiatives to integrate Wi-Fi more tightly with cellular. 6G could offer a more unified, converged network and Chapter 9, together with the concluding Chapter 16, explores these ideas more fully.

References

[1] Wikipedia Entry on Robert Metcalfe, [Online]. [Cited Jan 2024] Available from https://en.wikipedia.org/wiki/Robert_Metcalfe.

[2] IEEE, [Online]. [Cited Jan 2024] Available from: https://www.ieee.org/.

[3] Wi-Fi Alliance, [Online]. [Cited Jan 2024] Available from: https://www.wi-fi.org/.

[4] Coleman, D., *Wi-Fi 6 & 6E For Dummies®*. Hoboken, NJ:, David Coleman, John Wiley & Sons, Inc.; 2022. Available from: https://www.extremenetworks.com/resources/ebook/wi-fi-6-6e-for-dummies/.

[5] Sheth, J., Ramanna, V. K., and Dezfouli, B., 'Traffic Characterisation for Efficient TWT Scheduling in 802.11ax IoT Networks', *2023 IEEE Wireless Communications and Networking Conference (WCNC)*, Glasgow, United Kingdom, 2023, pp. 1–6, https://doi.org/10.1109/WCNC55385.2023.10119047.

[6] WBA Wi-FI 6 trials Report, 2020, [Online]. [Cited Jan 2024] Available from:https://wballiancec.wpenginepowered.com/wp-content/uploads/2020/11/WBA-Wi-Fi-6-Trials-report-2020_Final-v1.0-copy.pdf.

[7] Wi-Fi Alliance News Item, [Online]. [Cited Jan 2024] Available from: https://www.wi-fi.org/beacon/the-beacon/wi-fi-6-shipments-to-surpass-52-billion-by-2025.

[8] Wi-Fi Alliance Spectrum Tool, [Online]. [Cited Jan 2024] Available from: https://www.wi-fi.org/countries-enabling-wi-fi-in-6-ghz-wi-fi-6e.

[9] Ofcom Statement, 2023, [Online]. [Cited Jan 2024] Available from: https://www.ofcom.org.uk/news-centre/2023/sharing-6-ghz-spectrum-for-wi-fi-and-mobile.

[10] Vodafone Announcement, 2023, [Online]. [Cited Jan 2024] Available from: https://www.vodafone.com/news/technology/vodafone-tests-reveal-6ghz-spectrum-to-avoid-5g-capacity-crunch.

[11] Maldonado, R., Karstensen, A., Pocovi, G., *et al.*, 2021 'Comparing Wi-Fi 6 and 5G Downlink Performance for Industrial IoT', *IEEE Access* 2021; 9:86928–37, https://doi.org/10.1109/ACCESS.2021.3085896.

[12] Lopez-Perez, D., Garcia-Rodriguez, A., Galati-Giordano, L., Kasslin, M., and Doppler, K., 'IEEE 802.11be Extremely High Throughput: The Next Generation of Wi-Fi Technology Beyond 802.11ax', *IEEE Communications Magazine*, 2019;57(9):113–119, https://doi.org/10.1109/MCOM.001.1900338.

[13] Garcia-Rodriguez, A., López-Pérez, D., Galati-Giordano, L., and Geraci, G., 'IEEE 802.11be: Wi-Fi 7 Strikes Back,' *IEEE Communications Magazine*. 2021;59(4):102–8, https://doi.org/10.1109/MCOM.001.2000711.

[14] Khorov, E., Levitsky, I., and Akyildiz, I. F., 'Current Status and Directions of IEEE 802.11be, the Future Wi-Fi 7', *IEEE Access*. 2020;8;88664–88, https://doi.org/10.1109/ACCESS.2020.2993448.

[15] Deng, C., Fang, X., Han, X., *et al.*, 'IEEE 802.11be Wi-Fi 7: New Challenges and Opportunities', *IEEE Communications Surveys & Tutorials*. 2020;22 (4):2136–66.

[16] Yang, M. and Li, B., 'Survey and Perspective on Extremely High Throughput (EHT) WLAN—IEEE 802.11be', *Mobile Networks and Applications*. 2020;25(5):1765–80.

[17] Chen, C., Chen, X., Das, D., Akhmetov, D., and Cordeiro, C., 'Overview and Performance Evaluation of Wi-Fi7', *IEEE Communications Standards Magazine*. 2022;6(2):12–18.

[18] Bellalta, B., Carrascosa, M., Galati-Giordano, L., and Geraci, G., 'Delay Analysis of IEEE 802.11be Multi-Link Operation Under Finite Load', *IEEE Wireless Communications Letters*. 2023;12(4):595–99.

[19] Bellalta, B., 'An Experimental Study of Latency for IEEE 802.11be Multi-link Operation', *ICC 2022 – IEEE International Conference on Communications*, Seoul, Republic of Korea, 2022, pp. 2507–12, https://doi.org/10.1109/ICC45855.2022.9838765.

[20] Carrascosa-Zamacois, M., Geraci, G., Galati-Giordano, L., Jonsson A., and Bellalta B., 'Understanding Multi-link Operation in Wi-Fi7: Performance, Anomalies and Solutions', *2023 IEEE 34th Annual International Symposium on Personal, Indoor and Mobile Radio Communications (PIMRC)*, Toronto, ON, Canada, 2023, pp. 1–6, https://doi.org/10.1109/PIMRC56721.2023.10293865.

[21] Naik, G., Ogbe, D., and Park, J.-M. J., 'Can Wi-Fi 7 Support Real-Time Applications? On the Impact of Multi-Link Aggregation on Latency', *ICC*

2021 – IEEE International Conference on Communications, Montreal, QC, Canada, 2021, pp. 1–6, https://doi.org/10.1109/ICC42927.2021.9500256.
[22] 'Wi-Fi 7 Test', 2024, [Online]. [Cited Jan 2024] Available from: https://www.tomsguide.com/opinion/i-just-tested-the-worlds-first-wi-fi-7-router-and-these-are-5-things-i-learned.
[23] Minutes of the IEEE UHR Group Meeting, March 2023, 'IEEE 802.11-23/0480r0: UHR Proposed PAR', IEEE 802.11-23/0480r0, [Online]. [Cited Jan 2024] Available from: https://mentor.ieee.org/802.11/dcn/23/11-23-0480-00-0uhr-uhr-proposed-par.pdf.
[24] Reshef, E. and Cordeiro, C., 'Future Directions for Wi-Fi 8 and Beyond', *IEEE Communications Magazine*. 2022;60(10):50–55, https://doi.org/10.1109/MCOM.003.2200037.
[25] Galati-Giordano, L., Geraci, G., Carrascosa, M., and Bellalta, B., 'What Will Wi-Fi 8 Be? A Primer on IEEE 802.11bn Ultra High Reliability', 2023. [Online]. Available from: https://doi.org/10.48550/arXiv.2303.10442.
[26] Szott, S., Gawlowicz, P., Kosek-Szott, K., and Torres Gómez, J., 'Wi-Fi Meets ML: A Survey on Improving IEEE 802.11performance with Machine Learning', *IEEE Communications Surveys & Tutorials*. 2022;24(3):1843–93.
[27] Garcia-Rodriguez, A., Lopez-Perez, D., Galati-Giordano, L., and Geraci, G., 'IEEE802.11be: Wi-Fi7 Strikes Back', *IEEE Communications Magazine*. 2021;59(4):102–8.
[28] Technology News, [Online]. [Cited Jan 2024] Available from: https://www.pocnetwork.net/technology-news/so-what-ever-happened-to-60ghz-802-11ad-Wi-Fi/.
[29] Siklu Primer One mmWave, [Online]. [Cited Jan 2024] Available from: https://go.siklu.com/blog/60ghz.
[30] Sun, Y., Peng, M., Zhou, Y., Huang Y., and Mao, S., 'Application of Machine Learning in Wireless Networks: Key Techniques and Open Issues', *IEEE Communications Surveys & Tutorials*. 2019;21(4):3072–108, https://doi.org/10.1109/COMST.2019.2924243.
[31] Technology Blog, 2023, [Online]. [Cited Jan 2024] Available from: https://www.thefastmode.com/technology-solutions/23264-pharrowtech-launches-ieee-802-11ay-compliant-cmos-rf-chip.
[32] EE Wi-Fi Calling, [Online]. [Cited Jan 2024] Available from: https://ee.co.uk/help/help-new/getting-started-and-upgrading/using-your-phone-features/how-do-i-use-Wi-Fi-calling.
[33] Croitoru, A., Niculescu, D., and Raiciu, C., 'Towards Wi-Fi Mobility Without Fast Handover', *12th USENIX Symposium on Networked Systems Design and Implementation (NSDI' 15)*, 2015, Oakland USA, [Online]. [Cited Jan 2024] Available from: https://www.usenix.org/conference/nsdi15/technical-sessions/presentation/croitoru.
[34] Trinity Mobile Networks, [Online]. [Cited Jan 2024] Available from: https://trinitymobilenetworks.com/mvno/.
[35] Galinina, O., Pyattaev, A., Andreev, S., Dohler, M., and Koucheryavy, Y., '5G Multi-RAT LTE-WiFi Ultra-Dense Small Cells: Performance

Dynamics, Architecture, and Trends', *Selected Areas in Communications, IEEE Journal*. 2015;33(6):1224–40.

[36] 'Chunghwa Telecom Launches Exclusive Pioneering LWA 4G + Wi-Fi Services', [Online]. [Cited Jan 2024] Available from: https://www.cht.com.tw/en/home/cht/messages/2017/en-msg-170223-173325.

[37] Adib, F., and Kati, D., 'See through walls with Wi-Fi', 2013, *SIGCOMM '13: Proceedings of the ACM SIGCOMM 2013 conference on SIGCOMM 2013*, Hong Kong, China, pp. 75–86 [Online]. [Cited Jan 2024] Available from: https://doi.org/10.1145/2486001.2486039.

[38] Wi-Fi Sensing Guidelines, [Online]. [Cited Jan 2024] Available from: https://wballiance.com/resource/wi-fi-sensing-deployment-guidelines/.

[39] Tain, L., Santi, S., Seferegic, A., Ian, J., and Famaey, J., 'Wi-Fi HaLow for the Internet of Things: An Up-to-Date Survey on IEEE 802.11ah Research', *Journal of Network and Computer Applications*. 2021;182:103036. https://doi.org/10.1016/j.jnca.2021.103036.

Chapter 4

The future of the Internet

I predict the Internet will soon go spectacularly supernova and in 10 years we will have forgotten all about it.

– Robert Metcalfe, inventor of Ethernet, 1995

4.1 Introduction

This chapter examines the nature of the devices and the applications accessing the future Internet to understand the requirements and potential innovations unlocked by 6G and future Wi-Fi supported by the broadband access network and its capabilities. Understanding the needs of a broad range of application requirements will lead to understanding the requirement for 6G and where to draw the line to avoid a 'point case' setting an expensive requirement. Therefore, we examined a comprehensive body of academic and industry studies to find quantitative evidence to specify the requirements and the impact of these future Internet applications on the network.

In driving the future of the Internet, three essential requirements come to the forefront. First, gigabits per second connectivity is paramount, particularly for services like the metaverse, enabling seamless and immersive experiences. Second, low latency, as seen in Box 4.1, plays a vital role in the Tactile Internet, ensuring real-time responsiveness for applications that rely on tactile feedback. Third, there is a growing need for mass-content delivery, exemplified by the transition of traditional TV broadcast services to the Internet. Chapter 7 describes the importance of this, where today, only circa 15–20% of the most valuable spectrum below 3 GHz is used. Moving broadcast TV content to broadband would liberate the terrestrial TV spectrum, creating an additional 1600 MHz for 6G. These three dimensions collectively define the Internet's capacity to offer various services, often combining these requirements innovatively.

Since the Internet became widely available to the public in the mid-1990s, data transfer speed has been the primary metric for Internet services. Back then, people were excited to upgrade from V.32bis modems with 14.4 kbit/s to V.34 with 28.8 kbit/s. However, with the advent of gigabit access speeds, the time it takes for data to reach us or latency is becoming increasingly important. 5G has set a benchmark to achieve 1 ms round trip time (RTT) latency, although we shall see if 5G achieves this. The question is whether 6G should offer a 'normal' ten times improvement associated with each mobile generation and if this is genuinely needed apart from outlier use cases. With mobile networks leading the charge for revolutionary change, the question arises about how the fixed Internet responds in an era of convergence. We discuss our vision of convergence in Chapter 9. This chapter analyses the application requirements for latency and the need for further improvements. It also examines the causes of latency in mobile and fixed networks and its impact on applications.

Once we understand the applications' requirements and what causes latency, the chapter describes how to address those requirements.

This chapter starts by providing insights into the potential user experience and network requirements for the future Internet that may arise for 6G, future Wi-Fi and the fixed network. Industry 4.0, which has specific stretching requirements, is addressed in Chapter 5.

The application performance requirements and the ultra-use cases section analyse the requirements to meet the future Internet applications of The Tactile Internet, the metaverse and linear TV over the Internet. In this chapter, the Tactile Internet is about interactions with the physical world, and the metaverse is about interactions with virtual worlds.[1]

The Latency section compares real-world latencies to future Internet requirements and examines what causes latency. The Reliability section summarises the importance of network availability or reliability, the issues with measuring it and how reliability can be improved. Finally, this chapter concludes with important lessons and recommendations.

Box 4.1: Definition of latency

Latency or lag is the time it takes something to respond. In the context of network applications, we have three important terms:

1. End-to-end latency (E2E) defined by the 3GPP as '*The time that takes to transfer a given piece of information from a source to a destination, measured at the communication interface, from the moment it is transmitted by the source to the moment it is successfully received at the destination.*' Application latency requirements are usually expressed as an end-to-end latency.
2. Round trip time (RTT) is the time it takes for a packet to travel from a source to a destination and back again. Network latency is usually measured as an RTT because it is the easiest to measure without requiring source and destinations to have a synchronised clock. RTT is

[1]Some might define the metaverse as being a subset of the Tactile Internet while others may think the Tactile Internet is a subset of the metaverse. For this chapter we consider Augmented Reality to be part of The Tactile Internet.

sometimes called the 'ping time' named after the ping tool used to measure network RTTs.
3. System latency is the latency the application experiences and includes processing time plus network latency. Processing time depends on the application.

E2E latency may not be half the RTT because of network asymmetry, the forward and reverse paths may have different E2E latencies.

Whether E2E or RTT latency is most important depends on the application. For those apps that require feedback, e.g. control loops be they machines or humans, such as a tele-surgeon or a games player, then RTT is important.

The definition of E2E latency is ambiguous because it depends on the type of the 'piece of information' e.g. if it is a file, then the complete file must be transferred in many packets. The location of the 'communication interface' is also ambiguous as this could be an API in the device's operating system used by the application or a virtual interface on a virtual machine or a physical network interface on the device.

4.2 Application performance requirements and the ultra-use cases

Many have summarised 5G and 6G application performance requirements in a graph plotting the latency and bandwidth requirements for selected applications or provided a list of use cases and their requirements. Table 4.1 and Figures 4.1–4.3 show a compilation of application performance requirements taken from a survey of references [1–6] supplemented with analysis for this book explored in this chapter.

Many authors have produced spider diagrams showing how 6G should be *N* times better than 5G in the dimensions of data rate, energy efficiency, spectral efficiency, latency, connection density and reliability, and Figure 4.4 shows an example. These spider diagrams should be understood in that they are theoretical and real-world 5G performance is significantly worse, i.e. global median 5G download speed is 80 Mbit/s, upload speed 10 Mbit/s [7] and latency greater than 4 ms.

We challenge the concept that 6G needs to be *N* times better than 5G in many dimensions, as does some of the groupthink practice of inheriting requirements from earlier papers and documents without question. To this end, this section examines three critical classes of future Internet applications: the Tactile Internet, which requires low latency; the metaverse, which requires high bandwidth; and the migration of linear TV to the Internet when Digital Terrestrial TV is turned off to release 400 MHz of spectrum for 6G but requires mass video distribution over the Internet. This section describes each application class. We compare the performance of current network technologies and networks in the real world against future Internet requirements. It concludes that 5G's current performance ambitions meet all the future Internet requirements, although today's 5G, Wi-Fi and fixed networks fall-short of meeting these requirements in some situations.

Table 4.1 Application performance requirements

Category	Use case	Mobile/ fixed	End-to-end latency (ms)		Bandwidth (Mbit/s)		Minutes not available per year[2]	
			From	To	From	To	From	To
AR/VR/ Gaming	AR/VR motion-to-photon	F&M	5	17	5	1000	500	500
	Collaborative gaming	F&M	5	50		1000	500	500
Healthcare	General	F&M	5	20	2	100	500	5
	Accurate tactile feedback	F&M	5	5	0.01	2	500	5
	Tele-surgery	F	5	72	20	1000	5	0.5
Education & Culture		F&M	5	20		1000	500	500
Tactile Internet	Audio feedback	F&M	20	70	0.1	1	500	500
	Video feedback	F&M	30	85	2	18	500	500
	Uncalibrated tactile feedback	F&M	5	50			500	500
	Simulcast	F	20	100	2	18	500	500
Linear TV		F&M		1000	2	18	500	50
Automotive	Wireless roadside backhaul	F	10		1000	10000	5	0.5
	Sensor sharing	M	<20		0.02	25	5	0.5
	Platooning	M	<10	20	0.01		5	0.5
	Coop collision avoidance	M	<10	100	0.01		5	0.5
	Coop lane change	M	<10	100	0.01	5	5	0.5
	Emergency trajectory alignment	M	<3				5	0.5
	Remote driving info-sharing	M	<5				5	0.5
	Video sharing	M	<10		10	700	5	0.5
	Info sharing for automated driving	M	<100				5	0.5
	Telemetry offload	M		1000	0.01	1000	5	0.5

[2] Where availability for an application has not been specified then it has been assumed that the average residential broadband availability is sufficient.

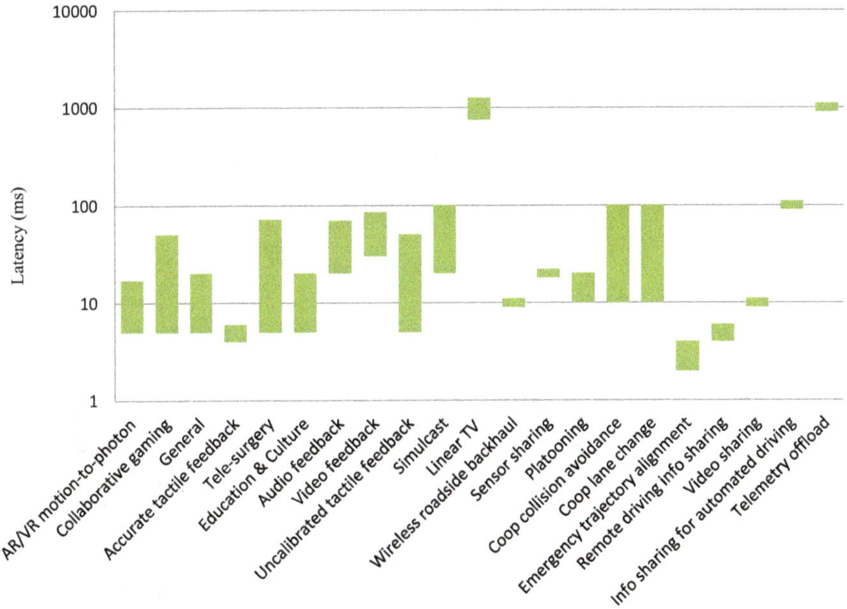

Figure 4.1 Latency requirements of use cases

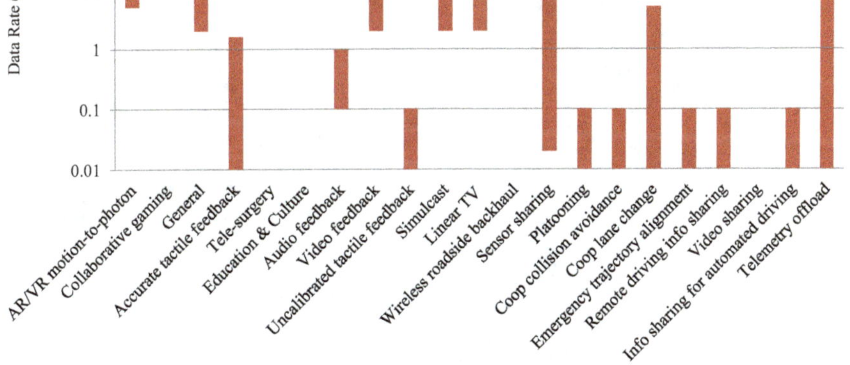

Figure 4.2 Bandwidth requirements of use cases

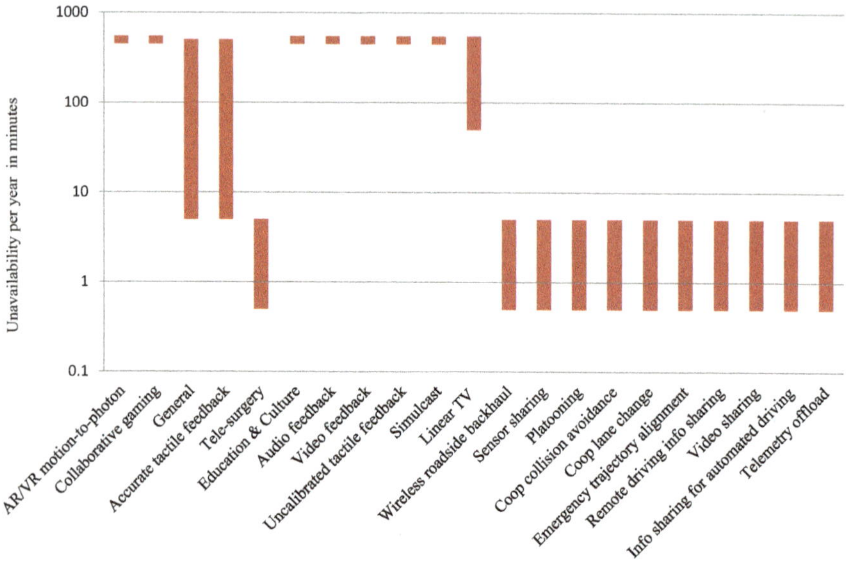

Figure 4.3 Availability requirements of use cases in minutes of unavailability per year

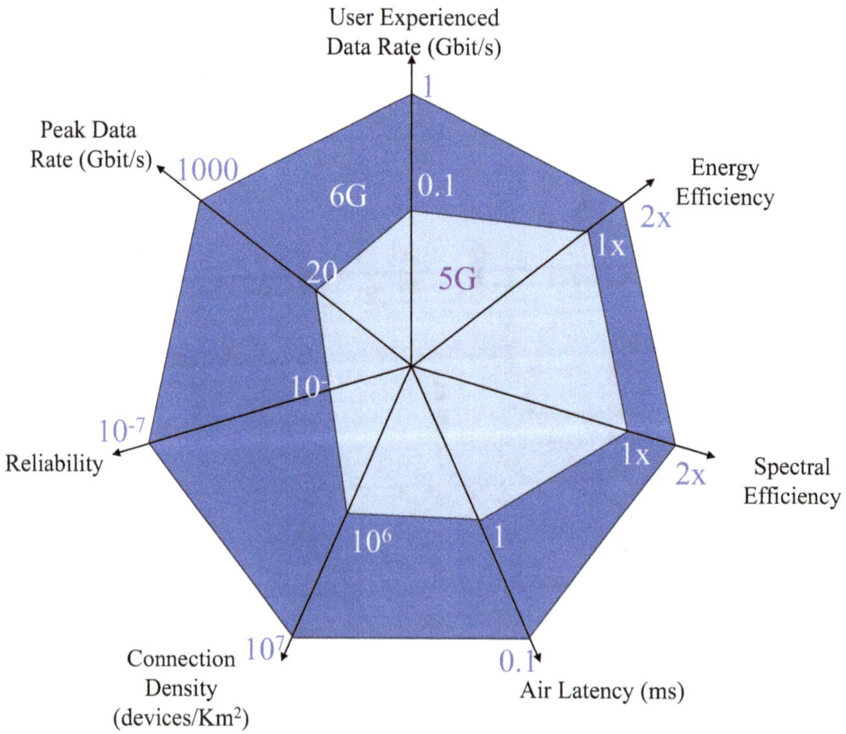

Figure 4.4 Example 5G vs. 6G Spider Diagram after Samsung 6G whitepaper

4.2.1 The Tactile Internet

Imagine being able to interact with real physical or virtual physical objects over the Internet, to touch, feel their surfaces, feel their weight, feel resistance, push, pull, or move them. The Tactile Internet covers these interactions between people, people and machines, and machines and machines. The IEEE [8] defines the Tactile Internet as '*A network, or a network of networks, for remotely accessing, perceiving, manipulating, or controlling real and virtual objects or processes in perceived real-time*'. 'Tactile Internet' was first coined in 2014 by Professor Fettweis, who characterised it as having approximately one-millisecond latency, high availability, reliability and security. The Tactile Internet would significantly increase the quality of our lives. Together with other technologies, it industrialises tele-healthcare, creates new ways of learning manual skills, allows workers to perform remotely from dangerous environments, saves over one million lives per year (World Health Organisation [9]) lost in road collisions, and creates a new genre of tactile gaming (Figure 4.5).

5G enables the Tactile Internet with a 1 ms RTT latency objective. Chapter 2 shows that 5G struggles to achieve 4 ms latency. If 5G delivers 1 ms latency, what is the role of fixed networks, particularly with Wi-Fi, to meet their comparable Tactile Internet requirements? Understanding the needs of a broader range of application requirements will lead to understanding the requirement for 6G and where to draw the line to avoid a 'point case' setting an expensive requirement. In this chapter, we explore specific performance requirements for these applications.

The focus of the Tactile Internet human-centric applications is on haptic communications. Haptic technology creates an experience of touch using electro-mechanical means. Everyday haptic communication is shaking hands with someone as a basis of trust. Perhaps shaking hands with someone over the Internet could be called haptic telecommunication, but this emerging concept is already littered with terms such as kinaesthetic[3] communication, 3D touch, cyber-physical systems, sensorimotor communication, and embodied interaction. 'Tactile Internet' extends beyond human interactions to include industry, robotics, road traffic, smart power grids, and anything to control real or virtual objects in real-time.

Implementing the Tactile Internet requires understanding how the latency between taking an action and feeling the effect of that action impacts how humans perceive and interact with objects. It also requires understanding how latency impacts task performance and control system stability for humans and machines.

Let us examine a few Tactile Internet applications and their requirements, then surmise the impact on society, industry, mobile and fixed networks.

4.2.1.1 Healthcare

Telemedical applications include the ability of doctors to perform tele-diagnosis, surgeons to perform tele-surgery, and physiotherapists to perform tele-rehabilitation. The COVID-19 pandemic quickly forced the adoption of

[3]Kinesthetics, in American English.

Figure 4.5 Haptic glove with tactile, force, and vibration feedback and motion tracking by HaptX, Inc. Image courtesy of HaptX.

telemedicine. Pre-COVID pandemic, telemedicine accounted for a single-figure percentage of patient visits. During the 'lockdown', telemedicine accounted for most patient visits. Many of us used telemedicine during the pandemic and experienced the shortfalls of Zoom diagnostics and physiotherapy; however, Dan Zeltzer's [10] study discovered that telemedicine during 'lockdown' in Israel reduced costs with no evidence of adverse outcomes. The UK National Health Service (NHS) has introduced 'Virtual wards' [11]. The NHS Virtual Wards enable patients to get care at home, including care homes, rather than hospitals. The virtual ward can use remote monitoring apps, video calls, wearables, and medical devices such as pulse oximeters and digital stethoscopes, supplemented with face-to-face care at home. The NHS virtual wards treated 100,000 patients in 2022 [12]. These initial steps show the value of telemedicine, eHealthcare or Healthcare 4.0 using a Tactile Internet that can provide high-fidelity video, audio, medical telemetry and tactile or haptic information.

The adoption of massive phased arrays in 6G for mmWave communications opens the potential use of these as computerised tomography (CT) scanners, which could bring significant remote diagnostic capabilities using domestic 6G user equipment, for example, the mmWave scanner in the iPhone 6G could be used for vital signs monitoring [13] and to detect skin cancers and microwave scanners can be used to detect breast and lung cancers [14]. Cable Lab's 'The Near Future: A Better Place' video [15] illustrates the potential telemedicine applications and their benefits, ranging from the use of in-home robotics, sharing medical wearables data with physicians, VR for consultations, and home sensors for monitoring the situation of vulnerable patients.

Telemedicine transforms medical care in hospitals; imagine Star Trek-like 'sick bay' medical beds where patients' vital signs are monitored wirelessly by AI, increasing patient comfort and safety as issues are detected early. There may not be an Emergency Medical Hologram, but a robot with tactile sensors allowing specialist consultants to examine patients remotely, reducing time spent travelling. Tele-surgery (Figure 4.6) brings expert surgeons to every hospital where the trend in western medicine is for hospitals to be increasingly specialised. The surgeon will have access to high-resolution images and real-time scans overlaid and supplemented by AI. Robotic surgery increases precision and reduces physical stress on the surgeons, e.g. it reduces time spent standing and reduces hand shaking. Telesurgery requires an incredibly reliable network and bandwidths commensurate with XR in the 20–1000 Mbit/s range, but the latency requirement is less obvious. If the surgeon requires tactile feedback from the robot, then [16] established that hard surfaces start to feel softer if latency exceeds 5.5 ms. I'm not a surgeon, but I imagine mistaking hard tissue for soft tissue may not have a good outcome. However, Kalyana Veluvolu [17], who studies robotics specifically for surgery, implies that a 72 ms delay generated by the filter required to remove surgeons' physiological tremors or handshakes is acceptable. Another feasible future is that AI will be used for surgery [18], making tele-surgery redundant.

The benefits of the Tactile Internet and 6G for healthcare may be significant, but we should consider that healthcare is a conservative, safety-first, risk-averse industry, which would delay the adoption of this technology until it is mature and reliable.

4.2.1.2 Education and training

We learn using our five senses, but with traditional remote learning, we are limited to sight and sound. The Tactile Internet allows the sense of touch to become involved. For example, students can use the Tactile Internet to create virtual laboratories to conduct experiments in a safe and controlled environment. Virtual laboratories would benefit students who do not have access to physical laboratories, such as those who live in rural areas or attend schools with limited resources. The

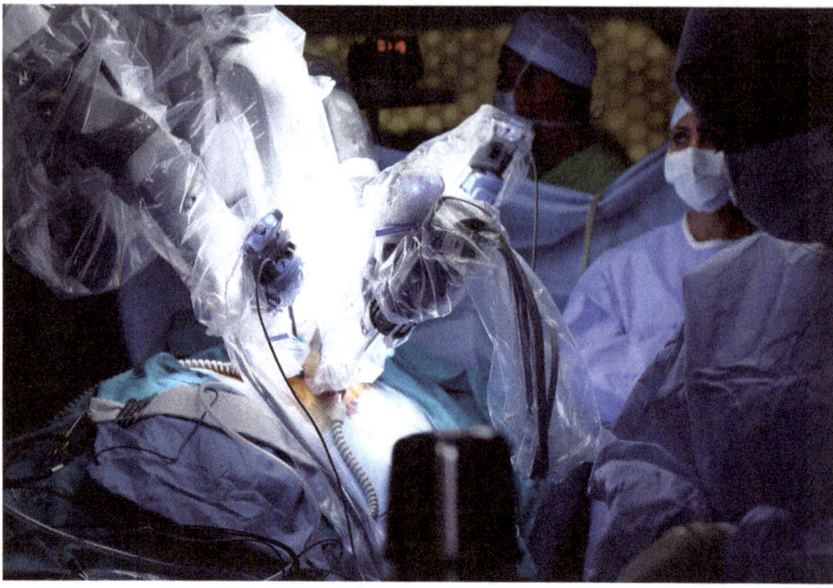

Figure 4.6 Robotic surgery. Image used under licence from Shutterstock.com.

Tactile Internet could also enable students to take virtual field trips to places they would not otherwise be able to visit, such as the International Space Station or the Great Barrier Reef, allowing a 'hands-on' experience. The Tactile Internet allows students to learn about diverse cultures and environments in a more immersive and engaging manner. Hamza-Lup [19] demonstrated that using haptic feedback for learning about hydraulics in high schools improved student engagement and the percentage of correct answers from 47% to 62%. The 'haptic overlay of student and teacher' takes this further with the, almost sci-fi, concept of a student being able to feel and follow a teacher's movements when executing a task and the teacher being able to feel and follow the student's efforts. Haptic overlay technology[4] improves remote learning of sports, manual technical tasks, medical procedures, playing musical instruments, and much more. With the Tactile Internet communications technology, electro-mechanical haptic gloves, exoskeletons, and VR headsets achieve a haptic overlay. Another feasible alternative to haptic overlay is the use of AI. The student learns skills by interacting with an AI model of the skill, which is much more straightforward than the mechanics of remote human-to-human interaction.

The Tactile Internet would enable dangerous manual tasks to be learned from a safe distance, although robotics might mean humans can avoid dangerous tasks altogether.

[4]The term 'haptic overlay' also applies to human computer interaction devices e.g. a flat touch screen display where buttons and dials can be felt.

Performance requirements for education and training differ from XR requirements because haptic force feedback accuracy is task-specific. The general rule of 0.74% error in force feedback per millisecond of latency covers any task requiring accurate haptic feedback of force. We should expect to find the amount of error acceptable to a music teacher different to a hobbyist wood turner. If the application is not concerned about the amount of force feedback but only some sense of haptic feedback, then latency in the 5–50 ms range should be acceptable.

4.2.1.3 Robotics and telepresence

Robots and the Tactile Internet allow people to perform tasks remotely and safely in dangerous environments, e.g. firefighting, construction and maintenance, decommissioning nuclear reactors, military conflicts, mining, offshore construction, and space satellite maintenance (Figure 4.7).

The robots are, in effect, providing a telepresence service. Wireless communications must provide low latency, high bandwidth, and high-reliability communications in potentially hostile radio environments. The wireless connectivity must work reliably in the face of deep concrete and metallic shielding, electromagnetic interference and potentially deliberate jamming, which may be challenging for any wireless system. Some of the high-risk environments also lack good telecommunications infrastructure, e.g. offshore. Robotic telepresence has significant commercial value in a limited range of scenarios, even if all these requirements cannot be met.

Figure 4.7 The Marseille firefighters [80] with a selection of their firefighting robots

Examining the scenarios mentioned above, super-human strength rather than the delicate touches of a surgeon or a musician is more critical, reducing the need for accurate tactile feedback and relaxing the low latency requirements. In some of the above scenarios, bandwidth requirements are higher than standard Augmented Reality because super-human visual acuity is needed, e.g. infra-red vision, multiple camera angles and zooms, RADAR and LiDAR.[5]

Robots and machine tools controlled by computers in industrial scenarios have speeds and reaction times over a thousand times faster than humans, requiring specialist protocols to meet demanding requirements. The 'Industry 4.0' Chapter 5 addresses this.

4.2.1.4 Augmented reality

Augmented Reality (AR) combines the real world and computer-generated content. AR must facilitate real-time interaction with the real world, so synchronisation with the real world is a crucial driver of the AR systems' technical requirements. For AR to be practical, the equipment users must wear and carry must be lightweight and robust for field use. AR glasses should be similar to regular glasses, and applications should run on smartphones. AR contact lens prototypes have even been developed [20]. The lightweight and low power requirements of field use mean there is limited computational power available in the glasses or the smartphone, so there may be a need to move the computations to the Cloud, especially AR applications, which require sophisticated manipulation of 3D images. Accurate synchronisation with the real world requires an Edge Cloud and low-latency communication. AR latency and bandwidth requirements are use case specific, so a small selection of use cases are examined.

For highly dynamic environments requiring accurate synchronisation of the augmentation video or images, such as a firefighter searching for injured people in a smoke-filled building, the challenge is to ensure the user does not suffer motion or cybersickness caused by a lag between where the user is looking, and the augmented visuals received. Studies suggest that latency should be less than a range between 5 and 17 ms to avoid cybersickness; see Adelstein [21] and Jerald [22]. The computation complexity required in this scenario is high. Images from multiple sensors with different points of view, for example, the mmWave scanner, a thermal imaging device[6] and the camera, are mounted at different positions on the body, must be synchronised, manipulated in 3D and aligned to a common point of view, so a nearby Extreme Edge Cloud (see Section 4.5.1) on the fire engine may be required. This scenario implies that the images from multiple sensors must be acquired, transmitted to the Extreme Edge Cloud on the fire engine, processed and transmitted back to the smartphone, and rendered onto the AR glasses in approximately 5 ms, as shown in Figure 4.8.

[5]LiDAR is an abbreviation of 'light detection and ranging' used to determine the distance an object is by measuring the time for a reflected laser to return to the receiver. LiDAR scanners were in the iPhone 12 Pro and more recent models, see Chapter 14.

[6]Today's thermal imaging devices for firefighters are handheld so a good AR solution should free up the firefighters' hands.

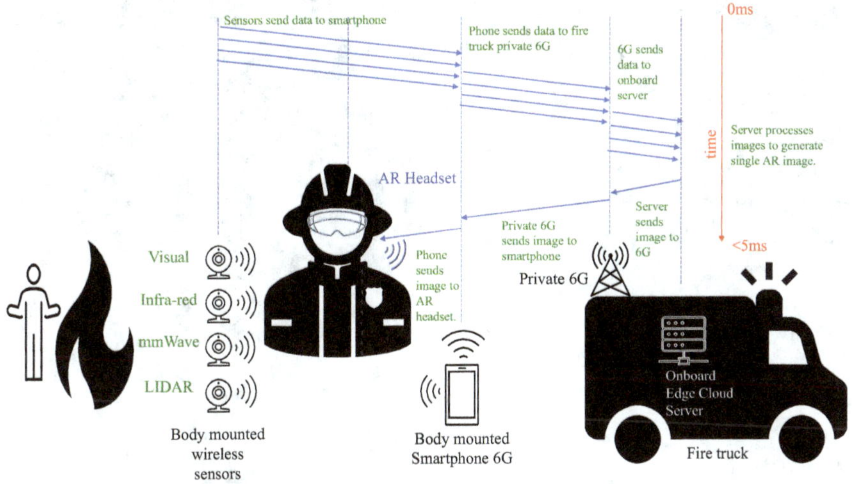

Figure 4.8 AR firefighting scenario

Bandwidth requirements are high, ranging from circa 20 Mbit/s to 40 Gbit/s, dependent on video resolution, frame rate and codecs[7] used, as the solution must transmit three or four high-definition images to the fire engine and return one high-definition image for every firefighter. In this scenario, we cannot assume that the building's internal Wi-Fi is working, but the fire engine may have its own private 6G network base station.[8]

For less dynamic environments, for example, a technician repairing a complex piece of equipment, as shown in Figure 4.9, the AR user can wait for the system to take a still image, process it, retrieve the information, and display it appropriately positioned. The augmented information may be textual or a relatively simple graphic; hence, bandwidth requirements for display are low.

Imagine a simulcast scenario where people view a live sports event in a stadium and may like to see a different view of the event on their tablet or smartphone. The stadium has many cameras, and everyone can choose their preferred viewpoint. The latency between the naturally visible event and the alternative viewpoint displayed on the screen should not be perceptible. Less than 20 ms, Britannica [23], latency is sufficient for people to not perceive the AR video as not simultaneous with actual vision.[9] The bandwidth density required in a stadium with tens of thousands of people watching live videos is very demanding. For example, Wembley Stadium in the United Kingdom has 90,000 seats with an HD video feed to a smartphone of ~5–10 Mbit/s, equating to 450–900 Gbit/s. Alternatively, the

[7]Some video codecs have a considerable latency e.g. >200 ms, so would not be suited to this application.
[8]The UK Emergency Services Network as well as using nomadic 4G base stations in Blue Light vehicles can also use an in vehicle LTE mobile router to create a Wi-Fi bubble to connect body-worn cameras etc.
[9]Or in other words at <20 ms latency people would perceive the AR video and the actual vision to be simultaneous.

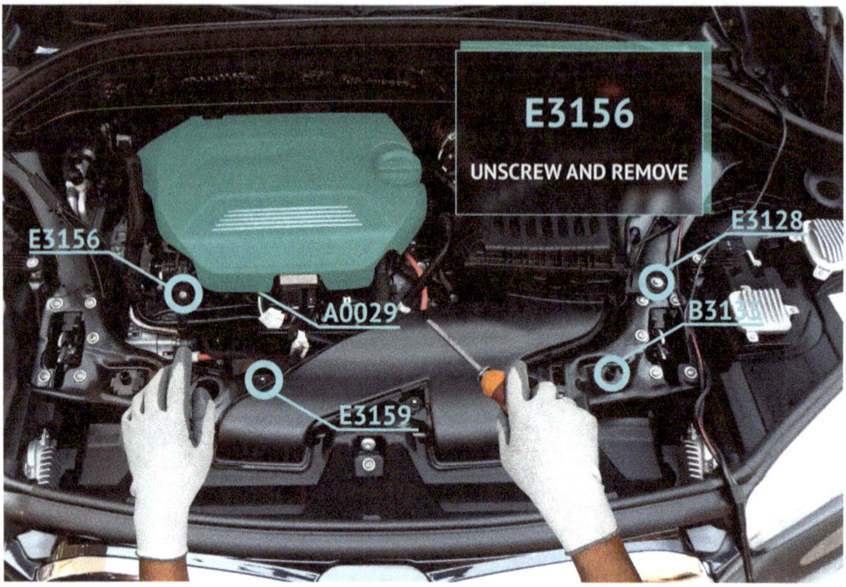

Figure 4.9 Use of AR for maintenance tasks. Image used under licence from Shutterstock.com.

feed from each video camera could be multicast, which for 100 cameras would only require ∼500–1000 Mbit/s.

4.2.1.5 Gaming

Online video gaming, where multiple players interact with each other, is now prevalent, with the most popular games having approximately one billion downloads. The Tactile Internet may enable a whole new genre of tactile gaming where gamers can more physically interact with each other and the virtual environment. Traditional online video gamers are sensitive to latency. Liu [24] showed a linear relationship between players' scores and latency. Players with 25 ms latency scored 20% more than players with 125 ms latency. Similarly, players with 25 ms latency reported a 20% improvement in Quality of Experience against those with a 125 ms latency. In general, online video games require latency of under 50 ms, with the more skilled players being the most sensitive to latency [25,26]. Tactile gaming is more sensitive to latency; it requires 5 ms latency for accurate haptic interaction; otherwise, latency less than 50 ms should be sufficient.

4.2.1.6 Road traffic

The World Economic Forum estimated that traffic congestion costs the United States nearly $87 billion in 2018 [27], and the World Health Organisation [28] estimates that 1.3 million people die yearly from road traffic crashes. The potential savings in lives and congestion costs from automating road traffic management could be significant. Many automobiles have autonomous collision avoidance and intelligent cruise control but are limited to the vehicle's onboard sensors. Networking would allow vehicles to see around corners and coordinate actions with other vehicles. Vehicle-to-vehicle (V2V) communications can improve collision avoidance and enable cooperative traffic optimisation, such as platooning, but this can be achieved with existing V2V standards and does not require 6G.

The US Department of Transport (DOT) withdrew its rule to mandate V2V in November 2023 due to the lack of adoption by car manufacturers. The USA FCC reallocated the 5.9 GHz spectrum used by V2V and permitted the transition of the spectrum to cellular vehicle-to-everything (C-V2X) technology. 5G C-V2X is specified in 3GPP Release 16. 5G C-V2X supports direct or 'sidelink' communications between cars, which can operate with or without cellular coverage. 5G C-V2X also supports standard vehicle-to-base station communications. The European Union standards use protocols based on IEEE 802.11p. Japan is exploring C-V2X and trialling its Wireless Safety Messages standard in the 755–764 MHz band. China adopted standards similar to those of Europe but has invested in LTE-V2X and is trialling C-V2X. South Korea adopted C-V2X as its standard in 2021. V2V standards and regulations are evolving; meanwhile, autonomous collision prevention systems that do not depend on all vehicles to have V2V prevail using onboard sensors.

MIT researchers [29] have found that humans need about 390–600 ms to detect and react to road hazards, so we expect there is plenty of scope for automated systems to do much better. The estimation of latency requirements for collision avoidance systems ranges from 10 to 100 ms; see ITU's paper [30] and Fu [6]. The variation is due to different assumptions. ETSI TR 102 638 [31] states pre-crash warnings require 50 ms RTT latency, while the US Department of Transport DOT HS 809 859 [32] states 20 ms E2E latency. While the benefits of vehicle-to-vehicle communication for collision avoidance are clear, especially at blind intersections, the benefits of using vehicle-to-infrastructure or vehicle-to-edge cloud communication for collision avoidance are unclear compared to the existing state of the art.

The '5G Automotive Vision' [33] states that an automated overtake on a two-way road requires 10 ms latency.[10] Cooperative collision avoidance requires a trajectory handshake of 100 ms with 10 ms status updates in case of deviations from the agreed trajectories. High-density platooning requires 10 ms.[11] 'See-through' view sharing requires 10 Mbit/s and a 'bird's-eye' view requires 4 × 10 Mbit/s. Again, the benefits of this compared to state-of-the-art autonomous vehicles without cooperation via a 5G network are unclear.

[10]This assumes that a car's maximum steering frequency is 10 Hz and requires an oversampling factor of 10 leading to an overall update cycle of 10 ms. This suggests an approach where the steering servo control loop is extended wirelessly from the car whereas an alternative architecture would be to keep the servo control within the car and signal the planned trajectory to the car wirelessly which would work with a much longer latency. This is another example of motion and trajectory control explored in more detail in the Industry 4.0 Chapter 5.

[11]Again based on the 10 Hz frequency response of directional controllers.

An autonomous car travelling at 100 km/h takes 57 m to stop (human drivers require ~98 m). If it takes 100 ms before the car system responds, that's equivalent to travelling 2.8 m or 5% of the stopping distance. It seems unlikely that the Tactile Internet communication with the Cloud is involved in an emergency stop and communication to cars tailgating the first car that brakes could be by non-5/6G V2V communication. In reality, cars now use radars to detect when emergency braking is required, which does not involve V2V communication. More extended distance communication, say to the Cloud, would be used for road traffic planning and control purposes and not require low latency communications. Compare this scenario to how air traffic control works today and is barely more sophisticated than the best satnav systems.

Bandwidth requirement estimates for vehicles vary wildly according to scenarios. Where most of the vehicle's computation is inside, communications with other vehicles are telemetry, with data rates often below 10 kbit/s [6]. Kathy Winter, vice president of the Automated Driving Division at Intel, states that autonomous vehicles collect four terabytes of data in approximately 90 min of driving [34], equivalent to a raw data rate of nearly 60 Gbit/s. The vehicle compresses much of that data, but how much needs keeping in a real or virtual black box? Do AVs store just telemetry; does the telemetry need to be transmitted to road traffic management systems and insurance companies? Will some diagnostics or legal reasons exist for storing or transmitting the raw data? In the unlikely scenario, AVs transmit Gbit/s of data while moving at high speed in densely populated areas, which would be a significant challenge for 6G and a significant cost to the AV owner.

Positioning accuracy demands 30 cm accuracy for lane change assistance and vehicle platooning [6] and 10 cm [33] to avoid vulnerable road users. Modern cars achieve this using cameras.

Innovations in onboard car sensors and computer systems essentially achieve all the aims of a more complex network-based system at a lower cost, which is already happening. 5G/6G has a role in car communications, such as vehicle tracking and traffic updates.

The key economic issue is that the car buyer pays for a better car with advanced autonomous safety features. The alternative is a massive public investment in telecommunications infrastructure to help car companies.

4.2.2 The metaverse

There is a vast range of predictions about the metaverse; some speculate the metaverse could become somewhere we spend hours a day, far removed from today's occasional Augmented Reality (AR) and Virtual Reality (VR) uses, while other pundits think the metaverse is already dead [35–37]. McKinsey says it could generate up to $5 trillion by 2030 [38], with people spending six hours a day on average in the metaverse, while Gartner predicts people will spend one hour a day in the metaverse by 2026 [39]. Walsh's argument [36] that people who use the metaverse 'don't date' reminds me of my director in the mid-1990s telling me that the Internet was like CB radio, 'it'll be dead this time next year', or when I demonstrated Amazon to a well-known clothes and food retailer that said they wouldn't want to damage their brand by being associated with the Internet. We should not think of the metaverse as a singular thing owned by one or a few companies but as a tool, the equivalent of a hi-tech web browser. In my view, the metaverse will transform the use of the Internet like it was by the web browser 20 years ago.

The metaverse is more than a virtual gaming world and 3D video conferencing; it has other practical applications. Cyber security operations or control rooms are today dominated by multiple physical screens, and there is a physical limitation to the number of screens that can present information. With a virtual cyber security operations centre [40] (Figure 4.10), the analysts can create their customised control room, with as many 'screens' as desired, readily rearranged to ease the analysts' tasks.

With 3D imaging and gesture-based user interfaces, the virtual operations room has become what many of us may remember from the movie Minority Report. We may never be able to have genuine 3D holograms as Pierra-Alexande Blanche [41], in his paper on the challenges of holographic displays, estimates that they require 6.6×10^{15} FLOPS, raw uncompressed data rates of 3×10^{15} bit/s, and 1.6×10^{12} phase pixels and not available until circa 2100 far beyond 6G. Feasible 3D imaging is limited to headsets, Pepper's ghost, spinning mirrors or projection onto a particular film stuck to the glass. Much research continues to improve holograms, such as metasurface holography [42].

Figure 4.10 Using the metaverse to analyse data. Image used under licence from Shutterstock.com.

A key question is, what network performance is required to create the metaverse? Estimates vary from tens of megabits per second at ∼100 ms latency to a few gigabits per second at ∼10 ms latency. The exact performance required depends on the depth of immersion and interaction. We expect performance requirements to increase as metaverse technology matures. However, all these estimated performance levels are within the potential of today's Wi-Fi and fibre-to-the-home networks.

It is worth analysing metaverse bandwidth requirements given the wide variations in predictions. In the following analysis, we examine the biological bandwidth of human sight and compare that to cameras, highly efficient video coding and foveated imaging.

One starting point to estimate the bandwidth required for the metaverse is to examine how much bandwidth there is between the eye and the human brain. Koch *et al.* [43] report ∼875 kbit/s per eye for guinea pigs. The human retina has ten times as many ganglion cells, which implies a human eye sends about 10 Mbit/s [44] to the human brain. Interestingly, from a communications point of view, average jitters between 40 and 15 ms in the signals from the guinea pig's eye to the brain were measured.

Another approach suggests that a headset receives an 8K ultra-high-definition 360-degree image, from which the headset selects the appropriate field of view depending on where the person is looking. The raw bandwidth output from an 8K camera[12] is ∼2.5 Gbit/s [45]. A 360-degree view requires Multiple 8K cameras. The number of cameras depends on lens focal length and sensor size, but for simplicity, let's assume four cameras are required, which equates to ∼10 Gbit/s of raw data. That's one thousand times more than the human eye requires, so the raw data has much redundancy. Compression reduces the data rate, such as high-efficiency video coding (HEVC); Sugito *et al.* [46] states, '*The experimental results confirm that the bitrate required for 8K 120-Hz videos is 85–110 Mbit/s, which is equivalent to the practical bitrate for 8K 60-Hz videos*'. Multiplying this by four, assuming four cameras for 360-degree vision, gives 340–440 Mbit/s.

Further compression can be achieved by only sending hi-res images for the central scene the viewer is looking at with lower resolution images for the peripheral vision. This method is called 'foveated imaging'. Kortum [47] reports that foveated image coding can reduce the bandwidth required by a factor of 18.8. Applying foveated imaging to a single 8K HEVC video feed gives us a bit rate of 4.5–5.6 Mbit/s. This bitrate is a similar order of magnitude reported by the University of Pennsylvania School of Medicine as generated by the human eye. Table 4.2 summarises the above findings.

Foveated imaging requires that the head and eye positioning information from the headset is sent into the Edge Cloud for the clipped and compressed image to be computed and returned to the headset; this requires latency in the 5–17 ms range to avoid cybersickness. Delivering ∼5 Mbit/s at low latency <5 ms may exceed the cost of delivering a ∼400 Mbit/s full image with 20 ms latency.

[12]V-Raptor 35.4 Megapixel, 8192 × 4230 @ 120 GPS, 16 bit redcode raw.

Table 4.2 Comparison of visual bandwidths

Scenario	Bandwidth
Guinea pig eye	875 kbit/s
Human eye	10 Mbit/s
8K Camera	2.5 Gbit/s
8K HEVC	110 Mbit/s
8K Foveated imaging	5.6 Mbit/s

4.2.3 Linear TV over the Internet

Chapter 7 describes how there is almost 2 GHz of additional spectrum for 6G in the low-band and mid-band frequencies (30–3000 MHz) if used efficiently. Digital Terrestrial TV (DTT) is the first piece to look at. Hence, the future of Digital Terrestrial TV is uncertain. In the United States, the United Kingdom, and other European countries, there is typically over 400 MHz of spectrum in the UHF band between 470 and 890 MHz used by TV, which would benefit mobile broadband services. Governments must balance the utility of ubiquitous free-to-air public service broadcasting TV, especially for people without Internet access, against the utility of the spectrum for mobile broadband services. From 2017 to 2020, the United States reallocated 84 MHz of spectrum, 614–698 MHz, of which 70 MHz went to mobile broadband services [48]. The UK government extended the DTT spectrum licenses in 2021 until at least 2034 but can trigger a revocation clause from 2030. In 2022, Tim Davie, Director-General of the BBC, gave a speech to the Royal Television Society in which he said, '*For the BBC, Internet-only distribution is an opportunity to connect more deeply with our audiences and to provide them with better services and choice than broadcast allows*' [49]. Tim Davie proposed that the BBC could switch off DTT by the end of 2030 to focus on online content and streaming. In its whitepaper [50] on the future of DTT, the UK government said, '*The government will ask Ofcom to continue to track changes to DTT viewing and to undertake an early review on market changes that may affect the future of content distribution before the end of 2025*'.

TV viewing behaviour is changing. In a 2022 survey, Ofcom's 'Media Nations 2022' survey [51] found that 16–24-year-olds watch seven times less broadcast TV than those over 65. Also, 90% of 18 to 24-year-olds go to streaming and social video services before considering watching broadcast TV. In August 2022, the USA's Nielsen survey [52] showed that streaming TV surpassed broadcast TV for a share of viewers' time.

What will be the impact of the extra traffic generated by all DTT broadcast TV migrating to streaming video over the Internet? Taking the United Kingdom as an example, Barb [53] shows that 58 million people watched broadcast TV with an average of 165 daily viewing minutes. Openreach, a UK provider of fixed broadband infrastructure, data [54] shows that the average amount of traffic an Openreach customer consumed was 9 Gigabytes daily in 2020. Adding 165 minutes more video per day equals between 2.5 and 22 GB of data daily, an extra 27% (standard definition video only) to 250% (ultra-high-definition video) more traffic dependent on video resolution. Given the trend towards UHD TV sets, Cisco estimates 66% of TV sets are UHD, so switching all DTT broadcast traffic to unicast streaming services could triple the amount of traffic on the Internet. However, a significant amount of linear TV viewing would transition to handheld devices with lower than UHD resolution, mitigating additional traffic created. It would be reasonable to assume this transition will happen over many years, and on the Internet this growth level is not unprecedented. Internet traffic tripled between 2010 and 2014 [55]. DTT switch-off could benefit from significantly upgrading Internet infrastructure to deliver more video content more efficiently. DTT switch-off should not impact 5G and Wi-Fi as TV streaming is not latency-sensitive, and the peak bandwidth of UHD video is within 5G and Wi-Fi capabilities.

Today, 65% of Internet traffic is video [56], Sandvine [56] reports video usage grew 24% in 2022, and Nielsen [52] reports streaming volume minutes increased 22.6% in the United States. The trend is clear that streaming video traffic dominates Internet traffic, and its importance will only increase. It seems only a question of time before governments will see the utility of the 470–890 MHz spectrum being greater for mobile broadband services than DTT and public broadcast services. Whether governments proactively reallocate the spectrum used by DTT to mobile broadband services, a policy the United States and the United Kingdom are implementing, or follow behind the broadcasters and consumer trends, the outcome will be the same.

4.3 Latency

You have enough bandwidth, but your application is running too slow or not working, possibly due to too much network latency.

It may appear to be splitting hairs to debate the latency requirements of applications within a few milliseconds, but delivering low-latency communications is expensive. One millisecond Ultra-Reliable Low-Latency Communications compared to 10 ms latency enhanced Mobile Broadband requires as much as 60 [57] times more spectrum for 5G and 100 times as many edge compute sites for ubiquitous coverage.[13] Milliseconds matter and make the difference between an application being technically and commercially viable or not. Unfortunately, there is a wide range of maximum latency recommendations across the scientific literature and many active in setting the telecommunications agendas have opted for a safe but expensive RTT 1 ms target with many speculating sub-1 ms is required for 6G. Much more research specifically focused on Tactile Internet applications and the scientific measurement of the impact of latency on tasks is required. At this point, we would assert that less than 5 ms end-to-end latency is required only for accurate haptic or tactile feedback, while less than 20 ms meets most other Tactile Internet requirements.

[13]Industry 4.0 edge compute site requirements will not be as severe as there will, in general, be only a few factory or office sites.

Multi-modal communication requires multiple media feeds to be accurately synchronised; this must not be confused with requiring low latency. Achieving synchronisation over high latency connections can use existing techniques for synchronising clocks.

Designers must take a holistic systems approach to designing the overall Tactile Internet applications to function within the limitations of the communications technology. For example, data rates can be reduced by compression and focusing the fidelity where it is required. Designers should accommodate network failure by allowing some fail-safe autonomy where safety is an issue. In hostile environments, e.g. fires or collapsed buildings, designers should not assume that any internal building Wi-Fi facilities are working.

This chapter has explored the critical performance requirements for the Tactile Internet, the metaverse and linear TV over the Internet. This exploration has revealed some common fundamental application components or 'building blocks' requirements that can be applied to build multiple applications and scenarios, summarised in Table 4.3.

Table 4.3 Application components performance requirements

Application components	**System latency**	**Band width**	**References**
Accurate Tactile Feedback	5 ms	*Tens kbit/s*	[16,17,81]
Uncalibrated tactile feedback	<50 ms	*Tens kbit/s*	[82]
Audio feedback	<70 ms	300 kbit/s–1.5 Mbit/s	[82]
Visual feedback	<85 ms	3 Mbit/s–2.5 Gbit/s	[82]
AR/VR headset/glasses movement sync	5–17 ms	<1 Mbit/s for augmentation	
		10–100 Mbit/s for video	[21,22]
AR screen sync with live	<20 ms	3.5 Mbit/s for smartphone	[23,82]
Gaming without accurate tactile feedback	<50 ms	4 Mbit/s per player	[25,26]
Vehicle platooning @ 100 km/h <1m spacing	<20 ms	Tens kbit/s per vehicle	[6,33]
Vehicle platooning @ 100 km/h <3m spacing	<100 ms	Tens kbit/s per vehicle	
Headset raw UHD video	<17 ms	10 Gbit/s	[45]
Headset compressed UHD video	<17 ms	440 Mbit/s	[46]
Headset compressed foveated	<5 ms	5.6 Mbit/s	[47]
Linear TV over the Internet	<1s	22 GB/day/house	[53]
Raw vehicle sensor data	<20 ms	2.6 TB/h	[34]
Vehicle telemetry	<20 ms	10 kbit/s	[6]

4.3.1 Latency conclusion

System latency is the latency the application experiences and includes processing time plus network latency. Processing time depends on the application. For many applications, the processing time is minimal compared to network latency; hence, the target network round-trip time is approximately the same as the system latency except for the compression of real-time video with state-of-the-art video codecs, creating two to three milliseconds of latency [58,59]. AR or VR applications requiring real-time video reduces the required network latency to approximately two milliseconds.

Applications involving humans, e.g. Tactile Internet, have relatively relaxed latency and bandwidths required because human reaction times are slow and the human nervous system has low bandwidth. Architectural decisions influence latency requirements for non-human applications, such as vehicle platooning, for example, setting the safe distance between cars and whether velocity information is carried in a communications protocol or detected by radar. The 'Industry 4.0' Chapter 5 looks at the requirements for automation, which are much more challenging for 6G because automated machines have much quicker reactions than humans. It is also worth reflecting on the network latencies enterprises want. The Spirent report 'Cutting Through the Edge Computing Hype' [60] reports on an interview of 150 enterprises in North America, mainly in the manufacturing and construction sectors, which found that more than half wanted guaranteed latencies, but only 3% wanted less than 10 ms latency, 26% wanted less than 20 ms.

In conclusion, looking at these requirements, achieving an RTT 1 ms latency, as per 5G URLLC requirements, is also sufficient for 6G, although the latency of 5G enhanced mobile broadband is more relevant to consumer applications.

4.3.2 Real-world latencies

The previous sub-section explored the latency requirements of various applications and use cases. Are existing fixed and wireless technologies, such as 4G, 5G, Wi-Fi, FTTC, FTTH and DOCSIS, addressing these latency requirements, or does 6G have to improve the low latency performance of today's technologies significantly?

Figure 4.11 shows some real-world measurements of round-trip time network latency using the ping tool, which measures round-trip times from London, Washington DC and New Delhi; this shows the real-world latencies are slower than the theoretical ones due to distance only.[14]

The latency between some key cities in the United Kingdom and latency from an FTTC 50 Mbit/s DSL connection was measured and overlaid on a map of the United Kingdom, as shown in Figure 4.12. Black text is inter-city latencies, and green text is measurements from an FTTC access near Ipswich. The red text shows the latency between an FTTC premises and the nearest central office with a router. The measurements show that measured latency is twice as high as the theoretical propagation delay due to distance; however, most UK FTTC and FTTH achieve less than 20 ms ping latency to London.

[14] This may be that international routing has a higher route factor than 1.4.

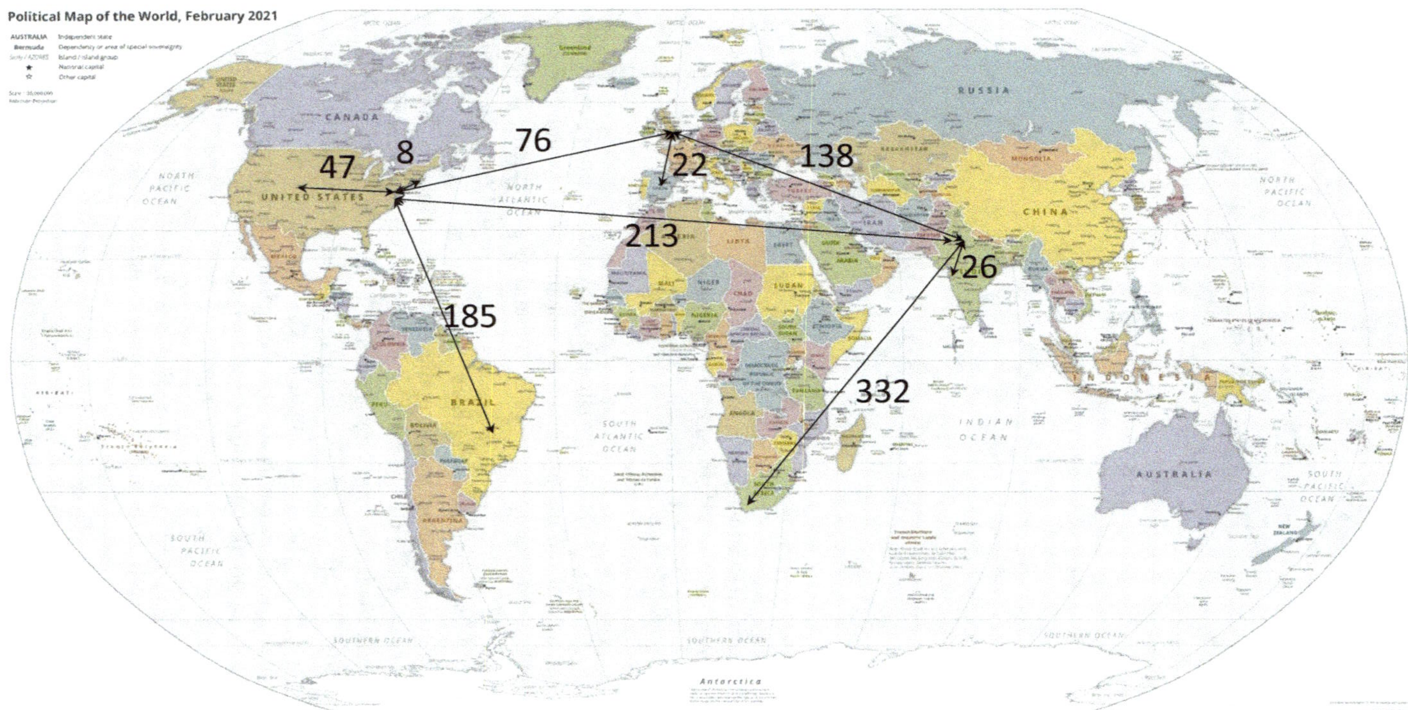

Figure 4.11 *Example real network RTT measurements in milliseconds This image is a work of a Central Intelligence Agency employee, taken or made as part of that person's official duties. As a Work of the United States Government, this image or media is in the public domain in the United States.*

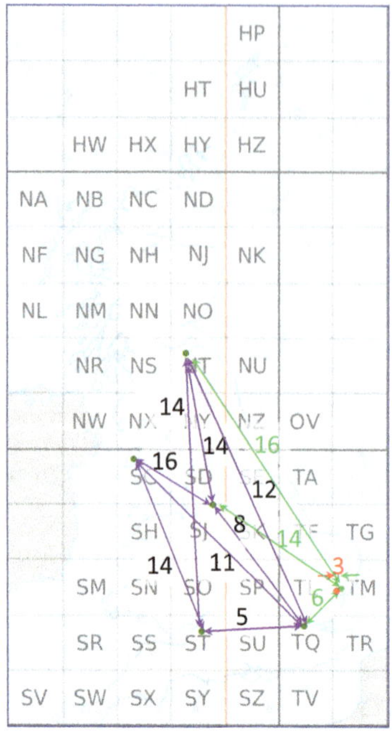

Figure 4.12 Measured UK inter-city and access latencies in ms. Ordnance Survey grid of UK from Wikipedia [79] (Creative Commons license CC BY-SA 3.0).

The USA's FCC has a 'Measuring Broadband America' program [61], which measures the latency of mobile and fixed broadband. From which the FCC has produced a report summarising the fixed broadband measurements [62] which in 2022 reported, '*Cable latencies ranged between 11 and 19 ms, fiber latencies between 9ms to 14ms, and DSL between 18 and 61ms.*' The FCC measured latency as the round-trip time between the consumer's home and the closest measurement server. The FCC has not published a similar report for mobile broadband but has made the measurement data available [63], which shows a median latency of 82 ms and is also helpful for the distribution of latencies. Figures 4.13 and 4.14 show the cumulative distribution of round-trip-time latencies measured in milliseconds for the 'Measuring Broadband America' for Mobile Broadband in 2020. Less than 25% of tests measured less than 50 ms, while 10% of measurements exceeded 200 ms, with 1% of tests exceeding 1 s. Less than 1% of tests achieved the 20 ms or less required for the Tactile Internet and the metaverse despite 10% of US customers connecting to 5G on average in 2020.[15]

[15]In 2020 22% of T-Mobile, 10% of AT&T, 14% of Sprint and 0.4% of Verizon customers connected to 5G on average [64].

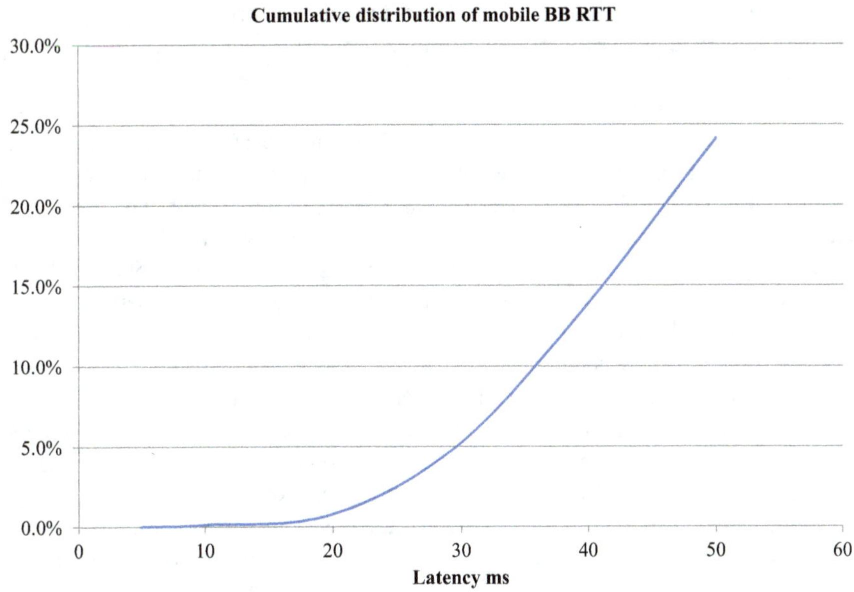

Figure 4.13 *Measuring Mobile Broadband America Latencies Distribution for <50 ms*

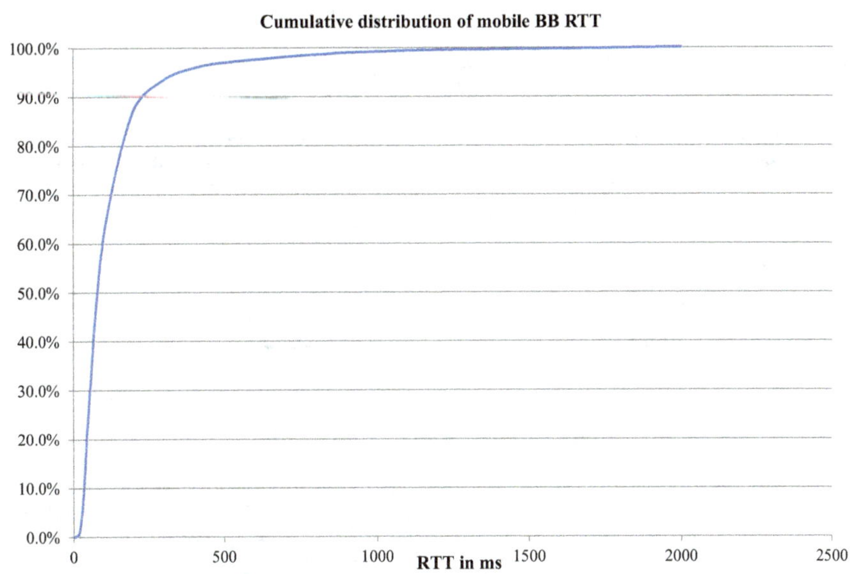

Figure 4.14 *Measuring Mobile Broadband America Latencies Distribution*

The UK's telecommunication regulator, Ofcom, performs similar monitoring of home broadband performance and a summary of the results for March 2023 [65] is shown in Table 4.4.

The Spirent Report [60] covers real-world testing of latencies to 5G Mobile Edge Computers (MEC) and shows RTT (ping) latencies in the range of 4–20 ms when downloading data and RTT latencies of 9–28 ms when uploading data for Seattle, Chicago, New York and Tokyo. The difference between the Spirent and 'Measuring Broadband America' tests is the location of the servers or test points being pinged on the Internet, i.e. Spirent tested to much physically closer MEC servers. 5G with MEC mostly achieves less than 20 ms latency as required for the future Internet.

Wi-Fi latency in the real world depends on the Wi-Fi traffic load, the number of users, the amount of interference, especially from other overlapping Wi-Fi networks and the quality of the Wi-Fi Access Point scheduler. Qualcomm [66] measured the latency of Wi-Fi 6, the latest available Wi-Fi technology in 2021, in several scenarios, summarised in Table 4.5. In an ideal single access point with a single user with no interference, a latency of two to three milliseconds is expected, illustrating the significant difference between the theoretical and the real-world performance of Wi-Fi.

Figure 4.15 shows a summary of the real-world latency performances for various technologies measured from 2021 to 2023. These results show that many of today's network technologies can support the future Internet now. The difference between the USA's and UK's results for similar technologies shows the impact of the size of the country on latencies; this illustrates the need for Edge Cloud and indicates how near to customers it needs to be deployed. It also illustrates the need

Table 4.4 Summary of OFCOM home broadband performance measurements in March 2023 for UK broadband networks

Technology	Median 24-h latency (ms)
Openreach fibre (PON)	6.9
UK Cable	12.5
FTTC	10.1
ADSL2+	24.2

Table 4.5 Wi-Fi 6 real-world latencies

Scenario	OFDM mode RTT latency ms upload/download	OFDMA mode RTT latency ms upload/download
Busy home	76/15	28/9
Office	70/53	54/25
Classroom	5875/452	66/31

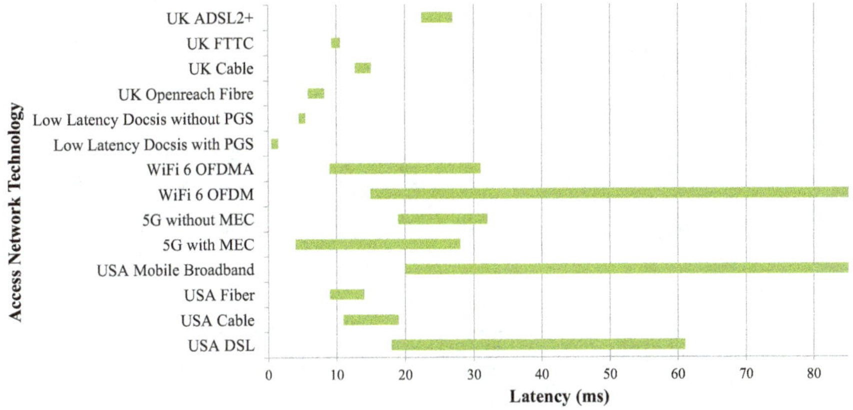

Figure 4.15 Real world performance of network technologies

for further improvements in wireless technology performance in the real world, where today, the worst-case latencies experienced by Wi-Fi 6 and 5G are not good enough for the future Internet.

A global Tactile Internet, where, for example, someone in South Africa could interact tactilely with someone in New York, is not possible simply because of the limitation of the speed of light.

4.3.3 Where does network latency come from

The causes of latency outside the access network, whether Wi-Fi, DSL, Cable, Fibre or 5/6G, must be addressed to enable the Tactile Internet. Improvements across the entire end-to-end service are required. This section examines the various causes of network latency and addresses how those causes can be mitigated for 6G.

Figure 4.16 shows a breakdown of the typical end-to-end latency for 5G Tactile Internet applications. The tactile input and visual output delays are typical of today's smartphone input devices and display screens, which would have to be significantly faster in the 'iPhone 6G' to enable the Tactile Internet. The other latencies shown are composed of the following factors[16]:

1. The distance between sender and receiver.
2. Packet queueing delays.
3. Coding or adaptation delays.
4. Media acquisition delays.
5. Multiple handshakes between endpoints.
6. Packet switching delays.

Figure 4.17 shows the latency ratio due to the distance between sender and receiver, or propagation latency, and packet queuing delays for various distances

[16]There are other causes of latency; for a complete analysis see 'Reducing Internet Latency: A Survey of Techniques and their Merits' [68].

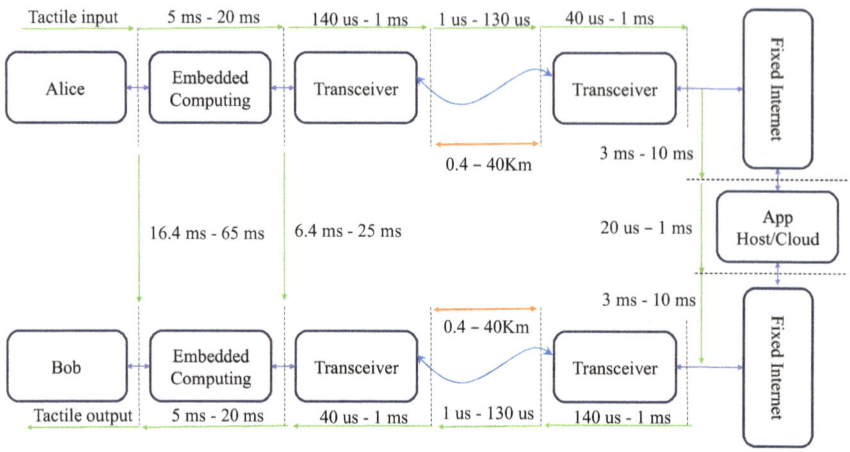

Figure 4.16 Breakdown of typical end to end latency for a Tactile Internet app (after Fettweis [67])

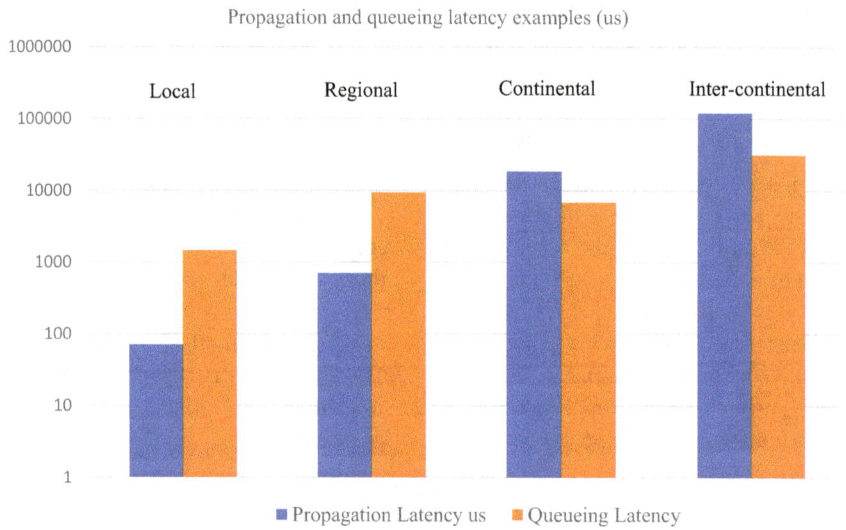

Figure 4.17 Propagation and queueing latency examples

estimated from the measurements shown in Figure 4.11. As the distance between sender and receiver increases, the proportion of delay due to propagation latency increases. Figures 4.17–4.19 show typical values for causes of latency in a local network connection, including handshakes, which have the impact of multiplying network latency. Packet switching delays are negligible at less than 10 microseconds for an average Internet connection.

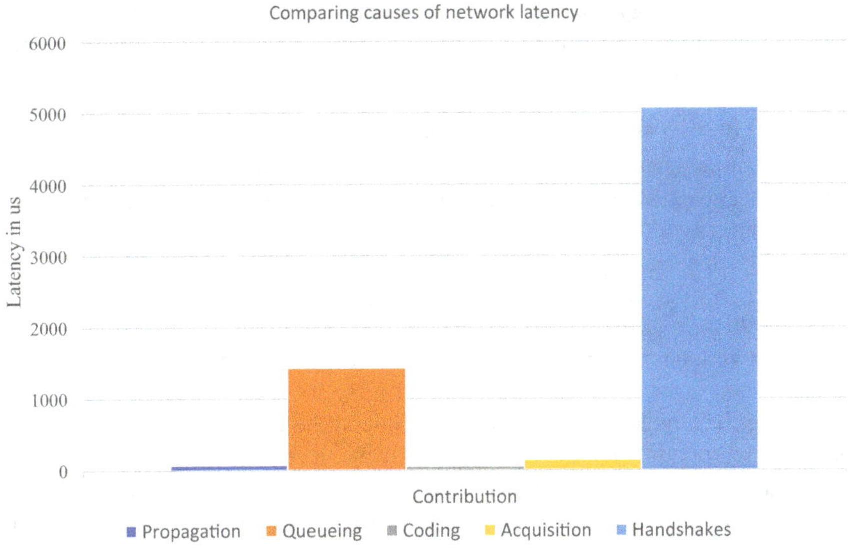

Figure 4.18 Comparing causes of network latency for a typical local connection including handshakes.

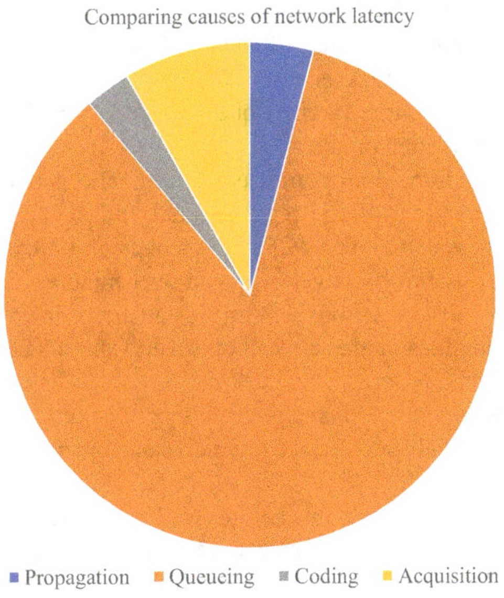

Figure 4.19 Pie chart comparing causes of network latency for a typical local connection.

Simply increasing the bandwidth of your connection to the Internet does not fix these latency causes. The causes of latency and potential methods to reduce latency are described below.

4.3.3.1 The distance between sender and receiver

The propagation delay between the sender and receiver follows the network's cable routes and associated transition time. Most applications and protocols require a response, so network latency is measured as a round-trip time (RTT). Figure 4.20 compares latency propagation distances, applications' latency requirements, and the number of edge or central cloud sites required to meet the requirements. It shows, by the size of the red circles centred on London, the theoretical areas covered for network latency times of 1, 5, 20 and 50 ms, considering cable and network routing.

Taking the United Kingdom as an example at the country level, if a propagation in fibre round-trip-time of one millisecond is required, then covering the United Kingdom in 100 km^2, with the sender of the data in the centre of the square and the receiver anywhere in the square, would achieve this. That is, the maximum propagation latency from the centre of a 100 km^2 would be from the centre of the square to a corner of the square, which would be 1 ms. By coincidence, The UK Ordnance Survey has a grid covering the United Kingdom in 59,100 km^2, as shown in Figure 4.21, implying 59 Edge Cloud data centres would be required to cover the United Kingdom for a 1 ms latency.

4.3.3.2 Packet queueing delays

Packet queueing is a significant cause of latency. In general, it contributes the most to latency. A router queues packets whenever it receives more traffic than it can send to the traffic's destination. Internet queueing theory and modelling are complex; books have been written on the subject, e.g. [69], it is made more challenging to model due to the non-random nature of some traffic patterns,[17] for example, some applications generate traffic in regular bursts and interact with the protocols in servers' operating systems, e.g. TCP, to discover and adapt to the maximum speed of connections. The traffic level averaged over minutes might be well within the capacity limits of the connections, but traffic can arrive in very short bursts, less than a second in length, and the arrival of several short bursts simultaneously causes a queueing delay.

There are many methods of managing the length of packet queues in network devices, which broadly fall into Active Queue Management (AQM) and multiple queues.

Active Queue Management manages queues to achieve target queue loss and latency requirements. It can vary from randomly dropping packets from anywhere in the queue when the queue is too long, which signals the applications to slow down, or it can use a more sophisticated algorithm to select key packets to drop. CableLabs [70] demonstrated AQM reduced latency under high loads from

[17]Most queueing theories assume a stochastic or random arrival of packets.

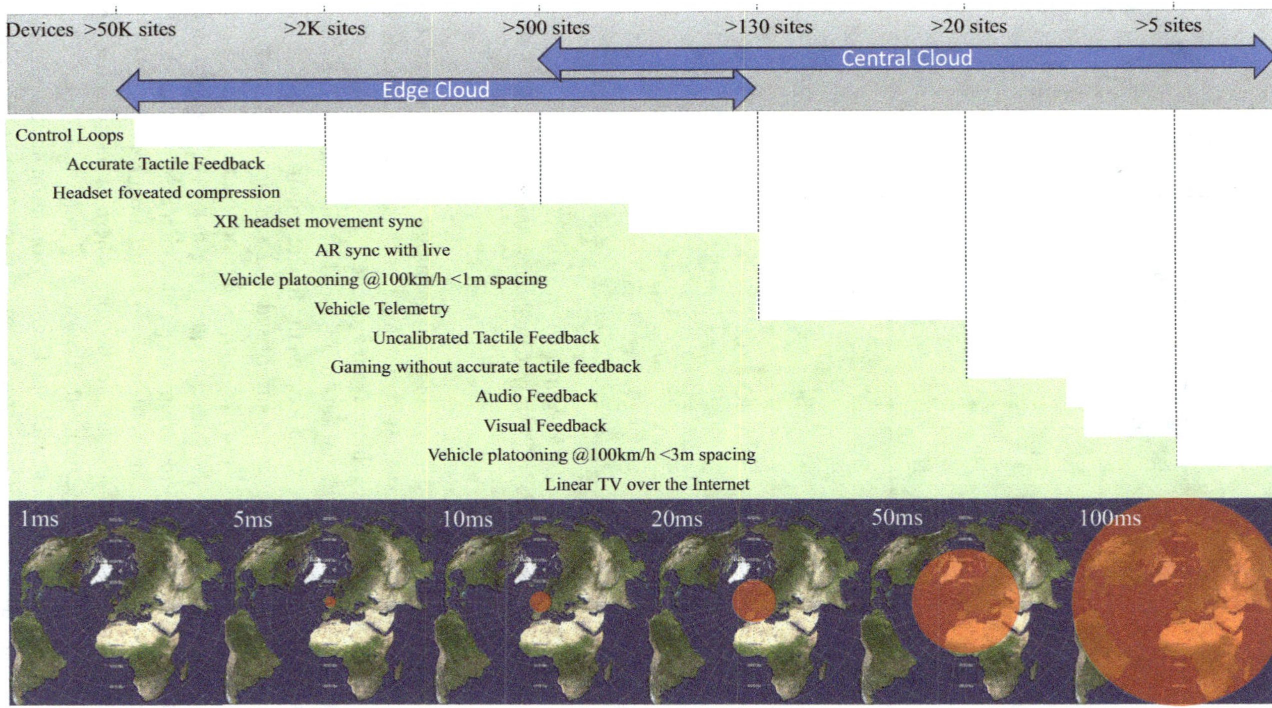

Figure 4.20 Propagation distances for RTT targets v. application requirements & number of global sites required. Azimuthan equidistance chart by Jacobo Tarrio, Azimuthal Equidistant Chart Generator (tarrio.org) Creative Commons Attribution 4.0 International License. The base Earth image comes from NASA's Visible Earth collection.

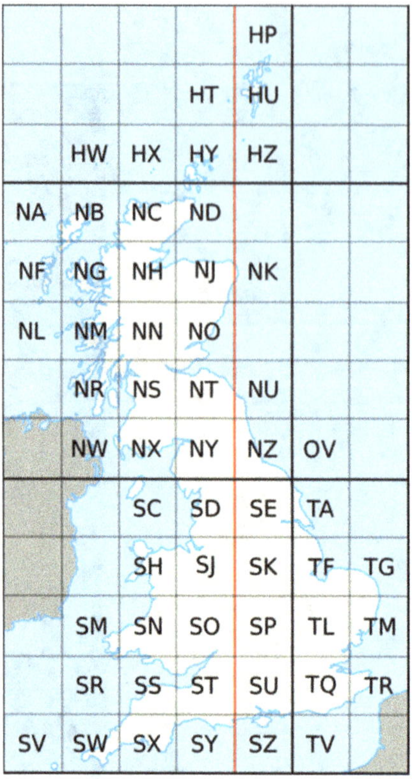

Figure 4.21 *UK Ordnance Survey grid equivalent to 1 ms RTT squares with 100 km^2. From Wikipedia [79] (Creative Commons license CC BY-SA 3.0).*

~250 to 32 ms for DOCSIS 3.1. A highway analogy for AQMctive Queue Management does not readily exist, as removing a packet from a queue is more straightforward than removing a vehicle from a queue.

Multiple queues are where the packets from applications requiring low latency are queued separately from the rest; so long as the traffic requiring low latencies is small compared to the rest, it can be prioritised and rapidly forwarded. In the highway analogy, multiple queues are like high occupancy vehicle (HOV) lanes.

A significant insight is that the cause of queueing delays is the behaviour of applications that try to find the maximum speed at which they can send data by sending data increasingly faster until routers begin to drop packets. RFC9330 [71] addresses this directly by specifying an architecture where excessive queueing is avoided by signalling the applications to slow down using a form of Explicit Congestion Notification (ECN) in packets. Highway on-ramp traffic signals are roughly analogous to Explicit Congestion Notification. Davide Brunello [72] showed that L4S can reduce the latency of video frames delivery by ~40% in an L4S over 5G network simulation.

4.3.3.3 Coding or adaptation delays

In most digital telecommunications, the user's data must be converted into an analogue signal and transmitted over radio, fibre or cables. The analogue signal, representing the user's data, is spread over time to minimise the impact of impulse noise and other interference. The user's data may be reordered and interleaved, and forward error correction (FEC) codes may be added. The receiver must store the received signals until it has received enough data to check the forward error correction code and return it to its original order. This process can add significant delays; for DSL using interleaving, it can add an 8 ms delay. There is a trade-off between robustness to noise or interference and latency. Making the transfer of data more robust increases latency. However, reducing robustness may lead to data loss, requiring the application to retransmit data, thus increasing latency.

4.3.3.4 Media acquisition delays

Where the communications technology depends on a shared medium, e.g. using the same frequency in radio or a shared cable or optical fibre for Passive Optical Networks (PONs, the standard technology used for FTTH), then the sender must wait until no one else is using the medium before sending. There are several Medium Access Control (MAC) methods available. Fixed assignments (TDMA, FDMA) provide access to predefined slots in time or frequency, giving predictable access latency. Demand assignment uses requests for access or capacity to generate allocations; the latency also includes the time to send and process the requests. DOCSIS uses a request-grant mechanism. Random access does not require coordination, but latency depends on load; if there is a collision, more than one sender transmitting simultaneously, both must resend their data. Early versions of Wi-Fi and Ethernet used random access.

Low Latency DOCSIS [70] has a proactive grant service where traffic requiring low latency can have timeslots proactively scheduled regularly; CableLabs [70] demonstrated this reduced latency from 4.7 to 0.9 ms. Earlier Wi-Fi and cellular technologies used OFDM (orthogonal frequency-division multiplexing) but Wi-Fi 6 and 5G use OFDMA (orthogonal frequency-division multiple access) to allow more users to have parallel access to the same spectrum by dividing it into smaller units. Table 4.5 shows the impact of this.

4.3.3.5 Multiple handshakes between end points

Before applications can pass data, endpoints must perform multiple handshakes, which causes delays in transactions or applications. On the Internet, a protocol ensures that the received data is the same as the transmitted data; the protocol must handle packet loss and reordering. It also must adjust the speed at which packets are transmitted to avoid loss due to congestion. TCP traditionally handles these tasks, and QUIC is an IETF protocol that improves on the performance of TCP.

TCP requires multiple round trips to establish a secure encrypted connection, as shown in Figure 4.22.

QUIC exchanges information in only one round trip to set up a secure connection. QUIC also includes other changes to improve overall latency and

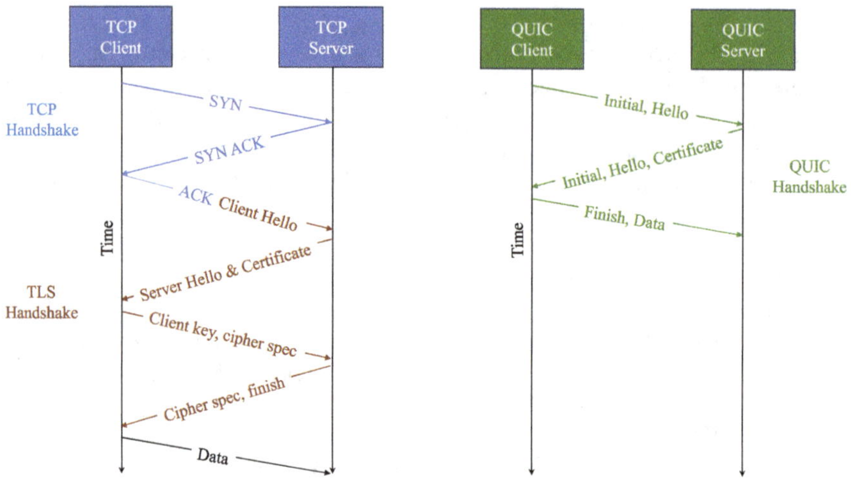

Figure 4.22 Comparison of TCP with QUIC handshakes

throughput. QUIC can also handle a network change, e.g. from a 5G connection to a Wi-Fi connection, by using a unique conlection identifier to the server regardless of the network connection. The IETF designed QUIC to be deployable and evolvable [73]. Google has driven QUIC deployment, and many Internet browsers are QUIC-capable, but QUIC still only makes up a small percentage of Internet traffic. As of February 2023, 8.7% of websites [74] use QUIC.

4.3.3.6 Packet switching delays

Network routers receive packets, determine their destination, and forward them along the appropriate link, which seems simple. However, in practice, routers may have to be aware of hundreds of thousands of destinations, receive and send packets on hundreds of links, and manage the queueing and prioritising of packets at speeds up to a million packets per second. Many assume that this complex task adds significant latency and that every router, therefore, adds significant latency, but engineers have highly optimised routers over many years; they use highly advanced specialised silicon chips, and as a result, a modern router can switch each packet in a few microseconds if not nanoseconds.

It is the impact of packet queueing and the time it takes to read in and send out a packet at the speed of the router's connections, the serialisation delay, that causes each router transversed to add latency.

4.4 Reliability

You have enough bandwidth and low enough latency for your essential applications, but is the network reliable enough?

Service providers deem the network service unavailable when it does not meet the communication requirements of the applications, e.g. packets are delivered too late. Measuring service availability as the amount of time it is unavailable can be deceptive because it depends on the time the service is required. For example, an industrial application may need to run non-stop all year, but remote surgery may only last a few hours; if both can tolerate a sub-second outage, the industrial application appears to be much more demanding. It seems wrong, however, that communications for remote surgery would be less available than communications for an industrial application; this is an illusion caused by measuring service availability as the time service is unavailable.

High availability is achieved by duplicating network equipment and ensuring duplicate paths across the network take separate paths. Designers should consider electric power feeds; in some cases, these need to be duplicated and take separate paths to the network devices that require power. The main impact of high availability is network costs. The technological impact is small in that protocols need to be implemented to detect failed equipment or paths and switch traffic to working equipment or paths. Unavailability of less than 5 min a year usually requires duplicated and diverse network paths. The fundamental choice at a network architectural level is whether to permit single points of failure or to duplicate or triplicate network equipment and paths. Network equipment duplication increases redundancy; having redundant equipment is good for availability but bad for economics. Duplicating cellular masts, which may fail due to weather events or attacks, is very expensive. It is impossible for a public network service to make redundancy design decisions per application.

The availability requirement depends upon the application components used in a scenario. For example, education or surgery uses tactile feedback, and the latter would require higher availability. System architecture decisions for particular scenarios can influence network requirements; for example should autonomous robots be designed to fail safely when network connectivity fails?

Packet loss should not be confused with service availability. A lost packet does not count towards service unavailability so long as the number of lost packets, measured over a specific period, does not exceed the required service level. Many applications are not impacted by lost packets because the operating system of the device the application uses resends lost packets, or the radio system detects and resends lost packets. A resent packet has higher latency and hence jitter – the variation in packet latency increases. Packet loss may not impact mission-critical applications like surgery if the lost packet is retransmitted within the latency target time. A latency-sensitive application like a VR headset may experience the loss of an entire video frame if a packet is lost, but that is unlikely to impact the user so long as packet loss is infrequent. In an industrial control scenario, packet loss may lead to an error in machine tooling, which could be unacceptable.

4.5 How to address the application requirements

This section further addresses future Internet application requirements, how Edge Cloud could reduce latency, and how edge caches could enable the Internet to handle Digital Terrestrial TV switch-off. It explains why 6G cannot depend on a quick evolution of Internet Protocols to address new requirements.

4.5.1 Edge cloud to reduce latency

The Cloud is a term many are familiar with, and we now use it knowingly to store our photos, perhaps unknowingly when we use smartphone apps. Cloud, or Central Cloud or Public Cloud, is a shared pool of configurable computing resources that the public can use on-demand via the Internet. By default, we assume the Cloud to be a central cloud located in a few (<1000) large (>50,000 servers) data centres. Well-known Public Cloud providers include Apple, Amazon Web Services, Microsoft Azure, Google Cloud Platform and Oracle Cloud. The main benefit of the Cloud is that it is an on-demand service; it can quickly scale up to deal with surges in demand, e.g. when a famous performer's tickets for a concert go on sale or when we had to work from home during the pandemic, and it can quickly scale down to save costs when not required. Edge Cloud is the concept of locating utility compute infrastructure close to the user to reduce latency.

The difference in broadband latency performance of DSL, fibre and cable technologies in the United States and the United Kingdom, shown in Table 4.5, illustrates the impact of distance. Most of the United Kingdom can achieve less than 20 ms broadband latencies for most services implemented in cloud datacentres located centrally, e.g. near London, but less than 20 ms is impossible in the United States due to its geographic size. The United Kingdom is about the same size as the state of Oregon, which is the USA's ninth largest state by area, which implies the United States requires many more distributed Edge Cloud sites to achieve less than 20 ms latency than the United Kingdom. For example, Microsoft Azure has 25 edge locations in the United States and two in the United Kingdom. Verizon is providing Mobile Edge Computing (MEC, a standardised version of Edge Cloud also known as Multi-Access Edge Computing, although in more recent press releases, the 'M' has been dropped) at 19 sites in the United States [75] and claims 75% of the US population is within 150 miles of an (M)EC site. For a wide area of public service, such as would be required for the Tactile Internet, the impact of MEC on 5G latencies is shown in [60] to reduce latency by ~15 ms. With a future Internet requirement of less than 20 ms latency, deploying Edge Cloud in these small numbers appears feasible. Appendix A gives the formula to calculate the theoretical number of sites required to meet a target latency for a public service area. For a private 5G or 6G service, typically to serve an industrial site, it is simple to consider an Edge Cloud on that site. The number of sites required for a public service is inversely proportional to the square of the target latency reduction, e.g. to half the latency requires quadrupling the number of sites. For example, for the United Kingdom, for a target RTT of 1 ms, 89 sites would be required, but a target of 0.1 ms RTT latency requires 8920 sites. For the United States, with a surface area of 9.83 million sq. km, to serve the 5G 1 ms low latency target would require at least, in theory, 960 sites (this is assuming no latency due to the 5G radio interface, which is of course not true), a 0.5 ms RTT latency would require at least 3841 sites.

To meet the 5G 1 ms latency target requires many Edge Cloud sites, which may not be commercially viable, although we should note that both fixed PSTN services and Cellular services require a minimum number of sites due to the physical characteristics of analogue voice signals in telephone cables or radio wave propagation. Hence, the legacy PSTN has ~5000 telephone exchange or central offices in the United Kingdom and 25,000 in the United States, more than enough sites to host Edge Cloud to meet a 5G 1 ms latency. Similarly, there are ~400,000 cellular masts in the United States, some of which may be suitable for hosting Edge Cloud servers, although cellular mast sites have limited space.

Extreme Edge Cloud is a term used to describe cloud computing hosted either on the end user's equipment (UE) or very close to the end user. The difference between a user's app running on a smartphone or on-site computer and Extreme Edge Cloud is that the latter is coordinated and managed as part of the more extensive cloud infrastructure.

In summary, achieving lower latency, even the 5G 1 ms target, by increasing the number of Public Edge Cloud or MEC sites to reduce propagation delay increases the complexity and decreases the efficiency of the overall system by inversely proportional to the square of the target reduction in latency, e.g. halving the latency quadruples the system complexity. The exact impact on efficiency and costs requires further research, but lower target latency requirements for public services must be fully justified.

Private Edge Cloud sites do not suffer the consequences of ubiquitous universal coverage as severely because the Industry 4.0 organisation can deploy Private Edge Cloud facilities at its existing locations. The number of Edge Cloud facilities required is limited to the number of factory or office sites it uses and not the size of the geography or population it serves. Installing and operating Private Edge Cloud facilities at many sites involves cost and complexity; therefore, the target latency requirements must still be justified.

4.5.2 Edge content caches to reduce core Internet traffic

We have seen earlier that digital terrestrial TV switch-off could triple the amount of traffic carried on the Internet. Most Internet video traffic is served from content caches today [76], so upgrading caching infrastructure seems inevitable to handle the extra traffic generated by terrestrial TV switch-off. Chapter 12 proposes that if linear TV over the Internet tripled the core bandwidth requirements, virtual CDN edge caches could reduce that to only a 30% increase in traffic on the Internet core.

4.5.3 What about Internet Protocol evolution?

Several Internet Protocols are decades old and were developed based on historical assumptions that do not apply to 5G or 6G technology. There have been many proposals to re-architect the Internet to make it more efficient and faster, but new Internet Protocols are slow to standardise and deploy. For example, IPv6, developed by the Internet Engineering Task Force (IETF) to address the problem of IPv4 address exhaustion, became a draft IETF standard in 1998 and was ratified as Internet

Standards in 2017. Google tracks the number of its clients using IPv6, which in 2010 was 0.25%, in 2017 20% and at the start of 2023 only 39% [77]. Despite 3GPP Release 8 specifications mandating IPv6 in 2009, the European Commission created an IPv6 task force in 2001 and invested tens of millions of Euros in IPv6 R&D [78]. The problem of Internet Protocol evolution is often referred to as 'ossification' and blamed on the difficulty of upgrading service providers' enterprise and residential Internet equipment. Deploying new Internet Protocols can be quicker if they do not impact the legacy Internet equipment. For example, QUIC, a replacement of TCP, a protocol used by Internet endpoints (user equipment) to transfer data, was initially implemented in 2012, became an IETF Internet Draft in 2015 and an agreed standards draft in 2021 (not yet a fully ratified Internet Standard). By February 2023, only 8.7% of websites [74] used QUIC. The historical speed of Internet Protocol evolution should set the expectations for 6G. In the time it took the mobile industry to go from 2G to 5G, the Internet has evolved 39% of its fundamental packet protocol traffic by one generation.

4.6 Summary and conclusions

6G researchers have produced spider diagrams showing how 6G should be ten times better than 5G in data rate, energy efficiency, spectral efficiency, latency, connection density and reliability. We must challenge the groupthink concept that 6G needs to be ten times better than 5G in performance and conduct more scientific research on the real 6G performance requirements.

The analysis described in this chapter shows that 5G's current URLLC 1 ms and 10 Gbit/s performance ambitions meet all predicted future Internet requirements. However, today's real-world 5G, Wi-Fi and fixed networks fall short of meeting these requirements in some niche situations, e.g. remote control of robots in dangerous environments, transmitting raw uncompressed UHD 360 degrees video for VR requires more than 10 Gbit/s and accurate or calibrated remote tactile feel requires latencies less than 5 ms. As a reflection of the latency requirements, in an interview with 150 enterprises in North America, only 3% wanted less than 10 ms latency.

With no predicted mainstream Internet application requiring less than 5 ms latency or more than 1 Gbit/s bandwidth, today's fixed access network technologies, FTTH and Cable, have the speed and latency to deliver the predicted future Internet requirements. Regions of more extensive geographic spread, e.g. larger than the state of Oregon or the United Kingdom, require Edge Cloud (MEC) to meet the low latency application requirements.

For mobile application requirements, measurements have shown that 5G in the United States requires Mobile Edge Cloud to deliver low enough latency. For example, in the Measuring Broadband America program, less than 1% of tests achieved the 20 ms or less required for the Tactile Internet and metaverse, while Spirent measured most latencies at less than 20 ms with MEC.

Edge Cloud standards should be integral to future 6G standards to drive adoption, implementation and critical mass. 6G Edge Clouds is an opportunity to offer more than basic Infrastructure as a Service (IaaS) but, like the Hyperscaler clouds, offer more value-added services such as Platform as a Service (PaaS) and Software as a Service (SaaS). Standardising these services for 6G will allow users to roam the globe while bringing sufficient scale and users to justify the investment.

Switching all DTT broadcast traffic to unicast streaming services could give another 400 MHz spectrum for 6G use but triple Internet traffic. Implementing content caches at Edge Cloud locations would remove most of this traffic from the Internet core.

Wi-Fi 6 does not have low enough latency in more challenging locations such as schools and busy offices, where latency exceeded 25 ms for downloads and 54 ms for uploads, but is adequate for domestic use, where latency was 9 ms for downloads and 28 ms for uploads. Wi-Fi 7 should have better latencies in all these scenarios.

5/6G does not operate in isolation, and some supplementary Internet Protocol methods, such as L4S and QUIC, can contribute to significant latency reduction, but the Internet has historically been slow in adopting new protocols.

Where robots are, in effect, providing a telepresence service in hazardous environments, wireless communications must provide low latency, high bandwidth, and high-reliability communications in potentially radio-hostile environments. Wireless connectivity must work reliably in the face of deep concrete and metallic shielding, electromagnetic interference and potentially deliberate jamming, which may be a challenge for any wireless technology.

Designers must take a holistic systems approach to design the overall Tactile Internet applications to function within the limitations of the communications technology. Data rates can be reduced by compression and focusing the fidelity where it is required. Designers should accommodate network failure by allowing fail-safe autonomy where safety is an issue.

Figure 4.23 summarises the future of the Internet and the iPhone 6G, as presented in this and Chapters 9 and 15.

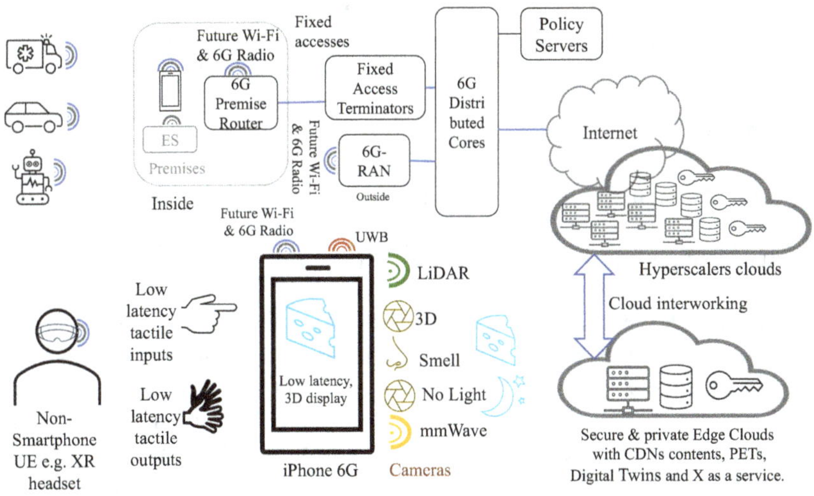

Figure 4.23 Future of the internet and the iPhone 6G

Appendix A Formula for number of MEC sites required

The theoretical number of MEC sites required is given by the surface area of the region to be serviced at the target latency divided by the area of the maximum size of the square that the target latency allows.

If the square is of length L, then its maximum latency is given by the distance H from the middle of the square to its corner.

$$H^2 = (L/2)^2 + (L/2)^2$$

$$H = L/2^{1/2}$$

For target latency, T

$$T \geq HR/C$$

where R = Route factor and C = speed of light in fibre. I usually use $R = 1.4$ and $C = 2 \times 10^8$ m/s.

$$L \leq TC2^{1/2}/2R$$

Given a region of area A, then the number of MEC sites required is

$$N \geq 2AR^2/T^2C^2$$

For the United Kingdom, with a surface area of 248,531 km^2, this requires 24 sites for a 1 ms latency. However, as can be seen from Figure 4.21, due to the shape of the United Kingdom, 91,100 km^2 are required to cover its total expanse, a surface area of 910,000 km^2. The above formula for servicing a 910,000 km^2 region at 1 ms latency gives 89. I recommend that area A is the rectangle size required to cover the region to be served rather than the exact land area.

References

[1] Spirent Communications Inc. Cutting Through the Edge Computing Hype. 2022.

[2] 3GPP. Study on Communication for Automation in Vertical Domains. 2020. Report No.: 3GPP TR 22.804.

[3] Parvez I, Rahmatic A, Guvenc I, Sarwat AI, and Dai H. A Survey on Low Latency Towards 5G: RAN, Core Network and Caching Solutions. *IEEE Communications Surveys & Tutorials*. 2018; 20(4): 3098–130.

[4] Kiesel R, Jakob F, Vollmer T, and Schmitt RH. Evaluation of ICT for Networked Control Systems of Latency-Critical Applications in Production. *Procedia CIRP*. 2022; 112: 238–44.

[5] Fanibhare V, Sarkar N, and Al-Anbuky A. A Survey of the Tactile Internet: Design Issues and Challenges, Applications, and Future Directions. *Electronics*. 2021; 10(17): 2171.

[6] Fu Y, Li C, Yu FR, Luan TH, and Zhang Y. Survey of Driving Safety with Sensing, Vehicular Communications, and Artificial Intelligence-Based Collision Avoidance. *IEEE Transactions on Intelligent Transportation Systems*. 2022; 23(7): 6142–63.

[7] Speedcheck. Speedcheck.org. [Online]. 2024 [cited 2024 January 23]. Available from: https://www.speedcheck.org/5g-index/.

[8] IEEE. IEEE Draft Standard for Tactile Internet: Application Scenarios, Definitions and Terminology, Architecture, Functions, and Technical Assumptions. 2016. Report No.: IEEE P1918.1.

[9] World Health Organisation. [Online]. 2022 [cited 2023 March 15]. Available from: https://www.who.int/news-room/fact-sheets/detail/road-traffic-injuries#:~:text=Key%20facts%201%20The%20United%20Nations%20General%20Assembly,a%20result%20of%20road%20traffic%20crashes.%20More%20items.

[10] Zeltzer D, Einav L, Rashba J, and Balicer RD. The Impact of Increased Access to Telemedicine. *National Bureau of Economic Research*. 2021 July.

[11] NHS.UK. [Online]. 2023 [cited 2023 March 20]. Available from: https://www.england.nhs.uk/virtual-wards/.

[12] NHS.UK. [Online]. 2023 [cited 2023 March 20]. Available from: https://www.england.nhs.uk/2023/03/world-leading-nhs-virtual-wards-treat-100000-patients-in-a-year/.

[13] Paterniani G, Sgreccia D, Davoli A, Guerzoni G, Di Viesti P, and Valenti Ac. Radar-Based Monitoring of Vital Signs: A Tutorial Overview. *Proceedings of the IEEE*. 2023; 111(3): 277–317.

[14] Mirbeik-Sabzevari A and Tavassolian N. Tumor Detection Using Millimeter-Wave Technology: Differentiating Between Benign Lesions and Cancer Tissues. *IEEE Microwave Magazine*. 2019; 20(8): 30–43.

[15] Cable Labs. [YouTube Video]. 2017 [cited 2023 May 5]. Available from: https://youtu.be/k0KIqRAoGBk.

[16] Kajimoto H and Hachisu T. Vibration Feedback Latency Affects Material Perception During Rod Tapping Interactions. *IEEE Transactions on Haptics*. 2017; 10(2): 288–95.

[17] Veluvolu KC and Ang WT. Estimation of Physiological Tremor from Accelerometers for Real-Time Applications. *Sensors*. 2011; 11: 3020–36.

[18] Zhou XU, Guo Y, Shen M, and Yang GZ. Application of Artificial Intelligence in Surgery. *Frontiers of Medicine*. 2020; 14: 412–30.

[19] F. G. Hamza-Lup. Feel the Pressure: E-learning Systems with Haptic Feedback. In *2008 Symposium on Haptic Interfaces for Virtual Environment and Teleoperator Systems*; 2008. p. 445–50.

[20] Perry TS. spectrum.ieee.org. [Online]. 2020 [cited 2023 March 20]. Available from: https://spectrum.ieee.org/ar-in-a-contact-lens-its-the-real-deal.

[21] Adelstein BD, Lee TG, and Ellis SR. Head Tracking Latency in Virtual Environments. In *Proceedings of the Human Factors and Ergonomics Society 47th Annual Meeting*; 2003. p. 2083–87.
[22] Jerald J and Whitton M. Relating Scene-Motion Thresholds to Latency Thresholds for. In *Proc IEEE Virtual Real Conf.* 2009; 2009.
[23] Britannica. Britannica.com. [Online]. [cited 2023 March 17]. Available from: https://www.britannica.com/science/time-perception/Perception-of-sequence-and-duration.
[24] Liu S, Claypool M, Kuwahara A, Sherman J, and Scovell J. Lower is Better? The Effects of Local Latencies on Competitive First-person Shooter Game Players. In *2021 CHI Conference on Human Factors in Computing Systems*; 2021. p. 1–12.
[25] Claypool M and Claypool K. Latency and Player Actions in Online Games. *Association for Computing Machinery*. 2006; 49(11): 40–5.
[26] Amin R, Jackson F, Gilbert JE, Martin J, and Shaw T. Assessing the Impact of Latency and Jitter on the Perceived Quality of Call of Duty Modern Warfare 2. *Lecture Notes in Computer Science*. 2013; 8006: 97–106.
[27] World Economic Forum. https://www.weforum.org. [Online]. 2019 [cited 2023 March 15]. Available from: https://www.weforum.org/agenda/2019/03/traffic-congestion-cost-the-us-economy-nearly-87-billion-in-2018/.
[28] World Health Organisation. https://www.who.int. [Online]. 2022 [cited 2023 March 15]. Available from: https://www.who.int/news-room/fact-sheets/detail/road-traffic-injuries.
[29] Matheson R. MIT News. [Online]. 2019 [cited 2023 March 15]. Available from: https://news.mit.edu/2019/how-fast-humans-react-car-hazards-0807.
[30] ITU. The Tactile Internet – ITU-T Technology Watch Report. ITU-T; 2014.
[31] ETSI. Intelligent Transport Systems (ITS); Vehicular Communications; Basic Set of Applications; Definitions. ETSI, ETSI Technical Committee Intelligent Transport System (ITS); 2009. Report No.: ETSI TR 102 638 V1.1.1 (2009-06).
[32] The CAMP Vehicle Safety Communications Consortium. Vehicle Safety Communications Project Task 3 Final Report Identify Intelligent Vehicle Safety Applications Enabled by DSRC. Washington, DC: US Department of Transportation (USDOT), National Highway Traffic Safety Administration; 2005. Report No.: DOT HS 809 859.
[33] 5G-PPP. 5G Automotive Vision. 5G infrastructure Public Private Partnership; 2015.
[34] Winter K. Businesswire.com. [Online]. 2017 [cited 2023 March 16]. Available from: https://www.businesswire.com/news/home/20170414005225/en/Intel-Editorial-For-Self-Driving-Cars-There%E2%80%99s-Big-Meaning-Behind-One-Big-Number-4-Terabytes
[35] Swaby M. BBC News. [Online]. 2023 [cited 2023 March 16]. Available from: https://www.bbc.co.uk/news/newsbeat-64862006

[36] Walsh JD. Why Facebook's Metaverse Is Dead on Arrival. [Online]. 2021 [cited 2023 March 16]. Available from: https://nymag.com/intelligencer/2021/11/why-facebooks-metaverse-is-dead-on-arrival.html

[37] Kunert P. Most Metaverse Business Projects Will be Dead by 2025. [Online]. 2022 [cited 2023 March 16]. Available from: https://www.theregister.com/2022/10/24/metaverse_dead_by_2025/

[38] McKinsey & Company. Value Creation in the Metaverse. McKinsey & Company; 2022

[39] Gartner. Gartner.com. [Online]. 2022 [cited 2023 March 16]. Available from: https://www.gartner.com/en/newsroom/press-releases/2022-02-07-gartner-predicts-25-percent-of-people-will-spend-at-least-one-hour-per-day-in-the-metaverse-by-2026.

[40] East Anglian Daily Times. eadt.co.uk. [Online]. 2019 [cited 2023 March 21]. Available from: https://www.eadt.co.uk/news/business/21449245.new-virtual-universe-suffolk-frontline-britains-war-cybercrime/.

[41] Blanche PA. Holography, and the Future of 3D Display. *Light: Advanced Manufacturing*. 2021; 2(4): 446–59.

[42] Huang L, Zhang S, and Zentgraf T. Metasurface Holography: From Fundamentals to Applications. *Nanophotonics*. 2019; 7(6): 1169–90.

[43] Koch K, McLean J, Segev R, *et al.* How Much the Eye Tells the Brain. *Current Biology*. 2006; 16(14): 1428–34.

[44] ScienceDaily. Penn Researchers Calculate How Much the Eye Tells the Brain. [Online]. 2006 [cited 2023 April 4]. Available from: https://www.sciencedaily.com/releases/2006/07/060726180933.htm.

[45] RED. red.com. [Online]. [cited 2023 March 21]. Available from: https://www.red.com/v-raptor.

[46] Sugito Y, Iwasaki S, Chida K, *et al.* Video Bit-Rate Requirements for 8K 120-Hz HEVC/H.265 Temporal Scalable Coding: Experimental Study Based on 8K Subjective Evaluations. *APSIPA Transactions on Signal and Information Processing*. 2020; 9: 1–17.

[47] Kortum P and Geisler WS. Implementation of a Foveated Image Coding System for Image Bandwidth Reduction. *Proceedings Volume 2657, Human Vision and Electronic Imaging*. 1996; 2657.

[48] Shure. shuer.com. [Online]. 2022 [cited 2023 March 22]. Available from: https://service.shure.com/s/article/600-mhz-auction-resource-info?language=en_US.

[49] BBC. BBC Media Centre. [Online]. 2022 [cited 2023 March 23]. Available from: https://www.bbc.co.uk/mediacentre/speeches/2022/tim-davie-director-general-royal-television-society/.

[50] UK Government – Department for Culture Media & Sport. Up Next – The Government's Vision for the Broadcasting Sector. UK Government, Department for Culture, Media & Sport; 2022.

[51] Ofcom. [Online]. 2022 [cited 2023 March 23]. Available from: https://www.ofcom.org.uk/media-use-and-attitudes/media-habits-adults/media-nations-2022.

[52] Nielsen. [Online]. 2022 [cited 2023 March 23]. Available from: https://www.nielsen.com/insights/2022/streaming-claims-largest-piece-of-tv-viewing-pie-in-july/.
[53] Barb. Barb Monthly Viewing Summary. [Online]. 2022 [cited 2023 March 24]. Available from: https://www.barb.co.uk/monthly-viewing-summary/.
[54] Openreach. UK Broadband Usage More than Doubled in 2020. [Online]. 2020 [cited 2023 March 24]. Available from: https://www.openreach.com/news/uk-broadband-usage-more-than-doubled-in-2020—driven-by-live-sport-online-gaming-and-home-working/.
[55] Sumits A. The History and Future of Internet Traffic. [Online]. 2015 [cited 2023 March 27]. Available from: https://blogs.cisco.com/sp/the-history-and-future-of-internet-traffic.
[56] Schwartz S. Prnewswire. [Online]. 2023 [cited 2023 March 22]. Available from: https://www.prnewswire.com/news-releases/sandvines-2023-global-internet-phenomena-report-shows-24-jump-in-video-traffic-with-netflix-volume-overtaking-youtube-301723445.html.
[57] Pocovi G, Kolding T, and Pedersen KI. On the Cost of Achieving Downlink Ultra-Reliable Low-Latency Communications in 5G Networks. *IEEE Access*. 2022; 10: 29506–13.
[58] Mody M, Swami P, and Shastry P. Ultra-Low Latency Video Codec for Video Conferencing. In *2014 IEEE International Conference on Electronics, Computing and Communication Technologies*; 2014; Bangalore. 1–5.
[59] Stabernack DB. H.264 Ultra Low Latency Video Codec. [Online]. [cited 2023 April 6]. Available from: https://www.hhi.fraunhofer.de/fileadmin/PDF/VCA/ESG/Flyer_ULLVC-2011-09_web.pdf.
[60] Spirent. Cutting Through the Edge Computing Hype. [Online]. 2022 [cited 2023 May 9]. Available from: https://www.spirent.com/assets/report-cutting-through-the-edge-computing-hype.
[61] Federal Communications Commission. Measuring Broadband America. [Online]. 2021 [cited 2023 April 12]. Available from: https://www.fcc.gov/general/measuring-broadband-america.
[62] Federal Communications Commission. Twelfth Measuring Broadband America Fixed Broadband Report. Federal Communications Commission, *Office of Engineering and Technology*; 2022.
[63] Federal Communications Commission. https://www.fcc.gov. [Online]. 2022 [cited 2023 April 12]. Available from: https://www.fcc.gov/reports-research/reports/measuring-broadband-america/measuring-broadband-america-mobile-data.
[64] Fogg I. 5G User Experience Report. [Online]. 2020 [cited 2023 April 12]. Available from: https://www.opensignal.com/reports/2020/06/usa/mobile-network-experience-5g.
[65] OFCOM. UK Home Broadband Performance. OFCOM; 2023.
[66] Vegt Rd. Reduced Latency Benefits of Wi-Fi 6 OFDMA. [Online]. 2021 [cited 2023 May 9]. Available from: https://www.wi-fi.org/beacon/rolf-de-vegt/reduced-latency-benefits-of-wi-fi-6-ofdma.

[67] Fettweis GP. http://www.5gsummit.org/. [Online]. [cited 2024 January 25]. Available from: http://www.5gsummit.org/docs/slides/Gerhard-Fettweis-5GSummit-SiliconValley-11162015.pdf.

[68] Briscoe B, Brunstrom A, Petlund A, Hayes D, Ros D, and Tsang IJ. Reducing Internet Latency: A Survey of Techniques and their Merits. *IEEE Communications Surveys & Tutorials*. 2014; 18(3): 2149–96.

[69] Giambene G. *Queuing Theory and Telecommunications*: Springer Cham; 2021.

[70] White G, Sundaresan K, and Briscoe B. *Low Latency DOCSIS: Technology. CableLabs*; 2019.

[71] Briscoe B, De Scheppe E, Bagnulo K, and White GM. *Low Latency, Low Loss, and Scalable Throughput (L4S) Internet Service: Architecture*. IETF, 2023. Available from: https://www.rfc-editor.org/rfc/rfc9330.html.

[72] Brunello D. *L4S in 5G Networks*. Stockholm: School of Electrical Engineering and Computer Science, Kth Royal Institute of Technology; 2020.

[73] Kühlewind M and Trammell B. Applicability of the QUIC Transport Protocol. [Online]. 2022 [cited 2023 February 14]. Available from: https://quicwg.org/ops-drafts/draft-ietf-quic-applicability.html.

[74] W3Techs. Usage Statistics of QUIC for Websites. [Online]. 2023 [cited 2023 February 14]. Available from: https://w3techs.com/technologies/details/ce-quic.

[75] Ashraf C. Verizon and AWS Expand Edge Computing to 19 U.S. Metro Areas. [Online]. 2022 [cited 2023 May 24]. Available from: https://www.verizon.com/about/news/verizon-aws-expand-edge-computing-19-us-metro-areas.

[76] Lynn T. *Managing Distributed Cloud Applications and Infrastructure*. Cham: Palgrave Macmillan; 2020.

[77] Google. Google IPv6. [Online]. 2023 [cited 2023 February 14]. Available from: https://www.google.com/intl/en/ipv6/statistics.html.

[78] Palet J. https://meetings.ripe.net/. [Online]. 2003 [cited 2023 February 14]. Available from: https://ripe45.ripe.net/presentations/ripe45-plenary-ipv6-strategy.pdf.

[79] Wikipedia. https://en.wikipedia.org/. [Online]. 2010 [cited 2024 January 25]. Available from: https://en.wikipedia.org/wiki/Ordnance_Survey_National_Grid#/media/File:British_National_Grid.svg.

[80] [Online]. [cited 2023 March 20]. Available from: https://www.marinspompiersdemarseille.com/pages/eos-appui-robotise.

[81] Leib R, Rubin I, and Nisky I. Force Feedback Delay Affects Perception of Stiffness but Not Action, and the Effect Depends on the Hand Used but Not on the Handedness. *Journal of Neurophysiology* 2018; 120(2): 781–94.

[82] Kaaresojaand T, Brewster S, and Lantz V. Towards the Temporally Perfect Virtual Button: Touch-Feedback Simultaneity and Perceived Quality in Mobile Touchscreen Press Interactions. *ACM Transactions on Applied Perception*. 2014; 11(2): 1–25.

Chapter 5

Opportunities for 6G in Industry 4.0 and roles for Wi-Fi

Technologies that are emerging today will soon be shaping the world tomorrow and well into the future – with impacts to economies and to society at large. Now that we are well into the Fourth Industrial Revolution, it's critical that we discuss and ensure that humanity is served by these new innovations so that we can continue to prosper.

– Mariette DiChristina, Editor-in-Chief of Scientific American, and chair of the IEEE Emerging Technologies Steering Committee

5.1 Introduction

Mass production has a significant role in shaping the modern world. It has driven economic growth and development, creating affordable goods that have increased our standard of living and spurred technological innovation. However, industry and society must address the challenges of a scarcity of resources, the impact of industry on the environment, the disruption of global supply chains and labour shortages. Industry 4.0 aims to address all these issues.

Industry 4.0 [1,2] is the fourth industrial revolution, integrating digital technologies, such as artificial intelligence, cloud computing, big data and the Internet of Things into the manufacturing and service sectors. Telecommunications is essential to Industry 4.0, and the industrial market sectors are essential for future MNO revenues; hence, 6G should have an Industry 4.0 focus. Industry 4.0 is more than factory automation or applying digital technologies in single organisations or locations, e.g. in a factory or an office. It is about using digital technologies across the supply chains across multiple organisations. Industry 4.0 includes AI processing data gathered from factory machines to predict when they may fail or need maintenance. It can track and optimise logistics and use digital twins that model and test optimised industrial processes using real-time data from the real world. The aim of Industry 4.0 is to improve flexibility, use of resources, costs, worker support and quality of industrial production and logistics.

Industry 4.0 should provide enough value to justify investments in 6G. McKinsey [3] reports that Industry 4.0 can improve manufacturing productivity by 15–30%. The global trade in manufactured goods in 2020 was US$12.1 trillion [4]; hence, Industry 4.0 productivity improvements could be worth at least approximately US$1.8

trillion per annum. Industry 4.0 would not be possible without telecommunications, and if 6G could claim only 2%[1] of the value of the global productivity improvements, this would be worth US$36B per annum. According to Gartner [5], global 5G infrastructure investment is running at US$23B for 2022. If 6G investments are similar to 5G investments, it can be concluded that 6G need only claim 2% of the value of Industry 4.0 global productivity improvements to make the business case that Industry 4.0 alone could justify the investment in 6G.

Critical features of Industry 4.0 are smart factories that use robots and autonomous systems to monitor and optimise the production processes and can collaborate with other machines and people across the supply chain. Cyber-physical systems combine physical systems, such as factory machines, with software and data to create a system that can adapt to changing conditions and user preferences. Digital twins are virtual representations of the physical world used to simulate, test and optimise systems. Augmented reality provides people with guidance and support for various production or maintenance tasks.

Industry 4.0 is not a product but a concept that combines multiple technologies. The architectural tenants of Industry 4.0 are interoperability between multiple organisations and technologies, modularity, digital twins and flexibility. Industry 4.0 makes use of several types of technologies applied all along the value chains:

- The Internet, data communications and the Cloud.
- Machine learning and artificial intelligence.
- XR for human–machine interactions.
- Robotics and automation.
- Advanced engineering, e.g. additive manufacturing, 3D printing, renewable energy and nanoparticles.

Industry 4.0 technologies are already widely used by manufacturers. The MPI Group survey [6] of 445 global manufacturers with revenues greater than US $10 million in 2021 reports that more than half have implemented strategies to apply Industry 4.0 technologies to their processes and embed Industry 4.0 technologies in their products. Manufacturers have incorporated smart devices into 45% of their production processes and equipment. Today, Industry 4.0 uses a wide range of network technologies, including but not limited to wired industrial Ethernet, wired field buses, Wi-Fi, cellular, industrial wireless, e.g. WirelessHART, private LTE, 5G, satellite networks, low-power wide-area network (LPWAN), e.g. LoRaWAN. In the MPI Group survey, many manufacturers reported that their network infrastructure could not support Industry 4.0. More than 13% required significant network upgrades or overhauls, while 40% required some network upgrades for Industry 4.0. Adoption by small and medium-sized enterprises (SME) varies as they lack the financial resources and skills to invest in Industry 4.0 [7].

5G was 3GPP's first attempt to provide a common communications framework optimised for Industry 4.0. 5G introduced Ultra-Reliable and Low Latency Communications (URLLC), necessary for the control of machines and industrial

[1]Analysys Mason [38] claim connectivity revenue accounts for 9% of the IoT value chain.

processes, complemented by local edge computing to reduce the latency and risks associated with centralised clouds, network slices and private 5G networks for the security of data and performance, and a 5G core network to allow the overall management and orchestration of the network. 3GPP designed 5G, in the Industry 4.0 context, to control robots cooperating with each other or human workers, factory process control, transport and logistics, and the Industrial Internet of Things (IIoT).

The most advanced features of 5G will come late in its generation cycle, and we will see 6G evolve to fix 5G for Industry 4.0. This chapter gives you a better understanding of what is required for 6G to optimise it for most Industry 4.0 cases at the lowest cost.

For 6G, Industry 4.0 is where operations technology (OT) meets information technology (IT) meets telecommunications. MNOs have already experienced the disruptive impact of IT on telecommunications in the form of Network Functions Virtualisation (NFV) and Software Defined Networks (SDN). OT brings different protocols and objectives to the 5G/6G table; notably, OT is focused on safety, uptime and maintenance of machinery.

This chapter focuses on what Industry 4.0 means for 6G. It summarises Industry 4.0 and other interesting use cases to determine their communications requirements and 6G expectations. It examines if the value of Industry 4.0 is enough to pay for 6G alone. A feature of Industry 4.0 is communications across the supply chain, emphasising the value of extranets. What is the role of the network in the corresponding supply chains created? Is there a role for Wi-Fi, and could this replace 5G? What are Operations Technology Architects' requirements and expectations from 6G when designing Industry 4.0 solutions?

5.2 Industry 4.0 use cases

This section explores Industry 4.0 use cases and describes them and their communications requirements. Table 5.1 shows a compilation of application performance requirements taken from a survey of [8–14] supplemented with analysis for this book explored in this chapter. The 3GPP's 'Study on Communication for Automation in Vertical Domains' [10] provides input as it details the essential requirements, including performance, of nine vertical domains. Table 5.1 is colour-coded to show where 5G can meet those requirements: Green indicates that 5G meets that requirement, amber indicates that in some situations, 5G cannot meet the requirement, and red indicates that 5G cannot. 5G meets most use case requirements, which raises questions about the need for 6G to perform better. The Motion Control use case, which is the only use case that drives 6G to have lower latency than 5G can offer, is examined in the next section.

We must consider other performance requirements for many Industry 4.0 use cases, including positioning accuracy, see Chapter 14, wireless range, velocity of mobile devices, how often devices need to send data, reliability and availability. The EU Hexa-X project [15], a leading 6G research project, examined these criteria, and Wymeersch [16] summarises them, which Table 5.2 partly reproduces.

Table 5.1 Summary of Industry 4.0 performance requirements

Category	Use case	Latency (ms)	Data rate (Mbit/s)
Mobile robots	Coop motion control	1	200
	Cooperative driving	10	100
	Real-time video streaming	10	18
	Video-operated remote control	10	1500
Factory	Motion control Assembly robots or milling machines	0.25 4	16 2
	Mobile cranes	12	1
	Process automation	>50	<5
	Process automation monitoring	50	<5
	Control to control comms	<10	<5
	Discrete automation	10	<5
	Video-assisted app	10	25
	Sensor	10	
	Video		10
	Voice		
Transport and logistics	Field sensor/instrumentation Emergency safety notification		0.02
	Asset tracking		0.02
	Time-critical sensing	<30	0.02
	Remote drone operation	10	50
	Real-time control for discrete automation	<1	16

Table 5.2 Use case requirements proposed by Hexa-X (after Wymeersch [16])

Use case family	Accuracy (m)	Range (m)	Velocity (km/h)	Latency (ms)	Update rate (Hz)	Reliability (per cent)
Massive twinning						
Manufacturing	0.1–0.5	<50	<5	<100	>100	99.999
Smart city	1–5	<200	<50	<1000	>1	99
Cooperative robots						
Transport	<0.2	<200	<100	<10	>100	99–99.9
Industrial	<0.01	<200	<30	<10	>10	99.999
Trust zones						
Health care	0.01	0.1–10	<1	<1000	>10	
Web of sensors	0.1–1	<200	<300	<10	>100	99
Public safety	1–5	<200	<50	<1000	>1	95–99

5.2.1 Factories of the future

Factories of the future embody the Industry 4.0 technologies described in the introduction. From a communications perspective, they are particularly challenging. Motion control of robots, machine tools, and production lines requires network latency, jitter, determinism, and availability orders of magnitude better than the Tactile Internet presented in Chapter 4. Safety and security are paramount, so private networks must be used. Radio systems must integrate into a predominantly wired environment, which uses specialist protocols and production facilities with lifespans exceeding 20 years. Radio systems must operate in adverse propagation environments, with interference from heavy-duty electric machines and arc welders. Any network downtime leading to interruption of production could cause significant financial damage.

It is necessary to understand the fundamentals of motion control to understand why it has strict requirements for low latency. Figure 5.1 shows a generic motion controller; examples of plants are machine tools and robots.

The motion controller is formed from multiple nested control loops. The outermost loop compares the required position obtained from the trajectory planner or motion controller to the actual position, and any error results in a signal to change the speed of the motor. The speed controller compares the actual speed to the required speed and adjusts the torque accordingly, and the innermost loop monitors the torque current and adjusts the driving voltage to achieve the desired torque. These controllers generate corrections based on the proportional difference between the desired and actual variables. For example, if the plant needs to operate to an accuracy of 1 mm, the controller does not wait until the error exceeds 1 mm, but it compensates as soon as any deviation is detected. An analogy would be to compare this system to driving a car. The Satnav provides the trajectory plan while the driver adjusts the speed and position of the car, and the engine management system adjusts the fuel required. The inner loops must react much quicker than the outer loops. In control theory, the speed of the loops is not specified in seconds but in the frequency of the loop's bandwidth. The higher the frequency, the quicker it reacts to errors. The frequency of these control loops is application-specific, but 10–20 Hz is generally sufficient for general-purpose applications, as it provides reasonable control without being too slow. The range of 100–200 Hz is better for applications that require faster response times, such as machine tools or robots.

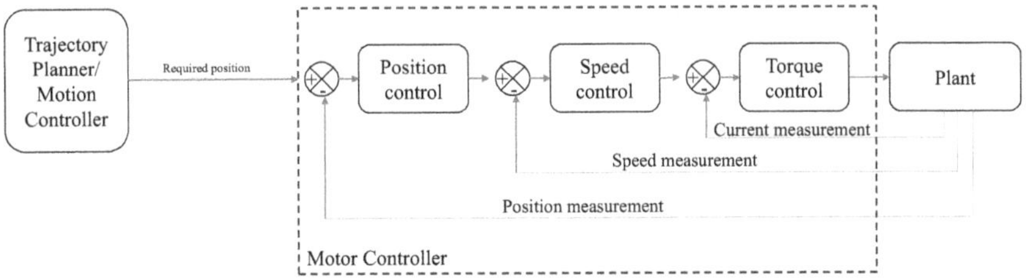

Figure 5.1 Generic motion and motor controller

While 1000–2000 Hz is best for applications that require swift response times, such as semiconductor manufacturing equipment.

Hundreds of motors, robots and machine tools may need to be synchronised in a factory. Unsynchronised robot or machine tool movements could lead to damaged products and, in the worst cases, damaged plants or injured workers. Synchronisation of motion is traditionally achieved with a powerful central Programmable Logic Controller (PLC), as shown in Figure 5.2.

The real-time network must deliver the required position or trajectory commands to all the motor controllers reliably and synchronised. Network latency in the order of 0.25–10 ms is required.

Figure 5.3 shows the trend to centralise the control loops driven by the ease with which new plants or axes[2] can be added and removed without requiring new motor controllers and improved precision through combining trajectory planning and motion control.

Trajectory planning can be optimised according to feedback from the control loops. This approach requires better than 5G ultra-low latency (<1 ms). Any network latency creates a phase shift in the control loop. Excessive phase shift makes the control loop unstable. The maximum phase shift allowed before the system becomes unstable is called phase margin. Typical servo loops operate with phase

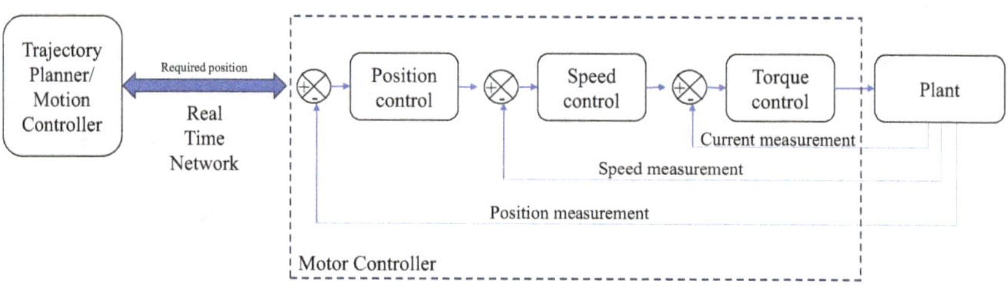

Figure 5.2 Generic motor controller using a centralised motion controller

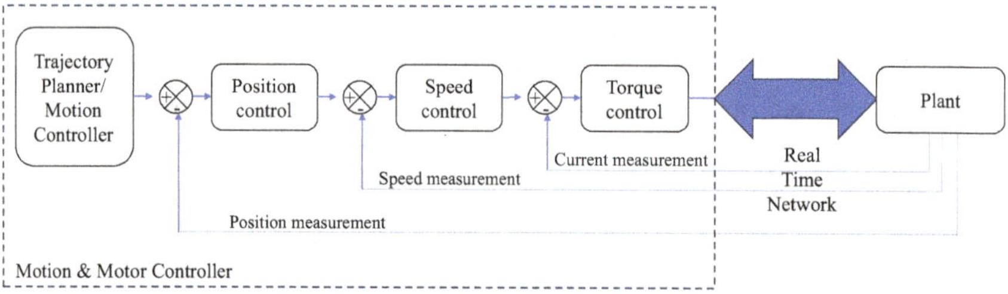

Figure 5.3 Generic centralised motion and motor controller

†Machines and robots have axes, plural of axis, about which tools, arms, wheels etc. rotate. Axes may be added or removed from machines as production jobs change which often requires motors to be added or removed to control the axes.

margins of 35°–45°, equating to approximately 5 ms at 20 Hz, 0.5 ms at 200 Hz and 50 μs at 2000 Hz. Network latency must be significantly less than the phase margin; hence, network latencies for the centralised motor controller must be 0.5 ms to 5μs, depending on the application.

The wired network technologies used in factories are called fieldbuses. The highest performance fieldbuses deliver packets with tens of microsecond latencies. Often, the round-trip time of fieldbuses is called cycle time because it works in a strictly cyclic and deterministic manner, such that during one cycle time, the motion controller sends data to all plants and receives updates back from the plants. An example of a high-performance fieldbus is EtherCAT, which is designed for cycle times less than 100 μs and achieves less than 30 μs in practice. Cables and wires provide simple and reliable industrial connectivity but are prone to failure due to corrosion, especially at connectors, ingress of dust and moisture into connectors, shorts or open circuits caused by wear and tear or accidents.

In food and beverage production lines, cables are subject to and obstruct regular hygienic steam cleaning of equipment. Cable installation needs careful planning, ducts and inspection boxes need installing, terminals installed to facilitate future modifications and maintenance. According to Cabling Installation & Maintenance [17], '*The cost of cabling is a significant part – from 5 per cent to 20 per cent – of the overall communications costs. Both new cabling installations and upgrades can run from $2 to $5 per square foot and, for technologically intensive applications, increase to $50 per square foot.*' VW's Wolfsburg plant is the world's largest automobile factory, which covers 70 million square feet. It could cost $140M to upgrade or more than $350M to cable from scratch. The high cost of installing and maintaining cables motivates Operational Technology architects to consider wireless connectivity, at least for monitoring, if not control, of machines.

Wireless communications are essential for autonomous guided vehicles (AGVs) that typically move goods around a factory. The performance requirements for AGVs depend upon the motion controller architecture adopted. If all motion control is on board the AGV, then communication requirements are significantly more relaxed than if all motion control is centralised. The former approach loads complexity and costs into the AGV, while the latter challenges existing wireless technologies. Smart conveyor belts, or linear track systems, perform asynchronous operations on goods while the conveyor belt moves at high speed. A robot shuttle may move on the conveyor belt along with the goods, which is only made possible by wireless communications and power. A significant motivation when upgrading a factory to Industry 4.0 is that adding new wireless sensors to equipment is much easier than adding wired sensors.

The HMS 2023 survey of the industrial communications market [18] shows wireless communications is the fastest-growing segment, rising by 22% from 7 to 8% of the market. Wi-Fi dominates wireless systems with 5% of the total market. Other wireless technologies, including 5G, account for less than 2% of the market. Industrial Ethernet accounts for 68% of the market.

Motion control latency requirements depend on the application, the controlled machines, the tolerance requirements of the produced goods and the OT architects' choices. All wireless technologies are severely challenged if the OT architectural target is a wireless modular factory with centralised motion control for optimum high-precision tool performance.

An exciting application of wireless communications for ultra-low latency is replacing wired slip rings with Wi-Fi, the established protocol for high-speed low latency communications in factories. Robot arms and similar slip rings are required at rotational joints, e.g. wrists, to enable communication with devices and controllers beyond the rotational joint. Mechanical slip rings are prone to failure due to dust, wear and tear. Contactless slip rings are available [19] using wireless Ethernet connections across the slip ring gap at up to 1 Gbit/s, compatible with EtherCAT. mmWave (60 GHz unlicensed ISM band) solutions are also available [20], achieving 2 μs latency, which would be necessary if multiple rotating joints on a robot need to connect in series.

OT architects are aiming for modular factories, which can be quickly and easily reconfigured, so they are motivated to replace wired fieldbuses with wireless fieldbuses for faster reconfiguration of factories without the expense and time of removing and installing cables. However, power over Ethernet is also a technology that can reduce the amount of cabling required. Wireless solutions also require configuration and potential reconfiguration when the factory is reconfigured. Placement of Wi-Fi access points (APs) may have to be moved due to changes in radio propagation due to changes in the location of sizeable metallic equipment, potentially requiring a new site survey; even a private 5G network may need its software configuration changed or reconfiguration of antenna sectors by an expert. Introducing AI to automate 5G operations should help automate private 5G network reconfiguration.

If the factories of the future are to use 5G or 6G, one primary consideration is that there are few large factories. According to Eurostat [21], there are 15,724 manufacturing enterprises with more than 250 employees across the European Union's 27 countries. That's an average of 521 large manufacturing enterprises per country. This type of market is far from the MNO volume operations for consumer services and requires a complex systems [22] approach that systems integrators specialise in.

5.2.2 Logistics

Logistics is crucial to modern life. It ships goods and materials between producers and consumers, individuals and businesses, drives economic growth through its efficiencies and improves the quality of life by making essential goods and services accessible. The continued digitisation of logistics plays a vital role in improving its efficiency, and logistics' distributed and global nature requires global wireless communications.

Logistics is mostly about monitoring and tracking assets, which includes everything that goes in or out of an organisation, materials and products across the supply chain, including all the vehicles and containers used to carry them but also workers, contractors, laptops, smartphones and other tools.

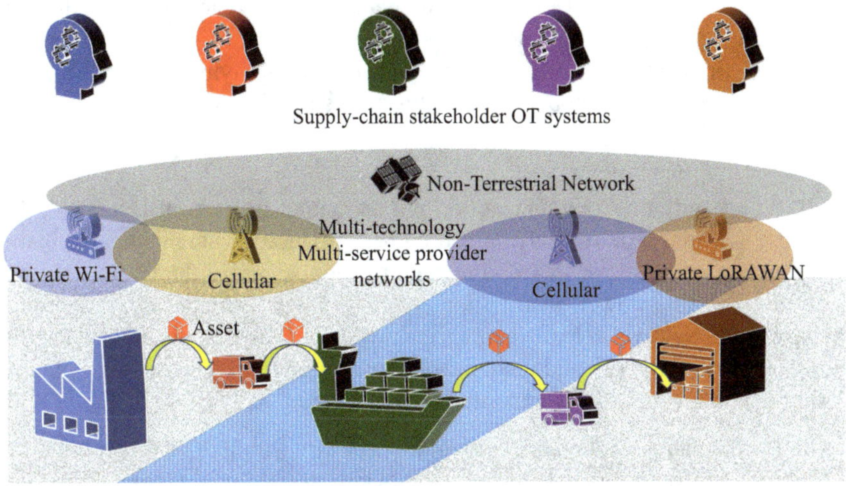

Figure 5.4 Industry 4.0 6G logistics challenge

Not all Industry 4.0 communications occur inside factories or buildings; logistics, where goods are shipped between the various companies in a supply chain, requires connectivity to a public network for real-time tracking, as shown in Figure 5.4. When the goods arrive at a factory, the goods' trackers may need to connect to the factory's private network for more accurate position tracking and automated incorporation into the inventory management system. Where public networks overlap with a private network, how are the IoT devices made to scan for a private network when they arrive on-site?

The 3GPP's 5G Access Network Discovery and Selection Function (ANDSF) is used to install network selection policies in UEs. However, the challenge is for the UE or IoT device associated with the goods to connect securely to the correct network, especially when it may visit multiple, potentially international locations, not all of which may be known at the start of its lifetime. Mutual trust between the asset and the private network is an issue. A malicious asset could be used as a trojan horse or a fake private network to make assets 'disappear'. The asset's owner may change as it moves through the supply chain. Does each owner negotiate multiple roaming agreements and set up interconnects with MNOs and the other organisations they need to connect to? Does the asset get reconfigured at every step in the supply chain, or is it a service provided by a third party globally that everyone in the supply chain agrees to use?

The efficiency of the logistics industry, transporting goods, could be significantly improved by applying technology. For example, 30% of heavy goods vehicles (HGV) miles travelled in the United Kingdom were empty [23]. In the European Union, it is 20% [24] and 15% in the United States [25]. Silva [26] claims a digital freight platform can reduce empty runs by 30%, saving 10% of haulage costs. However, sceptics point out [27] that it is a challenge for digital systems to allow for all the variables that can happen on a journey, e.g. accidents, roadworks, waiting times to unload due to adverse weather conditions or goods receivers not being ready. In the author's opinion, this area is ripe for applying AI and data about road traffic, weather and supply chain status across multiple organisations.

5.2.3 Other use cases

The 3GPP's 'Study on Communication for Automation in Vertical Domains' [10] details the essential requirements, including performance, of nine vertical domains for 5G: railbound mass transit, building automation, factory of the future, eHealth, smart city, electric-power distribution, central power generation, programme making and special events and smart agriculture, in a 196-page document. The list below summarises the key and unique requirements from these use cases, potential missing requirements and 6G opportunities:

- Railbound mass people transport requirements are within 5G's capabilities, i.e. lowest latency <100 ms, highest bandwidth 100 Mbit/s except for CCTV offload or archiving, which requires <1000 Mbit/s (an information fountain at train stations).
- Building automation requirements are not demanding. It does not mention that fire detection and suppression signalling should operate without an electricity supply in an emergency.
- For Smart Cities, the 3GPP study focuses on CCTV distribution as part of a Smart City. The 6G opportunity would be to use joint communications and sensing (JCAS), see Chapter 14, to give the Smart city valuable data, e.g. about how many people are where. mmWave sensing may be more beneficial than CCTV for some applications.
- For programme making, it looks at live event scenarios that require robust wireless communications with a mix of private and public networks. End-to-end latency for feedback to live performers (microphone to in-ear monitor) must be less than 4 ms, requiring the wireless system to have end-to-end latency <1 ms. Uncompressed UHD TV signals have very high bandwidth, up to 61 Gbit/s for digital cinema, but the 3GPP study assumes compression to <200 Mbit/s. Multiple uncompressed UHD TV feeds would exceed the capacity of 5G systems.
- For smart farming. The 3GPP study looks at the use cases of the control of an automated irrigation system containing valves and soil moisture, humidity and temperature sensors. And protection against animal poaching. The driving requirement identified is high-density connections (1 million per square kilometre). There is no mention of network coverage challenges in rural areas where non-terrestrial networks, see Chapter 10, could have a role. For example, John Deere, see DeGrasse [28], uses SpaceX's Starlink service to supplement LTE coverage on its autonomous farm machinery. The service provides connectivity to a cloud platform application that supplies farmers with a real-time view of their equipment, reports on their soil health and recommendations to optimise the use of fertiliser and other resources.

5.3 Industrial IoT

Sensors enable Industry 4.0 by measuring the many critical parameters of industrial or agricultural processes. The term IIoT describes networking sensors, instruments, tools, machines and other devices with computers for industrial applications. The definition of IIoT is wide-ranging and can encompass all of the telecommunications, cloud computing and data processing required for Industry 4.0, in which IIoT is synonymous with Industry 4.0 ICT infrastructure. This section focuses on the requirements of wireless sensors for IIoT.

IIoT devices must be more precise, robust in harsh environments, reliable and secure than consumer IoT devices, and usually connect to private networks rather than the Internet.[3] IIoT uses a wide range of wireless systems, including, but not limited to, 3GPP cellular solutions, licenced, unlicensed and shared spectrum such as CBRS, Wi-Fi, LoRaWAN, Zigbee, WirelessHART (IEEE 802.15.4), Wi-Fi HaLow (IEEE 802.11ah) and 3GPP's suite of NB-IoT, LTE-M EC-GSM[4] or 5 G's NR-Light. The choice of wireless system depends on the specific applications' requirements for range, data rate, data volumes, latency, reliability, security and power consumption. This variety of requirements has led to many wireless solutions as each is optimised for specific use cases and cellular IoT cannot generally compete with the wide range of non-cellular IoT systems. 6G does not reduce or simplify the options but may add more. While the total number of IoT devices are predicted to increase by 162% by 2028 [29], the cellular IoT market share is expected to fall with the 2G/3G switch-off. From Ericsson data, the cellular market share will drop from 26% of the IoT base in 2022 (2.7 billion out of 13.2 billion devices) to close to 19% in 2028 (5.4 billion out of 28.7 billion).

Table 5.3 compares a selection of IIoT wireless technologies ordered by the highest data rate first. The values in Table 5.3 are indicative as there are many variables in establishing these parameters. Data rate generally reduces with increasing range, and latency increases with increasing number of connected devices. Maximum ranges and maximum data rates will not be achieved simultaneously.

Low data rate (tens of kbit/s) and long-range devices (several km) could use LoRa, with a range of 5 km in urban environments and 10 km when in rural line of sight, with a 22 kbit/s maximum data rate falling to 250 bit/s at maximum range with latencies of several seconds. LoRaWAN uses spread spectrum modulation and unlicensed spectrum. Wi-Fi 6/7 is suitable for short-range high-bit-rate devices that are not power constrained, while Wi-Fi HaLow uses the 900 MHz licence except frequency and is intended for IoT devices with low energy consumption, a one-kilometre range and high density. Wi-Fi HaLow has data rates from over 15 Mbit/s to 150 kbit/s. Wi-Fi HaLow can support 8192 devices per AP but may be limited to a few thousand in practice. Zigbee and Bluetooth are suitable for short-range, low-bit-rate devices. Zigbee in the European Union can use 868–868.6 MHz so is ideal

[3]The Internet in IIoT is perhaps a misnomer, more accurately it should be the Industrial Networks of Things.

[4]With the planned closures of 3G EC-GSM will no longer be available.

Table 5.3 Comparison of some potential IIoT wireless technologies

Technology	Spectrum (GHz)	Bandwidth (MHz)	Data rate (Mbit/s)	Max range metres	Approx latency ms	Max number devices
Wi-Fi 7	ISM 2.4, 5, 6	320	46,000	120	<5	<255
Wi-Fi HaLow	ISM 0.9, 2.4	1	15	3000	5000	<8192
NR-RedCap	Licenced 5G	20	10		250	
Weightless	TV white space	8	16	10,000		
LTE-M	Licenced LTE 0.7–0.9	1.4	0.6	15,000	<15	<900 K/km^2
WirelessHART	ISM 2.4	3	0.25	250	2000	250
Zigbee	ISM 0.9, 2.4	2	0.25	100	100	<15 M
NB-IoT	Licenced LTE	0.2	<0.25	10,000	5000	<1 M/km^2
Bluetooth	ISM 2.4	1	0.25	100	34	7
Z-wave	ISM 0.9	0.2	0.1	100	30	232
Ingenu	ISM 2.4	1	0.03	48,000	10,000	
LoRaWAN	ISM 0.9, 0.16,0.43	0.5	0.027	10,000	10,000	<10,000
SIGFOX	ISM 0.9	0.2	0.001	50,000	>1000	

Notes:
1 IEEE 802.15.4 is the basis for Zigbee, ISA100.11a, WirelessHART, MiWi, 6LoWPAN, Thread, Matter and SNAP.
2 LoRaWAN is a protocol built on the top of LoRa wireless technology.

to use for accessing devices where Wi-Fi or BlueTooth cannot reach, e.g. the smart meter in the basement. 3GPP offers a range of cellular solutions tailored for IoT.

Before Release 17, 3GPP provided three different cellular protocols. NB-IoT is a subset of LTE which offers <180 kbit/s throughput, 10–15 km range from the cell site, high latency (10 s) using approximately 200 kHz of licenced spectrum and does not support handovers if the IoT device is mobile. LTE-M, or eMTC, is a modified version of LTE offering <1 Mbit/s throughput, 100 ms latency. EC-GSM, similar to LTE-M but for GSM networks, <240 kbit/s with 2 s latency. From 3GPP Release 16, NB-IoT can co-exist with 5G NR and will continue to evolve with 5G.

3GPP's massive machine-type communication use case demands long battery life, aiming for 10 years, for low data rate devices up to 15 km from cell sites and anticipates up to one million devices per square kilometre. Ericsson's evaluation of NB-IoT [30] shows it can support more than one million devices per square kilometre, with a three-second latency, and its performance drops off significantly at 1.2 million devices per square kilometre. In this evaluation, NB-IoT achieved an approximately 0.75 s latency at a low number of devices per square kilometre.

Wireless IoT technologies using an unlicensed spectrum must trade off increasing density with increasing latency. Bankov *et al.*'s [31] research shows LoRaWAN, with a high number of devices, approximately 5000, is capacity limited to two messages per day per device, making it unsuitable for many high-density applications. Bilgin and Gungor [32] concludes that ZigBee can only be used for low data rates and low-power applications without high-reliability requirements or real-time deadlines. Muller's study of WirelessHART shows it supports 50–100 devices with a greater than two-second latency, stating the number of devices is limited by bandwidth requirements and the possible joins of new devices. The HART communication foundation claims WirelessHART has reliability better than 99.73%.

3GPP Release 17 introduced Reduced Capability New Radio (RedCap NR) or NR-Light to address 'mid-tier IoT' devices requiring more bandwidth, up to 150 Mbit/s, and lower latency than NB-IoT. RedCap will continue to evolve with future 3GPP releases.

According to Vodafone [33], '*in the United Kingdom, there are 4,500*[5] *miles of roads with no cellular coverage and 29,000 miles of roads with partial coverage only*'. Iridium [34] claims that only 15% of the Earth has cellular coverage.[6] NTN-IoT or satellite IoT can connect IoT devices in areas without cellular coverage and out of range of other low-power wide-area network (LPWAN) technologies. 3GPP has completed a study of enhancements to NB-IoT for NTN use, and it is expected to be standardised in Release 18 (5G Advanced, due for completion in June 2024).

For any organisation building its own IIoT solution, there is a rich plethora of wireless technologies to choose from. The challenge is matching the applications' requirements to the wireless technology performance. Key factors to consider are range, latency, data rate, number of devices, cost, and battery life. A single organisation may require multiple wireless technologies. Most IIoT solutions require gateways to relay messages between IoT wireless systems and the organisation's wired private networks. Further, given that IIoT standards are still evolving, application-level gateways are required to convert IoT device data into a standard format. 3GPP's 4G NB-IoT, 5G RedCap NR and the future 5G Advanced NTN-IoT protocols will have a role to play but will not provide a complete IIoT solution.

6G could address the complexity of building IIoT solutions by filling the performance gaps in the 3GPP portfolio of IoT services to offer a single, easy-to-use 3GPP solution that works everywhere for most IIoT applications with a single interface standard and interworks with non-3GPP IIoT wireless networks. 6G should not be able to ignore the role of Wi-Fi for IIoT and should address Wi-Fi – 3GPP IoT

[5]262,000 miles of roads in the UK hence 1.7% do not have cellular coverage.

[6]Approximately 70% of the Earth's surface is water therefore half of all land does not have cellular coverage.

convergence. There will be many opportunities for system integrators and MNOs to offer IIoT-integrated services, but there is a need to understand specific market verticals' requirements.

6G's JCAS, see Chapter 14, complements the IIoT sensors and is a significant differentiator from 5G, enabling new sensing services that MNOs can sell to the industry.

5.4 Digital twins

Digital twins can be used to optimise products or systems, test new designs and processes, perform 'what if' analysis and predict the future state of systems. It is, therefore, central to Industry 4.0. The generation of measurements or data from the real world, often using IoT technologies, and communicating that to the digital twin is essential; hence, digital twins use the IIoT communications technologies.

The first international standard [35] definition of a digital twin in 2019 was '*digital asset on which services can be performed that provide value to an organisation*'. A digital asset was defined as '*data set describing an asset that is not necessarily physical*'. The more useful Wikipedia definition is '*a digital representation of an intended or actual real-world physical product, system, or process (a physical twin) that serves as the effectively indistinguishable digital counterpart of it for practical purposes, such as simulation, integration, testing, monitoring, and maintenance*'. A digital twin is more than a simulation or emulation of a system as it is updated with data, sometimes in real-time, from the real-world system. The defining characteristic of a digital twin is that a change in the physical system automatically leads to a change in the digital twin. A change in the digital twin could also be reflected as a change in the physical system, as shown in Figure 5.5.

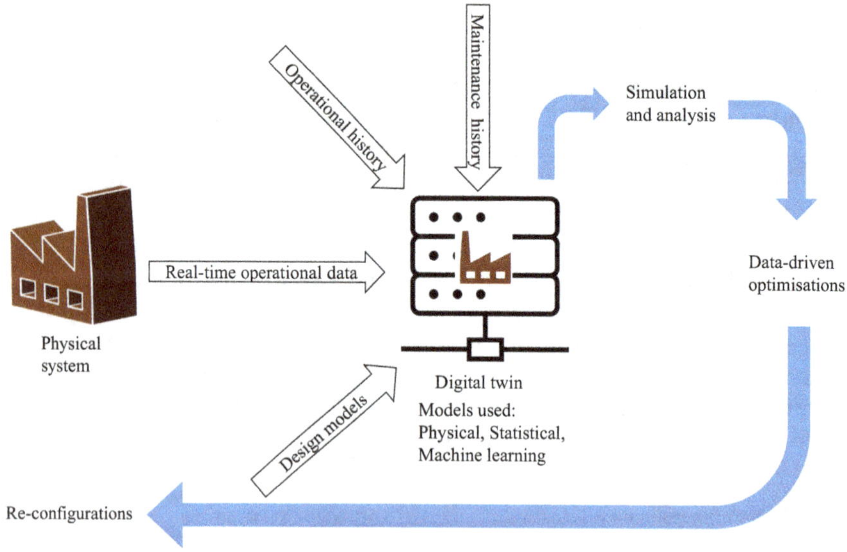

Figure 5.5 Digital twin overview

A digital twin is not a product that can be purchased 'off-the-shelf' but a system that the enterprise, or industry vertical expert, needs to define, specify, build and test. It involves the integration of many components.

Examining digital twins' challenges may reveal more value-added roles 6G could take in digital twins beyond simple wireless communications. Digital twin standards are under development in ISO, CEN, ITU and others to address the significant hurdle of lack of standardisation to the widespread adoption of digital twins. Without standards, each organisation building a digital twin would have to integrate data from many sources, each source using a proprietary approach. The methods and standards used to build digital twins are vertical-specific; for example, a smart city digital twin would use different techniques and standards than an automobile factory. The preferred telecommunications and wireless technologies are vertical-specific. There are no generic plug-and-play digital twin solutions; therefore, system integrators have a prominent role in building digital twins for large organisations, leaving scope for MNOs to partner with those developing SME digital twin products and services.

Key technology challenges impacting 6G are associated with collecting massive amounts of data from many distributed sources. The data must be securely transmitted and processed. Data integrity is vital as it indirectly drives physical systems via the digital twin. Incorrect data could lead to incorrect reconfiguration of the physical system, at best resulting in sub-optimal performance, at worse, damaging equipment or putting lives at risk. An IIoT system using low-cost wireless systems using unlicensed bandwidth to gather data that drives a digital twin that influences the organisation in real-time is at risk of critical failures. Its criticality may favour ultra-reliable 6G systems using licenced spectrum against lower-cost unlicensed spectrum systems, especially in the wide area.

The digital twin system, including its communications, must be dimensioned and scaled to handle peak data loads. Mechanisms must be in place to prevent data overload, which could swamp the network or the digital twin system. An Edge Cloud service would be well placed to manage digital twin data near its source, and a large-scale 6G network infrastructure would be better placed to handle the data that could overload SMEs' facilities.

Digital twins will be a valuable asset to many organisations, and MNOs could have a role in providing the ultra-reliable and secure processing of the IoT data that drives digital twins, exploiting partnerships with vertical solution providers for SMEs and leveraging MNOs' ultra-reliable radio and Edge Cloud infrastructures.

5.5 Private 6G networks

In a 2021 Deloitte global survey, 69% of networking executives regarded 5G as the most critical wireless technology for their business initiatives, and 58% Wi-Fi 6. That demonstrates a significant amount of interest and trust in the evolving private 5G networks technology, but are there any weaknesses in Private 5G networks that 6G addresses to satisfy Industry 4.0 requirements? This section analyses the

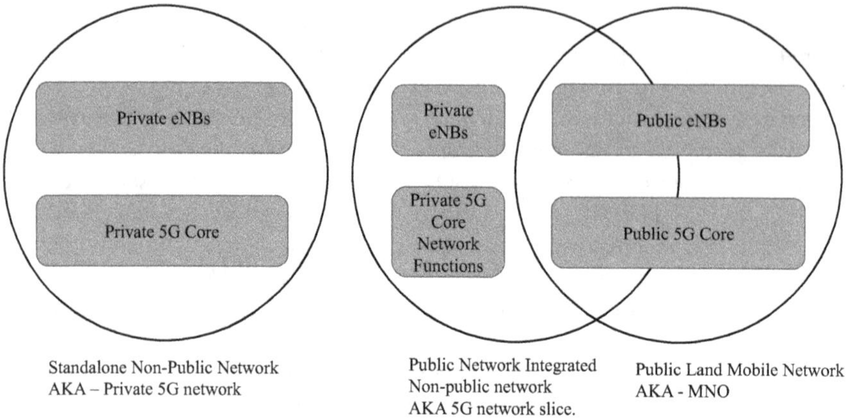

Figure 5.6 5G private versus public networks

challenges private 5G networks face to identify the issues 6G private network standards should address.

3GPP defines standalone non-public networks (SNPN) and, confusingly named, public network integrated non-public networks (PNI-NPN). An SNPN does not use network functions provided by a public land mobile network (PLMN) or MNO, while a PNI-NPN uses network functions provided by a PLMN. Figure 5.6 illustrates the relationship between these terms.

As usual, there are various options regarding which network functions are provided by the PLMN and which are standalone provided by the private organisation, including the PLMN using virtualisation techniques to provide private slices. Neutral hosts, see Chapter 2, can also provide private network facilities. 3GPP does not include non-3GPP network technologies in considering NPNs until Release 18. This book section loosely uses 'Private 5G Networks' to encompass all 3GPP's NPN types.

A significant step to making 5G Advanced more usable for Industry 4.0 is 3GPP Release 18, 5G Advanced, which supports non-3GPP access, e.g. Wi-Fi using N3IWF and TNGF (Chapter 11). Release 18 also supports mobility between Standalone NPNs without new network selection. It studies cooperation between 3GPP and non-3GPP management systems. It also addresses the security implications of mobility between Standalone NPNs and the support of non-3GPP technologies. Some of the logistics use case issues of managing connected assets roaming between multiple public and private networks will also be addressed in 3GPP Release 18.

Where a private network uses PNI-NPNs, i.e. a 5G network slice, the MNO must guarantee service levels and meet the security requirements of the private organisation. PNI-NPNs may require security certifications specific to the enterprise customer. These requirements are no different from the fixed VPN services many telcos offer, but the devil is in the detail of how MNOs are operated. In telcos that offer fixed VPN services and mobile/cellular services, the MNO and VPN operations are at arm's length with no overlap of systems or people. Most telcos offering fixed VPN services treat 5G as an access technology for backup, rapid deployment or temporary locations. Merging or hybridising the MNO and VPN operations is an involved undertaking. The MNO cannot sell Industry 4.0 solutions alone nor sell 5G or 6G as a separate part of an Industry 4.0 solution without partnering with system integrators.

Private networks are required for security, but conversely, a key feature of Industry 4.0 is that it needs connectivity across supply chains between multiple organisations. There must be interoperability between multiple private networks using different technologies at private sites and global public networks. Such arrangements are called extranets. The Automotive Network Exchange (ANX) is an example of an extranet managed by a third party to benefit USA automobile manufacturers. Some multinational companies run their own extranets to collaborate with their suppliers. Extranets are not a new idea; the ANX was built in 1995, but Industry 4.0 increases the geographic range, number of organisations and amount of data with IIoT. Extranets of old covered common standards for electronic document exchange, but for Industry 4.0, common standards are required for AI processing and digital twins. Extranets are commonly implemented using two technologies: either fixed MPLS VPN services or over-the-top secure tunnels, e.g. IPsec. Software-defined wide area network (SD-WAN) technologies use secure tunnels to build VPNs over-the-top of broadband and Internet services, and some SD-WAN vendors include the ability to automate extranet configuration. Extranets impose the following extra requirements on private networks:

- Increased configuration complexity.
- Handling overlapping private IP addresses.
- Multiple networks requiring common security and QoS policies.
- Supporting multiple vendors' products.

5G private network standards do not address extranets; they must be implemented as an over-the-top service that a third party provides.

An interesting example of the value of the 'over the top' part of extranet services is the ANX, which OpenText acquired for US$100 million in 2016.

If Private 5G networks become very popular, there is an issue with insufficient Mobile Network Codes (MNC), which are only three digits per country, to give every Private 5G network a unique identifier. There is a private mobile country code (999) to give a private identifier space, like private IPv4 addresses, but if a device must access multiple private 5G networks, there is no guarantee that a unique mobile network code would have been used for each network. Devices on Private 5G networks using the 999 mobile country code cannot roam on to public MNO networks (PLMNs) using the 999 MCC which may be limiting for logistics use cases. Those familiar with IPv4 private addresses can see the similarity of this situation with private IP networks starting 192.168. The lack of MNCs could become a similar problem to the exhaustion of IPv4 address space and the long and painful transition to IPv6.

Private 6G network standards should address:

1. Multiple wireless network technologies, especially Wi-Fi.[7]
2. Managing wireless connected assets roaming between multiple public and private networks.[7]
3. Extranets.
4. Globally unique identifiers to allow millions of private 6G networks per country.

Network operators must merge or hybridise their fixed network VPN and 6G private network operations to leverage the marketing and support of fixed VPNs into 6G private networks and offer seamless services between fixed and mobile VPNs. Network operators must be open to offering extranet services in cooperation with their competitors.

5.6 Wi-Fi versus cellular for Industry 4.0

There are roles for both private cellular networks and Wi-Fi in Industry 4.0, and the best choice depends on the location and use cases. Table 5.4 compares the main features of Wi-Fi and private 5G networks. The main driver of choice is the size of the organisation and site. A small SME site is usually best served by Wi-Fi, while private cellular networks best serve a sizeable multinational factory for critical operations supplemented by Wi-Fi for non-critical operations.

Enterprises might consider Wi-Fi for in-building Industry 4.0 applications, complemented by Wi-Fi HaLow for medium-range IIoT applications outside buildings, due to Ethernet's legacy installed base, existing availability, ease of use for smaller sites, high-performance potential and ease of upgrading to future Wi-Fi standards.

However, for larger enterprises, private 5G networks using cellular solutions look better. 5G Base stations have a wider coverage due to higher power and increased sensitivity of receivers. A single base station can also support more devices than a single Wi-Fi AP. Fewer base stations reduce costs and factory reconfiguration issues. Cellular has seamless handovers between base stations, inside and outside the factory and with public networks, easing integration with logistics use cases. Cellular's licenced or shared spectrum is more reliable than Wi-Fi's unlicensed spectrum.

A critical application of in-building Industry 4.0 wireless is controlling moving machines, e.g. AGVs. However, mobility support has not been addressed in Wi-Fi standards to date. Many Wi-Fi AP are required to cover a large factory. As a UE, e.g. an AGV, moves about the factory, it may need to switch between APs. However, a successful handover between APs takes approximately 0.5 s, which creates 'Wi-Fi blind spots' in the factory. 'Sticky AP' is also a common problem where the UE's connection quality must fall significantly before connecting to the better AP. These are not issues for cellular networks.

Even for non-mobile, fixed UEs, if an AP fails, then, dependent on the way the AP failed, it takes approximately 0.5 s before the UE connects to a working AP. In some

[7]Anticipated to be addressed in 5G Advanced Release 18.

Table 5.4 Comparison of Private 5G and Wi-Fi technologies for Industry 4.0

Feature	Private 5G network	Wi-Fi 7
Spectrum type	Licenced, shared or unlicenced	Unlicenced ISM
Spectrum bandwidth	100 MHz/channel	320 MHz/channel
Maximum transmit power	1 kW base station 1 W mobile	0.4–1 W AP
Intra-factory handover delay	Seamless	0.5 s
WAN handover	Available	UE driven
Maximum UE speed	500 kph	10–20 kph
Latency	1 ms @ 0.72 bit/s/Hz FDD 1 ms @ 0.5 bit/s/Hz TDD 1 ms @ 0.31 bit/s/Hz NR unlicenced	Typical 12 ms 1 ms @ 0.16 bit/s/Hz
Typical throughput	~1 Gbit/s	~1 Gbit/s
Range	Low and mid-band: 1–3 miles radius of base station. High-band: less than 1 mile radius	30 m per AP
Factory reconfiguration	Not impacted	New site survey and relocation of APs may be required
Reliability	Full self-healing	Requires UE to detect loss of IP connectivity to drive selection of new AP in some situations
Skills base	Enterprises have little experience and skills, so they must partner with a vendor or integrator	Large base of skilled technicians across the IT industry

cases, the failure will not self-heal, such as a failure in the IP network or AP's connection to the fixed network. Wi-Fi 8, probably available in 2028 and being developed by the IEEE Ultra-High Reliability Study Group, should address some of these shortfalls.

Wi-Fi 7 has mechanisms that constrain worst-case latency, with 99% of traffic having less than 12 ms latency, while, for Wi-Fi 6, this would be 75 ms, although the median latency of Wi-Fi 7 and Wi-Fi 6 is similar at 2 and 3 ms, respectively [36]. However, the low latency operation of Wi-Fi reduces its spectral efficiency, reducing the data throughput available. From a latency point of view, Wi-Fi 7 should satisfy all but the most stringent motion control and Industry 4.0 application requirements. While 5G URLLC promises 1 ms latency.

Wi-Fi 7 should be sufficient for Industry 4.0 applications and locations not requiring handover between APs or the outside world or very high availability, a scenario that applies to many SMEs. However, will Wi-Fi 8 and Wi-Fi HaLow be sufficient for in-building and near-building Industry 4.0 applications for larger enterprises and MNCs? Wi-Fi 8 must deliver on its promise of ultra-high reliability, including for roaming devices, to have wide acceptance as an Industry 4.0 technology for larger enterprises and MNCs. While private 5G networks are superior in many aspects, they are a new technology with a higher entry barrier for many organisations.

As we have read, there are roles for both private cellular networks and Wi-Fi in Industry 4.0, and the best choice depends on the location and use cases. The main driver of choice is the size of the organisation and site. For example, Tecknexus [37] describes a use case for a US automaker with a 1.5 square miles campus with driverless 'yard trucks' or AGVs. The automaker deployed a Private 5G network solution using a quarter of the APs of Wi-Fi and saved 40% costs. However, the challenge of installing a private 5G network in this case required a site survey to install 18 5G APs. Connectivity and handover problems had to be optimised, which required expert tuning of network parameters.

5.7 Edge Cloud

Industry 4.0 digitalisation requires lots of cost-efficient computer processing and data storage. Although Cloud computing services offer this, the Cloud data centres may be too far from the factory or farm for latency-sensitive applications. Edge Cloud offers the flexibility and costs of computing and storage in the Cloud but at sites closer to where the data is produced and consumed to reduce latency.

For Industry 4.0 applications, the Edge Cloud is in the factory or campus, or 'on-premises', and used to run applications that require low latency, e.g. trajectory control real-time digital twins. On-premises Edge Cloud may also be a requirement for those organisations that do not want to ship sensitive data off-site. The organisations Private 5G core could run on the organisation's Edge Cloud or the organisation's Edge Cloud could leverage spare compute on the Private 5G core infrastructure. An MNO has fewer opportunities to offer private Edge Cloud services on-premises as they will not typically have an on-premises presence.

The distribution and orchestration of compute workloads between on-premises Edge Cloud, centralised Clouds, and public Edge Clouds has all the challenges described in the '6G for Edge Cloud' section of Chapter 11, 'Radical Core Network'.

Chapters 4 and 11 examine Edge Cloud in detail, describing how 6G could catalyse a more general-purpose Edge Cloud.

5.8 Artificial intelligence

AI is responsible for the transformative improvements in efficiency and automation in Industry 4.0. It can optimise production processes, perform automated quality control, control robots and perform real-time data-driven decision-making.

AI has a significant role in 6G networks, as described in Chapter 13, but are there any synergies between Industry 4.0 AI and 6G AI that would naturally lead to 6G networks offering AI services for Industry 4.0?

Factories of the future need to be modular and flexible. Replacing fixed cables with wireless solutions helps achieve this, but only if the wireless network is modular and flexible. AI will be required to automate the design and operation of 6G networks, as described in Chapter 13; therefore, there is a need for the systems driving the reconfiguration of the factory to communicate requirements to the AI systems, automating the operation and reconfiguration of the 6G network. The 6G service providers become part of the supply chain of the Industry 4.0 organisation.

6G networks will deploy AI to perform network diagnostics. However, some network faults will be caused by UE or operations equipment. Network AI could inform the Industry 4.0 organisation AI of network diagnostics pointing to operations equipment faults. A more advanced solution would allow the network diagnostics AI and the operations diagnostics AI to work together to locate and resolve faults.[8] An even more advanced solution would be for the Network diagnostics AI to predict network equipment failure, often indicated by changes in analogue parameters such as equipment temperature, power consumption or electrical noise levels, and inform the operations AI so that proactive actions can be taken to minimise the impact of any network downtime.

A cyber security attack on the Industry 4.0 organisation may be detected by the 6G network security AI significantly before the organisation becomes aware of an attack. In some cases, the 6G security AI may mitigate the attack, but in other cases, it may require cooperation with the operations security systems to minimise operational impacts.

Localising assets or positioning information of mobile devices is vital for Industry 4.0. Positioning information can be obtained from 5G networks with an accuracy of a few metres, but this may not be accurate enough for some Industry 4.0 applications that require centimetre accuracy. The Industry 4.0 organisation may use its UE-based localisation technologies. AI can combine positioning information from the 6G network and the UE to improve the reliability and accuracy of positioning information, see Chapter 14.

IIoT could generate vast amounts of data that may be best processed closer to the sources at the network's edge. In Industry 4.0, much of the IIoT data would be processed by AI; in some cases, it is feasible to distribute AI processing across a network. The IIoT data could be processed by an AI distributed across the 6G networks using the compute resources of the 6G networks.

5.9 Summary and conclusions

Operational technologies are focused on safety, uptime and maintenance of machinery. 5G and Wi-Fi 7 are not Industry 4.0 ready because they lack proven

[8]A task that is performed by network and customer technicians today.

reliability. 5G and Wi-Fi 7 do not fully address cross-technology interoperability and cross-organisation integration capability, which are vital for Industry 4.0. A plethora of telecommunications options does not necessarily lead to the reliability, implementation and maintenance modularity required to give the flexibility needed for Industry 4.0. If 5G Advanced and Wi-Fi 8 address these issues, then 6G could be a repackaging of standards to make them more consumable by industry, especially OT architects. Industry 4.0 is not a product that MNOs can sell. MNOs must partner with system integrators to sell 5G/6G products as part of Industry 4.0 solutions.

The architectural tenets of Industry 4.0 are interoperability between multiple organisations and technologies, modularity, digital twins and flexibility. 5G has the same tenets for its internal construction, but viewed from the Industry 4.0 architect's point of view, it does not support multiple network technologies, it does not address extranets, the modularity options are obscure and not plug and play, it appears more ossified than flexible, especially concerning spectrum.

3GPP Release 18 (5G Advanced) should address some of these concerns as it will support non-3GPP access to Private Networks (Standalone Non-Public Networks) via the N3IWF and TNGF, including adding lists of permitted SNPNs and Wi-Fi APs to UEs. Incorporating Ethernet into 5G Advanced Private Networks is a critical strategy to enable Industry 4.0 to use 5G.

From a digital twins perspective, will MNOs give Industry 4.0 organisations access to a digital twin model of their networks to become part of the connected supply chain? Can the OT architect model the choices between different MNOs' services and other network technologies?

Industry 4.0 should provide enough value to justify investments in 6G. Gregolinska [3] reports that Industry 4.0 can improve manufacturing productivity by 15–30%. The global trade in manufactured goods in 2020 was US$12.1 trillion [4]; hence, Industry 4.0 productivity improvements could be worth at least approximately US$1.8 trillion per annum. Industry 4.0 would not be possible without telecommunications, and if 6G could claim only 2%[9] of the value of the global productivity improvements, this would be worth US$36B per annum. According to Gartner [5], global 5G infrastructure investment is running at US$23B for 2022. If 6G investments are similar to 5G investments, it can be concluded that 6G need only claim 2% of the value of Industry 4.0 global productivity improvements to make the business case that Industry 4.0 alone could justify the investment in 6G. Additionally, the Market Research Future [38] predicts revenues from 5G will be worth US$229B by 2032, so if 6G achieves only 2% of the value of Industry 4.0, it contributes to 16% of MNO's forecasted 5G revenues. There is scope to invest in 6G if it solves Industry 4.0 issues with 5G.

Reliability is critical, and in the author's opinion, 5G protocols are incompatible with the cloud-native architecture implied by the service-based architecture

[9]Analysys Mason [39] claim connectivity revenue accounts for 9% of the IoT value chain.

approach; this makes 5G core SBA implementations fragile. Only the test of time will prove the 5G cores' reliability.

5G Advanced must deliver better interoperability with non-3GPP wireless systems, especially Wi-Fi, Ethernet fieldbuses and the plethora of IIoT protocols. 5G Advanced must also deliver on NTN-IoT to track assets on 85% of the Earth's surface with no cellular coverage. Wi-Fi 8 must also deliver on its ultra-high reliability promise to be useful for Industry 4.0.

The logistics use case of asset tracking and management requires assets to log in to the correct network as it transverses a global supply chain; sometimes, that network may be public or private 5G or Wi-Fi or another IIoT technology. Mutual authentication is key as assets could host a trojan horse to infiltrate a private network or a network may be faked to misappropriate the asset. 5G Advanced will look at some of these issues, but it is unlikely to address them all.

There are some use cases where 5G will not have sufficient performance. For example, the wireless transmission of live uncompressed UHD, 61 Gbit/s per camera, is a niche application where 5G does not have a high enough data rate. 5G does not have low enough latency for the wireless control of servo loops for high-precision machine tools, which requires latencies of 5 μs to 0.5 ms. OT architects are motivated to replace wired fieldbus technologies with wireless to reduce the time and cost it takes to reconfigure a factory, in a mega-factory, the cost of cabling could exceed US$100M. The OT architect also wants to centralise motion and trajectory control within a factory to optimise them together for better precision. There may be a role for mmWave in factories for super-ultra-low latency and super-ultra-high reliability, but that may be beyond 6G or Wi-Fi 8 timescales.

In the vision of a reconfigurable factory network, reconfiguration must be automated. If Wi-Fi or 5G requires a new site survey and re-optimisation of the placement of Wi-Fi or 5G APs, this obstructs the automated reconfiguration of the factory. There is also a bottleneck if 5G networks require a 5G expert to reconfigure or tune the network after a factory reconfiguration. Network configuration needs to be automated for the reconfigurable factory use case.

5G NB-IoT is designed for a density of 1 million devices per square kilometre but allows only one thousand mobile networks per country. If 5G Private networks and IIoT are successful, we should consider millions of private mobile networks per country.

For the MNOs, there are challenges and opportunities. The challenge is for MNOs to exploit synergies with fixed VPN operations for private 5G networks. Opportunities include being part of integrated ecosystems for digital twins and AI solutions, especially for SMEs. They must work with other MNOs and vendors to build Industry 4.0 extranets.

Industry 4.0 is about seamlessly working with others, which 3GPP and MNOs need to do to make their services attractive to the industry and unlock the value of Private 5G networks, even if it means devaluing their legacy spectrum in the short term.

Let us consider the task of the OT architects when asked to architect an Industry 4.0 implementation; they must select the technologies to build solutions that connect the global multi-organisational supply chain, use wireless to build a flexible, modular factory and integrate with legacy technologies. 5G provides more

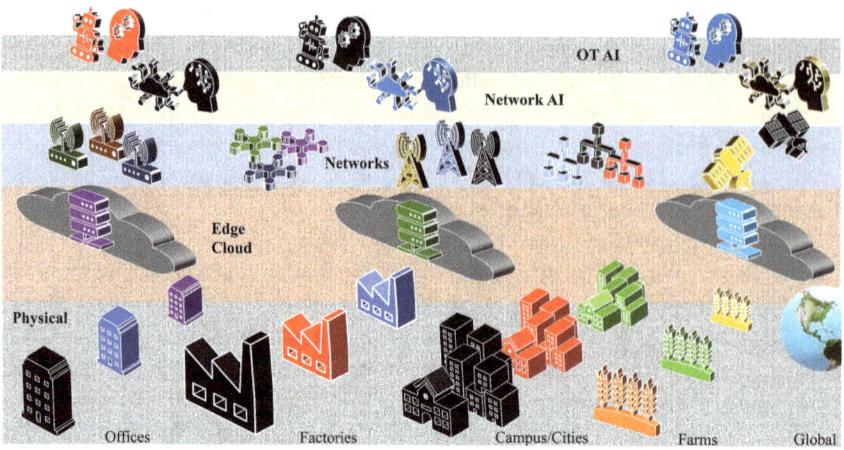

Figure 5.7 Industry 4.0 6G ecosystem

tools for the OT architect's kit bag, but no more. The OT architects need vertically integrated blueprints and an open ecosystem that does not lock them into product vendors or network operators.

Figure 5.7 shows the 6G ecosystem for Industry 4.0, which must cover use cases ranging from small offices to global coverage, where supply chains cross multiple organisations and types of organisations, multiple network technologies, especially wireless, will be used. The network service provider AI will have to manage the complexity of these networks and cooperate with the operational technology AI, which will be instrumental in delivering the efficiencies promised by Industry 4.0. Private and public Edge Cloud infrastructure provides the underlying computing power to support all this.

For Industry 4.0, 6G must be more than 5G by

1. Recognising that IIoT standards are highly optimised for specific use cases, making it difficult for cellular solutions to broadly compete; hence, a network of networks approach is required where multiple non-cellular IIoT and cellular IoT standards can be networked together via gateways.
2. Being more reliable. 6G should use stateless protocols and examine a 'coreless' strategy to make the system truly 'cloud native' and highly reliable.
3. Having proven reliability. Proven reliability may only come about by use in live Industry 4.0 deployments, but extensive testing of failure scenarios helps, e.g. testing what happens when nodes and links in the 6G network fail.
4. Implementing JCAS as described in Chapter 14.
5. Being more enterprise-focused:
   (a) Support mixed technology private networks with minimal cost overheads.
   (b) Address extranet scenarios.
   (c) Adopt an interworking first philosophy.
   (d) Have automated network configuration and reconfiguration.

   (e) Connect Network AI to Operations AI.
   (f) Increase the mobile network code space.
   (g) Allow private networking roaming onto public networks.
6. Having a long operational life, especially for long-lived infrastructure investments, e.g. emergency services networks, mega-factories, whose lifespan may be measured in several decades.

The six themes of 6G, introduced in Chapter 1, of network of networks, sustainability, global service coverage, extreme experience, trustworthiness and connecting intelligence are exactly the themes required to address Industry 4.0 telecommunications requirements. The value for MNOs to engage with Industry 4.0 and 6G technologies is clear with Industry 4.0 predicted to improve global productivity costs by approximately US$1.8 trillion per annum.

References

[1] IBM. IBM.com. [Online]. 2024 [cited 2024 Feb 2]. Available from: https://www.ibm.com/topics/industry-4-0.

[2] Mckinsey & Company. mckinsey.com. [Online]. 2022 [cited 2024 Feb 2]. Available from: https://www.mckinsey.com/featured-insights/mckinsey-explainers/what-are-industry-4-0-the-fourth-industrial-revolution-and-4ir.

[3] Gregolinska E, Khanam R, Lefort F, and Parthasathy P. Capturing the True Value of Industry 4.0. [Online]. 2022 [cited 2023 August 31]. Available from: https://www.mckinsey.com/capabilities/operations/our-insights/capturing-the-true-value-of-industry-four-point-zero.

[4] World Trade Organisation. World Trade Statistical Review 2021. [Online]. 2021 [cited 2023 August 31]. Available from: https://www.wto.org/english/res_e/statis_e/wts2021_e/wts2021_e.pdf.

[5] Goasduff L. Gartner Forecasts Worldwide 5G Network Infrastructure Revenue to Grow 39% in 2021. [Online]. 2021 [cited 2023 August 31]. Available from: https://www.gartner.com/en/newsroom/press-releases/2021-08-04-gartner-forecasts-worldwide-5g-network-infrastrucutre-revenue-to-grow-39pc-in-2021.

[6] The MPI Group. https://mpi-group.com/. [Online]. 2021 [cited 2024 Feb 2]. Available from: https://mpi-group.com/mpi-industry-4-0-study-2021/.

[7] Agostini L and Nosella A. The Adoption of Industry 4.0 Technologies in SMEs: Results of an International Study. *Management Decision*. 2019; 58(4): 625–643.

[8] Nokia. *Convergence of Operational and Communications Technology*.

[9] Spirent Communications Inc. Cutting Through the Edge Computing Hype. 2022.

[10] 3GPP. Study on Communication for Automation in Vertical Domains. 3GPP, SA1; 2020. Report No.: 3GPP TR 22.804.

[11] Parvez I, Rahmatic A, Guvenc I, Sarwat AI, and Dai. H. A Survey on Low Latency Towards 5G: RAN, Core Network and Caching Solutions. *IEEE Communications Surveys & Tutorials*. 2018; 20(4): 3098–3130.

[12] Kiesel R, Jakob F, Vollmer T, and Schmitt RH. Evaluation of ICT for Networked Control Systems of Latency-Critical Applications in Production. *Procedia CIRP*. 2022; 112: 238–244.

[13] Fanibhare V, Sarkar N, and Al-Anbuky A. A Survey of the Tactile Internet: Design Issues and Challenges, Applications, and Future Directions. *Electronics*. 2021; 10(17): 2171.

[14] Fu Y, Li C, Yu FR, Luan TH, and Zhang Y. Survey of Driving Safety with Sensing, Vehicular Communications, and Artificial Intelligence-Based Collision Avoidance. *IEEE Transactions on Intelligent Transportation Systems*. 2022; 23(7): 6142–6163.

[15] Hexa-X Consortium. https://hexa-x.eu/. [Online]. [cited 2024 Feb 8]. Available from: https://hexa-x.eu/.

[16] Wymeersch H, Shrestha D, Lima CD, *et al.* Integration of Communication and Sensing in 6G: A Joint Industrial and Academic Perspective. *IEEE 32nd Annual International Symposium on Personal, Indoor and Mobile Radio Communications (PIMRC)*. 2021: 1–7.

[17] Cabling Installation & Maintenance. Costing Out a Cabling System. [Online]. 1995 [cited 2023 Sep 6]. Available from: https://www.cablinginstall.com/design-install/cabling-installation/article/16469357/costing-out-a-cabling-system.

[18] Drives & Controls. drivesncontrols.com. [Online]. 2023 [cited 2023 Oct 3]. Available from: https://drivesncontrols.com/news/fullstory.php/aid/7359/Industrial_Ethernet_exceeds_two_thirds_of_comms_market.html.

[19] Everaxis. Robots & Machinery Slip Rings & Rotary Joints. [Online]. 2023 [cited 2023 Sep 4]. Available from: https://everaxis.com/markets/industrial/robots-and-machinery/.

[20] Analog Devices. Industrial Wireless Connectivity Technology. [Online]. 2023 [cited 2023 Sep 4]. Available from: https://www.analog.com/en/solutions/industrial-automation-technology/industrial-connectivity-technology/industrial-wireless-connectivity.html.

[21] Eurostat. https://ec.europa.eu. [Online]. 2021 [cited 2024 Feb 2]. Available from: https://ec.europa.eu/eurostat/databrowser/view/sbs_sc_ovw__custom_9633693/default/table?lang=en&page=time:2021.

[22] Moore GA. Innovation and Business Architecture. In Moore GA. *Dealing with Darwin: How Great Companies Innovate at Every Phase of their Evolution*. Portfolio; 2005. p. 30–55.

[23] UK Department for Transport. Domestic Road Freight Statistics: July 2021 to June 2022. [Online]. 2023 [cited 2023 August 31]. Available from: https://www.gov.uk/government/statistics/domestic-road-freight-statistics-july-2021-to-june-2022/domestic-road-freight-statistics-july-2021-to-june-2022.

[24] Eurostat. A Fifth of Road Freight Kilometres by Empty Vehicles. [Online]. 2021 [cited 2023 August 31]. Available from: https://ec.europa.eu/eurostat/web/products-eurostat-news/-/ddn-20211210-1.

[25] Alex Leslie DM. An Analysis of the Operational Costs of Trucking. [Online]. 2022 [cited 2023 August 31]. Available from: https://truckingresearch.org/wp-content/uploads/2022/08/ATRI-Operational-Cost-of-Trucking-2022.pdf.

[26] Silva J. Why Should You Partner with A Digital Freight Platform? [Online]. 2022 [cited 2023 August 31]. Available from: https://www.yourzeus.com/news/why-should-you-partner-with-a-digital-freight-platform.

[27] Latham K. Why Empty Lorries are a Problem and How to Fix It. [Online]. 2023 [cited 2023 August 31]. Available from: https://www.bbc.com/news/business-66461766.amp.

[28] DeGrasse M. fiercewireless.com. [Online]. 2024 [cited 2024 Feb 2]. Available from: https://www.fiercewireless.com/tech/john-deere-taps-starlink-connect-autonomous-tractors.

[29] RCR Wireless News. Total IoT Connections to Surge, but Cellular IoT to Lose Share with 2G/3G Switch-off. [Online]. 2023 [cited 2023 June 23]. Available from: https://www.rcrwireless.com/20230622/internet-of-things-4/total-iot-connections-to-surge-but-cellular-iot-to-lose-share-with-2g-3g-switch-off.

[30] Liberg O. LTE-M and NB-IoT Meet the 5G Performance Requirements. [Online]. 2018 [cited 2023 Sep 6]. Available from: https://www.ericsson.com/en/blog/2018/12/lte-m-and-nb-iot-meet-the-5g-performance-requirements.

[31] Bankov D, Khorov E, and Lyakhov A. On the limits of LoRaWAN channel access. *International Conference on Engineering and Telecommunication*. 2016: 10–19.

[32] Bilgin B and Gungor V. Performance Evaluations of ZigBee in Different Smart Grid Environments. *Computer Networks*. 2012; 56(8): 2196–2205.

[33] Vodafone. Satellite IoT. [Online]. 2023 [cited 2023 Sep 6]. Available from: https://www.vodafone.com/business/iot/managed-iot-connectivity/iot-satellite.

[34] Iridium. What is Satellite IoT and How is it Used? [Online]. 2021 [cited 2023 Sep 6]. Available from: https://www.iridium.com/blog/what-is-satellite-iot-and-how-is-it-used/.

[35] ISO/TC 184 Automation Systems and Integration. Automation Systems and Integration – Oil and Gas Interoperability – Part 1: Overview and Fundamental Principles. Standard. ISO, ISO/TC 184 Automation Systems and Integration; 2019. Report No.: ISO/TS 18101-1:2019.

[36] Garcia-Rodriguez A, López-Pérez D, Galati-Giordano L, and Geraci G. EEE 802.11be: Wi-Fi 7 Strikes Back. *IEEE Communications Magazine*. 2021; 59 (4): 102–8.

[37] Tecknexus. https://technexus.com. [Online]. 2023 [cited 2024 Feb 12]. Available from: https://tecknexus.com/5g-network/private-networks-5g-magazine-dec-2023-edition/5g-in-manufacturing-navigating-the-evolution-and-impact-of-advanced-connectivity/.

[38] Market Research Future. 5G Market Research Report Information. [Online]. 2023 [cited 2023 August 31]. Available from: https://www.marketresearchfuture.com/reports/5g-market-2630.

[39] Kasujee I and Rebbeck T. https://www.analysysmason.com. [Online]. 2023 [cited 2024 Feb 2]. Available from: https://www.analysysmason.com/research/content/articles/iot-value-chain-rdme0/.

Part II

6G – an introduction to standards and key enabling technologies

Chapter 6

6G – the definition begins with views from standards organisations, industry and academia

If you believe that what you're doing will have positive results, it will – even if it's not immediately obvious.

– Sophia Amoruso, Girlboss

6.1 Introduction

Since 2017, thousands of researchers and developers across multiple countries have been collaborating to develop a new mobile system for potential release in 2030 for the sixth generation. While we have yet to reach the midpoint of 5G, we are already beyond the beginning of the 6G journey with substantial research activity since 2017 in China with the Chinese IMT-2030 (6G) Promotion Group and with the European Hexa-X project, which began in 2018. In other regions, leading collaborations such as the North American Next G Alliance (NGA) have proposed concrete 6G technology roadmaps [1] with MOUs with the European Hexa-X project and Japan's Beyond 5G Promotion Consortium (B5GPC). There is also broad agreement on the key technologies and research areas for 6G worldwide. In this chapter, we consider the European Hexa-X project as an exemplar of the direction of 6G, recognising that there is also brilliant work in other regions following similar approaches and sharing their results. By considering the details of one of these mature programmes, Hexa-X, we will gain insight into what 6G is, its use cases, and the key underlying technologies. It will also explain whether it is solely an evolution from 5G Advance or whether it also brings about revolutionary features that excite the world.

The Third Generation Partnership Project (3GPP) is expected to submit specifications for 6G standards to the ITU in 2027, with commercialisation anticipated around 2030. 3GPP is an umbrella term for several regional standards organisations that develop protocols for mobile telecommunications and are responsible for creating the 6G proposals. Meanwhile, since the commercialisation of 5G in 2019/20, the standardisation of 5G-advanced is also in progress in 3GPP to further enhance 5G networks with Releases 18–20, continuing up to 2028. This is widely seen as the starting point for 6G.

5G addressed the broader digitalisation of Society and Industry as a communications technology within the novel paradigms in computing and networking, such as cloud computing, digital twins, artificial intelligence (AI), software-defined networks (SDN) and network function virtualisation (NFV). It is based on wider radio channel bandwidths with low latency and spectrum utilisation reaching towards 100 GHz to support new extreme data services. However, it underestimated the timescales of other elements required to address this paradigm change in society and industry as it only addressed the communications stack. Usually, the success of the early stages of 5G would influence the direction of 6G as it would address issues identified from these early deployments. Mobile operators are frustrated with 5G's apparent overpromising, leading to a credibility problem for further investment in the upcoming 6G. However, the slow introduction of core network technology, such as Standalone Core, which is necessary for many industrial use cases, is partly to blame for this, and some more advanced features only appeared in 5G advanced towards the end of the decade. Therefore, investing in 6G around 2030 to further improve key performance indicators such as speed, latency, reliability and number of supported IoT devices, while 5G still needs to prove itself in these areas, is a real credibility challenge. Furthermore, proposing massive enhancements to 5G in data speeds for new use cases like holography is equally challenging when this technology has yet to exist. Nevertheless, the business case for the broader digitalisation of society and industry is exceptionally compelling, as described in Chapter 5. As always, the question concerns timing, and not all paradigm changes fit neatly into mobile generation ten-year cycles each new decade. The 6G community is aware of the issues around 5G and how to address these. For example, while the standardisation of 5G was rushed to get an early release date, it had to rely on 4G for non-standalone cores, opening various deployment options. Each required testing, but only a few were deployed. 6G hopes to avoid this mistake.

Since 3G, when code-division multiple access (CDMA) was introduced as a global standard unifying the formerly divided technology landscape, it is natural to think of 6G as producing a single global standard. However, growing global tensions may return to different 6G standards for other regions.

This chapter focuses on the status of 6G, which is reaching a level of maturity definition. For example, the ITU report, 'Future Technology Trends of Terrestrial International Mobile Telecommunications Systems Towards 2030 and Beyond' [2], published in November 2022, will provide input for ITU recommendations for 6G. We review this document in the last chapter to support what is needed in the evolution and revolutionary roadmaps. Our views on the current state of 6G are based, in part, on an analysis of seminal white papers published by Ericsson [3] with additional comments from the authors. We also focus on the work of the EU Hexa-X project as an exemplar to show how a research group is advancing the 6G standards with its vision of 6G for the EU. This has broader applicability to other regions as they share similar visions and technology research for 6G. However, we also cover the work of other regional standards organisations to demonstrate this alignment.

In other chapters of this book, we have gone into greater depth on how we got there with a critical analysis of individual components of 6G, with suggestions on how to improve it dramatically and to paint a picture of what we believe will happen with Wi-Fi playing its role with 6G in the further digitalisation of Society and Industry which we feel is inevitable.

6.2 Third Generation Partnership Project

The 3GPP [4] is an umbrella term for seven national or regional telecommunication standards organisations as the primary members, who are referred to as 'organisational partners', and several other organisations as associate members, who are referred to as 'market representation partners'. Three 3GPP different work streams exist: radio access networks, services and systems aspects and core networks and terminals. 3GPP's best-known work is the development and maintenance of

- GSM and related 2G and 2.5G standards;
- UMTS and related 3G standards;
- LTE and related 4G standards;
- 5G NR and related 5G standards, including 5G Advanced;
- An evolved IP Multimedia Subsystem (IMS).

The market representation partners include Next Generation Mobile Networks (NGMN), GSMA and IPV6 forum. They can offer market advice to 3GPP to bring a consensus view from their markets but do not have the capability or authority to define standards within the 3GPP scope. Nevertheless, they are committed to all or part of the 3GPP scope. The seven 3GPP Organisational Partners are from Asia, Europe and North America. They aim to determine the general policy and strategy of 3GPP.

The Organisational Partners are shown in Table 6.1, and their location is shown in Figure 6.1.

Table 6.1 Organisation partners of 3GPP responsible for developing the 6G standards for the ITU [4]

Organisation	Country/ region	Website
Association of Radio Industries and Businesses (ARIB)	Japan	ARIB [5]
Alliance for Telecommunications and Industry Solutions (ATIS)	USA	ATIS [1]
China Communications Standards Association (CCSA)	China	CCSA [6]
European Telecommunications Standards Institute (ETSI)	Europe	ETSI [7]
Telecommunications Standards Development Society (TSDSI)	India	TSDSI [8]
Telecommunications Technology Association (TTA)	South Korea	TTA [9]
Telecommunication Technology Committee (TTC)	Japan	TTC [10]

Figure 6.1 6G technology and standards in regions

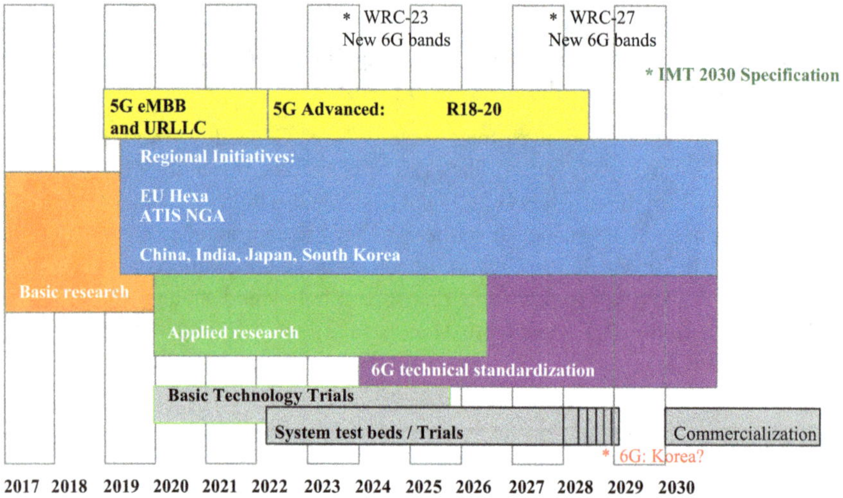

Figure 6.2 Timelines and activities for 6G (based on Ericsson and Nokia Diagrams)

6.3 When will 6G arrive and how – timelines

Ericsson believes that 3GPP will begin work on 6G requirements in 2024, with work on technical standardisation starting in 2025. The timelines for 6G are shown in Figure 6.2, with the various contributing elements, including the regional initiatives we review below. These include notable activities, including ongoing research and industrial projects, such as the second phase of the EU's Hexa-X project led by Nokia, which started in 2023 with a focus on the systemisation of 6G. In addition, many academic communities are focusing their research on enabling 6G and presenting at annual events in the United States, such as the Brooklyn 6G Summit and 6G@UT Forum. The Wireless Summit and key International conferences are held annually in the EU. These fuels the technology narratives for the 6G vision.

Of course, spectrum is critical, and the World Radiocommunication Conferences (WRC) and the ITU-R have decided on the frequency band options and the conditions for 6G, as shown in Figure 6.2. The WRC 2023 identified the upper 6 GHz band for IMT (spectrum for 6G), completed the harmonisation of the 3.5 GHz band for IMT, allocated the UHF band (470–694 MHz) to mobile and approved studies for WRC-27 on new bands.

On the future agenda item for WRC 2027 on IMT, WRC-23 approved the study of the bands 4400–4800 MHz, 7125–8500 MHz and 14.8–15.35 GHz for IMT. On the UHF band, the final decision of the WRC-23 was a secondary mobile allocation for Europe in all the band (470–694 MHz), while 11 countries in the Middle East allocated the band 614–694 MHz to mobile on a primary basis and identified the band for IMT. With a secondary mobile allocation, Europe will be able to deploy IMT once the decisions on the future use of the UHF band are made within Europe after the planned review of the TV broadcast usage of the band by 2025 [11].

While 3GPP will need to conclude on the standardisation requirements, the first 6G specifications are expected to be available in 3GPP Release 21 in the second half of 2028 [12].

The standardisation of a new cellular network to a single standard has become, since 3G, a global and collective effort of several standardisation bodies who work collaboratively and complementary towards a single standard. Figure 6.3 depicts the

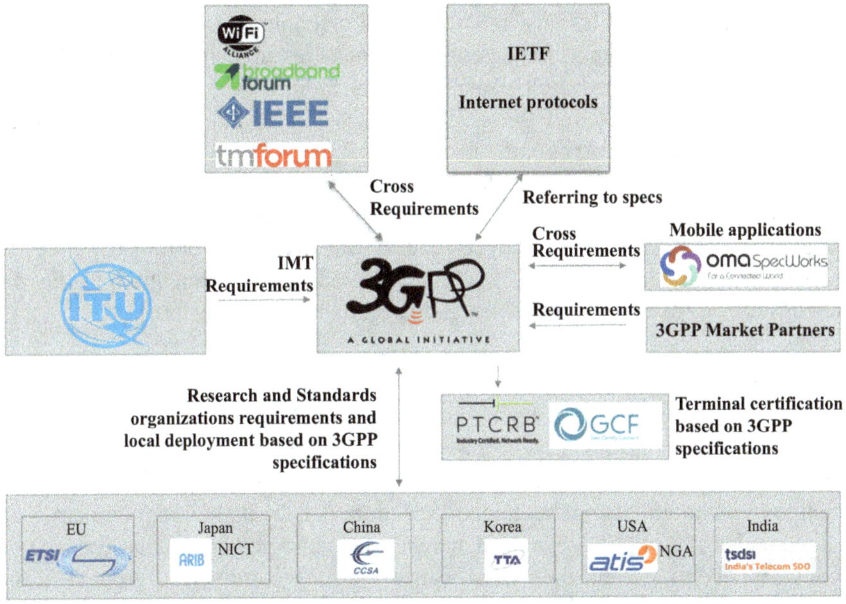

Figure 6.3 Structure of research and standardisation effort for 6G based on ITU diagram

standardisation efforts of 6G led by 3GPP and the ITU supported by major standardisation bodies in six regions, as shown in Figure 6.1. 3GPP combines the ITU vision requirements via the 'Future Technology Trends of Terrestrial International Mobile Telecommunications Systems Towards 2030 and Beyond' [2] with 3GPP market partners, research and standards organisations. There are also cross-reference requirements with other organisations, such as Wi-Fi Alliance, IEEE, IETF, etc.

6.4 Global trends

China has taken a strong position on developing 6G technology and currently holds the highest percentage of 6G patent filings at 40.3% as of 2023. The United States follows with 35.2%, Japan with 9.9%, Europe with 8.9% and South Korea with just over 4.2%, according to Williams IP Law Field [13]. Williams IP Law points out that experts had predicted that China's development of 6G-related technologies would slow down after the US government levelled sanctions against Huawei Technologies in 2019. However, China has pivoted from Huawei to universities and several state-run businesses. Nevertheless, Huawei continues to be a key player in this arena. A joint research project by Cyber Creative Institute and Nikkei Asia [14] in November 2022 discovered that approximately 20,000 patent applications have been filed on 6G-related technologies covering areas of AI, base stations, quantum technology and communications. South Korean and Chinese companies are amassing patents for metaverse hardware, with LG Electronics and Huawei Technologies climbing the ranks as the electronics sector looks beyond the smartphone [14].

According to a study by the Australian Strategic Policy Institute (ASPI) [15], in March 2023, China is leading in 37 out of 44 technology domains, of which 6G is included. The ASPI report indicates that Chinese researchers are responsible for 29.7% of high-impact research papers on new radio research, followed by the United States with 9.5% and the UK with 5.2%. In advanced optical research, China has a significant lead over the United States, with 37.7% compared to 12.8%. Although China and the United States are evenly matched in machine learning and natural language processing, China excels in other ICT domains, such as advanced data analytics, AI algorithms and cybersecurity. Finally, China dominates three out of four quantum research fields, including quantum communication, while the United States leads in small satellites and space launch systems [16].

6.4.1 European Union

The Hexa-X project [17], led by Nokia and Ericsson, is the European Commission's flagship initiative to research 6G technology to drive its overall 6G vision to connect the physical, digital and human worlds through future wireless technology and architectural research, see Figure 6.4 for its key requirements and characteristics. The EU believes that wireless technologies are crucial for today's society and economy, and their significance will only continue to grow with the steady evolution of 5G, which will enable new ecosystems and services. Their project aims to create unique use cases and scenarios for 6G, develop fundamental

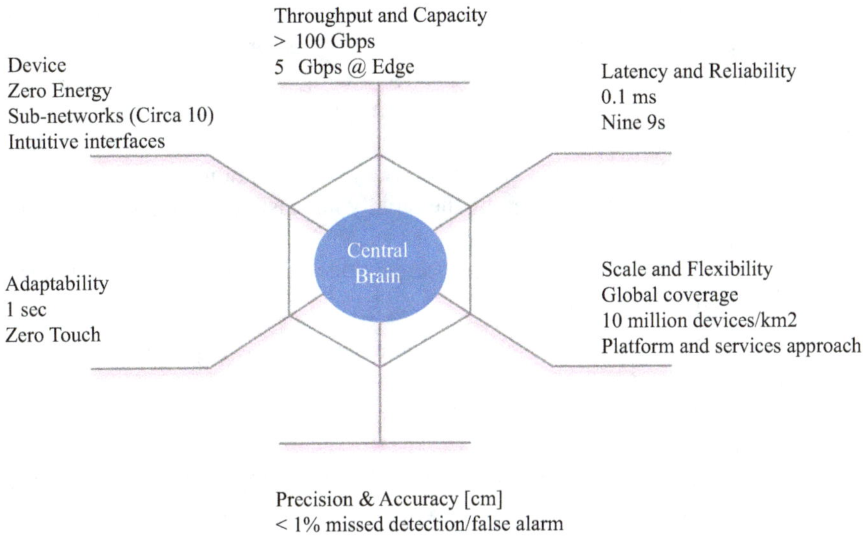

Figure 6.4 Hexa-X key requirements and characteristics of 6G based on reference [21]

6G technologies and define a new architecture for an intelligent fabric that integrates critical 6G technology enablers [18].

> *Its key aim is to develop smart communication components, systems, and networks for 6G, following an evolutionary path from 5G through further enhancements of 5G advanced technology and a more revolutionary path by investigating the benefits of promising technological enablers.*
>
> *European Commission [19]*

This book aims to understand this critically and examine wireless technologies' role more generally.

The European Commission has granted the Hexa-X project funding for the EU's Horizon 2020 research and innovation programme. It initially brings together 25 key industry stakeholders from across Europe. These stakeholders represent future connectivity solutions, including network vendors, communication service providers (CSPs), technology providers, verticals and the most prominent European communications research institutes.

In October 2022, the EU announced 35 new projects worth €250 million in 6 G-related Research and development under the EU's Smart Network and Services Joint Undertaking (SNS JU) [19]. Expanding the Hexa-X phase II to 44 stakeholders.

In particular, the EU initiated the Hexa-X project to address the upcoming opportunities and challenges related to growth and sustainability that the world will face in 2030. In a post-pandemic world, proactive measures are crucial to tackle the issues of green deal efficiency, digital inclusion and health and safety assurance.

The Hexa-X vision is developed to connect the physical, digital and human worlds through future wireless technology and architectural research. The Hexa-X vision aims to create an x-enabler fabric of connected intelligence, networks of networks, sustainability, global service coverage, extreme experience and trustworthiness [20]. The importance of wireless technologies for our society and economy cannot be overstated, especially with the advent of 5G and its evolution, which will enable new ecosystems and services driven by the rapidly increasing traffic and trillions of devices.

The ambitions of the Hexa-X project include developing key technology enablers in the areas of the following:

- Fundamentally, new radio access technologies at high frequencies and high-resolution localisation and sensing.
- Connected intelligence through AI-driven air interface and governance for future networks.
- 6G architectural enablers for network disaggregation and dynamic dependability.

Like other regional research, Hexa-X recognises that developing 6G networks requires global support and collaboration. Therefore, it will strive for openness and cooperation between the European and international research community, standardisation bodies and policymakers. This will be achieved through public workshops, joint whitepapers and active participation in significant events, which will help explain the broader consensus of the global 6G community.

6.4.2 The United Kingdom of Great Britain and Northern Ireland (UK)

In December 2022, the UK government announced its plan to boost research in the 5G and 6G technology [22]. As part of a £110 million telecoms R&D package, the government has granted £28 million to the Universities of York, Bristol and Surrey to develop next-generation 6G network technology. These universities will collaborate with major telecom companies such as Nokia, Ericsson and Samsung to design and build networks of the future, such as 6G. The UK government aims to enhance the UK's position as a global leader in telecoms research following Ericsson and Samsung's recent decision to establish cutting-edge 6G research centres in the United Kingdom. Additionally, a £80 million fund is set to launch a state-of-the-art UK Telecoms Lab for testing network equipment.

As part of the overall funding of £110 m, the United Kingdom has partnered with the Republic of Korea to accelerate the deployment of Open RAN and associated technologies. This joint project, which will receive more than £3 million with a UK contribution of £1.2 m

Within the UK government funding of University research, coordinated through UK Research and Innovation (UKRI) and the Engineering and Physical Sciences Research Council (EPSRC), there is an additional £100 M in funding to push to a position at the 'leading edge of future telecoms'. The investment will parallel the government's 6G strategy and include funding to establish a series of Future Telecoms Research Hubs.

The United States and the United Kingdom have announced plans to develop a detailed science and technology partnership agreement, including a provision for strategic collaboration on developing the 6G technology [23].

6.4.3 USA

In 2022, in anticipation of 6G, the Federal Communications Commission (FCC) granted a ten-year special category of authorisation called the Spectrum Horizon License. This license permits experimentation with frequencies ranging from 95 GHz to 3 THz. Additionally, to ensure smooth and effective implementation of spectrum in 6G, the FCC has established an expert group.

Keysight Technologies claims to be the first company to obtain an experimental license for 246 and 275.5 GHz for 6G research. They are optimistic that advancements in the sub-THz spectrum will enable the development of extended reality, immersive telepresence, digital twins that combine real and virtual environments, and computer technology and wearables that generate human–machine interactions.

Meanwhile, the Alliance for Telecommunications Industry Solutions (ATIS), the US leading standards organisation, established the Next Gen Alliance (NGA) in 2022 to help secure its leadership for 6G. This is a private consultative body for next-generation telecommunications, which includes leading Universities, large mobile operators such as AT&T, T-Mobile and Verizon with others, BS and UE equipment manufacturers such as Qualcomm, Samsung, Nokia and Apple and Open RAN equipment manufacturers such as HP, Dell, etc., with platform companies such as Microsoft and Google. Together, they will participate in the US 6G technology leadership, vision and roadmap development. They have published their first white paper, 'Next G Alliance Report: Roadmap to 6G' [1]. This report addresses all aspects from markets to technologies to networks for 6G.

The US vision from this report is

The International Mobile Telecommunications (IMT) systems for 2030 and beyond will be developed as a global standard to better serve the communication needs in every continent of the world. The Next G Alliance has identified six audacious goals that describe top priorities for North America's contribution and leadership in these future global standards, deployments, products, operations, and services. These priorities contemplate both the societal and economic needs across North America and the technology strengths that North America will contribute to the rest of the world.

Their report identifies six goals:

1. *Trust, Security and Resilience should be advanced such that future networks are fully trusted by people, businesses and governments to be resilient, secure, privacy-preserving, safe, reliable and available under all circumstances.*

2. *An enhanced Digital World Experience consists of multi-sensory experiences to enable transformative forms of human collaboration, as well as human-machine and machine-machine interactions that will transform work, education and entertainment, thereby improving quality of life and creating significant economic value.*
3. *Cost-Efficient Solutions should span all aspects of the network architecture, including devices, wireless access, cell-site backhaul, overall distribution and energy consumption. These must be improved for delivering services in a variety of environments, including urban, rural and suburban, while also supporting increased data speed and the services that are expected for future networks.*
4. *Distributed Cloud and Communications Systems built on virtualisation technologies will increase flexibility, performance and resiliency for key use cases such as mixed reality, URLLC applications, interactive gaming and multi-sensory applications.*
5. *An AI-Native Network is needed to increase the robustness, performance and efficiencies of wireless and cloud technologies against more diverse traffic types, ultra-dense deployment topologies and more challenging spectrum situations.*
6. *Sustainability related to energy efficiency and the environment must be at the forefront of decisions throughout the life cycle, toward a goal of achieving IMT carbon neutral by 2040. Advances will fundamentally change how electricity is used to support next-generation communications and computer networks, while strengthening the role that information technology plays in protecting the environment.*

Taken from the 'Next G Alliance Report: Roadmap to 6G' [1]

This aligns exceptionally well with the EU's research activities in the Hexa-X project. The ATIS Next G Alliance's research priorities strongly focus on delivering 'digital equity' in response to the US government's calls to overcome digital divides within the country. Similarly, this is also an essential aspect in other regions as well. The ATIS has signed MOUs with the EU's 6G IA – Smart Networks and Services Industry Association (6G-IA), the European Industry and Research for next-generation Networks and Services, which is leading the Hexa-X phase 2 project and Japan's Beyond 5G Promotion Consortium (B5GPC). This helps to align the various worldwide research activities on 6G.

Another important 6G policy group, founded by the former Google CEO Eric Schmidt, The Special Competitive Studies Project (SCSP), has released a national action plan for its views on US leadership in the Advanced Networks Field [24]. It recognises that the United States needs major producers of end-to-end telecom solutions but also recognises its strengths in cloud, software and satellites. It contains a list of policies directed at regaining a leading global position beyond 5G networks.

U.S. leaders must think beyond the recent policy focus on 5G networks to encompass other elements of the connectivity stack, including low-earth orbit (LEO) satellites, data centres, and potential leapfrog technologies, like free space optical communications and networks (FSONs),

Reference: SCSP National Plan [24]

6.4.4 China

In 2017, the Chinese government, led by the Ministry of Science and Technology, became the first to initiate research on 6G technology. To support this research, a $425M programme was launched and set to run from 2019 to 2027. China has taken a proactive stance on this program and has filed the highest number of 6G patents, accounting for 40.3% of all filings as of 2023.

Huawei, which started its 6G research in 2017, plans to launch 6G products in 2030. It emphasised its 6G infrastructure vision via the Smart World forum hosted in April 2022 [25]. It also researches 6G satellite communication by launching three low-orbit research satellites. From its 6G Horizon white paper published in 2022, its vision is broadly like others for 6G: ***6G will serve as a distributed neural network that provides communication links to fuse the physical and cyber worlds.***

As other regions do that, 6G will build on the foundations of 5G in eMBB, URLLC and mMTC, with further enhancements helping to accelerate the digital transformation of vertical industries.

Here, Huawei sees 6G serving as a distributed neural network that provides links with integrated communication, sensing and computing capabilities to fuse the physical, biological and cyber worlds, ushering in an era of true Intelligence of Everything [26]. Its six pillars of 6G, see Figure 6.5, correspond exceptionally well with research in other regions, such as the EU hexa-X and the US Next G Alliance. They also see the need for a native AI air interface and Integrated non-terrestrial networks (NTNs) using LEO satellites as identified in the Hexa-X project. Importantly, they see sensing, combined with enhancements to eMBB, URLLC and mMTC, under AI, as a necessary feature to fully digitalise society and industry.

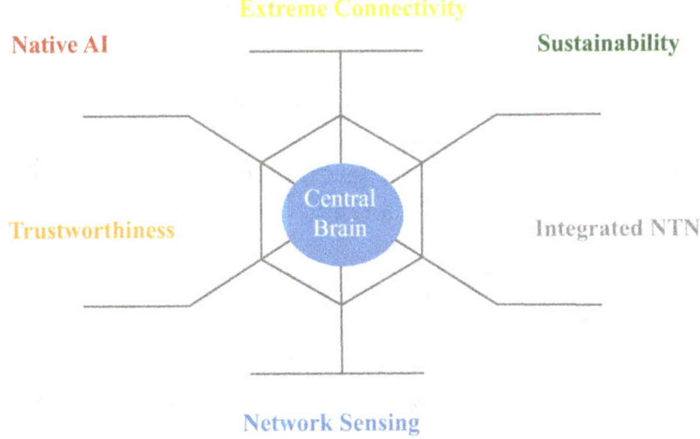

Figure 6.5 Huawei 6 Pillars of 6G (redrawn from reference [26])

6.4.5 South Korea – 6G in 2028

South Korea's Ministry of Science and ICT (MSIT) plans to commercialise an initial 6G network service in 2028, two years earlier than its initial schedule, the Korean newspaper *Aju Business Daily* reported in February 2023. The ministry said that it has decided to advance the launch of the 6G service 'as the country cannot achieve industrial innovation without having a global competitive edge in the 6G field'. Also, the ministry plans to hold a 'pre-6G vision fest' in 2026 to demonstrate the outcomes of the country's research in the 6G field. The event will invite global telecom firms, IT companies, standard experts and government officials from other countries [27].

The government announced the K-Network 2030 strategy to boost private–public cooperation to develop 6G technologies, innovate around software-based next-generation mobile networks and strengthen the network supply chain. The ministry has launched a feasibility study for research and development on core 6G technologies with KRW 625.3 billion ($481.7 million) funding to produce materials, components and equipment for future 6G networks locally [28]. This budget represents a tripling of the budget announced in 2020.

In May 2021, the United States and South Korea agreed to encourage joint R&D on emerging technology, including 6G [29]. South Korea plans to promote collaborative studies on core 6G technologies and spectrum, including 11 studies with the United States, one with China and two with Finland. The country's 5G Forum will sign MoUs for 6G collaboration with organisations in the private sector, like the Next G Alliance in the United States.

Specifically, MSIT outlined its strategic technologies that include (i) Tbits/s-capable wireless and optical communication for maximum 1 Tbit/s speeds; (ii) associated with this aim is terahertz RF components and spectrum model for bands between 100 and 300 GHz; (iii) space mobile and satellite communications to help expand support altitude to 10 km above ground; (iv) end-to-end ultra-precision networking for 1/10 latency compared to 5G; (v) intelligent wireless access and network with a focus on applying AI to all sections of the network; and (vi) technology for constant network quality monitoring for 5G focused on embedded security [30].

6.4.6 Japan

Japan has established the 'Beyond 5G promotion Strategy' for 6G-based technologies for 2030 to help recover its competitiveness in the mobile telecommunications sector. According to Nikkei Asia, Japan plans to create a $450 million fund for 6G research to foster research activities into future 6G communications systems.

The National Institute of Information and Communications Technology will oversee the 6G fund, providing financial support for the research and development of 6G technology over several years. Japanese companies NTT Docomo, NTT, NEC, Fujitsu and Finnish telecom equipment maker Nokia have announced plans to conduct experimental trials for new mobile communication technologies in preparation for the commercial launch of 6G services by 2030 [31].

The Beyond 5G consortium published its white paper in September 2022 as its roadmap to the 6G [32]. Like the US Next G Alliance, this describes market drivers, technology trends, capabilities and KPIs. Like others, they see applications of virtual reality, mixed reality, augmented reality and the metaverse as essential drivers. They also have extensively analysed the requirements from many sectors of industries and businesses with associated 6G requirements.

6.4.6.1 Summary of Japan KPIs

While following similar areas of research to other regions, they have uniquely specified 6G KPIs:

- Latencies: the requirement for sub-ms latencies, such as 100 μs for motion control of machinery, but in the prominent other applications requiring only a minimum of 1 ms.
- Ultra-fast and large capacity reaching 100–200 Gbit/s speeds for holographic communications and volumetric video but below 100 Gbit/s for other prominent applications.
- Ultra-security and resiliency require 10^{-7} for remote surgery.
- Positioning and sensing accuracy of 1–2 cm.
- Tens of millions of devices/km^2 and 100% land coverage.
- Ultra-low power consumption.
- Others include distributed learning and inference functions, inter-device interfaces, evacuation instructions that can be received at 1000 km/h and NTN nodes that connect to other NTN nodes.

6.5 The EU Hexa-X and Hexa-X phase II – representative for 6G end-to-end research

While 3GPP is expected to submit 6G standards specifications to the ITU in 2027 to meet technical performance with commercialisation around 2030, we are nevertheless reaching a point where the direction of 6G is well defined and where there is broad consensus on its direction as can be seen in the previous section on global trends. First, there is unanimous agreement on building 6G on the 5G's foundations of eMBB, URLLC and mMTC with further enhancements. The debate we cover in Chapter 16 concerns how much these need to change. Second, the major new global research themes for 6G are essentially the same. Each region may articulate its vision in slightly different ways, but the totality of the research areas is the same. For example, the six pillars of Huawei's 6G research align well with those of the EU Hexa-X project, ATIS Next G Alliance and Japan's B5GPC. An analysis of these would see in one form or another the key themes of (1) Trust and security, (2) sustainability including 'zero energy' devices for increased IoT roles, (3) bringing together the Physical, Human (biological) and Cyber (Digital) worlds through a multi-sensory experience, (4) An implementation of Cloud and virtualisation technologies with AI and ML, (5) addressing digital equity/digital divide with an emphasis on satellite technologies, and (6) extreme experience where future 6G systems access the spectrum above 100 GHz for extreme data rates, >100 Gbit/s and can provide cm-level positioning and sensing.

To understand this and how governmental research, with Industry and academia, can influence and direct standards bodies, we cover one of these projects in detail, the EU Hexa-X project. The motivation for the EU 6G flagship project, Hexa-X, is based on the recognition that the past four decades have driven continuous wireless innovations, primarily focused on the needs of people and their markets, transforming society and the lives of billions. Since the era of 4G, for the first time, the focus for 5G has also been to deliver a digital infrastructure supporting machine-to-machine communications for professional services and vertical sectors. 5G is intended to pave the way for a broader digitalisation and transformation of critical industries such as manufacturing, mining, transportation, logistics, commerce, health, smart cities and public safety, as described in our chapters on The Future of the Internet and Industry 4.0. Therefore, its effects will be profound even if it does not align well with neat decade cycles of mobile generations.

By 2030, according to Hexa-X, human intelligence will be augmented by being tightly coupled with network and digital technologies. User interfaces will change. It recognises that demanding services such as extended reality (AR/VR at scale) will present consumer and business interfaces to the network, services and intelligence. They expect further development in AI that will enable machines to transform data into reasoning and decisions, assisting humans in better understanding and navigating our world. Hence, the belief is that connectivity services will require bit rates of up to 100 Gbit/s, depending on the context and place of use. The current domestic and industrial machines will evolve into swarms of multi-purpose robots and drones equipped with new man–machine haptic and thought interfaces that can be controlled from anywhere.

> *Such a transformation will undoubtedly generate unprecedented economic opportunities and societal challenges as we move towards the 2030 timeframe; moreover, it will call for a fundamental shift in the way mobile networks are designed.*
>
> *Reference: About Hexa-X*

In Figure 6.6, taken from the Hexa-X vision, the EU project believes that everyday experiences will be enriched by the seamless unification of the physical, digital and human worlds, achieved through the new network and device technologies.

Interestingly, the Hexa-X project aims to resolve a significant issue that arose during the 3G era. Although sustainability was a crucial requirement, the success of mobile technology led to substantial growth in data traffic, data centres and the number of devices that consume more energy. However, there are some savings, such as using video conferencing to reduce travel. Nevertheless, the goal of 6G is to reconcile this by achieving the highest standards for energy efficiency, security, deployment and operation while promoting sustainable growth.

On the theme of reduced energy consumption, in Chapter 9, we explore the concept of fixed-mobile convergence and its potential benefits. One of these benefits is the support of 'Inside-Out Networks' – an alternative to the current 'outside-in'

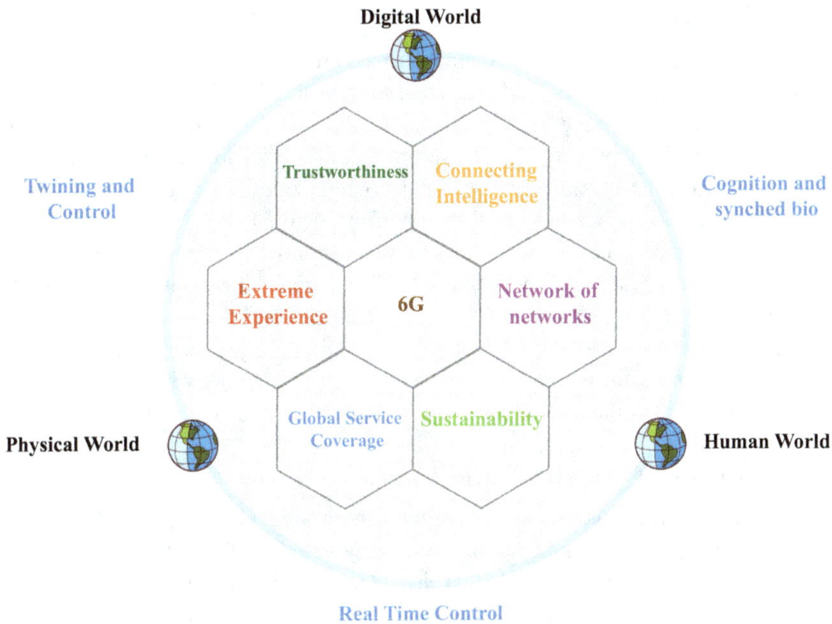

Figure 6.6 Hexa-X Vision (redrawn using Designed by Freepik)

approach used in all cellular networks. This new approach to network deployment can improve the user experience and reduce energy consumption. The authors suggest that using spectrum below the '5 GHz' threshold for wide-area cellular deployment, focusing on indoor use, could revolutionise small cell deployment within buildings. Additionally, incorporating dual-band coverage of the lower mid-band spectrum could lead to developing 'inside-out' networks.

6.5.1 Hexa-X Technology Pull

The emergence of advanced artificial intelligence (AI) technologies, radio access beyond 100 GHz, network virtualisation and disaggregation concepts offer excellent potential for enhancing wireless networks. Although some technologies are still at a low Technology Readiness Level (TRL), Hexa-X plans to investigate how they could influence the overall system architecture.

6.5.2 Society and Industry Pull

Hexa-X aims to provide sustainable digital solutions and support the United Nations (UN) and European Sustainable Development Goals (SDGs). The goal is to digitise various industry sectors, improve economic efficiency and resilience, encourage sustainable growth, generate meaningful job opportunities and promote the transformation of Europe into a robust circular economy.

6.5.3 Hexa-X Consortium

At the start of Hexa-X in 2021, it aimed to carefully select participants with the necessary backgrounds and critical mass to establish the foundations of 6G systems. Through a flagship research project, they seek to place the EU at the forefront of B5G/6G research and development. The Consortium incorporates top representatives from vendors, operators, the IT industry and high-tech companies to achieve this. Together, they want to create an ecosystem capable of performing the main objectives of Hexa-X and taking on a global leadership role. The Consortium comprises 25 organisations from nine EU countries, including multinational industries, innovative SMEs and distinguished research labs and universities. This ensures a broad level of internationally acknowledged technical expertise, guaranteeing high-quality contributions and solutions to the challenges of Hexa-X. The project is led by two network vendors, Nokia and Ericsson. Nokia is the overall lead, while Ericsson acts as the technical manager.

6.5.4 Hexa-X 6G research challenges

The Hexa-X initiative, comprising 25 participants, has developed a shared vision for the future of wireless communication, which they call 6G. The vision, illustrated in Figure 6.7, considers various factors that drive technological advancements and combines essential technical elements into an ‘x-enabler fabric’. This fabric acts as a foundation or modular architecture for developing the vision. The Hexa-X vision consists of three interdependent worlds: human intelligence and values, a digital world of information and a physical world of processes. The seamless integration of these worlds through real-time interactions is critical to overcoming future challenges. The Hexa-X vision identifies six primary research challenges that must be addressed to establish the technical foundation for the wireless systems of the B5G/6G era.

6.5.4.1 Connecting intelligence

In the future, 6G technology will be important in deploying intelligence throughout society. It will provide a framework for supporting and enhancing AI and machine

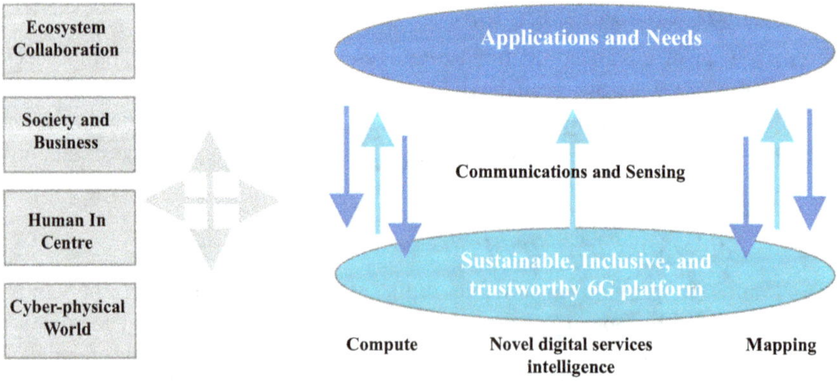

Figure 6.7 Hexa-X 6G platform (redrawn from reference [34])

learning technologies, enabling real-time trustworthy control. This will help improve efficiency and service experience, focusing on including human input in the process.

6.5.4.2 Network of networks[1]

The 6G network will combine various resources, such as communication, data and AI processing, to connect on different levels, from within the body to indoor locations, data centres and wide area networks. This will create a vast digital ecosystem that grows increasingly intelligent, complex and diverse, ultimately becoming a network of networks. The 6G network will support various needs and connectivity options while being cost-effective and flexible. It will promote business and economic growth and address societal challenges such as sustainable development, health, safety and digital inequality.

6.5.4.3 Sustainability

The development of 6G technology aims to improve the efficiency of digital networks and reduce their environmental impact. This will involve a complete overhaul of the resource chains used in wireless networks. In addition, 6G will enable real-time monitoring and analysis of the physical world, promoting sustainability from multiple perspectives, including environmental and economic factors.

6.5.4.4 Global service coverage

In developing 6G technology, digital inclusion should be a top priority. This means finding cost-effective solutions that provide global service coverage, even in remote areas such as rural regions or across oceans and vast land masses. By doing so, we can enable new services and businesses to promote economic growth and reduce the digital divide. Additionally, implementing 6G technology in uncovered areas can improve safety and operational efficiency.

6.5.4.5 Extreme experience

The upcoming 6G technology will offer incredibly high data transfer rates, ranging from hundreds of gigabits per second to a few terabits per second. The latency will be so low that it will be imperceptible, while the capacity will seem infinite. Additionally, it will provide precision in localisation and sensing, improving network performance beyond what is possible with 5G. This will unlock the commercial value of new technologies in the GHz–THz range, enhance the experience of services, such as fully immersive communication or remote control at scale, and accelerate the pace of digitisation.

6.5.4.6 Trustworthiness

The goal of 6G is to maintain the privacy and security of end-to-end communications while ensuring data privacy, operational resilience and security. This will help

[1]We discuss the implications of this theme in Chapter 9 – 6G Convergence, where we argue that if it adopts non-cellular networks, it could accelerate the digitalisation of society and industry faster and create significant opportunities for mobile carriers.

build trust in wireless networks and their applications for consumers and enterprises. It also aligns with European security, trust and privacy protection values. It supports the EU sovereignty goal of creating an open, trustworthy and more democratic Europe in the digital age.

The X-enabler fabric, an open, modular and flexible framework, will be developed as a foundation to integrate and weave together the technical enablers that address the six research challenges. This will be done by both the Hexa-X project and other 6G projects. The realisation of a new network generation takes about ten years. Hexa-X aims to lead the global Research and Innovation efforts towards 6G by laying a solid foundation for the network of 2030. To achieve this goal, the project will develop long-term strategic roadmaps based on research outputs obtained within the Hexa-X project and from other EU 6G projects.

The Hexa-X project understands that to fully embrace its vision for 6G networks, it is necessary to expand the fundamental network design paradigm from being mainly performance- to both performance- and value-oriented. This means considering intangible yet important human and societal needs such as sustainability, trust and inclusion. To achieve this, a new class of evaluation criteria called ***Key Value Indicators (KVIs)*** must be developed, understood and adopted in the network design process.

6.5.5 Hexa-X-II – second phase of Hexa-X

The Smart Network and Services Joint Undertaking (SNSJU) Hexa-X-II project [33] is the second phase of the European 6G flagship initiative (Hexa-X), which started in October 2022, within the SNSJU €900 million publicly funded research programme on next-generation networks with matching funding from industry.

This new phase will expand the list of Hexa-X partners to 44 organisations. Their task is to create a pre-standardised platform and system view, forming the basis for future inputs into 6G standardisation. The project will continue the tracks of the original Hexa-X, which laid its foundation for the global communication network of the 2030s by developing the 6G vision and basic concepts, including candidate key technology enablers.

Hexa-X-II has advanced from research to systemisation, early validation and proof of concept. This progress includes the 6G key enablers that connect the human, physical and digital worlds to advanced technology readiness. This readiness includes validated technology, providing critical aspects of modules, protocols, interfaces and data. Hexa-X-II aims to design a system blueprint to create a sustainable, inclusive and trustworthy 6G platform that can meet the future needs of society and businesses, as shown in Figure 6.7.

Hexa-X-II will address the implementation of the 6G platform by

- Defining use cases, services and requirements, ensuring the value for society
- Designing the platform and system to provide a global impact on 6G development.
- Assuring technology readiness in critical areas, ensuring EU strategic autonomy.

6.5.6 Hexa-X-II deliverables

The Hexa-X-II focuses on the following key deliverables with new results available at Hexa-X.EU/deliverables/

- D1.2 Expanded 6G vision, use cases and societal values – including sustainability, security and spectrum aspects.
- D1.3 Targets and requirements for 6G – initial E2E architecture.
- D2.3 Radio models and enabling techniques towards ultra-high data rate links and capacity in 6G.
- D3.3 Final models and measurements for localisation and sensing.
- D4.3 AI-driven communication and computation co-design: final solutions.
- D5.3 Final 6G architectural enablers and technological solutions.
- D6.3 Final evaluation of service management and orchestration mechanisms.
- D7.3 Special-purpose functionalities: final solutions.

The organisation and time scales for the three phases of SNS JU are shown in Figure 6.8.

6.5.7 Review of Hexa-X and Hera-X II progress

The 6G series workshop by Hexa-X and Hexa-X-II (6 June 2023) presented the latest deliverables and progress on both projects [34] with conference tracks on 'Wireless, Optical, and Satellite Networks', 'Network Softwarisation', 'RAS – Radio Access and Softwarisation' and '6G Visions and Sustainability', within a consolidated view on the 6G research from the major European players. Here, we present the key research directions and their status, which will confirm the status and direction of at least the early stages of 6G.

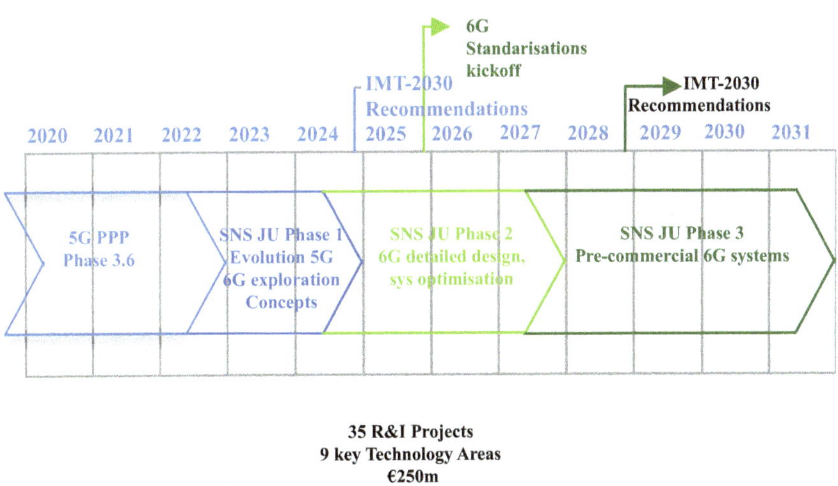

Figure 6.8 Hexa-X Phase II – organisation and time scales for the three phases of SNS JU (redrawn from reference [33])

Here, we include some key highlights from the workshop while all the presentations are available at [34].

6.5.8 What will 6G be?

Figure 6.9, from the workshop's Erik Dahlman Ericsson keynote presentation, shows the 6G focus areas. In these, it builds on those in 5G by expanding eMBB for immersive communications, URLLC for critical communications requiring low latency and reliability, and mMTC for massive communications. It also continues the theme of wireless communications everywhere, which started in 5G with NTN. Integrated sensing is one of the key differentiators between 5G and 6G in beyond-communications, which will build on AI and digital twins for new services in 6G, all under the umbrella of sustainability and trust aiming upwards offering sustainable digital solutions that support the United Nations (UN) and European SDGs. Its key value drivers include an ×10 capacity increase with a 50% power reduction compared with 5G. Meanwhile, digital inclusion aims to address accessibility and affordability. For security and privacy, it recognises that increasing these aspects will require higher control.

Regarding the 6G spectrum, Hexa-X sees possible use of the 7–16 GHz and above 100 GHz regions, sharing legacy 5G bands in the low and mid bands with licensed spectrum and local licenses and unlicensed spectrum.[2]

Nokia points towards the importance of MIMO in these new 6G bands and the advance of moving to higher frequencies, as we described in Chapter 7 on

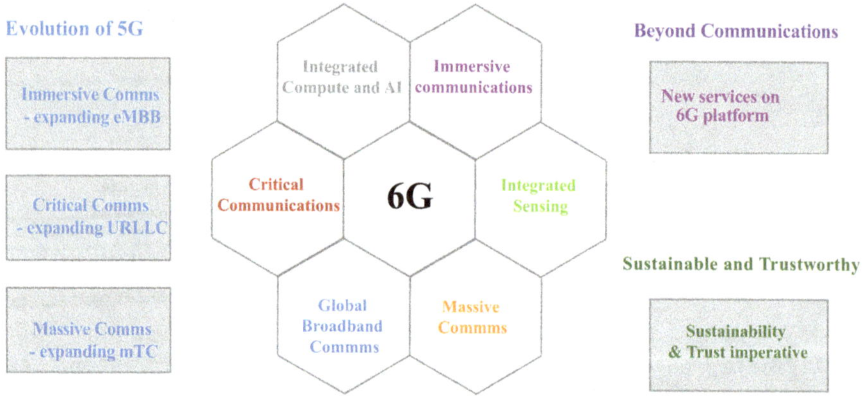

Figure 6.9 6G focus areas (based on Erik Dahlman, Ericsson keynote from reference [34], June 2023)

[2]IMT (spectrum for 6G) see Section 6.3. The WRC 2023 identified the upper 6 GHz band for IMT (spectrum for 6G), completed the harmonisation of the 3.5 GHz band for IMT, allocated the UHF band (470–694 MHz) to mobile and approved studies for WRC-27 on new bands. On the future agenda item for WRC 2027 on IMT, WRC-23 approved the study of the bands 4400–4800 MHz, 7125–8500 MHz, 14.8–15.35 GHz for IMT.

mmWave. They gave an example of 1024 MIMO at 7 GHz with 28 bit/s/Hz. They also featured work on this technology's power efficiency, achieving a 50% saving relative to a 64-transmitter MIMO. Nokia also points to their work on a native AI air interface to not only improve capacity but also help to reduce power consumption. The sub-THz communication scenarios focus on 10–100 Gbit/s for digital twins manufacturing, fixed wireless access (FWA), wireless fronthaul fully merged cyber-physical worlds and potentially holographic communications.[3] The range limits of these scenarios cover 10 m for holographic communications, 10–100 m for manufacturing and fully merged cyber-physical worlds, and up to 200 m for FWA.

The Nokia keynote presentation – What will 6G be? by Harish Viswanathan, initially covers the Immersive Experience use cases as shown in Figure 6.10 for Industry, Enterprise and Consumer. In Figure 6.11, Analysys Mason summarises the evolution of Metaverse services from 2023 to 2030 with critical performance parameters evolution from today's KPI of 5–10 Mbit/s and 50–125 ms latency to true Metaverse use cases requiring 1 Gbit/s+ and <10 ms in 2030 with 6G. When you reach True metaverse use cases with unlimited users and cloud processing, you need speeds more significant than that in 5G, namely 1 Gbit/s and latency of

Figure 6.10 Defining experience of the 6G Era (based on Harish Viswanathan, Nokia Bell Labs, keynote from reference [34], June 2023)

[3]We discuss the latter viability of true holographic communications in Chapter 16 where we are deeply sceptical.

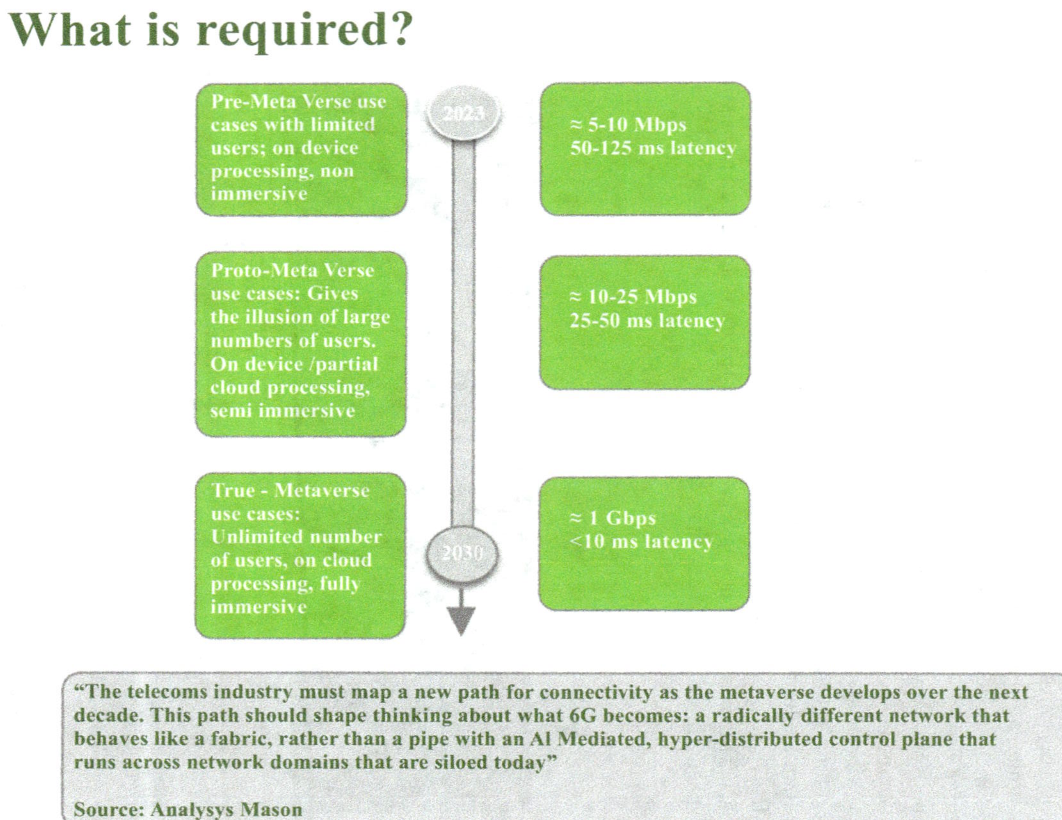

Figure 6.11 What is required? (based on Harish Viswanathan, Nokia Bell Labs, keynote, from reference [34], June 2023)

<10 ms.[4] This crystallises the critical challenge for the mobile industry in realising success from initial 5G and then 6G over this decade. Namely,

> *The telecoms industry must map a new path for connectivity as the metaverse develops over the decade. This path should shape thinking about what 6G becomes: a radically different network that behaves like a fabric, rather than a pipe, with an AI mediated, hype-distributed control plane that runs across network domains that are siloed today.*
>
> *Source: Analysys Mason*

One of the revolutionary aspects of 6G is sensing (JCAS), as shown in Figure 6.12. Sensing is used to bridge the digital and physical worlds. We cover this in greater detail in Chapter 14.

Nokia also points out the importance of the 6G network using AI to be cognitive with automation, while Chapter 13 describes AI for 6G.

6.6 Future technology trends International Mobile Telecommunications (IMT) towards 2030 and beyond

In November 2022, the ITU released a report titled 'Future Technology Trends International Mobile Telecommunications (IMT) Towards 2030 and Beyond' [2],

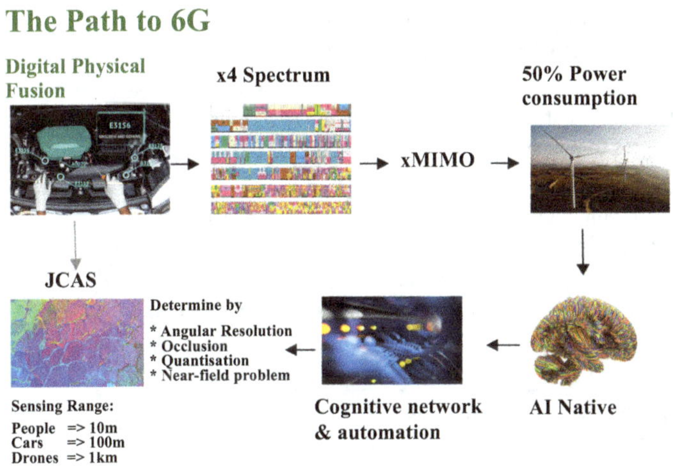

Figure 6.12 The path to 6G (based on Harish Viswanathan, Nokia Bell Labs, keynote, from reference [34], June 2023)

[4]A sceptic might point out that this is within the performance levels of Wi-Fi/FTTH networks today, as we have previously discussed in Chapters 3 and 4.

which outlines its vision for 6G technologies based on its views of the new 6G services. This report, in collaboration with other organisations, such as the Next Generation Mobile Networks Alliance (NGMN), which includes 59 contributions from operators, vendors and research parties, plays a significant role in defining the requirements for the future mobile network based on its views of services.

It supports the 6G research community consensus about the technologies and use cases, as shown in Tables 6.2 and 6.3; it identifies newly emerging technology trends for the development of 6G, including native AI technology – which refers to both AI-enabled air interface design and radio network for AI services, integrated sensing and communication, sub-Tera Hertz (THz) transmission, extreme MIMO and reconfigurable intelligent surfaces (RIS), enhanced trustworthiness with distributed ledger and quantum technology, as well as the interconnection between terrestrial networks (TNs) and NTNs.

Table 6.2 Technology trends adopted in ITU-R FTT Report M.2516 (based on reference [2])

Emerging technology trends and enablers:
• Technologies for AI-native communications. • Technologies for integrated sensing and communication. • Technologies to support the convergence of communication and computing architecture. • Technologies for device-to-device communications. • Technologies to efficiently utilise spectrum. • Technologies to enhance energy efficiency and low power consumption. • Technologies to natively support real-time services/communications. • Technologies to enhance trustworthiness.
Technologies to enhance the radio interface:
• Advanced modulation, coding and multiple access schemes. • Advanced antenna technologies. • In-band full-duplex communications. • Multiple physical dimension transmission (RIS, holographic radio, OAM) • THz communications. • Technologies to support ultra-high accuracy positioning.
Technology enablers to enhance the radio network:
• RAN slicing. • Technologies to support resilient and soft networks for guaranteed QoS. • New RAN architecture. • Technologies to support the digital twin network. • Technologies for interconnection with NTNs. • Support for ultra-dense radio network deployments. • Technologies to enhance RAN infrastructure sharing.

Table 6.3 Typical use cases for the six usage scenarios of 6G (IMT-2030)

Usage scenarios and typical use cases
Immersive Communication • Communication for immersive XR, remote multi-sensory telepresence and holographic communications • Mixed traffic of video, audio and other environmental data in a time-synchronised manner • Standalone support of voice
Massive Communication • Expanded and new applications such as in smart cities, transportation, logistics, health, energy, environmental monitoring and agriculture • Applications requiring a variety of IoT devices without batteries or with long-life batteries.
Hyper Reliable and Low-Latency Communication • Communications in an industrial environment for full automation, control and operation • Facilitating applications such as robotic interactions, emergency services, telemedicine and monitoring for electrical power transmission and distribution
Ubiquitous Connectivity • IoT communication • Mobile broadband communication
Integrated AI and Communication • 6G-assisted automated driving. • Autonomous collaboration between devices for medical assistance applications • Offloading of heavy computation operations across devices and networks
Integrated Sensing and Communication • 6G-assisted navigation. • Activity detection and movement tracking (e.g. posture/gesture recognition, fall detection, vehicle/pedestrian detection) • Environmental monitoring (e.g. rain/pollution detection) • Provision of sensing data/information on surroundings for AI, XR and digital twin applications (e.g. environment reconstruction, sensing fusion)

6.6.1 Usage scenarios

Six major usage scenarios have been defined for 6G, depicted in Figure 6.13 by the hexagon. This extends from the triangle, which features those from 5G (eMBB, mMTC, URLLC) with further enhancements. The circle around the hexagon contains four aspects: sustainability, ubiquitous intelligence, security/privacy/resilience and connecting the unconnected [36]. The services are depicted in Table 6.3.

In Chapter 16, the authors critically review the extended use cases to the original three use case types in 5G: enhanced Mobile Broadband (eMBB), machine-type communications (mMTC) and ultra-reliability low-latency communications. Generally, this allows us to specify the envisioned 6G KPIs in a more realistic light to avoid over-engineering the next generation as part of the conclusions to the book. **It will present an insight for mobile operators to see what is technically and commercially practical.**

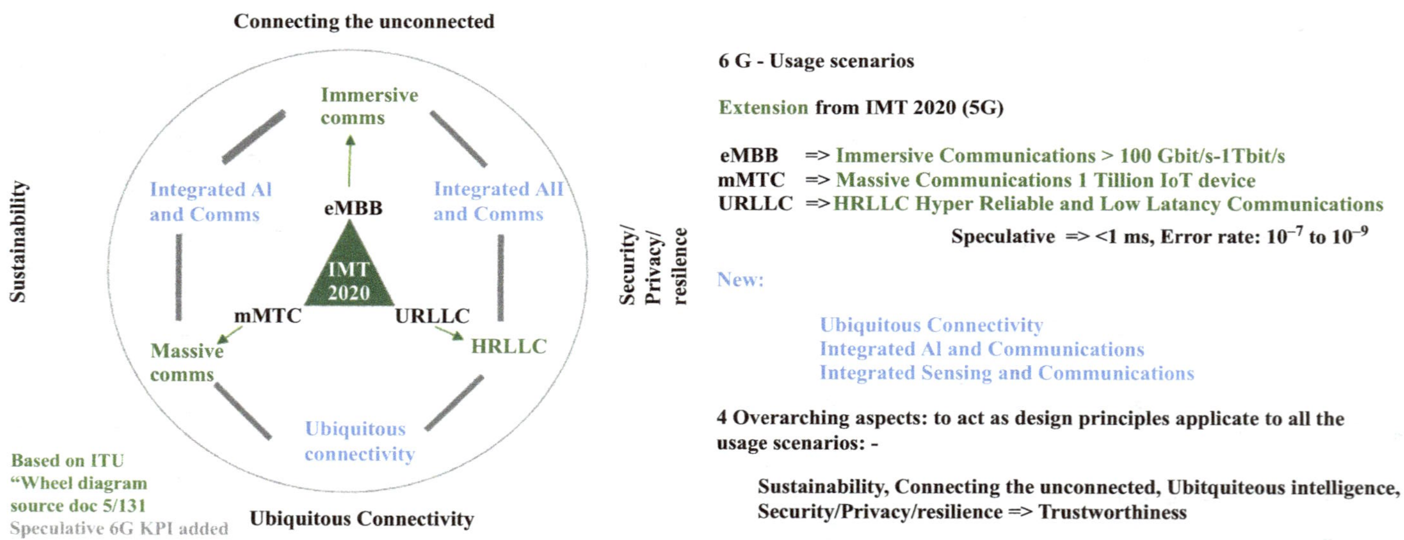

Figure 6.13 'Wheel diagram' of 6G use cases ITU. ITU-R, Draft New Recommendation [36]

6.7 Ten key takeaways on 6G from published white papers

Hexa-X and Hexa-X II have established a firm direction for 6G for the EU contribution to 3GPP, broadly shared amongst other regional partners with contributions to 3GPP. This has been achieved based on international consensus through MOUs, such as that with the USA ATIS, and presentations at international conferences. Nevertheless, reviewing key 6G white papers from the other six regions helps to contrast any difference with the Hexa-X vision and plans or provides additional definitions of the services, networks and technologies. Patrik Persson *et al.*, from Ericsson, have produced an excellent analysis, Nine Takeaways from Early 6G Research [3], of crucial 6G white papers published across the industry. We have reviewed these here with additional recent white papers. (To avoid confusion, we use Ericsson headings in these takeaways.) The consensus at this stage is highly 'directional' into the likely 6G standard or at least most of it.

6.7.1 Takeaway one: lessons from 5G for 6G

According to the SK Telecom white paper on the 6G [37], there are important lessons to be learned from the development of 5G that should influence the direction of 6G. The paper notes that while many exciting services were proposed for 5G, such as autonomous driving, UAM, XR, holograms and digital twins, they still need to meet their expectations, even now, as we approach the mid-way point of 5G. They identify other issues, including device form factor constraints, device and service technology immaturity, sizeable industrial equipment purchasing cycles, low or absent market demand and policy/regulation issues. This means a significant gap existed between the public's expectations and reality. SK believes that to successfully settle innovation services in a 6G era, all the ecosystem participants must work together to prepare the environment and 6G technology.[5]

6.7.2 Takeaway two: sustainability goals will be crucial to 6G use case development

Perhaps it is not surprising that this comes out top in all the various white papers because of the importance of climate change and the contributions the mobile industry can make with 6G to the UN Net Zero objective and broader sustainability. However, there have been similar goals in 4G and 5G. Yet, the total energy used by the mobile industry increased over that time even though one of 5G's achievements was reducing the energy requirement per bit. In particular, for such an important goal, Williams *et al.* [38] highlight that:

- The literature on green 5G and future mobile networks is dominated by small-scale, single-technology assessments.

[5]This is why we have included a critical analysis of the ITU report mentioned in Section 16.6 in Chapter 16.

- A lack of up-to-date, publicly available whole network level assessments of the energy use implications of 5G
- That the embodied energy use and indirect energy use effects of 5G have been largely overlooked in this literature.
- Insufficient attention has been paid to 5G-driven user behaviour changes and the prevention of rebound effects.

For instance, the Next Generation Mobile Networks Alliance (NGMN) industry body has suggested that 5G will be required to improve network energy efficiency by ×2000 [39] as an indication of the challenge needed.

The Hexa-X project is centred on developing 6G technology, addressing challenges arising from the ageing population and increasing global urbanisation. They call this shift moving from '5G engineering' to '6G humanity'. The US ATIS Next G Alliance aims to show that 6G applications will bring substantial economic and societal benefits, including improved cost efficiency, affordability, access and digital equity.

As the importance of climate-related action and sustainability issues continue to rise, Ericsson emphasises that all 6G use cases must prioritise technical and policy solutions that promote sustainability and net zero carbon emissions. To achieve this, the NGMN Alliance recommended monitoring the overall sustainability value of each service or application, which should be considered an essential step in the development of future 6G. However, how this can be enforced needs to be clarified. While a new IoT service, e.g. may initially look niche, it could grow significantly. If it had a considerable energy requirement per user, then its growth could be significant for the overall energy it uses. Additionally, Ericsson believes that ensuring high network energy performance will remain a critical design factor in future network platforms. This can be achieved by reducing node energy usage to close to zero when not carrying traffic and improving scalability by adapting to rapid traffic variations.

6.7.2.1 Energy harvesting

Energy harvesting is a fascinating concept that helps reach Net Zero in wireless systems, particularly in low-power IoT. Devices that harvest energy rely on solar, thermal, RF and piezoelectric sources to store energy that can power or reduce the power needs of wireless transmitters with typically 70% energy conversion efficiencies. Among these devices, solar cells using photovoltaic (PV) technology have the highest power density and output, while energy-harvesting systems based on radio frequency energy take advantage of the proliferation of radio transmitters worldwide. Statista forecasts that by 2025, over 18 billion mobile devices will be in use. Energy storage is often done using rechargeable batteries; however, storing energy in a capacitor is more practical for powering IoT tracking devices. Capacitors have unlimited charge–discharge cycles, and rechargeable batteries will wear out over time and eventually need replacing.

Nevertheless, it offers exciting IoT use cases where the battery requirements have been limiting, such as for the Bio Internet of Things. Now, body monitor devices can function without batteries. Ericsson believes that energy harvesting will become one of the key technology enablers for supporting a massive number of embedded devices in 6G. Deploying such 'zero energy devices' will remove existing use case limitations associated with battery replacement or charging requirements. However, energy harvesting is a new technology that needs to be clarified to the extent that this will influence wireless systems in general or IoT use. For example, powering a cell site would require solar cells with many square metres of space. Only relatively small amounts of power, with 70% conversion efficiency, can be achieved from thermal, RF and piezoelectric sources, limiting any access point's data rate and range. However, this technology could be used to augment existing electrical supplies.

6.7.3 Takeaway three: 6G will support trillions of embeddable devices?

In the coming years, various use cases, such as Digital Twins, Smart Cities, Industry 4.0 and the Internet of Everything, will require input from many embedded sensors. Additionally, they will need the ability to send information to actuators. To support this, it is argued that 6G will need to support trillions of embeddable devices with trustworthy connections that ***are always available***. This ambition is effectively orders of magnitude higher than the KPI in 5G of support of IoT devices (10^6/km). Addressing low-cost deployment and the energy supply are the two fundamental aspects that need consideration for success.

The Vivo Communications Research Institute Field [35] in China believes that Extremely Low Power Communication (ELPC) will be one of the critical technologies required to achieve ubiquitous connectivity and enable the interconnection between the physical and digital worlds. Their ambition for this technology is wider than short-range IoT. They advance wide-area scenarios, which include logistics and warehousing, environmental monitoring, smart agriculture, railroad operation and maintenance, powerline inspection and industrial IoT, which mirror 5G use cases. Vivo has proposed two possible network architectures for such ELPC device use cases. In the first option, the mobile network operator would provide a reader and proxies to the extremely low-power device to access the core network, meaning the device does not need to have NAS protocol stacks, reducing energy consumption. In the second option, ELPC devices do not access the core network. The core network only transfers data between the ELPC devices and an application server. The application server records the entity that transfers data to the ELPC power device and forwards the downlink data destined for the standard power device to the entity.

To put the challenge of achieving these ambitions in perspective, it is worth considering the performance of BlueTooth Low Energy (BLE), which was designed for low energy use and found many applications in health care, for example, and it probably meets many use cases requirements for 'Trillions of IoT' in turns of its range and data speed. These BLE devices can be powered by a 10-year life battery today with a capacity of approximately 0.75 Wh. To achieve a 'BlueTooth-type performance' with a range of up to 10–30 m and 100 kbit/s, using 2.4 GHz, these devices are usually in sleep mode, only waking up periodically because the radio and processors take a peak current of up to 10 mA. Requiring ***always-on*** and, hence, low latency would increase the power requirement by many orders of magnitude and challenge the ability of energy harvesting to power these alone.

One possible solution for the radio alone to reduce its power requirement would be to relocate this type of use to even lower frequencies, saving about half the energy requirement of a BLE device. However, this presents an enormous multiple-access problem given the concentration of a trillion devices, 10–100 devices per metre. More sophisticated multi-access schemes would place a much greater load on the processor, latency and power requirement. A simple calculation reveals another way to consider the scale of the challenge: Assuming only 1 mWh for an always-on device, which would significantly improve the performance of BLE – A trillion IoT would need almost 9 TWh per year. This is close to an additional third increase in worldwide electricity consumption, which was 25 TWh in 2022, according to Statista. Hence, the challenge of energy harvesting with a 70% conversion efficiency is approximately equal to the global challenge of Net Zero – Ten years earlier!

However, wide-area applications would require considerably greater than 1 mWh energy consumption even using low-frequency bands not considered in 6G use cases. For example, your mobile probably has a transmit power of 100 mW at around 2.4 GHz.

Ericsson has identified several research challenges before zero-energy or extremely low-power devices become a reality. These challenges include energy harvesting and storage and designing a system that can potentially handle orders of potentially higher IoT devices than 5G. It is necessary to develop new physical-layer designs since traditional transmission schemes may not be feasible due to the minuscule amounts of energy that can be harvested.

6.7.4 Takeaway four: network resilience will be a crucial design element of 6G systems

As identified for 5G for industrial and societal use cases, ensuring high network reliability, availability and resilience is necessary and will remain a key factor for 6G. We covered the KPIs for different use cases in Chapter 5 on Industrial 4.0. As in 5G and Wi-Fi 7/8, radio resilience can be improved by providing several overlapping techniques such as capacity, coverage redundancy, diversity of connectivity and medium access control. Achieving the highest network reliability, availability and resilience level for the most critical use cases will always remain a challenge for a wireless system, and some of the essential instances of using I4 will use cables! While 6G focuses on wireless solutions, it is crucial to recognise that other solutions work better in critical use cases which do not require mobile access.

For network resilience, we consider the centralised 5G Core and the statefulness of UE, gNB to Core (mainly AMF) a significant weakness in the 5G architecture, as it is only partially compatible with a cloud-native architecture. This results in the 5G service needing help to take full advantage of the high availability associated with cloud-native implementations. The loss of some network functions can lead to session drops. For 6G, it is recommended to review the design of 5G protocols that are not stateless.

Nevertheless, according to Ericsson, it is essential to approach network resilience from various angles. One way to achieve this is by creating a distributed architecture to avoid centralising all information and risks with a select few. Additionally, Nokia Bell Labs suggests that developing 6G sub-networks can guarantee fast data rates, minimal latency and reliable connections for life-critical communication. This can be achieved by implementing 6G security and resilience features on all devices in the sub-network. The EU Hexa-X points out that allowing regional network portions to continue operation is essential even when central functions may fail. Ensuring the continuation of critical use cases, e.g. emergency health care, is vital. We believe this could be taken further to a coreless 6G network architecture, using traditional Internet distribution mechanisms like DNS and newer mechanisms like distributed ledgers, distributed hash tables and consensus algorithms to highly distribute all network functions for super-ultra-high availability.

In ensuring resilience against dynamic changes in traffic load and radio environments of 6G systems, the emergence of AI and real-time analytics is expected to play a significant role. Huawei refers to this phenomenon as 'smart resilience' [40]. They believe that by leveraging situation awareness and big data analytics, 6G systems can identify, avoid, or transfer risks.

6.7.5 Takeaway five: 6G network architecture will be more adaptable and dynamic

Several organisations conducting research, including Nokia Bell Labs, Ericsson and NTT Docomo, a Japanese mobile phone operator, have predicted that 6G networks must be more adaptable and responsive to overcome deployment costs, energy consumption, network expansion and management challenges. NTT Docomo has identified several factors driving Advanced 5G and 6G development, including a significant increase in traffic from advanced cyber-physical systems, growth in wearable devices, the rapid implementation of new services in response to market changes and the need for robust defence against advanced cyber-attacks.

Nokia Bell Labs predicts that 6G systems will have varying technology and business models, forming a 'network of networks' that includes sub-networks, RANaaS and connectivity across different layers using new multi-connectivity techniques.

We believe that Industry 4.0 will be held back without blueprints that combine multiple wireless and wired technologies across multiple organisations in carefully controlled extranets with standardised vertical industry information models. This reflects our first takeaway: all Industry 4.0 service delivery areas must be addressed.

Nokia Bell Labs highlights advances in network slicing and virtualisation. Slices can be configured with dedicated software stacks for different types of flows, leading to more granular disaggregation of RAN functions.

Ericsson, Nokia Bell Labs, NTT Docomo and others emphasise the importance of cloud-native architectures and simplification. One crucial step towards this goal is separating essential, standardised business interfaces from implementation and deployment aspects. This approach allows for efficient implementation and leverages increasingly virtualised and cloud-native infrastructure, enabling different deployment decisions for various types of networks.

6.7.6 Takeaway six: 6G sensing capabilities

Sensing capabilities in their broadest forms is one of the revolutionary ideas in 6G, representing a significant new direction from 5G, albeit its standards included this capability. It can offer accurate spatial mapping through detailed sensing and high-precision positioning technologies. This is one of the key pillars that will bridge the digital and physical world enshrined in digital twins (we will discuss this in more detail in Chapter 14). This means that 6G networks can map physical surroundings accurately using radar-like technologies. Sensing can be achieved by observing the characteristics of received signals used for communication or by sending additional signals and observing their reflections on objects. Ericsson believes that reusing cellular systems for sensing can result in both a more cost-efficient sensing system and broader coverage than what can be provided by dedicated sensing systems. Using mmWave with millimetre and sub-millimetre wavelengths will significantly enhance the accuracy and pave the way for novel use of this technology than simple ultra-speed communications. In addition, including mmWave capabilities in UEs and networks will open new sensing applications, e.g. remote biometric monitoring.

There are various potential applications for enhancing network performance and offering innovative sensing services to users and applications. For instance, Ericsson has suggested use cases such as environmental modelling, road traffic detection and alarm detection systems. Huawei has identified various use cases, including high-precision and tracking, multi-robot collaboration and AI-based semantic localisation with context awareness and dynamic address resolution. Additionally, simultaneous improvements in multiple areas can be achieved.

Significant technological advancements will be necessary across multiple areas to achieve complete immersion and joint network communication during the 6G era. Nokia Bell Labs has highlighted the importance of improving AI/ML technologies and developing new software and knowledge systems that can effectively interpret the information gathered by networks. This information can then be fed into digital twins, enabling the wireless industry to develop applications and services that can use these data. Ericsson collaborated with NXP semiconductors to investigate potential new use cases for a network where communication and sensing functionalities are fully integrated into the same transmission/reception nodes [41].

6.7.7 Takeaway seven: 6G extreme performance

To create a more advanced telepresence experience and minimise the need for travel, it is recommended that 6G technology enhances the critical aspects of 5G. This encompasses faster speeds, higher throughput, decreased latency of less than one millisecond[6] and uninterrupted multi-access service continuity across various land, sea and air networks. This aligns with the fifth research challenge of the Hexa-X list in Section 6.5.4.

In addition to the Hexa-X objective, extreme performance is also taken to include '100%' coverage with terrestrial and Non-Terrestrial Networks (NTN), a theme also in 5G. This is best described as global satellite coverage focusing on outside environments but covering the planet's land and sea masses. This amplifies the primary aims of 5G. Interestingly, in August 2022, Starlink Operator SpaceX and T-Mobile announced the delivery of space-to-ground service to mobile phones in areas not covered by T-Mobile's cellular network. Given the size of the USA's land mass, this implies it is cheaper than a build-out of their cellular network covering the last few per cent of the area coverage where the number of calls is very low. Similarly, Apple has included satellite coverage for emergency texts in their iPhone and iPhone 15 smartphones for two years.

Moving towards '100%' global coverage primarily focuses on new use cases such as emergency access and remote control of autonomous robots and drones. In other parts of the world, NTN could be essential to address the digital divide, provided there are affordable solutions. However, we explore these use cases in Chapter 10, and achieving affordable prices looks challenging for broadband access. For example, using cm and mm wavelengths in 5G primarily opened the doors for mobile operators to offer FWA using a 3GPP standard; in most developed countries, populations already have access to circa 90% coverage with high-speed fixed broadband networks. The remaining, circa 10%, is addressed by various technologies.

mmWave technology in the 100–300 GHz range for 6G has the potential to offer 100 Gbit/s plus access network speeds with low latency. Due to the reduced bandwidth utilisation, these speeds may be required to provide ultra-low latency (Chapters 4 and 5).

According to Ericsson [3], achieving high performance in densely populated areas requires more than advanced access-link technology. It will also be necessary to implement packet fronthaul and modern wireless transport technologies like relay and mesh networking and further integrate access and backhaul functions.

To achieve the exceptional performance required for 6G, new spectrum bands and radio technologies will be necessary, including holographic beamforming, advanced duplexing technologies and gigantic massive MIMO technology. The Next G Alliance identifies massive MIMO as a crucial factor in achieving high data rates and broad coverage for 6G, and Nokia Bell Labs recommends the use of multi-user MIMO in mmWave bands to enable widespread multi-user massive MIMO and efficiently manage network density.

[6]However, in our analysis of the latency requirements for most industry use cases, we found that less than 1 ms is an outlier; we covered this in Chapter 16. Applications that require sub-ms latencies have ethernet solutions available and do not require mobility.

6.7.8 Takeaway eight: 6G networks will have the ability to learn and act autonomously

Enabling and supporting the expected 6G services scale and versatility will require new network intelligence and autonomy levels. Ericsson points out that this paradigm shift will likely occur gradually over the coming years, resulting in 6G fully cognitive networks that can observe, reason, acquire new knowledge and act autonomously. These new cognitive networks will enable energy efficiencies identified in the second takeaway and produce optimal performance with high service availability specified for industrial or critical applications.

The NGMN describes 'seamless hyper-automation' as a complete automation framework that enables CSPs to manage their services, networks and business/policy domains automatically. This framework would require end-to-end system visibility and full integration of AI functionality to achieve fully automated life-cycle management.

In their recent white paper '6G Position Statement – An Operators View' [42], they state

> *NGMN Alliance (NGMN) believes that 6G is the graceful evolution of communication networks into the 2030s, delivering compelling new services and capabilities for customers whilst maintaining essential offerings such as voice. 6G will build on, and extend beyond, our existing 5G ecosystem to foster new innovations which deliver value to customers and simplify network operation.*
>
> *Concurrent to this journey towards the 6G era is the development of network disaggregation and an open, interoperable cloud-native architecture.*

Experts in the field of 6G research agree that AI will play a crucial role in developing 6G systems. AI must be utilised in every aspect of the network's design to create self-sufficient and perceptive networks. This will allow the network to adapt to changes in its surroundings and learn from past experiences. Therefore, integrating AI as a service and a core component of the 6G system will be essential. We cover AI for 6G in Chapter 13.

Furthermore, AI could also be used to optimise radio codecs on the fly to adjust automatically to a changing environment or application's requirements. AI could also create higher-layer protocols, e.g. the MAC layer, that adapt to new requirements without requiring new standards. Nokia Bell Labs has been researching a native AI air interface.

As we continue to move towards the future, the upcoming 6G networks are expected to provide AI-as-a-service (AIaaS). This will facilitate applications while utilising high-quality data derived from the vast amount of raw network intelligence generated at the edge. In a previous chapter, we discussed how Huawei views distributed learning and inference as a critical use case for AIaaS. They believe feeding vast amounts of data into deep learning algorithms can create deep neural network (DNN) models for each application. Huawei also predicts that AIaaS for distributed learning and inference applications in 6G networks will play a crucial role in meeting the future real-time and large-scale learning and inference requirements of various industries and society.

We believe that distributed AI using the Edge Compute infrastructure supporting the network could offer a disruptive AIaaS, with the benefits of reducing the volume of IoT data uploaded, protecting privacy, and quicker reactions to events, which could compete with the Hyperscalers AIaaS.

A revolutionary idea is that AI could make standards bodies redundant in the long term, significantly increasing the velocity of deploying new services [40].

6.7.9 Takeaway nine: an integrated network computing fabric will fuel 6G network evolution

Future 6G use cases, such as the Internet of Senses and Cyber-Physical Systems, will require new capabilities beyond the connectivity [43]. As a result, 6G systems will be created to provide an all-in-one network compute fabric that will turn the network into a widely connected compute and storage platform. This will help efficiently handle application components while giving a sense of locality. The network compute fabric relies on real-time infrastructure, unified data access and intelligent operations to simplify serviceability.

Some distributed compute algorithms could be implemented in network hardware, accelerating some computations by orders of magnitude.

The evolution of the compute and storage paradigm forms a central theme in early 6G research. In their earlier research on 6G, Nokia Bell Labs identified it as one of the essential dimensions driving the design of the new communication system.

The Next G Alliance agrees that the convergence of mobile communications and cloud computing will be one of the critical drivers of network evolution. This will result in a 6G system that provides a wide-area cloud with ubiquitous computing across devices, network nodes and data centres.

The upcoming 6G system is set to revolutionise the mobile industry by expanding its capabilities and services beyond communication and what 5G offers today. This transformation is expected to pave the way for new service subscription models such as Everything as a Service (XaaS) for mobile users, device vendors, application providers, cloud service providers and other stakeholders. This paradigm shift will allow mobile devices to perform computing and data processing tasks in addition to traditional communication functionalities.

Hexa-X believes that delegating workloads to powerful nodes in the network will be essential for a stable closed-loop system that involves capturing measurements, processing, issuing an actuation policy and implementing it. This is the requirement driving the interest in sub-1 ms latency. In its white paper, it identifies three use cases that can potentially benefit from the Containers as a Service (CaaS) concept: industrial maintenance setting, where computing workloads need to be processed in a reliable and timely fashion; remote data collection and processing; and multi-player gaming, where complex computer games may be processed on computational resources in the network, addressing high computation load requirements and meeting low-latency requirements.

Ericsson argues that to create a compute and storage infrastructure of this magnitude, it is necessary to collaborate with a diverse group of stakeholders who work within a globally federated ecosystem. This group should include network and cloud providers, application developers, service providers and device and equipment vendors. This collaboration is vital to unlock the full potential of 6G systems and enable anticipated innovation.

However, resource management has many research challenges, such as how computing workloads are distributed across a heterogeneous network and security, especially if unqualified code can run on the network. New compute resource paradigms more suited to wide-area distribution across individually low compute power nodes are required.

6.7.10 Takeaway ten: 6G will be built to ensure trustworthiness in a new age

With the advent of 6G systems, the architecture of communication networks is undergoing significant changes. However, these changes also bring new and complex cybersecurity challenges that must be addressed. In particular, the threat analysis of future 6G use cases will have to consider various new factors, such as the potential use of on-body sensors and actuators, the extensive use of AI and advanced 3D+audio content spoofing. Furthermore, the emergence of quantum computing, combined with recent advancements in data privacy and cryptography, makes the technology area even more complicated, thus adding further challenges to existing ones.

Consequently, early 6G research has focused on ensuring the trustworthiness and dependability of future 6G systems by strengthening security controls for well-known threats and disruptions and exploring new aspects. The ability to withstand, detect, respond to and recover from attacks and unintentional disturbances is a cornerstone in designing trustworthy systems.

Ericsson believes that the key elements for developing trustworthy 6G systems are confidential computing, secure identities and protocols, service availability, security assurance and defence. These elements will need to be further improved in the years ahead. Confidential computing can protect future cloud users' privacy and enhance future network slices' security through cryptographic isolation.

We have proposed that 6G networks could use privacy-enhancing technologies (PETs) to optimise them across MNOs, and 6G customers could use them as a service; see Chapter 12 on 6G PETs.

Based in Japan, the Beyond 5G Promotion Consortium (B5G) has identified various critical components for a future cyber-resilience framework. These include AI security, automated software creation, quantum-safe cryptography, physical layer security and jamming protection. Besides confidential computing, B5G has recognised future technologies such as multiparty computation, federated learning, artificial data synthesis using digital twins and homomorphic encryption as crucial in the privacy paradigm.

The Next G Alliance also states that new tools will be needed to allow the 6G system to learn, detect and respond to threats autonomously. One of those tools will be AI, which the Next G Alliance has identified as pivotal in ensuring network trustworthiness.

To tackle the increasing risk of sophisticated cyber-attacks and breaches, Hexa-X suggests the development of new and adequate security and privacy schemes. This involves leveraging AI to anticipate potential threats and identifying and automatically resolving attacks caused by classical or AI-based methods. Additionally, Hexa-X emphasises embedding resilience and security-enabled trustworthiness in future network technologies' software and hardware implementations.

6.8 6G candidate spectrum

6G is exploring more usage scenarios than 5 G's eMBB, URLLC and mMTC and is expected to have more diverse bands, including lower, upper mid-band and sub-THz. For 6G's extreme performance, existing reframing and additional spectrum are necessary. Ericsson believes that the sub-1 GHz frequency bands will remain required for comprehensive area coverage even in the 6G era. At the same time, the mid-band spectrum will continue to address wide-area use cases that require capacity. Spectrum in the mmWave range will continue providing ultra-high capacity in crowded environments. Ericsson believes new scope in the centimetric (7–15 GHz) will be essential to enabling mobile high capacity 6G use cases, while others have suggested 7–24 GHz, while a complementary sub-THz (100–300 GHz) range will help to deliver required speeds beyond 100 Gbit/s and extremely low latencies of 6G niche use cases.

Nokia Bell Labs has identified the sub-THz range for future backhaul networks. Narrow beam point-to-point communication in these bands can free up spectrum for access in mmWave bands.

An interesting idea from Nokia Bell Labs for the future of spectrum management is that spectrum assignment could go from a static split between operators and services towards much more dynamic AI-based spectrum access in time, frequency and space.

Figure 6.14 illustrates a significant shift in spectrum usage for mobile systems. The frequency spectrum below approximately 5 GHz can be utilised in wide-area cellular networks, providing uninterrupted handover for various applications and services. 5G uses the 3.5 GHz band; however, it has been proposed to extend the capacity by using the C-band 3.6–4.2 GHz, so some 5G cells may use this band towards the end of the decade. However, this is not viable for frequencies above approximately 5 GHz due to the high costs associated with increasing cell density for wide-area use and the challenges of finding sufficient small cell sites. The other factor that comes into play is that in-building coverage falls significantly as frequencies increase. Moving to the cm bands for 6G, 7–24 GHz, will be necessary for even higher capacities for 6G. However, their use case will look like a WLAN or a cluster of WLAN access points for coverage within a building with no significant external wide area coverage, hence the spectrum dichotomy we discuss in the final chapter. Therefore, the higher 6G bands will be used predominately for indoor and campus-like scenarios like sports stadiums.

Furthermore, the main advantages of using phased arrays in MIMO become advantageous with system-on-a-chip (SoC) implements above circa 60 GHz, where

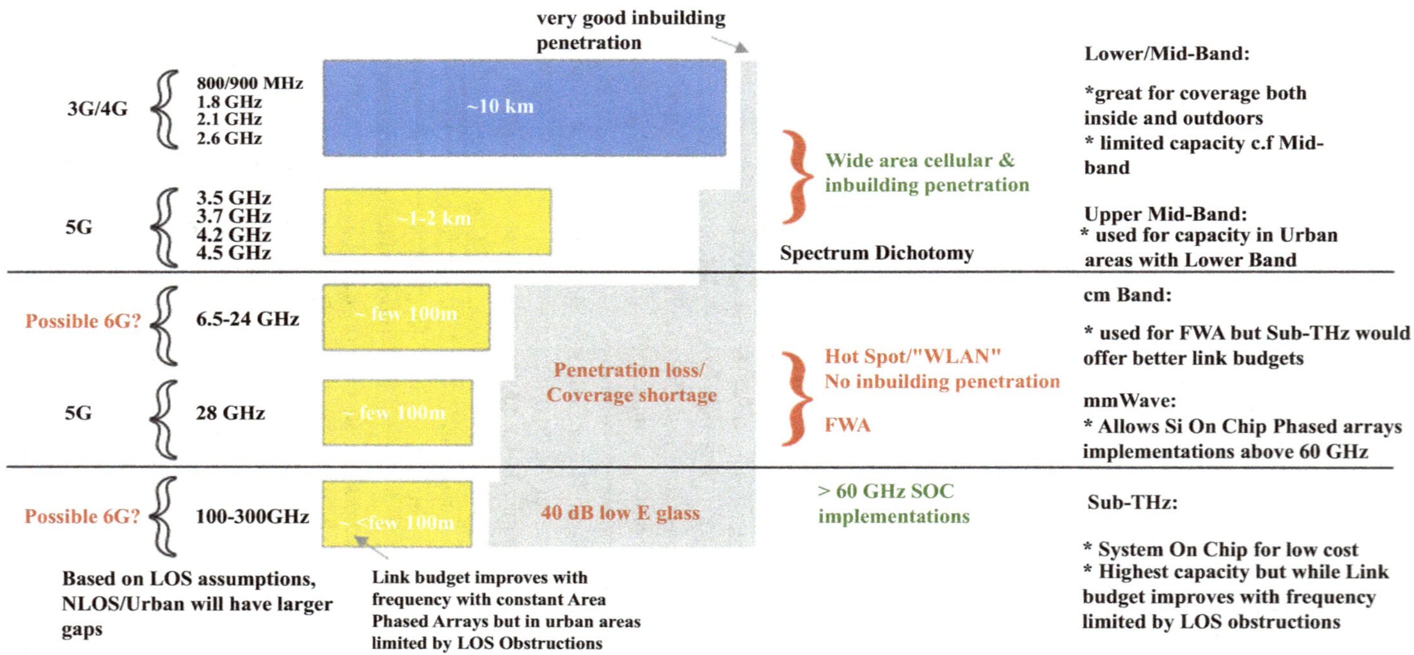

Figure 6.14 Use of spectrum in cellular systems and their implications

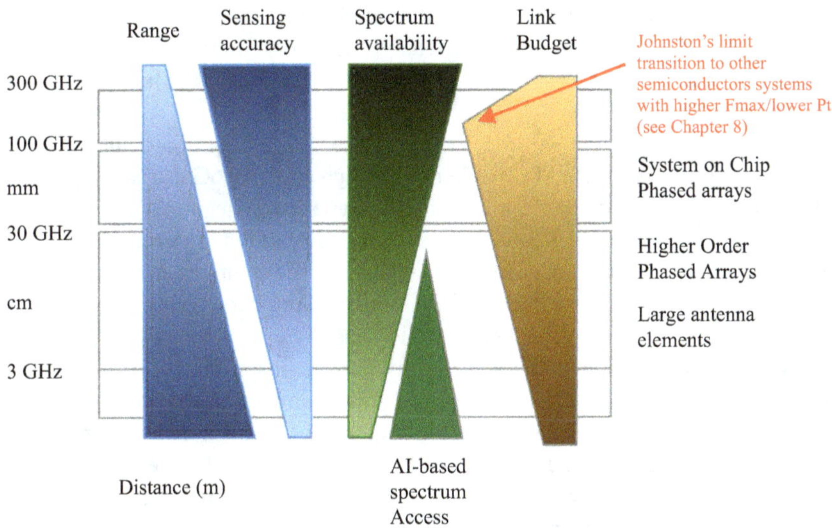

Figure 6.15 Nokia Spectrum options for 6G (re-drawn from EU's 6G Project Hexa-X with modifications [44])

the antenna arrays can be integrated with the electronics for low-cost implementations in a single IC, as we explained in Chapter 8, on mmWave.

Nokia has an interesting perception of spectrum use in 6G [44] (Figure 6.15). This includes the additional dimensions to Figure 6.14 required for sensing with spectrum availability with MIMO antennas, which we discuss in Chapter 14 for position accuracy.

6.9 Summary and conclusions

Collective efforts to research and define 6G are ongoing, as shown in Figures 6.2 and 6.3, with broad agreement emerging on themes and technologies. We have covered continuing research and industrial projects in detail within the second phase of the EU's Hexa-X project [Hexa-X-II] as representative of this. This has now focused on the systemisation of 6G. Parallel to this, there is ongoing ITU-R work about spectrum processes and vision with KPI/KVP activities.

The present 6G 'directional vision' is emerging, where 5G's eMBB, URLLC and mMTC technologies will further evolve. In contrast, AI and sensor convergence technologies will likely be applied to communications. The 6G network is set to grow significantly to learn and act autonomously. Its core implementation could be based on the 5G Core used in 5G advance with SDN and NFV technologies.

6G networks will be needed beyond just providing connectivity and, e.g. using digital twins for spatial mapping based on sensor measurements. With the advent of technologies like the Internet of Senses and Cyber-Physical Systems, 6G systems will be designed to offer a much more integrated network compute fabric. This will transform the network into a pervasive, globally interconnected platform optimised for efficiently handling application components while giving the impression of locality.

The spectrum for 6G will be more diverse. cm Wave, particularly in the 6.5–24 GHz region, is planned to enable mobile high capacity 6G use cases, probably focused mainly on indoor applications or canvas-type coverage. mmWave in the sub-THz region, 100–300 GHz, is intended to address 100+ Gbits/s speeds for extreme communication needs, but it may find uses other than for communications, such as in sensing. Over time, this technology could find revolutionary applications in spectrometry, such as scanning your meat for allergens.

While a '100%' global coverage network appears inspirational, its primary use cases are likely for emergency services, control of autonomous robots and drones, etc., and extension of mobile coverage to vast areas of land where it is uneconomic to deploy cellular networks because of the low traffic density. Its use is outside with a clear view of the sky. However, using this to address the global digital divide will depend on affordability in monthly access costs and UE implementations, as we cover in Chapter 10, but it does not look promising. The present 6G standards need to address these commercial issues.

While it is claimed that 6G will support Trillions of embedded devices for the IoT, this faces immense challenges in energy harvesting and multi-access schemes. Nevertheless, sustainability, and in particular, the reduction of energy in 6G networks, is of paramount importance. However, the SK Korea white paper warning could be remarkably prescient about the expectation gap a 6G era could create to reality if it does not address the other critical factors in developing and delivering new services other than just the communications stack.

6.10 Related reading for the latest updates to 6G including Videos

- Visit Ericsson's 6G resource Centre: [45]
- Visit Next G Alliance 6G library: [46]
- Visit Nokia's vision for the 6G era: [47]
- Visit Huawei 6G: The Next Horizon: [26]
- Visit Hexa-X: [48]
- Visit Library of white papers – University of Oulu 6G white paper Library: [49]

6.11 List of seminal white papers

- Ericsson's 6G white paper: Connecting a cyber-Physical World: [50]
- NGMN – 6G Position statement (An Operator's View) v1.0: [42]
- Visit Qualcomm white paper on 6G: [51]
- SK Telecom 6G white paper: [52]

- Huawei's 6G: The Next Horizon white papers: [25,26]
- 5GIA: European Vision for 6G network ecosystem white paper: [53]
- Beyond 5G Promotion Consortium white paper – Message to the 2030s: [54]
- Hexa-X white paper – 6G Vision, value, use cases and technologies: [55]
- Hexa-X white paper – Expanded 6G vision, use cases and societal Values: [56]
- NGMN Alliance white paper – 6G Drivers and Vision: [57]
- NGMN Alliance white paper – 6G use cases and analysis: [58]
- Nokia white paper – Communications in the 6G era: [59]
- Nokia white paper – Technology Innovations for 6G system architecture: [60]
- Nokia white paper – Joint Design of Communications and Sensing for Beyond 5G and 6G systems: [61]
- NTT DoCoMo white paper – 5G evolution and 6G: [62]
- Vivo white paper – 6G services, capabilities and enabling technologies: [35]

References

[1] ATIS – Next Gen Alliance. Roadmap to 6G. [Online]. 2023. Available from: https://www.nextgalliance.org/white_papers/roadmap-to-6g/.

[2] ITU-R M.2516-0. Future Technology Trends International Mobile Telecommunications (IMT) Towards 2030 and Beyond. [Online]. 2022. Available from: https://www.itu.int/dms_pub/itu-r/opb/rep/R-REP-M.2516-2022-PDF-E.pdf.

[3] Persson P, Arkko J, John W, Parkvall S, and Wikström G. What to Expect from 6G: Here are Nine Important Takeaways from Early Global Research. [Online]. Available from: https://www.ericsson.com/en/blog/2023/2/6g-early-research-global-takeaways.

[4] Wikipedia. 3GPP. [Online]. Available from: https://en.wikipedia.org/wiki/3GPP.

[5] Association of Radio Industries and Businesses. [Online]. Available from: https://www.arib.or.jp.

[6] China communications Standards Association. [Online]. Available from: https://www.ccsa.org.cn/english/.

[7] European Telecommunications Standards Institute. [Online]. Available from: https://www.etsi.org.

[8] TSDS. [Online]. Available from: https://tsdsi.in.

[9] TTA. [Online]. Available from: https://www.tta.or.kr/eng/index.do.

[10] Telecommunication Technology Committee (TTC). [Online]. Available from: https://www.ttc.or.jp/e.

[11] Telefonica. WRC-23 approves a clear roadmap for mobile services spectrum. [Online]. 2023. Available from: https://www.telefonica.com/en/communication-room/blog/wrc-23-approves-a-clear-roadmap-for-mobile-services-spectrum/.

[12] Nokia. Spectrum for 6G explained | Nokia. [Online]. Available from: https://www.nokia.com/about-us/newsroom/articles/spectrum-for-6G-explained/.

[13] Williams J. 6G Patents: China Continues Its Domination. [Online]. Available from: https://www.txpatentattorney.com/blog/6g-patents-china-continues-its-domination/.

[14] Nikkei Asia. LG, Samsung, Huawei build trove of metaverse patents: survey. [Online]. 2022]. Available from: https://asia.nikkei.com/Business/Technology/LG-Samsung-Huawei-build-trove-of-metaverse-patents-survey.
[15] Australian Strategic Policy Institute (ASPI). ASPI's Critical Technology Tracker – The global race for future power. [Online]. 2023. Available from: https://www.aspi.org.au/report/critical-technology-tracker.
[16] LightReading. China dominates research in 6G and optical – report. [Online]. 2023. Available from: https://www.lightreading.com/6g/china-dominates-research-in-6g-and-optical-report.
[17] EU Hexa-X project. [Online]. Available from: https://hexa-x.eu.
[18] Times E. EU Hexa-X Project Gets €12m to Develop 6G Vision. [Online]. Available from: https://www.eetimes.eu/eu-hexa-x-project-gets-e12m-to-develop-6g-vision/.
[19] EU. Europe scales up 6G research investments and selects 35 new projects worth €250 million. [Online]. 2022. Available from: https://digital-strategy.ec.europa.eu/en/news/europe-scales-6g-research-investments-and-selects-35-new-projects-worth-eu250-million.
[20] EU. A flagship for B5G/6G vision and intelligent fabric of technology enablers connecting human, physical, and digital worlds. [Online]. Available from: https://cordis.europa.eu/project/id/101015956.
[21] Review EI. Nokia to Lead the EU's 6G Project Hexa-X. [Online]. 2020. Available from: https://energyindustryreview.com/tech/nokia-to-lead-the-eus-6g-project-hexa-x/.
[22] UK Government. UK to Accelerate Research on 5G and 6G Technology as Part of £110 Million Telecoms R and D Package. [Online]. 2022. Available from: https://www.gov.uk/government/news/uk-to-accelerate-research-on-5g-and-6g-technology-as-part-of-110-million-telecoms-r-and-d-package.
[23] Fierce Wireless. U.S., U.K. Team to Tackle 6G. [Online]. 2021. Available from: https://www.fiercewireless.com/tech/u-s-u-k-team-to-tackle-6g.
[24] SCSP. National Action Plan for US. [Online]. Available from: https://www.scsp.ai/wp-content/uploads/2023/05/Final-National-Action-Plan-for-U.S.-Advantage-in-Advanced-Networks.pdf.
[25] Huawei. Defining 6G. [Online]. 2021. Available from: https://www.huawei.com/en/huaweitech/future-technologies/6g-the-next-horizon.
[26] Huawei. Future Technologies – 6G: The Next Horizon White Paper. [Online]. 2022 [cited January 2022]. Available from: https://www.huawei.com/en/huaweitech/future-technologies/6g-white-paper.
[27] rcrwireless.com. South Korea Aims to Launch First Commercial 6G Networks in 2028. [Online]. Available from: https://www.rcrwireless.com/20230224/6g/south-korea-aims-to-launch-first-commercial-6g-networks-in-2028.
[28] Digital Times Asia. South Korea K-Network 2030 Strategy (Part 1). [Online]. 2023. Available from: https://www.digitimes.com/news/a20230719RS400/digitimes-research.html&chid=2.
[29] The White House. FACT SHEET: United States – Republic of Korea Partnership. [Online]. 2021. Available from: https://www.whitehouse.gov/briefing-room/statements-releases/2021/05/21/fact-sheet-united-states-republic-of-korea-partnership/.

[30] Wireless F. South Korea kickstarts 6G plans. [Online]. 2021. Available from: https://www.fiercewireless.com/tech/south-korea-kickstarts-6g-plans.

[31] RCR Wireless News [Online]. 2022. Available from: https://www.rcrwireless.com/20221103/6g/japan-create-450-million-fund-6g-research-report.

[32] Consortium JB5P. [Online]. 2022. Available from: https://b5g.jp/doc/whitepaper_en_1-5.pdf.

[33] Smart networks and services Joint Undertaking (SNSJU). Home page. [Online]. Available from: https://smart-networks.europa.eu.

[34] Hexa-x II. [Online]. 2023. Available from: https://hexa-x-ii.eu/the-6g-series-workshop-by-hexa-x-and-hexa-x-ii/.

[35] Vivo. vivo Releases Third 6G White Paper: 6G Services, Capabilities and Enabling Technologies. [Online]. 2022 [cited 22 July 2022]. Available from: https://www.vivo.com/en/about-vivo/news/vivo-6g-services.

[36] ITU. ITU-R, DRAFT NEW RECOMMENDATION, Framework and Overall Objectives of the Future Development of IMT for 2030 and Beyond. [Online]. 2023. Available from: https://www.itu.int/en/ITU-R/study-groups/rsg5/rwp5d/imt-2030/Pages/default.aspx.

[37] SK Telecom. SKT Publishes *6G White Paper to Present Blueprint for Future Networks*. [Online]. 2023. Available from: https://www.sktelecom.com/en/press/press_detail.do?idx=1575.

[38] Williams L, Sovacool B, and Foxon T. The Energy Use Implications of 5G: Reviewing Whole Network Operational Energy, Embodied Energy, and Indirect Effects. *Renewable and Sustainable Energy Review*, 2022, 157, 112033.

[39] NGMN White paper: A deliverable by the NGNM. [Online]. Available from: https://ngmn.org/wp-content/uploads/NGMN_5G_White_Paper_V1_0.pdf.

[40] Liu F, Sun R, Wang D, Javali C, and Liu P. 6G Native Trustworthiness. Communications of HUAWEI RESEARCH | pp. 94–109, 2022 [cited September 2022. Available from: https://www.huawei.com/en/huaweitech/future-technologies/6g-native-trustworthiness.

[41] NXP. Joint Communication and Sensing in 6G Networks. [Online]. Available from: https://www.nxp.com/company/blog/joint-communication-and-sensing-in-6g-networks:BL-SENSING-IN-6G-NETWORKS.

[42] NGMN. 6G Position Statement – An operator's view. [Online]. 2023. Available from: https://www.ngmn.org/wp-content/uploads/NGMN_6G_Position_Statement.pdf.

[43] Ericsson CTO Erik Ekudden. [Online]. 2021. Available from: https://www.ericsson.com/en/reports-and-papers/ericsson-technology-review/articles/network-compute-fabric.

[44] Review EI. Nokia to Lead the EU's 6G Project Hexa-X. [Online]. 2020. Available from: https://energyindustryreview.com/tech/nokia-to-lead-the-eus-6g-project-hexa-x/.

[45] Ericsson. Follow the Journey to 6G. [Online]. Available from: https://www.ericsson.com/en/6g.

[46] Next G Alliance. 6G Library. [Online]. 2024. Available from: https://www.nextgalliance.org/6g-library/.

[47] Nokia. Nokia's Vision for the 6G Era. [Online]. 2024. Available from: https://www.nokia.com/about-us/newsroom/articles/nokias-vision-for-the-6g-era/.

[48] Hexa-X. EU Hexa-X. [Online]. Available from: https://hexa-x.eu.

[49] Oulu University. 6G Visions on paper. [Online]. Available from: https://www.6gflagship.com/white-papers/.

[50] Ericsson. 6G – Connecting a Cyber-Physical World. [Online]. Available from: https://www.ericsson.com/en/reports-and-papers/white-papers/a-research-outlook-towards-6g.

[51] Qualcomm. Vision, market drivers, and research directions on the path to 6G. [Online]. 2022 [cited December 2022]. Available from: https://www.qualcomm.com/content/dam/qcomm-martech/dm-assets/documents/Qualcomm-Whitepaper-Vision-market-drivers-and-research-directions-on-the-path-to-6G.pdf.

[52] SK Telecom. Press Release – SKT Publishes 6G White Paper to Present Blueprint for Future Networks. [Online]. 2023 [cited 10 August 2023]. Available from: https://www.sktelecom.com/en/press/press_detail.do?idx=1575.

[53] 5GIA. The 5G Infrastructure Association – European Vision for the 6G Network Ecosystem. [Online]. 2021 [cited 7 June 2021]. Available from: https://5g-ppp.eu/wp-content/uploads/2021/06/WhitePaper-6G-Europe.pdf.

[54] Beyond 5G Promotion Consortium. Output – White Paper. [Online]. 2024 [cited March 2024]. Available from: https://b5g.jp/en/output/.

[55] Hexa-X. 6G Overview: Vision, Value, Use Cases and Technologies. [Online]. 2021. Available from: https://hexa-x.eu/6g-overview-vision-value-use-cases-and-technologies-ieee-access-2021/.

[56] Hexa-X. Deliverable D1.2 Expanded 6G Vision, Use Cases and Societal Values. [Online]. 2021 [cited 30 April 2021]. Available from: https://hexa-x.eu/wp-content/uploads/2022/04/Hexa-X_D1.2_Edited.pdf.

[57] NGMN. 6G Drivers and Vision by NGMN Alliance. [Online]. 2021 [cited 19 April 2021]. Available from: https://www.ngmn.org/wp-content/uploads/NGMN-6G-Drivers-and-Vision-V1.0_final_New.pdf.

[58] NGMN. 6G Use Cases and Analysis. [Online]. 2022 [cited 22 February 2022]. Available from: https://www.ngmn.org/wp-content/uploads/220222-NGMN-6G-Use-Cases-and-Analysis-1.pdf.

[59] Nokia. Communications in the 6G Era. [Online]. 2020 [cited 9 September 2020]. Available from: https://www.bell-labs.com/institute/white-papers/communications-6g-era-white-paper/.

[60] Nokia. Technology Innovations for 6G System Architecture. [Online]. 2022 [cited 18 April 2022]. Available from: https://www.bell-labs.com/institute/white-papers/technology-innovations-for-6g-system-architecture/.

[61] Nokia. Joint Design of Communication and Sensing for Beyond 5G and 6G Systems. [Online]. 201 [cited 9 April 2021]. Available from: https://www.bell-labs.com/institute/white-papers/joint-design-communication-and-sensing-beyond-5g-and-6g-systems/.

[62] NTT DoCoMo. White Paper 5G Evolution and 6G. [Online]. 2022 [cited January 2022]. Available from: https://www.docomo.ne.jp/english/binary/pdf/corporate/technology/whitepaper_6g/DOCOMO_6G_White_PaperEN_v4.0.pdf.

Chapter 7
A radical approach to spectrum management

I do not think that radio waves that I have discovered have any practical application.

One cannot escape the feeling that these mathematical formulas have an independent existence and an intelligence of their own, that they are wiser than we are, wiser even than their discoverers.

– Heinrich Hertz

7.1 Introduction

The electromagnetic (EM) spectrum is a limited natural resource. Radio frequencies are part of the EM spectrum with an internationally agreed upper limit of 3 THz. However, for the past century, our exploitation of the radio spectrum has focused mainly on just the 10 GHz of this spectrum, with minimal use above 6 GHz and virtually none above 100 GHz. While each generation of mobile technology has required wider channel bandwidths for higher speed and greater capacity, this has inevitably driven us to use higher frequencies in the belief that there is little available spectrum at lower frequencies. But, as regulators are becoming increasingly aware that, in most places, the radio spectrum is underutilised, can we find ways to access additional bandwidth without having to climb the 'spectrum staircase?'

The era of popular radio use commenced in 1906 when Reginald Fessenden broadcasted an hour of talk and music for technical observers in Massachusetts, USA. The first commercial broadcasting station, KDKA, was established in Pittsburgh, USA, 14 years later. Until the 1980s, wireless use was primarily confined to the first 6 GHz of the EM spectrum for broadcast services, radio communications, mobile and Wi-Fi. However, the launch of satellite TV in the same decade pushed us beyond 2 GHz for popular broadcast and wireless services, with also the advent of 2G mobile and Wi-Fi. During this period, regulators worldwide allocated the entire 275 GHz of the EM spectrum to a range of commercial, engineering and scientific services, with minimal use over 95 GHz. Yet, human radio use remained mainly within the first 6 GHz, with only a fraction of the remaining 90 GHz utilised. The upper limit of the radio spectrum was legally defined as 275 GHz until recently. Still, it has now been extended to 3 THz to acknowledge the potential for scientific research in this region. Chapter 8 will delve into the spectrum requirements for 6G, including using mmWave spectrum above 100 GHz.

This chapter explores the management of the radio spectrum, the extent of the current underutilisation and the current inefficient use. In particular, virtually all radio systems use propagation models to establish the system range and performance and avoid and manage interference with other users. These band-specific models can lead to inefficiencies of between 50% and 70%. We propose a radical approach to improving spectrum efficiency using weak signal beacon technology for propagation reporting, avoiding using a propagation model. Using weak signal beacons implemented in the transmitters can deliver a paradigm shift in spectrum management, mainly when implemented within a shared spectrum system, like Citizens Broadband Radio Service (CBRS). This approach has the potential to massively improve spectrum efficiency, using real-time propagation measurements to assign resources and carry out real-time interference remediation efficiently. Using the best spectrum below 3 GHz for mobile could yield an additional 1.9 GHz for 6G alone. Reframing 3G and 4G bands while moving TV terrestrial broadcast to broadband could further enhance this figure to 2.5 GHz. This could allow most 6G use cases to be supported on existing infrastructure apart from those for extreme performance (10–100 Gbit/s wing cm and mmWave spectrum). Furthermore, this new approach can be applied across the whole EM spectrum but would have significant utility below 10 GHz.

This chapter shows how such a scheme could be applied to sharing the C-band between satellites and mobile users (arguably one of the most challenging cases to consider) before examining how the approach could be extended to the entirety of the radio spectrum. If this ambitious vision could be achieved, its economic impact would far dwarf any improvement in spectral efficiency arising from any new air-interface options that could be considered for 6G.

7.2 Radio spectrum and its management

It is important to understand that the radio spectrum is a highly regulated resource by government authorities. They are responsible for ensuring that it is used appropriately and licensed. Telecom regulators use auctions to determine the value of the spectrum, especially for mobile. The economic theory behind this goes beyond just generating revenue for governments. The main idea is that the people who bid the highest price will have the most significant economic utility for that spectrum. As a result, they will have a greater incentive to use it efficiently. However, regulators also recognise that wireless technologies like Wi-Fi can share the spectrum without a rigid ownership/management model and provide a country with significant economic value. They make spectrum available for both approaches, often creating tension between them.

An international body, the International Telecommunications Union, coordinates the national laws covering radio frequency use. This is important as radio waves cross international boundaries, and using incompatible radio technologies in the same band can cause interference. The ITU World Radio Conference coordinates this every four years, with the last in 2023 (WRC 2023).

Figure 7.1 shows the famous chart of the US radio spectrum allocation [2]. This covers the bands: 0–300 kHz, 300 kHz to 3 MHz, 3–30 MHz, 30–300 MHz,

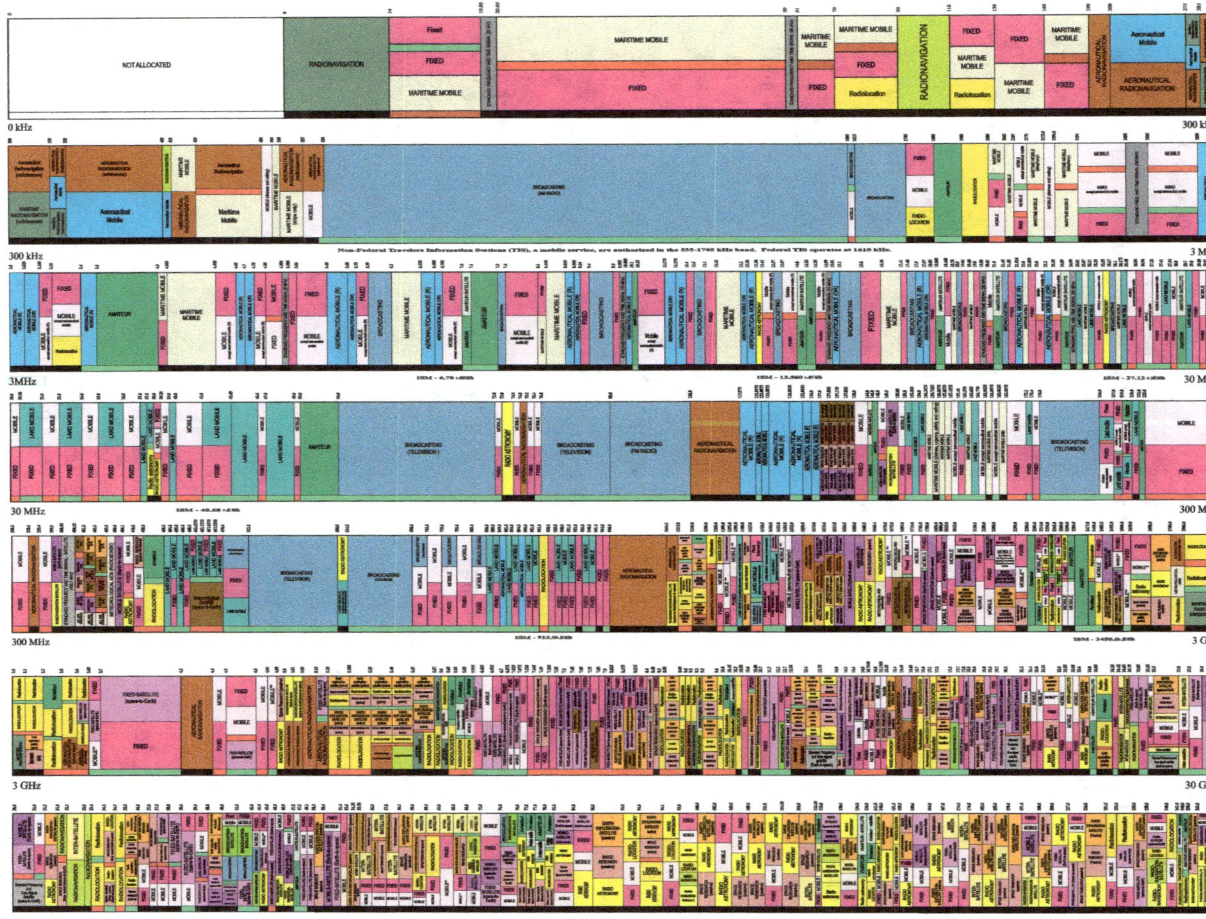

Figure 7.1 A 2016 chart of the US allocation of radio spectrum (source: NTIA [1])

300 MHz to 3 GHz; 3–GHz; and finally 30–300 GHz. This chart predates the change to the definition of radio spectrum up to 3 THz. Don't worry if you cannot see the details here, as that's my point – there is so much to it! However, the Internet reference will take you to the web page for the details of any specific band. A casual observer might conclude that this indicates a great success: the EM spectrum appears fully occupied. However, the vast majority of usage is currently in the bands below 10 GHz, with the spectrum above this largely unused in many parts of the world. For this reason, efforts to find additional spectrum for 6G primarily focus on the more lightly used higher frequencies, even though they will pose many practical problems when exploited for mobile applications.

When using the radio spectrum on mobile devices, the 3GPP mobile ecosystem marks off certain frequency bands reserved solely for mobile use. The 3GPP standards from the air interface, applied to both cellular base stations and devices, to the mobile core are only accessible with the specified mobile bands. Essentially, the mobile frequency band acts as a gatekeeper. This prevents, for example, Wi-Fi radios from being used with 3GPP-based networks.

However, there are examples where some of the more convenient lower frequency bands are beginning to be shared, albeit after protracted negotiations and financial compensation. For instance, the 3.7–4.2 GHz band (designated C-band and used by satellites globally) is now shared between mobile use for 5G and satellites in the United States.

Mobile network technologies take advantage of different characteristics of the radio spectrum, such as the following [3].

7.2.1 Low-band spectrum

Low-band (up to 1 GHz) has the following characteristics: it offers the most comprehensive economic coverage (fewer radio transmitters required) and better in-building coverage than higher frequencies. However, they inevitably need more channel bandwidth due to the high demand for these bands and their relatively low carrier frequency. Nevertheless, Ericsson believes it will play a role in future 6G systems at 700 MHz. Accordingly, low-band frequencies have been used heavily for commercial broadcast services (such as TV and radio), government communication services and cellular systems.

7.2.2 Mid-band spectrum

Mid-band (1–6 GHz) forms the basis of the most current cellular systems as it affords larger channel bandwidths for high-speed mobile data. It also accommodates the 2.4 and 5 GHz bands used by Wi-Fi.

For cellular use, frequencies up to approximately 3 GHz are suitable for macro cells covering distances of many kilometres. Above this limit, higher frequencies align better with micro-cellular use deployed in cities, typically covering a radius of a few hundred meters. Beyond 4 GHz, the cell densification requirement for wide area coverage is too expensive and only suitable for hot spot traffic areas in cities where the high concentration of users affords better economics.

7.2.3 High-band spectrum

High-band (>6 GHz) includes the cm bands, up to 30 GHz, and beyond that, the mmWave bands, which we discussed in Chapter 8. The high-band spectrum is interesting because it can offer extremely high channel bandwidths, supporting high data rates and significantly increased capacity. However, cm and mmWave do not readily penetrate physical structures such as buildings, so the present 'from outside to in' mode of mobile network deployment using external cells best suits the use of the Low band spectrum for in-building coverage. Later, we discuss a radical alternative approach called 'inside-out', using small cells within buildings, in Chapter 9.

5G was the first cellular system to start exploring the high-band frequencies (>6 GHz), using cm wavelengths (initially 24–29 GHz) to access channel bandwidths of 500 MHz or more. The high band will primarily be used for 5G up to 100 GHz, while 6G will likely exploit frequencies from 100 to 300 GHz to continue this strategy. It will also explore the use of 7–24 GHz, while the higher 6.5 GHz has already been allocated to 6G by the World Radio Conference 2023. Apart from using cm and mmWave for wireless local loops, the 5G handset use case for multi-Gbit/s services in these bands needs to be clarified (although these bands may find use in ultra-low latency applications, as described in Chapter 2.

7.3 The evolution of the use of the radio spectrum over five mobile generations

Each generation of mobile has a new radio air interface, which in turn requires a new access network infrastructure and new radios, modems, etc., all rolled out every decade. This was necessary for the first five generations as there were radical changes in the applications that mobile supported and the need to improve the spectral efficiency (bits/Hz) to conserve the amount of a scarce resource – spectrum. As the data speed and capacity requirements increase in each generation, this means expanding channel bandwidths, even given the improvements in spectral efficiency, as shown in Table 7.1. As the lower frequency band (<1 GHz), which supports much better coverage and building coverage in wide areas, is congested, this means inevitably moving up the 'spectrum staircase'. The low-band frequencies (<1 GHz) have already been used heavily for commercial broadcast services such as TV and Radio and government communication services. Even with spectrum re-farming, this does not help significantly as the lower bands have commensurate lower channel bandwidths. Consequently, 4G LTE uses 2.5 GHz, and 5G uses the 3.5 GHz mid-bands for high capacity in wide areas. For the first time, 5G also used cm-wave and mm-wave bands for high-speed applications that support multi-Gbit/s, allowing them to access even wider channels.

The prospects for 6G using greater parts of the Low band may be suitable for lower application speeds, such as voice, as terrestrial TV moves to fixed broadband during its lifetime. Ericsson believes that 6G will use 700 MHz. In many Western European countries, terrestrial TV uses coordinated frequencies at VHF and UHF (470–860 MHz). Similar coverage in the United States extends from 470 to

Table 7.1 The evolution of the spectrum over generations

Generation	Frequency bands	Bandwidths	TDD/FDD
1G (1980+)	900 MHz (Analogue)	25 kHz	FM
2G (1900+)	800, 850, 1900 MHz	<1.25 MHz	FDD
3G (2000+)	800, 850, 1900, 2100 MHz	1.25–5 MHz	FDD
4G (2010+)	600, 700, 850, 1700, 1900, 2100, 2300, 2500 MHz, unlicensed 5 GHz	5–20 MHz	TDD/FDD
5G (2020+)	600, 3500 MHz unlicensed 5 GHz and 6 GHz, 24 GHz, 26 GHz, 28 GHz, 29 GHz	5–100 MHz	TDD/FDD with mid and high bands being TDD
6G (2030+)	All of above+, 700 MHz, 6.5 GHz, 7–24 GHz, 100–300 GHz? + TV bands (<1 GHz)?	100–500 MHz	TDD?

884 MHz. This represents approximately 400 MHz of the spectrum, ideal for 6G with a national switch-off of this service. It is, however, difficult to predict if this will become available within the next 10 years. Still, with rapid penetration of Gbit/s broadband technologies such as Fibre to the Home (FTTH) and DOCSIS 4.0 Cable networks, this becomes inevitable as these broadband technologies can support multi-channels of 4K and 8K TV. Even one of the key TV satellite operators in Europe, SKY, announced in 2022 that it plans to offer its service on satellite and broadband with its new set-top box, eventually turning off its satellite service. The Internet-based approach offers their customers greater interactivity, enhancing their service offerings. So even if not available for the launch of 6G in 2032, it could become available for 6G during its life span.

In Table 7.1, 5G uses, apart from the 600 MHz band, the mid-band frequency band of 3.4–3.8 GHz for wide area coverage. It was also the first cellular system to start to explore the high-band frequencies (>6 GHz) with a move significantly into it with the use of cm wavelength (24–29 GHz), which to access channel bandwidths of 500 MHz or more.

The US FCC proposed in 2018 [4] the possible use of 70/80/90 GHz for 5G fixed wireless access (FWA), and the apparent conclusion is that 6G is likely to use frequencies above 100 GHz for this and other applications.

However, not long ago, the mobile industry viewed it as uneconomic to build commercial systems beyond circa 3.5 GHz because of the prohibitive costs of associated cell densification caused by the poorer propagation at these frequencies.

This created a dichotomy in spectrum use and the associated radio technologies. Frequencies below approximately 4 GHz have varying degrees of building penetration from outside deployment, allowing continuity of service from external networks deployment, 'outside-in'. These frequencies are suitable for widespread cellular deployment, while above this value, the costs of cell densification make this uneconomic. In comparison, frequencies above 30 GHz and certainly in the mmWave bands have virtually no building penetration and are suitable only for small cells. Therefore, their use inside buildings requires internal cellular network deployment with no corresponding outside coverage. Mobile operators who want to continue providing seamless handover need outside and inside networks for ultra-high-speed applications, which is uneconomic. This paradigm change for mobile operators is not well understood in discussions on 6G.

7.3.1 World Radio Conference 2023 (WRC 23)

WRC is organised by the International Telecommunication Union (ITU), a United Nations agency, and is usually held every four years. These treaty conferences have the power to change international regulations on the use of radio spectrum for the different services, including mobile, referred to at the ITU as International Mobile Technologies (IMT).

For the mobile industry, the WRC 23 [5] was critical to provide certainty of future spectrum availability for both 5G Advance and future 6G. More than 3900 delegates from 163 countries have delivered a successful agreement, identifying the upper 6 GHz band (6425–7125 MHz) for IMT, completing the harmonisation of the 3.5 GHz band for IMT, allocating the TV UHF band (470–694 MHz) to mobile and approving studies for WRC-27 on new bands.

Europe will study [6] in 2024–25 to determine the best usage of the upper 6 GHz band between Wi-Fi and IMT, including the option of a shared framework. A final decision is expected in 2026. Technical conditions have been established to protect satellite receivers in space from mobile base stations' aggregate interference.

On the UHF band used for TV today, the final decision of the WRC-23 was a secondary mobile allocation for Europe in all the band (470–694 MHz), while 11 countries in the Middle East allocated the band 614–694 MHz to mobile on a primary basis and identified the band for IMT.

With a secondary mobile allocation, Europe will be able to deploy IMT once the decisions on the future use of the UHF band are made within Europe after the planned review of the TV broadcast usage of the band by 2025.

On the future agenda item for WRC-27 on IMT (spectrum for 6G), the WRC-23 finally approved the study of the bands 4400–4800 MHz, 7125–8500 MHz, 14.8–15.35 GHz for IMT.

Identifying additional spectrum at WRC provides direction to manufacturers and operators, pointing to the concrete frequency bands on which to plan future developments. However, work at the regional level will now follow, and it will be some years before actual usage rights are issued.

7.4 How well is the radio spectrum used and managed?

Spectrum is arguably a scarce resource, particularly below 3 GHz, so what are the problems in general associated with its use and management? As Jenifer Manners points out in her book 'Spectrum Wars – The Rise of 5G and Beyond' [7],

> *Despite the apparent congestion of the spectrum resource, no one really knows the actual severity of congestion of the spectrum resource for several reasons. First, there is no international or national registry of what the spectrum resource is being used for. Even in heavily licensed bands, domestic regulators may not know the actual use of the licensed spectrum, including how much of the relevant band is used, and to what extent. This is compounded by the fact that no one has systematically reviewed every frequency band in the radio spectrum resource to determine what is in use. And even if such an analysis was performed, it is clear that even for the most intensely utilized bands, greater efficiencies of use could be made.*

In many respects, the ideal spectrum for cellular systems is the low-band (up to 1 GHz) and mid-band up to circa 3 GHz as they offer, in combination, wide area coverage, good building penetration and high capacity in a cell hierarchy of macro and microcells. But given its importance, how well is it used? This is not about the number of bits per Hertz a new air interface provides but instead how much is used on average across the country as if you were to carry out several spot frequency measurements, in many places, with a wide band antenna and a spectrum analyser to determine the average spectrum occupancy.

Over the past 20 years, spot spectrum occupancy measurements have been taken at various locations, including in the United States, Turkey and the United Kingdom. These measurements covered the critical 30–3000 MHz frequency range, which is heavily congested, and produced similar results. These indicate the efficiency of utilisation/occupation in that country and worldwide because these bands are harmonised internationally for identical use cases such as TV. Thus, the general agreement points to the overall effectiveness of international spectrum management.

Figure 7.2 shows some measurements carried out in Hull, a city in the North of England, UK, [8] in 2013, which aligns well with recent measurements in Ipswich,

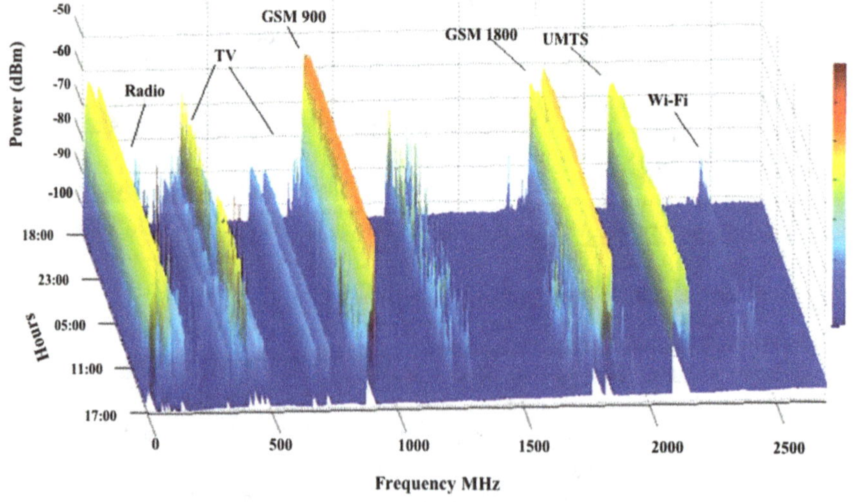

Figure 7.2 Measurement of the radio spectrum usage in Hull, UK, over 24 h [8]

Table 7.2 Analysis of the spectrum occupancy in bands across 180–2700 MHz in Hull, UK, and Konya, Turkey [8,10]

Hull, UK			**Konya, Turkey**	
Frequency band (MHz)	**Applications**	**Spectrum occupancy (%)**	**Frequency band (MHz)**	**Spectrum occupancy (%)**
			30–174	12.5
180–400	Military, FM, aerospace	6.9	174–230	16.0
			230–470	4.6
470–850	TV broadcast	13.5	470–790	8.5
880–960	Mobile GSM	32.9	890–960	4.1 and 91
1710–1880	Mobile	24.6	1710–1990	11.0
1900–2700	Mobile	9.5	2110–2170	88.3
	Mobile		2300–2500	2.9
Overall		**11.0**		**7.6**

a large town in the East of England by Suffolk University carried out by students under the supervision of Professor Peter Cochrane and Dr Felix Ngobigha [9]. The Hull measurements were carried out over 24 h to capture spontaneous use and indicate significant gaps of no use. For example, the overall spectrum occupancy for the 180–2700 MHz frequency range was only 11%. These spot measurements indicate poor radio spectrum utilisation even though it is fully allocated to different uses.

Meanwhile, Table 7.2 summarises spot measurements of spectrum occupancy and their band usage in Hull, UK, compared with Turkey to demonstrate similarities. In Konya, Turkey, in 2017, the overall spectrum occupancy over a slightly larger band of 30–3000 MHz was only 7.6%.

In the United States, in New York and Chicago, the overall usage was 13.1% and 17.4%, respectively, for the frequency range 30–3000 MHz [11,12]. None of these places shows significant use of the most valuable spectrum for commercial use, with a typical occupancy of around 15%. It is common for bands to have similar uses, radio technologies and deployment strategies for achieving nationwide coverage. In the 180–400 MHz frequency range, the UK and Konya have a similar occupancy rate of around 7%–8%, mainly utilised by government agencies. For TV broadcasting in the UHF bands, the occupancy rate ranges from 8% to 14%, depending on the number of broadcast channels assigned. The FM and TV broadcast bands are almost identical across most countries. While the mobile bands are globally harmonised for mobile roaming and scalability, they only make up approximately 10% of the first 3000 MHz. These bands have the highest occupancy rate, but their deployment heavily influences this conclusion in spot locations. This proves that the essential part of the EM spectrum for wireless communications has less than 20% utilisation/occupancy nationwide and may be as low as 11%. It is worth noting that external measurement campaigns do not effectively measure the internal use of radio, such as 2.4 GHz Wi-Fi. However, 2.4 GHz Wi-Fi only has a 60 MHz allocation.

These results on spectrum occupancy are likely a good representation of the situation nationwide. However, the overall occupancy of the spectrum between 30 and 3000 MHz at circa 15% is not an outstanding achievement for effective spectrum management, especially for the most crucial part of the EM spectrum. The current situation will likely stay the same as 5G, introduced after these results mainly use the 3.5 GHz band. Paradoxically, the perceived lack of spectrum has forced each mobile generation to use higher frequencies, leaving unused portions of the spectrum in most areas. There are potentially 1.9 GHz of additional spectrum available for 6G, below 3 GHz, assuming a new 80% occupation, with 20% (594 MHz) reserved for other uses, including previous mobile generations.

Of course, regulators are aware of this situation and are employing various means to address the problem. For example, in the United Kingdom, the Ofcom government strategy for the spectrum [13] states

> *Despite being finite, spectrum is also reusable. Notwithstanding progress in spectrum efficiency in particular frequency bands and locations, spectrum is still underutilised in many places and at many times. If this underutilised spectrum could be made available for use and reused intensively, additional value could be realised. Examples of opportunities to achieve this include more use of local and shared spectrum, automated and dynamic access to spectrum, reframing spectrum to accommodate higher value services and use of more agile and interference-resilient technologies.*

Now, of course, there are various explanations for this low occupancy. The broadcast networks for TV and radio deployments go back decades when we had simple analogue electronics and frequency division multiplexing for broadcast networks. Based on a simple hexagon propagation coverage model, one would expect utilisation of 1/7 or 14.2%, which aligns well with the UK figure of 13.5%.

In countries like the United States, the United Kingdom and various European nations, there is usually more than 400 MHz of spectrum available in the UHF band, which spans from 470 to 890 MHz. Recognising the importance of mobile usage, regulators have begun repurposing portions of the analogue TV bands, as witnessed by WRC 23. In 2017, the FCC incentive auction saw TV broadcasters selling off a portion of their TV spectrum, eventually acquired by mobile network operators. The FCC reorganised broadcasters into the remaining TV bands below Channel 37, freeing up 84 MHz of low-band spectrum. 70 MHz was designated for licensed mobile use, while 14 MHz was left for wireless microphones and unlicensed use. This auction generated $19.8 billion in revenue, with $10 billion going to the winning broadcasters and over $7 billion to the US Treasury for deficit reduction.

In comparison, $1.75 billion was also made available to repacked broadcasters to assist them in relocating to new channels. In March 2021, OFCOM in the UK announced plans to free up 80 MHz spectrum in the 700 MHz bands. This will be done following a four-year program to clear the band of its existing uses for digital terrestrial TV and wireless microphones.

Outside the broadcast use of spectrum, Governments are often some of the more significant users of the first 3 GHz spectrum, particularly for military communications systems. They guard their use conscientiously, which is why CBRS, described below, was so significant in that the US Navy was persuaded to share its use of the 3.5 GHz Band.

Another issue that can be seen clearly in Figure 7.2 is that gaps in spectrum occupancy do not simply arise from the use of legacy technologies but from the lack of localisation of the allocation of spectrum. This is particularly the case for the spectrum used by mobile operators. While auctions may seem efficient because they award spectrum to the highest bidder (as they have the most significant commodity for its use nationally), they are not required to use all their spectrum efficiently and have no requirement to share its use in places where it is not being fully exploited.

7.5 Implications of greater spectrum utilisation/occupation

Eventually, all linear TV could be delivered over the Internet, freeing up 400 MHz of high-value low-band spectrum for 6G, which would be an essential start to driving up the utilisation of the first 3 GHz of the EM spectrum. Perhaps a meaningful way to realise how significant this would be is through a heuristic consideration of this simple equation:

$$\text{Overall Mobile Capacity (bits)} = \text{Mobile spectrum (MHz)} \times$$
$$\text{Spectral efficiency (bits/Hz)} \times \text{Base Stations (function of Frequency)}.$$

The amount of spectrum available and the number of base stations deployed directly impact the overall capacity, as is well-known. All new mobile generations so far have used a new air interface, and the main driver for this has been to improve spectral efficiency, thereby driving up the overall mobile capacity. However, the spectral efficiency improvement of each new mobile air interface has slowed with each successive generation from 2G. Even though 4G LTE advanced offered circa 3–4 bits/Hz [14], 5G has extended this to around a theoretical figure of 5–8 bits/Hz. These quoted figures do not use MIMO, which has also shown smaller spectral improvements over time. The fundamental reason for this is that we are getting close to the Shannon limit, which describes the capacity of a wireless channel in terms of its bandwidth and the presence of noise.

However, through more efficient management of the radio spectrum, increasing the overall utilisation figure from circa 15% today to 85% in the future while allocating all this extra spectrum to mobile could yield 1955 MHz for mobile usage, up from circa 300 MHz today. Here, we assumed a lower figure of mobile use of 700 MHz. This would allow three national operators 650 MHz for 6G. This is beyond what is possible from years of R&D work on a new air interface for 6G!

However, because of the ‘spectrum staircase’, 5G is predominately deployed at 3.5 GHz, while 4G was deployed at 2.5 GHz and below. To understand the significance of better utilisation of the spectrum below 3 GHz – By using these frequencies, the costs of building a mobile network could be reduced significantly compared to using the 3.5 GHz bands, possibly by a factor of four times. At the same time, allocating the additional spectrum to an existing mobile network would increase capacity by over six times. Such advancements could bring about a fundamental paradigm change in mobile service.

7.6 Citizens Broadband Radio Service (CBRS) – democratising mobile spectrum

Let us describe an important initiative in improving spectrum utilisation. The FCC Citizen Broadband Radio Service (CBRS), launched in January 2019, represents the first mobile spectrum democratisation. It also represented the first attempt by a regulator to address some of the issues of spectrum management that leave large tranches of the spectrum under-utilised. To increase spectrum efficiency and meet the explosive growth of wireless data, the FCC established the CBRS in the 3.5 GHz radio band with a spectrum of 150 MHz (3.55–3.7 GHz). The US Department of Defense uses this band for its carrier ship radar systems, so a means had to be found to protect their use.

The CBRS radio band had previously been designated as a mobile band (band 48), which opened the opportunity for support in mobile handsets, tablets and laptops like any other mobile band. Nearly all new cellular smartphones in the US support the CBRS band today. While laptops and tablets from major suppliers also support this band. To put the 150 MHz in context, a typical national mobile operator for 4G would have a similar total spectrum holding of circa 150 MHz.

CBRS mobile spectrum is available on a sharing basis to anyone in the United States, not just mobile operators. The other key feature of this approach is the localisation of the spectrum. The auction licenses, 22,631, were the most ever auctioned in a single event. A total of 70 MHz of licensed spectrum was offered on a county basis across 3233 areas, each with seven 10 MHz blocks available. Before CBRS, choices for adding mobile spectrum required either acquiring it across a much wider area at a much higher cost during a spectrum auction or utilising unlicensed spectrum with little protection from interference from other users and dealing with the handset/device support issue. However, the CBRS spectrum changed that equation, allowing organisations to acquire and share unused spectrum with more benefits than previously unlicensed. Both the licensed and unlicensed components of CBRS are protected from interference, a key feature missing in Wi-Fi and very important to industrial wireless applications.

Spectrum in CBRS is dynamically shared locally in a three-tier priority access model (Figure 7.3). People do not buy spectrum as they have done previously for high service levels, but they can acquire higher priority access rights to its use. As governments often own high percentages of the valuable spectrum, this is also a

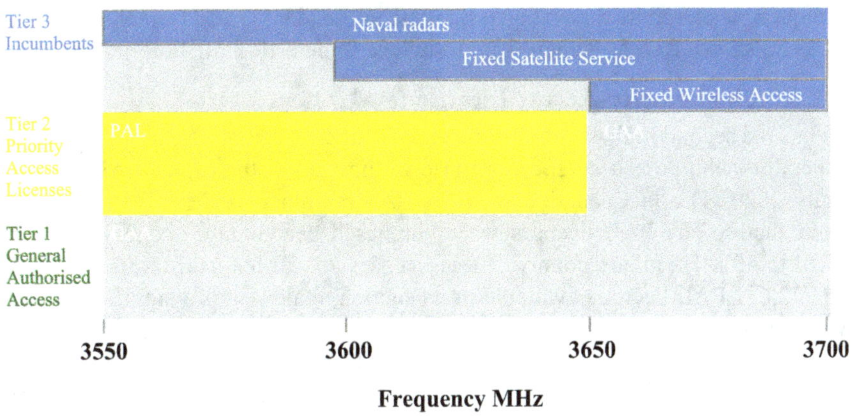

Figure 7.3 CBRS 3 tiers of users – managed by the SAS

way for the public to share its use with their government while protecting key users. The CBRS band supports cellular standards, including 4G (LTE) and 5G, but not Wi-Fi.

However, to fully understand this claim of democratisation, let us first review how the mobile ecosystem works and position the importance and nuances of a CBRS-like spectrum sharing as a future for 6G. As pointed out, the 3GPP mobile ecosystem builds a wall around itself by designating specific frequency bands for mobile use only. Only using these frequency bands allows access to all the 3GPP standards from the air interface to the core network. Effectively, the mobile frequency band is the Gatekeeper. Thus, CBRS allows non-mobile operators to access the 3GPP eco-system.

The CBRS band has been designated for sharing among three tiers of users: incumbent users, priority access license (PAL) users and general authorised access (GAA) users.

7.6.1 CBRS tiers of users

There are three tiers of users within the CBRS system, as shown in Figure 7.3.

7.6.1.1 The incumbent tier

This has the highest priority of use and is reserved for 'grandfathered users', traditionally using the 3550–3700 MHz frequency range. This group mainly includes the US Navy, which uses part of this band for their carrier ships' radar systems (requiring only a few MHz but not in a defined frequency location), commercial fixed satellite stations and broadband fixed wireless access operators. Approximately 20 US carrier ships use this band, mainly at sea, for most of the year. Nevertheless, as these are still the highest priority users of this band, a detection system has been built on the East and West coasts to detect the arrival of any of these ships (by detecting their radar transmissions). Furthermore, AI techniques pioneered by the National Institute of Standards and Technology (NIST) have been shown to offer superior performance in detecting offshore radars [15].

7.6.1.2 The Priority Access License (PAL) tier

This has the second highest priority tier. It effectively guarantees services for users of operators who purchased spectrum priority licenses at the CBRS PALs auction in July 2020. The FCC auctioned seven 10 MHz blocks in each county in the United States. The PAL licenses were purchased per county, and bidders were limited to 40 MHz in any county. These licenses are for ten-year terms. The FCC made 22,631 PAL licenses available in its auction, making this a significant move toward the localisation of the spectrum (i.e. removing the gaps seen in Figure 7.2).

The PAL auction was highly successful. Two hundred seventy-one bidders participated in 76 bidding rounds, with $4.5 billion spent on PAL licenses and nearly 23,000 new rights granted across more than 2300 counties in the United States. Verizon, one of the largest mobile operators, paid approximately $1.9 billion. Nevertheless, the range of companies that participated was very diverse, with mobile operators, organisations (e.g. Chevron, Texas A&M University, John Deere) and several Cable operators [who plan to use this spectrum with Mobile Virtual Network Operator (MVNO) arrangements] all bidding for licences. This demonstrates interest in private licensed networks.

In places where the PAL licenses were not taken up or fully allocated, this spectrum becomes available to all for free in the GAA tier, just like Wi-Fi, as described below.

7.6.1.3 The General Authorised Access (GAA) tier

This has the third highest priority. It is the 80 MHz of unlicensed spectrum in eight 10 MHz channels that users can access for free but is still a mobile band accessed by most smartphones, tablets and laptops. It has a similar status to Wi-Fi but with the protection of the Spectrum Access Sharing (SAS) system, which allocates channels to avoid any interference between access points, as explained below. Suppose there are no PAL users at a given location, or they are not entirely using their CBRS spectrum. In that case, the associated spectrum can be reallocated to GAA for free access while that situation remains. Interestingly, if you are an enterprise in a remote setting, this access scheme means you can rely entirely on GAA (i.e. no need to purchase licences) as there will be no contention with your neighbours.

7.6.2 CBRS is made for private wireless networks

It is hard to justify 5G as purely a consumer proposition as 4G LTE Advance would give consumers more broadband speed than they need, so 5G was in part positioned as the network for machines supporting Industry 4.0, as described in Chapter 5. 5G will provide protocols and standards to support low latency, higher reliability and resilience with options for Mobile Edge Compute with a Stand Alone (SA) Core necessary for high-performance Industry 4.0 use cases, which we cover in Chapter 12. It can offer a private mobile spectrum for exclusive use. And 6G, as an evolution play, is set to continue this. Many medium-to-large enterprises have long felt the need to access their private wireless network, which has led to the recent mobile proposition of private 5G networks. They have felt a lack of satisfaction with the mobile network quality provided by an 'outside-in' external network within their enterprise. Wi-Fi, which uses unlicensed spectrum, is not guaranteed against interference from other Wi-Fi deployments and is seen as unsuitable for critical manufacturing applications. Furthermore, many enterprises want to avoid having too many parallel networks for voice, data, IoT, etc. However, this desire for a private mobile network has never been fully realised, primarily due to the lack of spectrum ownership. This is why CBRS challenges the Private mobile 5G spectrum. Large enterprises may buy PAL licenses in the locations where they operate, while medium enterprises may find GAA sufficient. They can deploy a wireless network supporting all their applications from the Access network to their SA core.

7.6.3 CBRS – how it works

It is worth reviewing how CBRS works as it sets the scene for a more radical approach to sharing the spectrum. The central element of a CBRS network is the Spectrum Access System (SAS), certified by the FCC, whose primary function is to dynamically allocate a radio channel to the CBRS Access devices (CBRDs). This ensures no interference with neighbouring radios using a propagation model and frequency planning like that used in GSM networks. This will use conservative calculations to avoid any potential interference. In particular, it will not be aware of any propagation losses associated with placement in a building which acts as a shield.

In congested areas, the SAS will allocate the radio resources between PAL and GAA users, with priority given to the PAL users (as they have paid for priority access in that area).

The components of the CBRS system are shown in Figure 7.4. One of the critical inputs to the SAS for its decision-making is Environmental Sensing Capability (ESC), which detects an incumbent user, such as a US Navy carrier ship radar, who has the highest priority. The ESC was built for the US Navy and deployed on the East or West coasts, as their ships can approach the United States from any direction. Static incumbent users' locations are registered on an FCC database, so no ESC requirement is needed. These ESC detection networks were built before the launch of CBRS. They, in practice, detect not only the use of a ship radar system from many miles offshore but also its location within the CBRS band. The SAS can then dynamically reallocate users from that channel to other parts of the CBRS band under its priority access scheme. There is a provision in CBRS for contiguous channel aggregation with a Domain Proxy for large enterprise deployments.

7.6.4 How do CBRS private networks compare to Wi-Fi?

This is an important definition, particularly for critical industrial use cases. The CBRS band accommodates different cellular standards, including 4G (LTE) and 5G, but not Wi-Fi. Its spectrum is within the 3GPP band 48; therefore, it can use the 3GPP ecosystem of standards associated with its mobile status. This allows mobile

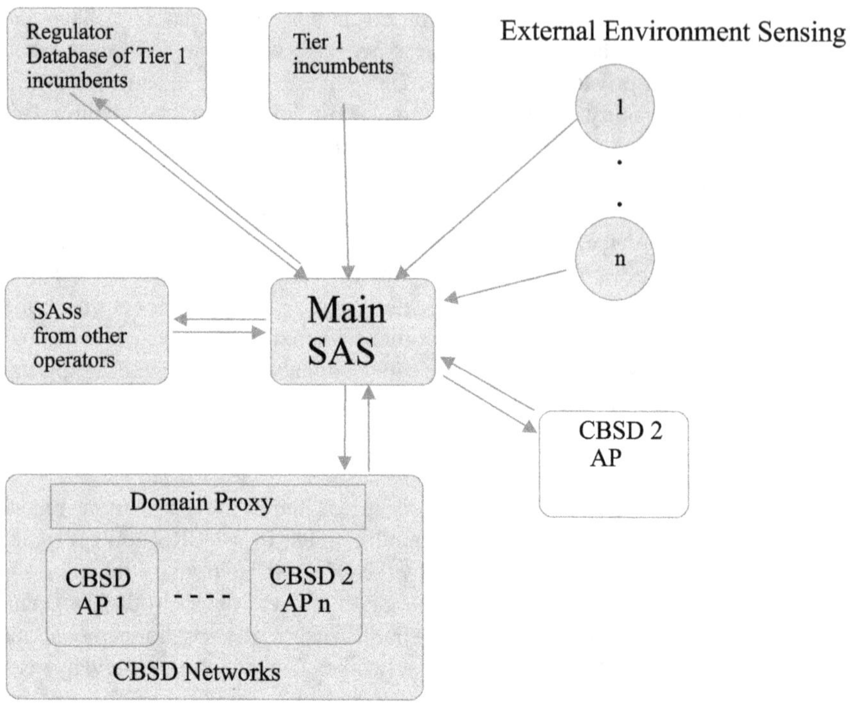

Figure 7.4 Elements of a CBRS spectrum sharing system

radios in devices, Mobile Edge Compute and a Standalone core within 5G for industrial applications for all CBRS channels.

Within the GAA tier, which could exist across the whole band if there were no PAL bidders in that location, it has a similar status to Wi-Fi use in that it supports free spectrum access. Nevertheless, SAS manages this with an associated yearly management fee from the installation's owner to the SAS provider. CBRS is meant to augment Wi-Fi, not replace it. So, how does this compare with Wi-Fi, and is it a replacement?

The three key differences are [16]:

7.6.4.1 Coverage

An indoor CBRS LTE AP operates in the 3.5 GHz spectrum, so propagation is better than the Wi-Fi 5 GHz, which is likely to be used in enterprises for similar higher bandwidth services. These will cover a roughly 4× larger area than a typical indoor 5 GHz Wi-Fi AP, while the SAS will coordinate its operating frequency to avoid interference.

7.6.4.2 Design

The CBRS system is part of a local mobile network that requires a cloud-based SAS, provided by a company like Federated Wireless, and an Evolved Packet Core (EPC) for LTE or a 5G SA Core. This is more complex than a Wi-Fi network used for access only. However, for manufacturing applications in Industry 4.0, the Standalone Core with possibly a MEC would be needed anyway.

7.6.4.3 Cost

The cost per access point of CBRS will likely be comparable to enterprise Wi-Fi. However, CBRS's greater coverage would lower the associated overall coverage cost in medium to large enterprises.

7.7 Can Wi-Fi be the basis of a mobile service?

Until now, every new generation of mobile technology has introduced a new radio air interface that utilises a specific portion of the mobile spectrum allocated for its service. Companies like Qualcomm produce modems and radios that operate within these designated frequency bands (reserved for mobile devices). Only mobile operators who have acquired the corresponding spectrum at great expense can utilise these bands. However, having made this investment, they can exploit the 3GPP standards and the resulting volume-driven equipment economies from the access network to the core.

Non-mobile operators have attempted to compete directly with this 3GPP ecosystem using the Wi-Fi spectrum but have yet to achieve any scale. To fully understand the issues, consider the case of British Telecommunications Plc (BT), which launched the BT Fusion phone bundled with its broadband services in 2007, see Figure 7.5, and withdrew a few years ago.

Figure 7.5 British Telecommunications Fusion Phone (Wikipedia)

This was based on a Motorola handset and had entirely seamless roaming between BT's home Wi-Fi network or BT Wi-Fi hot spots to an external mobile network using a MVNO arrangement with Vodafone. The idea for BT was to keep as much mobile traffic as possible on its broadband network, avoiding MVNO mobile roaming costs while offering its customers the same mobile experience. In many respects, this was one of the first examples of 'Convergence between fixed and mobile networks' and offered a better customer experience as it would have given them excellent in-building coverage in their homes, 10,000s of BT Wi-Fi hot spots and offices using Wi-Fi. However, because this did not use the mobile-designated frequency bands, a software client (universal mobile access – UMA) was needed to be embedded into the handset operating system. This client supported seamless handover to and from Wi-Fi and allowed direct access to the SIM within the device. This software approach is designated 'Southbound' as it is below the software interface between the device OS and its application layer and effectively requires a bespoke operating system. This limited the handset range to only three Motorola handsets on launch, using the bespoke operating system for each. While this succeeded technically, it was a commercial failure because of the limited handset range and the time delays associated with testing when introducing any new handset model. Furthermore, it required persuading other handset vendors to allow access to their operating system to embed the UMA client.

Since then, others have tried a more straightforward approach using Wi-Fi for mobile proposition using a similar MVNO model but with a downloadable client with a 'Northbound' interface to the device's operating system (i.e. just like any other smartphone app). While this means that any smartphone can now be used, it does not support seamless handover and requires the user to download the app and regularly turn it on. Further limitations include the lack of direct access to the SIM within the device and the vulnerability of this service to Wi-Fi or SIM card interface changes (thereby requiring constant monitoring with all the major device OS). Not surprisingly, there has not been any significant success. So, the wall around the mobile ecosystem system remains very high. However, the importance of voice calls has diminished in 4G, with mobile data and applications such as Skype and WhatsApp becoming increasingly popular, particularly for international calls. Nevertheless, 5G promises a new range of mobile services for machines using mobile devices and the Standalone Core network, and access to this will be through the mobile spectrum again! This model is likely to continue in 6G.

7.8 Spectrum inefficiencies from conventional propagation modelling

Today, virtually all cellular networks are planned from the perspective of 'outside-in' coverage. While perhaps upwards of 80% of mobile data is used inside buildings, the mobile networks attempt to provide an indoor service using an external cellular network, coping with an average of 15–20 dB of penetration loss. This has driven them to use high energy consumption MIMO base stations to compensate for this loss as data rates increase. As we move to even higher data speeds, this problem becomes even more challenging. However, inside buildings, people generally have access to Wi-Fi networks. Nevertheless, extending mobile coverage inside buildings is attractive for a mobile operator.

Mobile networks are planned using propagation modelling tools that attempt to predict the radio channel loss so that the mobile network can provide an excellent service to where its customers are. There are also two contradictory requirements in predicting the radio channel loss: Predicting radio path loss with high accuracy while being computationally efficient.

This is extremely difficult for indoor coverage. While the propagation planning tools accurately know the building's locations, floor area and height, they do not generally know the materials used in the construction and their effects on radio frequency penetration for indoor coverage. For example, there could be an enormous difference in the loss in window glass attenuation between clear glass and low emissivity (metalised) glass when using even mid-band frequencies, with virtually no penetration for cm or mmWave. A historical rule of thumb was to allow 17 dB for in-building loss for mid-band frequencies to achieve a level of in-building coverage. Of course, good propagation planning tools exist for in-building coverage for bespoke small cell or Wi-Fi coverage.

Even in the external environment, predicting the radio channel behaviour where networks are deployed is challenging. Models require constant updates during the year to consider seasonal changes in foliage, the construction or demolition of building infrastructure and other minor factors.

To illustrate the wide variations possible in radio channel loss, consider the predictions of three commonly used propagation models for path loss prediction: Free Space Path Loss (FSPL), COST 231 Hata Field and COST 231 Walfish Ikegami in the following section. In one case, using the COST models, the acceptable propagation distances range from 3.4 to 7 km for Hata and Walfish Ikegami, respectively. This would make an enormous difference in planning a cellular network. FSPL is not used in real-world situations as it is highly conservative.

Although the COST models were initially designed for frequencies up to 2 GHz, they have recently been extended to 3.5 GHz due to increased cellular frequency. However, the COST models are complex and require precise calibration with detailed measurement campaigns in a specific environment to produce accurate results. Typically, a calibrated model for a line-of-sight path will have a root mean square error of 6 dB.

7.9 Difficulties of mobile use of the satellite C-band

One of the most contentious issues for regulators is spectrum sharing, as it is desirable to drive up the efficient use of the EM spectrum and, within that, the use of a propagation model. Different user communities of the proposed shared spectrum will tend to pick a propagation model favouring their community interests. To illustrate this argument, consider the recent shared use of the satellite C-band (3.7–4.2 GHz for downlinks) for 5G mobile use in the United States. This band sits next to the CBRS band (3.55–3.7 GHz); in the future, it could provide a 650 MHz continuous spectrum for mobile use. These bands are extremely useful for 5G or future 6G microcells, which, to date, could only be found at much higher frequencies in the cm or mmWave bands.

However, this band is used globally for satellite communications. It is extensively used in the United States to distribute programming content for TV and radio broadcasts, live sports events and cable TV. It is a very informative exemplar case to consider because satellite systems are highly susceptible to any interference, with antennas with high gains of circa 40 dB and susceptible front-end electronics. This is sensitive to direct channel interference (co-channel) and out-of-band emissions (OOB) in adjacent channels and the aggregation effect of the total power across the whole of the C band, which could produce a change in the linearity of the front-end device if sufficiently high.

Let us consider the minimum protection distance for a satellite operating close to the C-band at 3.5 GHz to protect it from interference using different propagation models: The satellite has a 5° elevation and has an assumed gain profile vs. azimuth, as described in Table 7.3, with co-channel loss in dBs required between it and single CBRS base station operating at the height of 1.5 m with a transmitter power of 1 W.

Table 7.4 calculates the calculated protection distance requirement for each of the three models. While the FSPL model is unrealistic in this situation, yielding thousands of kilometres of distance compared to the COST231 models, it has been

Table 7.3 Gain profile of C-band Satellite vs azimuth for 5° elevation

Dish gain 5° elevation		**Total loss required**
Azimuth	**dB**	**Co-channel loss (dB)**
0	14.5	181
5	−5.9	160
10	−10	156
15	−10	156

Table 7.4 Protection distances between satellite and terrestrial radio using different propagation models

Azimuth (degrees)	**FSPL (km)**	**231 Hata (km)**	**231 WI (km)**
0	6940	3.4	7.4
5	662	1.1	2.1
10	413	0.9	1.7
15	413	0.9	1.7

cited in filings to the FCC for satellite protection. There is approximately a factor of two difference between the COST models in depicted protection distances with, in general, a protection distance of less than 10 km. We will let you guess which model establishes the protection distance. The FCC part 96 Protection Scheme for satellite earth station protection is 150 km, meaning you would not be allowed to site a base station within 150 km of a satellite earth station using the same spectrum band. This is cautious because there is no way of policing and remedying an interference situation should it arise. With initially over 5000 registered C-band earth stations in the US, each with a 150 km radius protection distance around each, this ruled out any effective sharing of the C-band (3.6–4.2 GHz) with any other wireless system.

Because of the importance of the C-band for mobile use (for both 5G today and 6G in the future) and the belief that there was no technical solution to sharing, the US Federal Communications Commission (FCC) announced an auction of part of it, 3.7–3.98 GHz (280 MHz bandwidth), for exclusive mobile use where those satellite operators using it would relocate to a higher part of the C-band. As a result, in the first phase, 100 MHz of the auctioned C-band spectrum was cleared in 46 top US markets by December 2021, while the remaining 180 MHz in these same 46 markets and the entire 280 MHz in the other markets will be cleared for use by 5G services by December 2023.

The FCC's C-band auction set a record, grossing $80.9 billion [17]. Later in this chapter, we proposed a technical solution allowing the entire C-band to be shared between mobile and satellites using a shared spectrum access system. This is based on real-time measurement of propagation losses between APs established using weak signal beacons whose power is so low it does not cause any interference during the propagation measurement, which could be implemented in a future CBRS system 'CBRS 2.0'

7.10 Spectrum sharing use – satellites and mobile

To grasp the most complex technical aspects of sharing spectrum, using a Spectrum Sharing System (SAS), it is illuminating to examine the specific situation of the C-band being shared between mobile and satellite use. This will also highlight the economic value of the spectrum. The solution found for this scenario can then be applied to other situations and serve as a universal architecture approach.

The extreme difficulty of sharing arises as the C-band 2 m satellite dish typically has a gain of 37 dBi and a highly directional gain profile of 1.3° half-power beam width (HPBW). As shown in Figure 7.6, any energy from a base station or a mobile device shown in blue could be coupled into the satellite dish if it is within its gain profile. The satellite also has an assumed gain profile outside its beam profile, significantly affecting its protection zones.

The Earth stations typically employ multiple satellite dishes, as shown in Figure 7.7, aligned to different satellites in orbit to obtain various programming sources.

In satellite installations, a low noise block downconverter (LNB) is the receiving device mounted on the satellite dish at its focus, which collects the radio

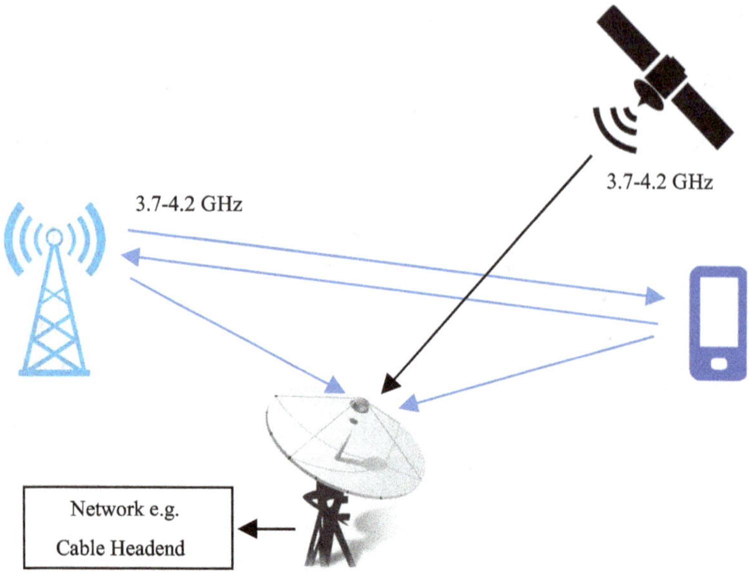

Figure 7.6 *Issues associated with shared use of the radio spectrum between mobile and satellite networks*

Figure 7.7 *Four C-band satellites dishes aligned to different satellites. The cylinders at the centre of each dish are the LNB.*

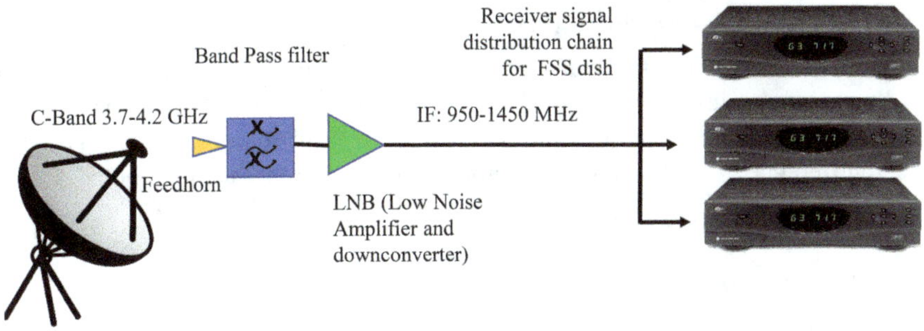

Figure 7.8 Electronics used in a satellite dish installation

waves from the dish and converts them to a signal (Figure 7.8). This contains a very low noise amplifier with a typical noise figure of only 0.3 dB. This operates at an effective thermal noise floor of only 80 K. It also contains a downconverter which converts its channel, within the C-band of 3.7–4.2 GHz, to a corresponding intermediate frequency band of 950–1450 MHz.

To avoid co-channel interference, any emission from a fixed terrestrial network within the gain profile of the fixed satellite stations (FSS) must be below a −129 dBm/MHz threshold, which is below the normal thermal noise of 174 dB/Hz at 290 K (space has a lower operating temperature). To meet this condition, the SAS managing shared use of the spectrum would employ a conservative propagation modelling tool based on line-of-sight calculations that could require a protection distance of tens of km (based on the calculations of Table 7.3 with a suitable safety margin).

Before the auction for most of the C-band in the United States, there were around 5000 registered C-band FSS as of July 2018. These were primarily registered for 'Full Arc' and 'Full Bandwidth' and, therefore, could legally change direction across the entire arc and access the C-band as a whole spectrum. However, in practice, most always kept their orientation direction the same as they were aligned to a particular geostationary satellite. Large installations contained many satellites, all aligned to different satellites for different content (Figure 7.7). Similarly, each satellite is tuned to a specific frequency for its content, and while that may change, typically, only some of the channels are used in each FSS. This can lead to the protection of frequencies that are not in use. The reason for this type of registration was that it was straightforward to register each FSS as full arc and full bandwidth for no additional cost. All of this demonstrates how intractable the problem is for sharing.

Another issue that emerged before the auction was that the FCC database was inadequate and did not reflect where the incumbents' satellite earth stations were. So, the FCC set a 90-day window for companies to re-register their earth stations to develop a better record.

Furthermore, additional interference calculations are needed for two other forms of interference: out-of-band transmitter emissions and low noise block oversaturation:

7.10.1 Out-of-band emissions

In addition to the desired spectrum for the channel, adjacent band transmitters emit low-level signals outside their intended frequency range. These OOBE, as they are called, can appear as co-channel interference to an earth station receiver. To mitigate interference, filtering could be implemented within the transmitter or the receiver at a cost.

7.10.2 Low noise block oversaturation

Many LNBs in service are not equipped with any filtering for the adjacent bands, such as from the upper part of the CBRS band (3650–3700 MHz band). A filter could be fitted between the antenna and the LNB. Still, its loss would affect the C-band performance directly, while its performance so close to the C-band may be limited such that a strong signal in the 3650–3700 MHz segment appearing at an LNB input could oversaturate the amplifier, causing reception problems for the entire 3700–4200 MHz band.

7.10.3 Implications for populations

The overall implication of all these calculations is to create exclusion zones around the FSS sites where terrestrial wireless network use is not permitted in bands adjacent to and in-band of the FSS spectrum because the Full Arc registration of these is circular, which are in the centres of the population as they are used to receive content for Cable Networks.

This would rule out terrestrial wireless use on most of America's east and west coasts. Even a 5 km radius exclusion zone around each FSS would reduce population coverage to less than 25%.

It is reasonable to observe that while this is the most challenging example to consider for spectrum sharing due to the extreme sensitivity of FSS from interference, it also establishes the issues of sharing more generally. Conservative calculations are used because there is no remedy to interference should it arise. An intelligent observer might suggest that 500 MHz of prime 5G spectrum is denied to more than 75% of the US population under this regime because the FSSs have no system which reports the critical operating parameters, even though there are fewer than 6000 sites. The key parameters, such as the satellite direction, elevation, channel of reception and antenna dish size, which were used in the SAS calculations, would yield more accurate results. Nevertheless, the database was inaccurate when simple parameters, such as at the earth station location, were registered. All of this led, in January 2021, to the US Federal Communications Commission (FCC) announcing the conclusion of a C-band auction for operating licenses in the frequency range 3.7–3.98 GHz (280 MHz bandwidth), as mentioned earlier, where this part of the C-band was clear for mobile use.

7.11 'CBRS 2.0' for 6G – a new radical approach to spectrum management

Earlier in Section 7.3, we quoted Jenifer Manners' spectrum management analysis, in which the radio spectrum is effectively fully allocated while parts may be fully congested; nevertheless, no one knows to what extent it is used. Furthermore, other parts of the radio spectrum allocated for a particular use are not or only partially used. In practice, the overall spectrum occupancy in the valuable part, 30–3000 MHz, is likely to be only between 10 and 20% nationally.

CBRS was an excellent start to show how the spectrum can be shared for mobile use and extended beyond traditional mobile operators. It also pioneered the localisation of the use of the spectrum. Still, a new radical approach is required to efficiently utilise this valuable finite resource – Enabling 6G with a broader selection of spectrum choices from low-band through to high-band and extending the use cases for 6G from outside networks to inside buildings.

To achieve that, we now need to address the following key points:

7.11.1 A registry of use

There should be an automatic registry of spectrum users with their locations and the characteristics of their systems. From a spectrum management viewpoint, this would provide valuable insight into the actual use of the spectrum band for policy decisions. However, this would be valuable for technical design to help design more efficient systems, avoiding interference as a mobile operator would do with their cellular deployment.

One reason for the lengthy description of CBRS here is that it paves the way for the future of spectrum management. In CBRS, the SAS is aware of the transmitters' location and characteristics, such as transmitter power and channels of operations.

7.11.2 Maximising spectrum efficiency

Maximising spectrum efficiency in spectrum management involves not only the air interface design but also filling the available spectrum in all locations according to local demand and avoiding interference between systems sharing the same frequency band. This approach can arguably make orders of magnitude gains in economic value.

As we saw earlier, the low-band and mid-band are two of the most valuable frequency bands because they combine characteristics such as wide area coverage, reasonably good in-building penetration and reasonable bandwidths available for most services. However, only circa 15% of these bands are utilised. This drives new generations, such as 6G, to use progressively higher frequency bands that don't share these favourable characteristics to the same extent.

CBRS was an excellent start in maximising spectrum efficiency by creating local licensed use at a county level. In the original CBRS proposal, it was proposed to do this at an even finer granularity, such as at the postcode level, with even greater utility, but that would have significantly increased the complexity of the auction process. Yet, Wi-Fi arguably carries circa 80% of mobile data, while cellular only the remainder, and is based on free spectrum without any auction. So, if all registered users were GAA, introducing a CBRS at a postcode level granularity could be achieved! This is similar to the Wi-Fi free-use model but with a SAS to maximise spectrum efficiency and avoid interference.

Planning tools for propagation have limitations in accurately modelling the natural world and its frequent changes. Although CBRS was a good starting point for many of these issues, it still faced challenges as it relied on traditional propagation models with conservative measures to protect other systems.

7.11.3 Compliance and enforcement

Spectrum regulators are responsible for overseeing the use of radio spectrum and ensuring that all users follow the rules and regulations related to its use. They often achieve compliance by communicating clearly with large multinational companies' operators. However, in certain circumstances, they may need to investigate potential violations and take enforcement action to ensure compliance with the law. This can be extremely difficult in the case of intermittent interference.

Guard bands separate different spectrum users, which impacts overall efficiency, particularly if the spectrum is widely shared. They also define the emission mask of the signal to ensure that even very low signal levels do not impact other users.

While this system has been undoubtedly successful, this is probably due partly to today's massively inefficient spectrum use, creating significant gaps in its utilisation. If the spectrum were filled with different users, this would make the potential for interference much more likely. However, there is no way of detecting this automatically. Today, this involves compliance with the regulator and teams of people sent out to try and establish the causes of the interference.

7.11.4 Encouraging innovation

While CBRS created a 150 MHz spectrum in blocks of 10 MHz, it allowed users to concatenate these individual blocks to create wider channels. Thus, it can be used for 4G LTE and 5G today. This approach is essential as it creates flexibility for new systems and innovation. The appropriate channelisation for the spectrum would depend on the carrier frequency. For example, at mmWave frequencies above 100 GHz, the channel could be at least 20 GHz for 100 Gbit/s systems, while in the low-band, they could be much smaller, perhaps 1 MHz.

7.11.5 Supporting legacy use

No change to spectrum management will come overnight, and any new scheme would have to deal with legacy for decades of use for existing systems with their embedded users. However, a CBRS 2.0 which addresses the above arguments would be a revolution in spectrum management and could be applied to 6G, reframing existing bands to take advantage of this approach. Terrestrial TV is a good example, and regulators are progressively reducing the spectrum used for TV in favour of cellular. However, can these points be realised in a technical system? If not, then this is academic. The following sections describe the realisation of this approach.

7.12 Weak signal propagation reporting (WSPR)

The foundation of a potential new approach to managing radio spectrums might have originated from the Amateur Radio community. The acronym WSPR (pronounced 'whisper') stands for Weak Signal Propagation Reporter. This protocol is employed in a computer program, K1JT [18], which amateur radio operators utilise to facilitate communication through weak signals. The protocol was crafted and initially coded by Professor Joe Taylor.

Did you know that Professor Taylor won the Nobel Prize for Physics? He discovered a new type of Pulsar, a highly magnetised rotating neutron star that emits electromagnetic radiation from its magnetic poles. This discovery has opened new possibilities for the study of gravitation waves. Using their observations of this Pulsar, Professor Taylor and Russell Hulse demonstrated the existence of gravitational radiation within 1% of the amount predicted by Einstein in his General Relativity Theory. Professor Taylor has been an amateur radio enthusiast since he was a teenager, which led him to explore the field of radio astronomy. He is renowned for his expertise in amateur radio weak signal communication and has been assigned the call sign K1JT by the Federal Communications Commission (FCC).

The K1JT programme is specifically designed to test propagation paths on the MF and HF radio bands by sending and receiving low-power transmissions. These transmissions contain the callsign, a character locator (Maidenhead Grid locator) and the transmitted power in dBm within a payload of only 50 bits. The measurements are made available on a website so people can see who and where they are communicating with. If the transmitter power and location were directly updated on the website, only the call sign ID would need to be transmitted.

Even with a signal-to-noise (S/N) ratio as low as −28 dB in a 2.5 kHz receiver bandwidth, the K1JT programme can directly decode signals without averaging. The minus sign for the S/N indicates the level below the noise floor, so −28 dB indicates it is 1/630th of the noise level.

With averaging over 30 min, the decoding can be improved by a further 6 dB. WSPR signals are only 6 Hz wide with 4-FSK modulation and transmitted within a voice channel of 2.5 kHz. (The channel bandwidth for noise calculations is 2.5 kHz.) This is remarkable. The signal is less than 1/1000 of the thermal noise in a 2.5 kHz bandwidth channel used for Amateur Radio communications. However, this beacon requires a very accurate alignment of the receiver's local oscillator (0.5 Hz) with the transmitters for successful and fast detection. Amateur Radio

Stations with Internet access automatically upload their reception reports to a central database called WSPRnet. The inclusion of location and power in the payload is to cover the case where the receiving station does not have Internet access; otherwise, the S/N could be further improved.

With this system, it is possible to transmit signals successfully over thousands of miles with a transmitter power of only 1 W, which is not that dissimilar to your Wi-Fi Access point. Since the original launch in 2008, Joe Taylor has worked on further improvements, with the latest being FT8 – Franke-Taylor design, which uses 8-FSK modulation to provide faster decoding of the weak signal in 50 Hz signal bandwidth for situations of fast fading at the expensive of a few dBs with 15 s averaging.

When using a technology like a Weak Signal John Taylor (WSJT) beacon, a transmitter power of 1 W can achieve a link budget of 198–204 dB. While amateur radio uses an omnidirectional antenna to achieve these results, most other wireless systems use directional antennas, which can further extend the link budget by its antenna gain. Typically, this link budget exceeds the signal link budget by 30–40 dB, making it possible to use it as a probe for radio system transmission initiation without affecting the use of existing users!

For the general use of propagation reporting and interference control for spectrum management, we need to design a beacon system [19] that can:

- carry necessary information, such as a unique Radio Access ID used for identification
- support the extremely low SNR
- is measurable at the receiver
- only requires practical and low-cost transmitters and receivers support a multiple access scheme for a large number of potential transmitter beacons.

Remember that only the beacon ID transmission is necessary and sufficient for such a system. At the same time, its location and power can be transmitted over the Internet or through another channel to a SAS controller database. This conserves the information payload, enhancing the link budget. However, in applications where the radio AP is mobile, the additional fields should include the location and power, as in K1JT. Additionally, this technology allows for direct assessment of signal aggregation within the pass band or adjacent bands.

7.12.1 Weak signal system design – a revolutionary approach to spectrum management

Here, we describe a revolutionary approach to spectrum management that overcomes all of the issues of traditional approaches using weak signal radio beacons. To move beyond the advances made by CBRS in sharing band 48 for mobile to a more generalised solution for spectrum sharing in low-band, mid-band and high-band, it is proposed to use weak signal radio beacons to overcome the disadvantages of other approaches. These include achieving three important features not found previously: an accurate propagation measurement, identification of users causing interference, and a means to remedy any interference in real-time.

The first aim is to explicitly address the deficiencies of using conservative planning tools by taking real-time measurements of the radio environment rather than calculations. This would remove arguments between different users of the shared spectrum about which propagation model accurately describes real-world behaviour. It would also extend the use of propagation measurements into spaces not modelled well (e.g. inside buildings).

The second aim would be to identify any interference source and, thirdly, a tool to control its interference in real-time, should it arise. This would address the use of highly conservative protection criteria required by different types of spectrum users, such as satellite ground stations sharing the spectrum with mobiles. While this type of interference should never occur in a well-designed system, a means to control it would further alleviate any concerns.

The radio beacon could contain, directly or through association, four fundamental pieces of information: The ID of the transmitter, its location, its transmitter power and the channel it is using. However, the only necessary piece of information that needs to be sent is the ID of the transmitter, which is assigned before its use, as the other pieces of associated information can be transmitted over the fixed network to which the Access point is connected. This is an example of using Information Theory to reduce the information contained in the beacon to improve the link budget and extend its range. Its data can occupy a minimal bandwidth. Of course, it is necessary to ensure that the radio beacon itself is not a cause of interference, so it needs to be below the noise floor of the radio channel yet detectable with the right approach!

The radio beacon is first transmitted on installation and then periodically in a maintenance cycle. Any receiving device would detect the beacon by measuring its magnitude and relay this information to a central SAS in real-time. Measures of the signal strength would yield valuable real-time data about the propagation environment from the transmitter's location to the receivers, within listening range, with known locations of both devices. Effectively, these are being used to calibrate the SAS initial propagation planning algorithm in real time and can automatically consider any temporal changes in the environment with periodic measurements initiated by the SAS.

The initial use of the radio beacon within a CBRS-type system is shown in Figure 7.9, which would use a similar CBRS architecture described in Figure 7.4. In addition to using radio beacons to initiate the service, these could be monitored periodically to maintain accurate information about the propagation environment and to develop, in effect, a real-time measurement campaign for any environmental changes. In this use, all the receivers within range, with a circa 200 dB link budget for the beacon detection, would report these received signal strengths to the SAS to calibrate the propagation model only briefly used on setup. Nevertheless, over time, it has built a 'perfect' propagation model for free, which could have other uses. Usually, the system link budget is at least 30 dB less than this, so these beacons appear within the channel noise floor, making no real difference. As they are transmitted on installation or as part of a scheduled maintenance cycle, there is no significant increase in the noise floor due to an aggregation effect of many simultaneous beacon transmissions. Whereas the original CBRS uses a propagation model to assign resources, this approach allows measurements to calibrate the propagation model.

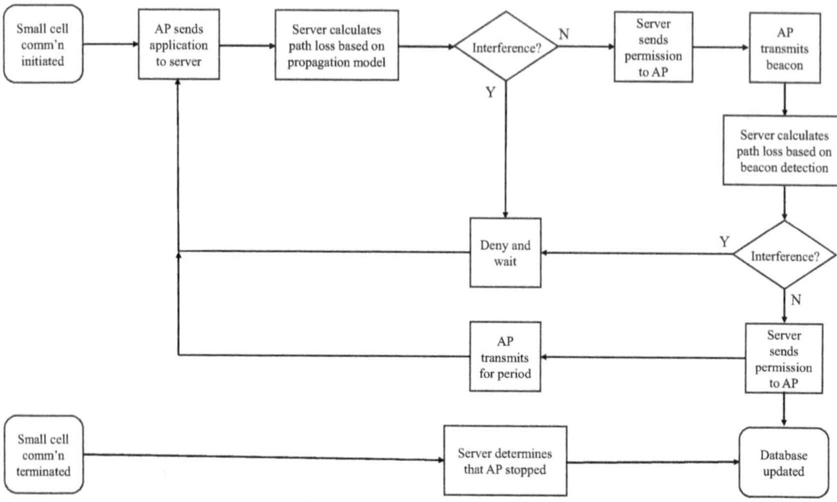

Figure 7.9 The process of initiating service with a radio beacon within a CBRS-2.0 type system [19–22]

To deal with any interference, a band user would automatically report the radio beacon ID to central SAS, which could take remedial action in real-time, such as reducing the interferer's transmitter power, re-assigning it to a different channel of use, or even closing its operation. This type of control is built into the present CBRS system for detecting radar systems on Navy carrier ships using the ESC network. Usually, this would not be occupied. However, this method would reassure and address any concerns of competing radio spectrum users. In addition, knowledge of all the locations of transmitters and their general usage would yield valuable information for spectrum management policies.

So, can we ensure that these beacons are not a new source of interference? Individually or in aggregation, what is needed seems impossible; a beacon that does everything required yet is practically indetectable! In the next section, we discuss a solution called weak signal propagation reporting, which can be applied to a CBRS-type shared spectrum system and was initially invented by the author to enable the sharing of the C-band by both mobiles and satellites [19–22].

7.12.2 Other schemes for weak signal reporting

Apart from WSJT, various modulation schemes can be used for weak signal reporting. These include Lora and Ingenu, which are commonly used in IoT applications. For instance, Ingenu utilises random phase multiple access (RPMA) [23] with a link budget of 180 dB in the 2.4 GHz band and a throughput of 19,000 bits/s/MHz. An Ingenu AP can reportedly cover an area of 300 square miles.

Since WSPR only requires the transmitter's ID, it does not need the throughput of a regular RPMA system; therefore, there are potential enhancements in receiver sensitivity and link budget for WSPR only. RPMA could achieve a link budget of 192 dB compared to 198 dB with K1JT on a like-for-like basis. One of the outstanding achievements of K1JT is designing the system using information theory to conserve the space symbols used to represent the minimum amount of information transmitted.

The bit space for the ID is important. We saw earlier that K1JT uses only 50 bits in its radio beacon to transmit the call sign, the locator, and the power level in dBm, where 28 bits are used for the call sign. In a system that supports only the Radio Beacon ID, 28 bits would support over 268 million transmitter locations, which is close to the number of the US population (circa 360 million). Thus, using 28 bits in a CBRS 2.0 system would yield close to a 3 dB improvement in link budget c.f. K1JT.

However, another viable option for weak signal reporting is implementing the Zadoff-Chu (ZC) sequences [24]. These sequences are already utilised in 4G and 5G cellular communication systems for unscheduled short message transmissions or initial network access on mobile devices. ZC sequences are a more recent method of spread-spectrum use in modern cellular technology. They have replaced PN and Walsh's sequences originally used in 3G cellular. A ZC sequence is a complex-valued mathematical sequence that, when applied to radio signals, creates an EM signal of constant amplitude.

A scheme of cellular ID reuse, as shown in Figure 7.10, could further reduce the bit space from 28 bits.

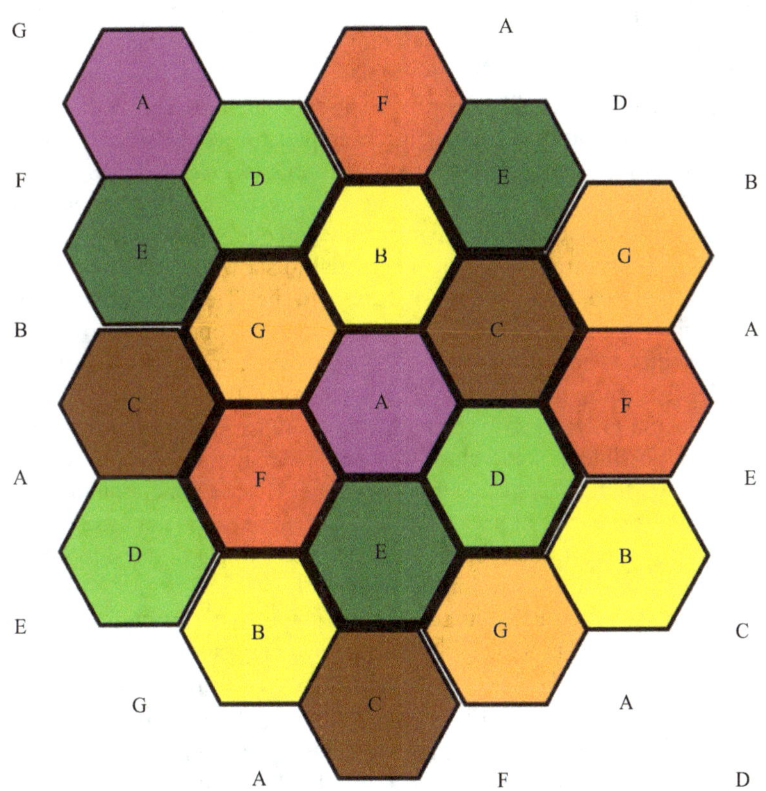

Figure 7.10 A cellular-type system with a 7-cell reuse cluster for the reuse of radio beacon IDs

An 18-bit word would allow seven-cell pattern reuse with 64 k RAPs per cell. This is like the idea of Frequency Division Duplex used in Cellular systems where blocks of frequencies are used per cell with a reuse repeat pattern of typically four. Furthermore, RAPs with the same ID could be differentiated using triangularisation by a network of beacon detectors.

ZC is currently utilised in both 4G and 5G systems, where this signal sequence is transmitted when the access device wakes up from sleep mode or performs a handover from one access point to another. If it were applied to radio beacons, it would function similarly. There is much to be gained in an optimum design of the radio beacon system for a CBRS 2.0 using methods like ZC, where the information payload is only circa 28 bits with an extremely low duty cycle and is amenable to averaging. So, circa 200 dB link budgets achieved in K1JT are achievable with a 1 W transmitter, while directional antennas could extend this further as the budget for K1JT was for an isotopic antenna.

7.13 The implications of C-band spectrum sharing with 'CBRS 2.0'

To showcase the potential of weak signal beacons for managing shared spectrum and propagation reporting, let us consider an example of how this approach could be used to share the C-band between mobiles and satellites. In Section 7.10, we discussed the significant challenges of sharing this satellite band with other radio users, so doing so would be a challenging exemplar for its use.

So far, we have described how the original architecture of CBRS, Figure 7.4, could be used to support a WSPR radio beacon, overcome the weakness of propagation modelling and identify and control any potential interference. However, to apply this to other spectrum bands, it is essential to show how these radio beacons could work in practice, which we must share with the 'Tier 1 incumbents' users.

The original architecture of CBRS, Figure 7.4, supports an Environmental Sensing Capability as an input to the SAS to identify the arrival of a US carrier ship radar and its frequency of operation. The SAS would then re-assign CBRS users that were potential interferers (to the high-priority radar system) to other channels. If that was not possible, it could close transmitters using the GAA tier to prioritise those in the PAL tier. Effectively, it creates an exclusion zone around the ship's location for the frequencies of use of its radar system only and not across the whole band/region. However, it does not monitor any interference to the ship's radar but relies on a conservative propagation planning tool.

In the case of C-band FSS, the registration process was a significant source of sharing problems. This allowed FSS operators to register for full arc and full bandwidth at every site's location, regardless. This was further compounded by the fact that many sites had incorrect locations in their registration. To fully realise the value of this 500 MHz of mid-band spectrum, an automated process is required for each satellite dish to record its GPS location, the directions of elevation and azimuth and its frequency of operation. Such a system would not be expensive and only have to be applied to circa 5000 locations in the United States to explore the C-band for sharing.

A 'CBRS 2.0 ESC' located at the FSS could detect the radio beacons of APs or mobile base stations sharing the same spectrum at the incumbent's site as inputs to its SAS for frequency planning purposes during its initialisation process. Thus, only APs within the narrow gain profile of the satellite dish would be detected if they were within range, leading to much more efficient spectrum use as most of the C-band spectrum would be unaffected and available for use. No propagation model is used; only real-time measurements of the environment are carried out by the SAS periodically after the initial registration process of the AP or base station. The weak signal beacon is an integral part of the AP transmission and could be implemented either in-band or separately in the adjacent guard band.

There is a coordination between the SAS and the EPCs with the mobile core networks, as shown in Figure 7.11. This would ensure no mobile use of frequencies identified at the FSS within exclusion zones around the FSS. This coordination role is essential to ensure that mobile handsets, UEs, are not also sources of interference as, unlike the base stations, they can move. The aggregate effect of mobiles associated with each base station can be modelled simply for interference calculations in the SAS. This approach can take advantage of the sectorisation of the mobile base station to ensure that while a sector covering the FSS may be denied

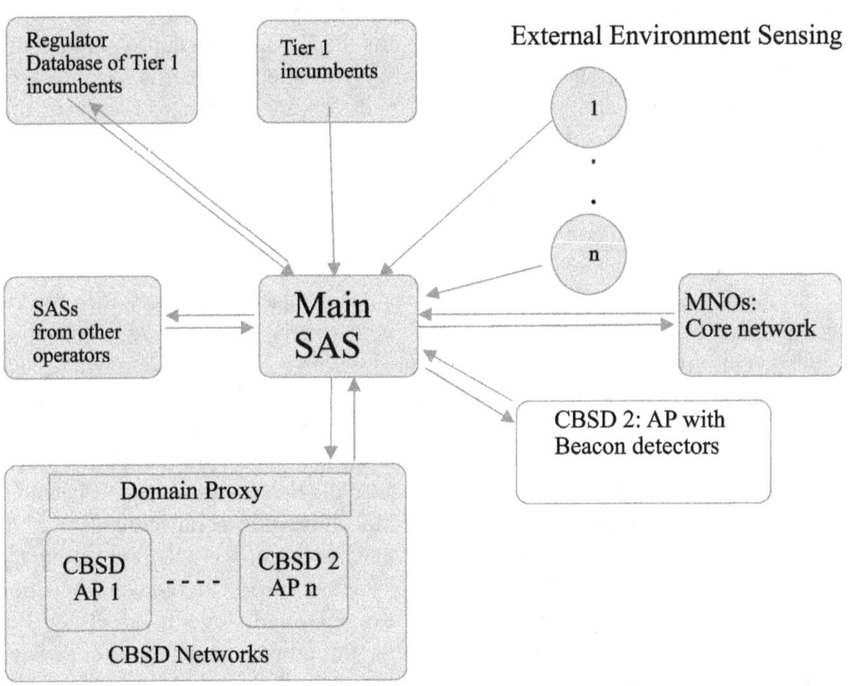

Figure 7.11 Diagram of CBRS 2.0 with interconnection to EPC of mobile network

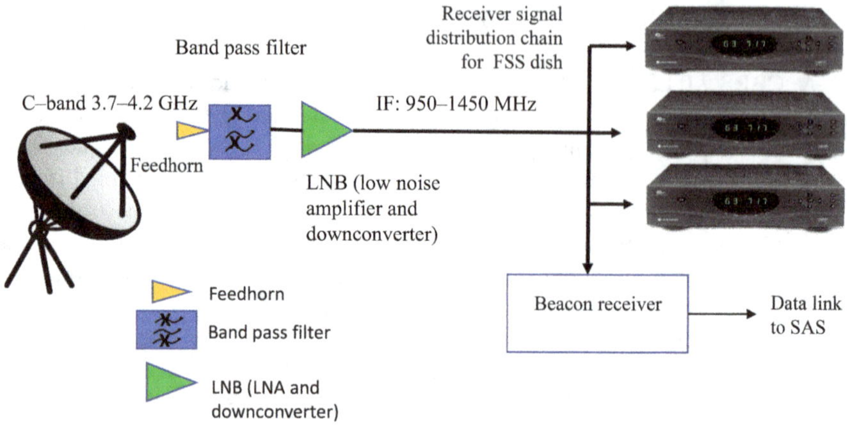

Figure 7.12 Radio beacon detection with fixed satellite service (FSS)

concurrent use of an active FSS channel, this would not apply to other sectors. In effect, CBRS 2.0 could create a tailored or bespoke exclusion zone for the incumbent's operation frequencies.

Figure 7.12 shows how a beacon detection system could be elegantly implemented within the electronics of the satellite dish. The signal from the satellite dish is fed to the LNB via its horn through an optional bandpass filter. The LNB contains the low noise amplifier and down converts the C-band channel to within its IF band of 950–1450 MHz. The beacon detector system directly benefits from the high gain of the directional satellite dish, extending its link budget range. Therefore, this system is not concerned with full arc or bandwidth protection as it is unnecessary for its successful operation.

Very occasionally, the FSS may wish to change its operating conditions by changing channels of operation or re-aligning to a new satellite. This is done with the SAS, which coordinates with the mobile EPCs.

As mentioned previously, radio beacons can be used to actively monitor the success of the frequency assignment and avoid overly conservative propagation planning. This approach could be implemented within a cellular system where each base station has its weak beacon ID, and reception by other base stations, or even mobiles, could be used to avoid using conservative propagation models.

Since the SAS now has access to accurate measurements from the transmitters to the FSS for all the transmitters within its 200+ dB link margin range, it becomes straightforward to measure and calculate total aggregate contributions to LNB blocking, first and second adjacent out-of-band-emissions (OOBE) meeting the attenuation limits as described in Table 7.5. The central SAS can remedy a situation where interference may be encountered, for example, by instructing an eNodeB or AP to change its frequency of operation and, through these changes, associated connected devices. There may also be a scenario where the noise floor of the FSS has increased and is approaching an unacceptable threshold due to the aggregation of individual mobile transmitters or radio APs. In this case, the SAS can instruct the

Table 7.5 Propagation loss limits on sources of interference

Dish gain 5° elevation		Total attenuation required dB			
Azimuth	dB	Co-channel	LNB blocking	1st adj OOBE	2nd adj OOBE
0	14.5	180.5	121.5	140.5	128.5
5	−5.9	160.1	101.1	120.1	108.1
10	−10.0	156.0	97.0	116.0	104.0
15	−10.0	156.0	97.0	116.0	104.0

Table 7.6 WSPR link budget with ZC

ZC budget (3.25 Mbit baud rate)	Value	Units
Transmitter power	25	dBm/MHz
1 W per 3.25 MHz bandwidth	3.25	MHz
Rx noise power (noise Fig. = 1.5 dB)	−112	dBm/MHz
Rx S/N	−30	dB
Rx antenna/satellite gain	37	dB
Max path loss	204	dB

devices near the FSS to reduce their transmitter power by small increments to bring the FSS noise floor down to the original acceptable limit.

7.13.1 Satellite WSPR link budget with ZC

Implementing the beacon detection system within the satellite system or a fixed Access system, as shown in Figure 7.12, directly benefits the link budget of the WSPR as it experiences the high gain of the satellite dish or FWA antenna.

The link budget for the ZC WSPR is shown in Table 7.6 for a 2 m satellite dish with a gain of 37 dB. With this, the ZC-based system can achieve a 204 dB link budget with a 3.25 Mbit/s baud rate, significantly higher than needed for a beacon ID. The link budget would dramatically increase with a ZC optimised for a beacon ID only.

7.13.2 Initial exclusion zones

Figure 7.13 shows the exclusion zone strategy to protect a C-band FSS [19] from mobile use. The tear-drop shape of the three zones around the FSS 1102 results from the satellite dish's gain directivity. The half-power (−3 dB) beamwidth is less than 2° for a typical large satellite dish of several metres in diameter. However, for ITU interference analysis using a conservative approach, the gain profile is assumed to extend over ±20°, and this gain profile is defined according to

$$\text{Gain (in dBi)} = 32 - 25 * \text{LOG}_{10}(\text{in degrees}) \text{ with 0 dBi at 20}$$

$$- \text{ degree angles around the centre.}$$

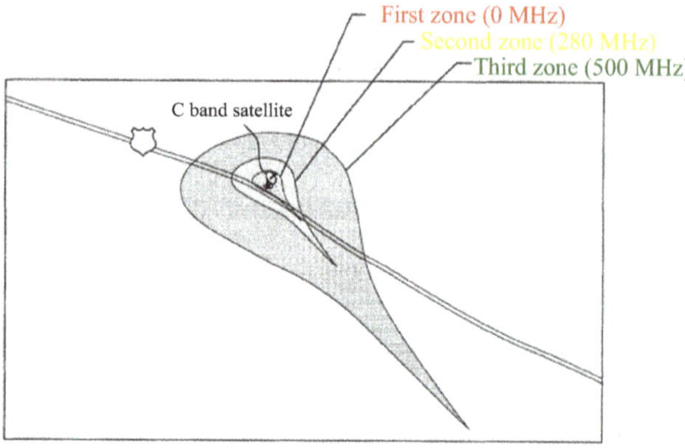

Figure 7.13 Initial exclusion zones around an FSS used by the SAS

Table 7.7 CBRS classification of limits for transmitter power

Device classification	dBm	dBm/MHz
AP – Category B: rural	47	37
AP – non-rural	40	30
AP – Category A	30	20
UE – end user	23	13

This leads to the circular area around the satellite dish apart from the peak associated with the dish's directivity. This gain profile equation thus follows the envelope of the satellite gain profile, which will have various peaks and troughs. Nevertheless, when implemented concerning the WSPR in a dynamic closed loop with the SAS, the actual gain of the satellite obtained from real-time measurements will be used. For example, a high-power transmitter directly behind the dish could be used without any direct effect.

The detailed calculations in reference [19] are based on the transmitter powers in Table 7.7, and the limits for co-channel, LNB, first and second channel OOBE detailed in Table 7.5 for a residential morphology below clutter (20 m) for the various transmitter powers.

In the first zone, Red, no AP or UE use is permitted. The circular region around the satellite dish has a typical radius of only 150 m. In the second zone, Yellow, the circular radius is now 300 m from the satellite dish, while 288 MHz of spectrum is available for transmitter powers of 1 W or less. In the third zone, Green, the circular area radius is now extended to a 750 m radius, and all of the 500 MHz is available for transmitter powers of 4 W or less. The range of powers described in Table 7.6 is available outside these zones. These ranges are orders of magnitude smaller than the current limits and would extend the use of the C-band on a sharing basis to circa 95% of the US population.

These exclusion zones would be used initially within the SAS to allocate resources to the AP or eNodeB before a weak signal probe is used to perform a real-time propagation loss measurement. Once these measurements have been made, the exclusion zones will be reduced and reshaped to accurately follow the satellite dish's actual gain profile. Furthermore, the indoor use of APs or eNodeBs would provide additional screening of the FSS.

7.13.3 Extension of the CBRS 2.0 to other low-band, mid-band and high-band sharing

As previously mentioned, CBRS has pioneered a new approach to spectrum sharing, allowing for the localisation of spectrum use while protecting incumbents and prioritising access to others. Nevertheless, it still bakes in using a propagation model, which makes it band-specific and carries all the conservative modelling assumptions that lead to significant inefficiencies in its application. For example, a study of the TV white spaces (TVWS) spectrum in New York found that 40%–75% of the TVWS spaces spectrum was lost primarily to the propagation modelling [25]. While CBRS has a mechanism to protect incumbents by controlling the channel used in APs, this relies on the external sensor network.

In the 'CBRS 2.0', we propose weak signal propagation reporting (WSPR), which uses actual propagation measurements between the transmitting and receiving nodes. This removes the need for a propagation model and makes the approach suitable for any frequency band. As the number of nodes increases in the network, the calibration measurements can grow exponentially to produce a real-world model that can adapt to temporal changes in the World. The use of AI would also significantly improve speed and accuracy in the development of this real-world model. This model could find other uses outside CBRS 2.0 systems.

The other key feature of 'CBRS 2.0' using WSPR is that it provides interference control using the ID from the beacon to identify a source of interference that is subsequently controlled. This removes the requirements for overprotection in such systems as satellites and brings massive gains in spectrum efficiency. This again removes the band-specific propagation models used for their protection. Sensitive installations such as an FSS would have an ESC to monitor their local environment with a connection to the SAS to report the beacons they observe. The massive link budget of WSPR allows them to see over the horizon for very efficient control of any interference.

It could be applied to TV white spaces (TVWS) where broadcasters transmit the TV signal and a WSPR beacon. Here, the TV signal has priority one access within a CBRS-like system. Access points using these TVWS bands could measure the TV signal beacon to determine its channel and signal strength. A SAS manages its channel allocation within the white spaces based on the signal strength measurement. In a sophisticated system, this can consider the propagation profile around it to avoid any sensitivity issues associated with a particular location.

These TVWS APs would also incorporate beacons for monitoring by other APs and vice versa, again under the local SAS. The functionality and operation are similar to the use described above for C-band satellites. In all these ways, a 'CBRS

2.0' with WSPR provides a generalised approach to spectrum sharing in any band, and its application to 6G or previous generations would be revolutionary.

7.13.4 Comments about cognitive radio alternatives

Previous attempts have been made to use cognitive radio techniques to utilise unused radio spectrum, particularly below 1 GHz. Therefore, in examining how the proposed WSPR technique within a CBRS-like architecture can be used as a generalised approach to managing and controlling the radio spectrum, it is illuminating to review some notable examples of spectrum sharing, such as TVWS.

7.13.4.1 TV white spaces (TVWS)

After the switchover from analogue to digital TV broadcast in the United States, TV broadcasting is limited to channels 2–51, in the VHF and UHF bands, 54–60 MHz and 692–698 MHz, respectively. The unused TV spectrum at any location, as shown in Figure 7.2, is designated as TVWS. The FCC allowed unlicensed operation in TV white space frequencies, subject to cognitive usage rules designed to protect the TV broadcasters (the primary users of these bands). Similarly, other countries, including the UK, followed this route. Ramjee *et al.* [25] from 2016 provides an excellent critique of the FCC TVWS.

In TVWS, unlicensed, secondary user devices were required to incorporate 'smart radio' features to protect TV broadcasters (primary users) from unlicensed devices interfering with home TV reception. This is particularly important as these frequencies have suitable propagation ranges and penetration of buildings compared to those using higher frequencies in the mid-band.

The FCC TVWS classified two functional categories: lower-power personal devices (100 mW or less), which could be mobile, and higher-power fixed devices (1 W of 4 W EIRP with 6 dBi), which could be used as base stations providing a wireless backhaul for the low-power devices. The low-power devices were subdivided into Mode 1, which gets its channel assignment from the fixed base station, or Mode 2, which has geo-location capability for channel assignment-based spectrum sensing.

The channel assignment is managed by access to a database that has implemented a conservative propagation model of the primary user's service areas with a protection region around this. It maintains records for all the authorised primary services within the TV bands. This is combined with the secondary users' high-power transmitter locations, EIRPs, antenna heights above ground, antenna pattern and tilt and the call sign (ID). As in the CBRS model, database administrators provide a chargeable service for channel assignments. It is a lookup table based on the geo-location for mode one devices with an accuracy of ±50 m. Spectrum sensing can be used instead of geo-location but requires device certification after testing to provide confidence.

7.13.4.2 TVWS issues

Ramjee *et al.* [25] describe the main issues (while the comments explain how WSPR and SAS solve this problem). These include

- Database inaccuracies, particularly in dense metro areas where it showed little or no TVWS availability. As we have described earlier, it is impossible to use a propagation model to accurately calculate the signal path loss at locations across a wide range of topologies ranging from urban to open spaces and external to internal and, at the same time, taking into account temporal effects such as seasons and building changes. Ramjee *et al.* quote a New York study where 40%–75% of the TVWS spectrum is 'lost' due to the conservative propagation-based database. *(As described above, WSPR avoids the use of a propagation model and its inherent inaccuracies.)*
- There is no consideration of secondary-to-secondary coexistence, which in Wi-Fi is handled elegantly by listening before talking (LBT).
  (while LBT is a good solution, the addition of the use of WSPR beacons and their receivers within the TVWS APs allows the SAS to build a more robust real-world local model for the management of the APs in high-density and complex environments)
- Spectrum sensing, where portable devices that incorporate this feature are limited to a maximum EIRP of 50 mW while being required to detect Digital TV signals at a sensitivity of −114 dBm. For a 6 MHz TV channel, as [25] points out, this requires an SNR in the range of −20 to −18 dB and is very challenging to achieve. This leads to lost TVWS spectrum use or potential interference.
  (The ability to detect WSPR beacons from the TV transmitter, combined with measuring the TV signal, can extend the SNR to 70−80dB!)
- Another concerning factor was the antenna height limit. A high-power device at a fixed location could cause interference, so the FCC adopted a maximum antenna height of 30 m above the ground. This had the side effect of limiting the base station range for backhaul, significantly increasing coverage costs.
  (WSPR avoids this issue and extends the use of this spectrum.)

These factors and other minor problems limit the market opportunity size significantly. Furthermore, when Microsoft wanted to allow unlicensed devices closer to the existing TV channels, the National Association of Broadcasters (NAB) opposed this over-interference concern (May 2021) as there were no means to remedy any potential interference.

It is argued that the proposed CBRS 2.0 approach would address all of these points if used for TVWS and, by extension, other wireless systems.

- There is no conservative propagation model limiting spectrum utilisation.
- There is a mechanism to identify any interference should it ever arise and remedy this so this removes the technical arguments made by competing users of the spectrum.
- The access point carries out the spectrum sensing (WSPR), eliminating the requirement for Mode 2 devices. Here, the aggregation of devices associated with the Access Point can be modelled simply in the SAS.

7.13.5 Link budget considerations of WSPR in a generalised spectrum sharing architecture

A vital feature of the WSPR approach is that it can achieve link budgets for the beacons that are orders of magnitude superior to the signal link budget. This allows the SAS to 'see over the horizon' when planning the frequency assignment based on accurate measurements. We saw two completely different use cases of beacon link budgets of circa 200 dB for WSPR, the first in MF and HF bands of 1.6–3.8 and 4–27.7 MHz, respectively, for amateur radio, while the second was an extension for use in the C-band of 3.7–4.2 GHz. In the first case, the amateur radio used an isotopic antenna, while in the second, the antenna was highly directional. Nevertheless, the intrinsic improvement in the link budget for WSPR arises from the extreme cases of perhaps requiring less than 50 bits of information representing all the key features of the signal transmission and the ability to use sophisticated coding, averaging and directional antennas, which is independent of frequency bands of use. So, while the signal propagation ranges enormously from 1000s of miles at HF and MF to perhaps a kilometre LOS at 4 GHz, the beacon maintained a circa 200 dB link budget. That is not to say that even a 200 dB link budget is required. In probably the most extreme case for spectrum sharing for C-band satellites, a worst-case 181.5 dB link budget is needed. Other use cases will require significantly less. Furthermore, as we move to higher frequencies where signal penetration of building becomes impossible, the beacon link budget requirements become more minor as the 'horizon' is the next cell.

7.14 Generalised architecture for management of low-band, mid-band and high-band

Let us discuss the architecture of CBRS 2, first presented in Figure 7.14, as a generalised approach to low-band, mid-band and high-band spectrum management, which is shown again below for convenience.

In all the bands, WSPR reporting is used to build a real-time propagation model based on a measurement that is periodically updated to reflect environmental changes. These beacons identify any interference and allow the central spectrum access system (SAS) to remedy it in real time. The nature of the real-time management process depends on the type of services and controls the Service Level Agreement (SLA) puts in place. The SAS also coordinates the frequencies it deploys with other Tier 1 incumbent licensed operators. In some cases, mobile operators have the agility to modify their frequency of operation. In other cases, they do not.

The access points that use the spectrum have in-line beacon detection via their antenna beam width and reporting to the SAS, as shown in Figure 7.12 in the case of satellites. This can be a generic device with a radio and modem for each frequency band. However, in some cases, such as radar installations, external sensing networks are required for this function, which reports to the SAS.

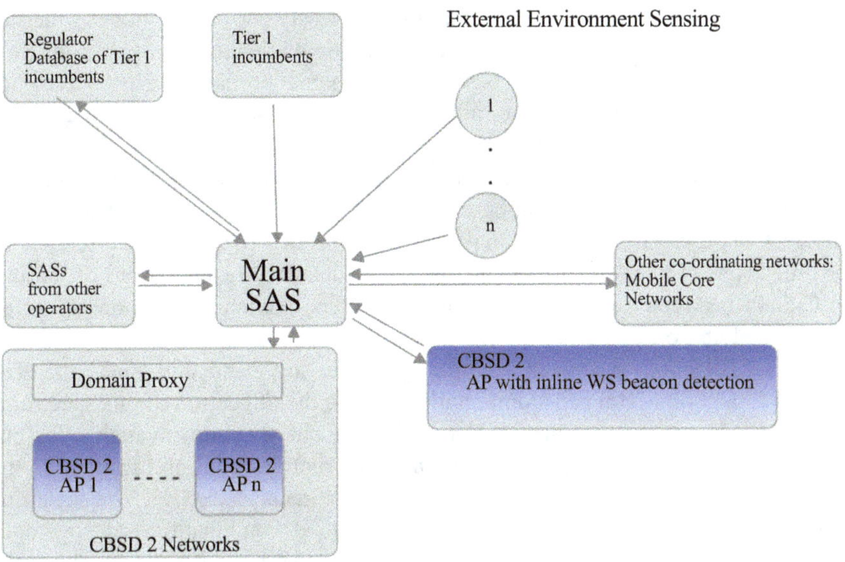

Figure 7.14 CBRS 2.0 architecture for low-, mid- and high-band

7.15 Summary and conclusions

Virtually every book or article discussing the radio spectrum refers to it as a scarce resource. Yet recently, we have added Terahertz to its legal definition in recognition that 3 THz is within human exploitation's reach. Chapter 8 of this book explores how we could reach the first 1 THz radio spectrum over the next decade, with 6G likely to use the sub-Terahertz spectrum in the 100–300 GHz range. So, is spectrum scarce in low-band (up to 1 GHz) and mid-band (1–6 GHz)?

In many countries, spot studies of spectrum occupancy in the 30–3000 MHz have found typical utilisations of between 10% and 20%, and there is good reason to believe that at a national level, the overall figure would lie in this range. Nevertheless, the need for increased channel bandwidth has driven us to seek ever higher use of the radio spectrum for each successive generation (the 'spectrum staircase'), with the perception that no spectrum is available at lower frequencies. Spectrum regulators recognise that despite the apparent congestion of the spectrum, no one knows what its utilisation is, while large areas of the country have low spectrum occupancy. The US CBRS is, in many ways, a revolutionary attempt to redress some of these issues by making mobile spectrum available at a county level and using a SAS to share the spectrum. It is a first step towards democratising the mobile spectrum, opening use to non-mobile operators, which will be increasingly crucial for Industry 4.0. However, it still bakes in a conservative propagation model within its design and cannot identify and control interference. The use of a propagation model has shown that it can 'lose' between 40% and 75% of the spectrum, so there is an extremely high price to pay for its use. In the case of sharing, this encourages discord as sharers of the spectrum endlessly debate the model's merits, and incumbent operators argue for greater protection while the new users want less. We have shown that a new system based on weak signal propagation measurements (reporting), first used in amateur radio and adapted within a CBRS-like architecture, could address all spectrum management issues. We call this 'CBRS 2.0'. For the first time, it will provide a means to share the spectrum widely with many different types of users, overcoming the concerns that have driven the adoption of conservative propagation models. In particular, the ability to address interference in real time is one of its most essential features. The detailed considerations of sharing the C-band between mobile and satellite providers exemplify its more comprehensive exploitation.

The 'CBRS 2.0' approach for managing the low-band and mid-band frequencies (30–3000 MHz) can significantly increase the utilisation of this spectrum. This approach can make nearly 2 GHz of additional spectrum available for 6G technology alone. Moreover, by reallocating other mobile bands and shifting TV to broadband, this figure could be increased to approximately 2.5 GHz. The spectrum below 3 GHz is the best for mobile, offering good combinations of coverage, channel widths and in-building penetration. It is worth noting that this approach, 'CBRS 2.0', can be implemented across the entire EM spectrum.

There is always a cost to change, and the FCC has led the way with auction design, using the proceeds to compensate existing spectrum users and cover the implementation costs.

In conclusion, spectrum is not a scarce resource but rather a resource unwisely used. In particular, enough spectrum is available for 5G and 6G below 3 GHz for most use cases apart from extreme performance, which uses cm and mmWave for circa 100 Gbit/s.

References

[1] NTIA. U.S. Radio Spectrum Allocation. [Online]. Available from: https://www.ntia.gov/page/united-states-frequency-allocation-chart.

[2] FCC, NTIA vow to fix spectrum allocation turmoil. [Online]. 2022. Available from: https://ww2.aip.org/fyi/2022/fcc-ntia-vow-fix-spectrum-allocation-turmoil.

[3] Nokia: Spectrum Explained. [Online]. Available from: https://www.nokia.com/about-us/newsroom/articles/spectrum-auctions-explained/.

[4] Federal Communications Commission (FCC): Notice of Proposed Rulemaking d(12). Federal Communications Commission (FCC). [Online]. 2018. Available from: https://docs.fcc.gov/public/attachments/FCC-18-21a1.pdf.

[5] Telefonica. WRC-23 Approves a Clear Roadmap for Mobile Services Spectrum. [Online]. 2023 [cited 26 December 2023]. Available from: https://www.telefonica.com/en/communication-room/blog/wrc-23-approves-a-clear-roadmap-for-mobile-services-spectrum/.

[6] Commission E. Radio Spectrum Policy Group. [Online]. 2023 [cited 25 October 2023]. Available from: https://radio-spectrum-policy-group.ec.europa.eu/system/files/2023-10/RSPG23-045final-Draft_RSPG_WP24_ and_ beyond_proposal.pdf.

[7] Manner JA. *Spectrum Wars: The Rise of 5G and Beyond*. Norwood, MA: Artech House; 2022.

[8] Mehdawi M, Fanan A, and Ammar M. Spectrum Occupancy Survey in HULL-UK For Cognitive Radio Applications: Measurement & Analysis. *International Journal of Scientific & Technology Research*, 2013; 2(4): 231–236.

[9] Cochrane P and Ngobigha F. Wireless Fit for the Internet of Things. *The Institute of Telecommunication Professionals*, 2022; 16(part 4): 9–15.

[10] Şeflek İ, and Yaldız E. Spectrum Occupancy Measurements at University Campus in Turkey. *International Journal of Electronics and Electrical Engineering*, 2017; 5(1): 1–6.

[11] Setoodeh P and Haykin S. *Fundamentals of Cognitive Radio*. 5th ed.: New York: Wiley; 2017.

[12] McHenry MA, Tenhula PA, McCloskey D, Roberson DA, and Hood CS. Chicago Spectrum Occupancy Measurements & Analysis and a Long-term Studies Proposal. *Proceedings of the First International Workshop on Technology and Policy for Accessing Spectrum (TAPAS '06)*. Association for Computing Machinery, New York, 2006. Available from: https://doi.org/10.1145/1234388.1234389.

[13] U.K Government – Department for Science, Innovation & Technology. [Online]. 2023. Available from: https://www.gov.uk/government/publications/spectrum-statement/spectrum-statement.

[14] Wikipedia. [Online]. Available from: https://en.wikipedia.org/wiki/Spectral_efficiency.

[15] Fierce Network. [Online]. 2019. Available from: https://www.fiercewireless.com/wireless/nist-taps-ai-for-better-radar-detection-3-5-ghz-band.

[16] Aruba Networking. [Online]. Available from: https://www.arubanetworks.com/en-gb/faq/what-is-cbrs/.

[17] FCC. FCC Announces Winning Bidders in C-band Auction. [Online]. 2021 [cited 24 February 2021]. Available from: https://www.fcc.gov/document/fcc-announces-winning-bidders-c-band-auction.

[18] Wikipedia. [Online]. Available from: https://en.wikipedia.org/wiki/WSPR_(amateur_radio_software).

[19] Smyth P, O'Connor J, Cheng L, and Winkelman E, Inventors; System and Methods for Fixed Satellite Protection Using Real-Time Measurements. *United States of America patent US* 10,574,343 B2. 2020 Feb 25.

[20] Smyth P, Cheng L, O'Connor J, Arendt S, Hamzeh B, and Winkelman E, Inventors; Systems and Methods for Beacon Detection Infrastructures. *United States of America patent US10*,367,577B2. 2019 July 30.

[21] Smyth P, Cheng L, O'Connor J, *et al.*, Inventors; Systems and Methods for Beacon Detection Infrastructures. *United States of America patent US 10*,581,514 B2. 2020 March 3.

[22] Smyth P, Poleti M, and O'Connor J, Inventors; Dynamic Frequency Planning in Shared Spectrum. *United States of America patent US10*,285,070 B2. 2019 May 7.

[23] Ingenu. *Random Phase Multiple Access*. [Online]. Available from: https://www.ingenu.com/technology/rpma/.

[24] Wikipedia. *Zadoff–Chu sequence*. [Online]. Available from: https://en.wikipedia.org/wiki/Zadoff–Chu_sequence.

[25] Ramjee R, Roy S, and Chintalapudi K. A Critique of FCC's TV White Space Regulations. *GetMobile*, 2016; 20(1): 20–25. Available from: https://www.microsoft.com/en-us/research/wp-content/uploads/2016/07/fcc.pdf.

Chapter 8

Millimetre and terahertz wave – 'optical fibres in air'

This 'telephone' has too many shortcomings to be seriously considered as a means of communication.

– William Orton, President of Western Union, 1876

8.1 Introduction

We are close to exhausting the valuable radio spectrum for mobile and wireless communications below 6 GHz with a rising tide of application speed in each new generation demanding greater channel bandwidth. One of the leading technology drivers supporting increased speed using this scarce resource has been improved spectral efficiency, but even here, we are very close to the Shannon limit. Recent trends in mobile and Wi-Fi generations have shown minor incremental spectral efficiency improvements in both the direct channel and the use of power-hungry multiple input and multiple output (MIMO) basc stations for mobile networks. Nevertheless, mobile data is still growing at a compound annual growth rate of 21% [1], which drives us up the spectrum staircase to use even higher frequencies.

This has driven the recent use of centimetre (cm) and millimetre waves (mmWave) towards 100 GHz for 5G, with an expectation that 6G will tap into frequencies above this. This expectation is based on emerging new applications requiring gigabit-per-second speeds (Gbit/s) or even greater. These mainly focus on new forms of imaging technology, such as virtual reality (VR) or Holographic displays. The former may require Gigabit-per-second speeds, while the latter is as high as several Terra bits per second, see Chapter 16. MmWave technology has potential non-telecom applications such as position sensing and detection of food allergens using future spectroscopes, as discussed in Chapter 14. In contrast, we discuss these technologies for the iPhone 6G in Chapter 15. Furthermore, mmWave is unaffected by darkness or smoke, offering new applications in rescue services or helping blind people. With speeds exceeding 100 Gbit/s, it could replace 'wiring' in ICs even operating at one of the absorption peaks at 380 GHz or for secure communications here.

One of the least understood benefits of mmWave is that the size of the antennas is so small that massive arrays of these can be directly integrated with all the electronics onto a single chip – a whole base station in a chip with a significant cost reduction.

Millimetre waves lie in the frequency region from 30 to 300 GHz with 10 mm to 1 mm wavelengths. Hence, terahertz waves are defined to cover 300 GHz–3 THz with associated wavelengths of 1 mm–100 μm. Historically, wireless regulation had ended at 275 GHz. In the United States, while frequency allocations went to as high as 275 GHz, the service rules ended at 95 GHz as there was virtually no use above this frequency. The World Radio Conference extended this to 3 THz in October 2019 [2] in recognition that we now had the potential to reach well beyond 275 GHz for future wireless applications, and the FCC was one of the first agencies at the vanguard of regulation of the mmWave spectrum [3]. That is not to say that anything like 3 THz will be exploited within the time scales for 6G.

While we have discussed the abundance of radio spectrum above 95 GHz with virtually no use there, the frequencies between 0.1 and 3 THz were referred to as the 'terahertz gap' before the 1980s [4]. Sengupta *et al.* present an excellent review paper in this reference. These frequencies were sandwiched between microwave use at one end and optical use at the other end. The gap existed then because efficient methods for generating and detecting signals in this range were lacking. This started to change in the 1980s with the first optical methods for generating and detecting THz signals. In the last two decades, there has been fantastic progress in enabling the electronic generation and detection of signals in this frequency range in compact IC scale electronics, helping close the first part of the gap [4] up to circa 1.5 THz. This was coupled with the recent discovery that cm wave frequencies could be used for cellular-like coverage, creating a new and important use case. However, the primary use has been for mobile operators to offer fixed wireless access (FWA), which competes with Cable and fibre to the home (FTTH) based on a single standard.

There is a common belief that using higher radio frequencies is worse, which has resulted in most usage concentrated in the first 6 GHz of the radio spectrum. However, this perception must be challenged despite the challenges as there is great potential in using mmWave and terahertz waves. This technology reminds me of the development of optical fibre technology in the 1980s, when there was worldwide interest and research, and the early uses were relatively modest compared to today's. Although mmWave and terahertz waves can approach the bandwidth of optical fibre, around 50 THz,[1] it may still take decades to demonstrate its full potential.

While mmWave offers the potential for an extraordinary increase in mobile and Wi-Fi communications speed, it is essential to understand the technical challenges and limitations that must be addressed over the next decade to make this suitable for use in the 2030s. To understand why anyone would even consider this viable against the background of the history of the wireless industry and a belief that lower frequencies offer better performance than high frequencies. While this

[1]https://physics.stackexchange.com/questions/56240/maximum-theoretical-bandwidth-of-fibre-optics provides an excellent discussion of the calculation.

might be part of the evolutionary roadmap of mobile and Wi-Fi communications, there are also revolutionary applications outside the communications domain for which this technology could be used. This chapter aims to cover a wide-ranging review of mmWave and THz applications, technologies, limitations and great expectations for the future.

8.2 The wireless spectrum to 1 THz and beyond

Are we facing a bandwidth crisis as the demands for mobile data increase each year and low-band and mid-band spectra become increasingly scarce? Actually no! Chapter 7 addressed how we could use the spectrum below 3 GHz more efficiently. Figure 8.1 shows the atmospheric absorption of electromagnetic waves up to 1 THz using the International Telecommunications Union model [5], which describes the specific attenuation for any pressure, temperature and humidity value. (This can be reproduced with a single-line code in MATLAB®.) Here, there are two curves. Blue shows a typical value of 7.5 g/m^3 water vapour compared to dry air, shown in brown.

These curves show that the whole 1 THz is available for potential use. However, there are absorption peaks across the entire EM spectrum due to individual resonance lines from oxygen and water vapour, together with small additional factors for the non-resonant Debye spectrum of oxygen below 10 GHz. You will see various attenuation peaks at 60, 119, 183, 325 and 380 GHz bands up to 400 GHz, together with others across the 1 THz bands, due to several different mechanisms associated with the summation of the individual resonance lines from oxygen and water vapour. The water content strongly influences the water-related peaks' peaks.

While 5G looks to exploit frequencies up to 100 GHz during its lifetime, 6G is likely to initially exploit frequencies in the range of 7–24 GHz in its first tranche as well as those frequencies close to wireless local area network (WLAN) use of 6 GHz as 6.425–7.125 GHz is a new IMT band following the World Radio Conference 2023 decision. In particular, 7.1–8.4 GHz is a candidate for the 6G global service rollout [6]. Nevertheless, mmWave above 100 GHz could offer superior performance and cost. This may emerge later as 6G is used for new use cases demanding multi-Gbit/s or high-precision sensing applications.

Focusing on the first 400 GHz of the spectrum with the standard water vapour curve, the attenuation only reaches 3 dB/km at 200 GHz and 20 dB/km at 400 GHz. Small cell applications covering a 100 m radius would only equate to 0.3 and 2 dB, respectively, of additional loss above free space loss with minimal effect on the system. The 77, 140 and 240 GHz bands only suffer 1 dB/km or less additional loss and are suitable for longer-range broadband fixed applications.

While it might look undesirable to operate at any of the high absorption peaks, even these bands are helpful for smaller cells or short-range applications where, in particular, security is a crucial requirement or the deployment requires a high degree of spatial reuse. For example, the 60 GHz band has been used for the WiGig

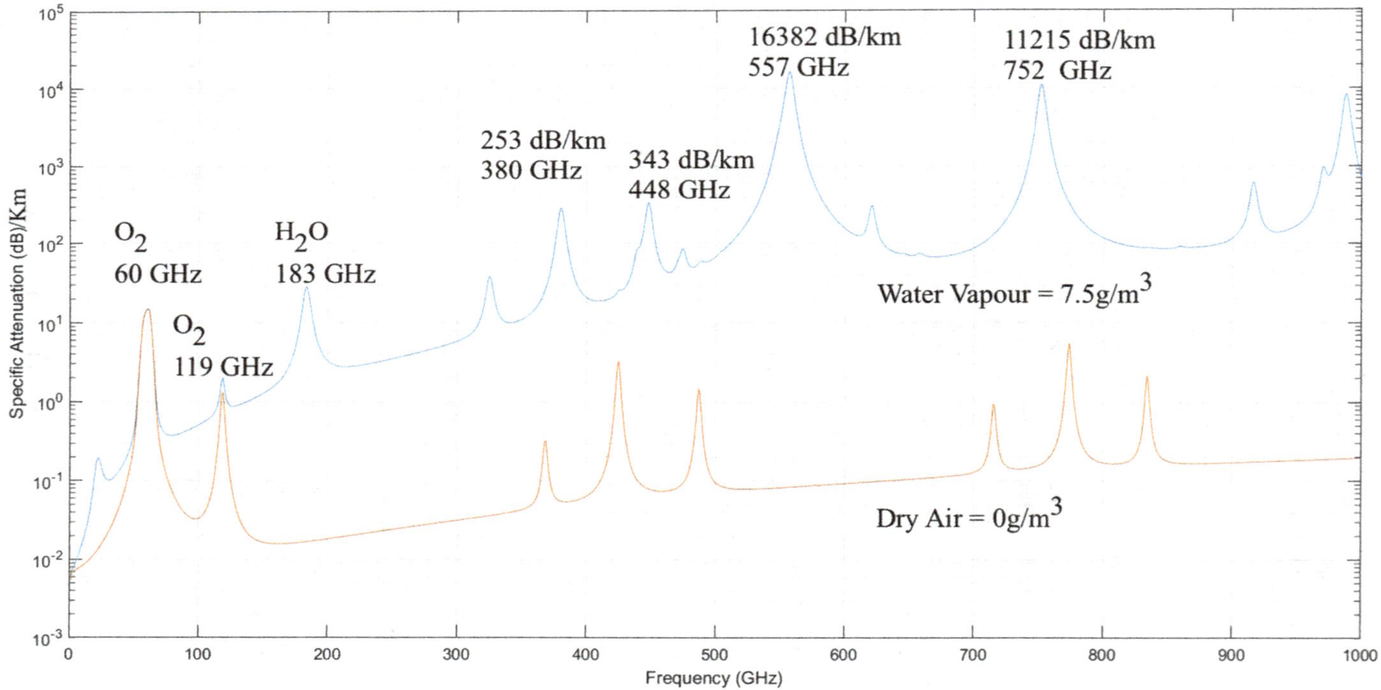

Figure 8.1 Atmospheric absorption of electromagnetic waves at sea level versus frequency for water vapour and dry air shows the additional path loss beyond free space propagation due to atmospheric absorption (MathLab simulation)

(Wi-Fi standard IEEE 802.11ad) standard to provide about 8 Gbit/s for short-range communications over 1–10 m. In addition, the 183, 325 and 380 GHz bands will likely replace the wiring harnesses, circuit boards and microwave waveguides in ICs with massive bandwidth channels, which will attenuate rapidly over a few meters. For example, the 380 GHz band reaches a peak of approximately 300 dB/km, while the band close to 550 GHz reaches over 10,000 dB/km! With just a focus on the first 400 GHz of the EM spectrum, there is 300 GHz available above 100 GHz for future mobile generations. Compare this with the concentration of use of only about 6 GHz over the last 120 years.

A decade ago, the cm wave was not considered 'mainstream' for cellular communications because of a misunderstanding of free space radio propagation law, limiting its primary use to point-to-point microwave systems and scientific sensor applications. The perception of the laws of physics shows that the loss from propagation increases at least quadratically with frequency and distance. For example, it is at least 900 times worse to increase the frequency of operation from today's low band, 1 GHz, to 30 GHz, and hence, the adverse ability to build a cellular communications system due to the high channel loss! This was compounded by the need for more availability of integrated electronic components that could transmit and receive mmWaves. This began to change in 2013 when Prof. Ted Rappaport from New York University published his groundbreaking paper [7], demonstrating cellular-like coverage with cm wave for a scenario of a city deployment of microcells. This eventually led to the use of cm wave in today's 5G and paved the way for even higher speeds in mmWave for 6G.

8.3 Wireless spectrum regulations change for mmWave and THz

Following the World Radio Conference 2019 decision to extend spectrum regulation to 3 THz [2] in recognition that we now had the potential to reach well beyond the previous 275 GHz limit for future wireless applications, the FCC was one of the first agencies at the vanguard of regulation of the mmWave spectrum [3]:

> *The FCC recognises innovators continue pushing technological boundaries in wireless communications. Frequency bands once considered unusable are now well within the range of modern communications systems. With their First Report and Order, the FCC provides new opportunities for innovators and experimenters to push those boundaries further and develop new equipment and applications for spectrum between 95 GHz and 3 THz.*

This is important as it sets the framework for innovation in the sub-THz and THz spectrum. The FCC offers experimental licenses for 95 GHz–3 THz – Spectrum Horizon ET Docket 18–21, while the rules on Licensed Spectrum will be deferred until there is sufficient technical and market data. Of course, other regulators and standards organisations, such as the European Telecommunication

Table 8.1 Operation proposed by the FCC

Frequency band (GHz)	Bandwidth (GHz)
116–123	7
174.8–182	7.2
185–190	5
244–246	2

Standards Institute, are actively reviewing this spectrum. In March 2019, the FCC made 21.2 GHz available for unlicensed use, shown in Table 8.1, and allowed investigational activities in the electromagnetic spectrum up to 3 THz.

Similarly, the National Telecommunications and Information Administration (NTIA) has set up a policy motion towards efficient spectrum policy for future US use and encouraged NTIA to lessen the restrictions on spectrum beyond 95 GHz. Meanwhile, the IEEE formed the IEEE 802.15.3d task force in 2017 for global Wi-Fi use at frequencies from 252 to 325 GHz, creating the first worldwide wireless communications standard for the 250–350 GHz frequency range, with a nominal new physical layer (PHY) data rate of 100 Gbit/s and channel bandwidths from 2 to 70 GHz. Meanwhile, the European Commission established the Radio Spectrum Policy Group to address THz-band communication management matters.

This demonstrates the international interest and moves by regulatory bodies to make available spectrum bands for future commercial exploitation.

8.3.1 Supporting research centres in mmWave and 6G

Outside regulation, essential institutions in this space include commercial research supported by key universities. Most notability the key universities include Prof. Ted Rappaport's NYU wireless research Centre, which focuses on mmWave propagation and channel sounding; Prof. Gerhard Fettweis' 5G/6G labs in Dresden, Germany which focuses now on 5G and 6G, including the applications of mmWave, Prof. Hua Wang from Georgia concentrates on the electronics and IC technologies for THz. The other key European research centres for 6G are in the UK at Surrey University under the leadership of Prof. Rahim Tafazolli and Finland at the University of Oulu. In the United States, the mmWave Coalition [8] is an important lobbying organisation for mmWave and, associated with this, information sharing amongst its members. Of course, there are many other excellent centres as well.

8.4 Transistor device performance – towards and beyond 1 THz

In the last two decades, there has been fantastic progress in enabling the electronic generation and detection of signals in the frequency range 0.1–3 THz, which were initially referred to as the 'terahertz gap' before the 1980s, sandwiched between microwave use at one end and optical use at the other end. Compact IC scale electronics are helping close the first half of this gap [4]. While the evolution path from 5G to 6G will likely explore the spectrum staircase initially from 100 to 200 GHz and then perhaps 200–400 GHz during its lifetime for communications applications, many of the new revolutionary uses of smartphones may be in imaging and sensor applications and spectroscopes, where frequencies well above 100 GHz will also be critical.

III–V semiconductor technologies such as indium phosphide (InP) heterojunction bipolar transistors (HBTs) and high-electron mobility transistors (HEMTS) have made operation to the midpoint of the terahertz gap on the electronic front possible. At the same time, gallium arsenide (GaAs) based Schottky diodes can generate power levels in the 100 μW to mW range. (Engineers familiar with RF design typically deal with RF signal output power levels of 100s of mW.)

Nevertheless, silicon-based electronics have been massively important in providing a platform for massive integration and extending frequency operation beyond 30 GHz. For example, as early as 2015, the world's largest single-chip phased array featuring 256 transmitter elements outputting 43 dBm (EIRP) at 60 GHz was built in the Towergate SiGe BiCMOS for 802.11 ad [9]. This phased array system-on-a-chip (SoC) was implemented in an IC chip measuring 41.6 mm by 41.6 mm. SiGe has also been used in the mid-terahertz gap (1 THz) to make a phased array of 42 elements using a 130 nm SiGe process [10]. This chip occupied only one mm^2 area and demonstrated the potential for massive phased arrays.

Although Si or SiGe is predominantly used for IC design for mmWave, it is worth noting that gallium nitride (GaN) has remarkably high-power handling capabilities, as we described in Section 8.5.3, which is crucial for high-power amplifiers/phased arrays used in 5G cm bands and beyond into the lower sub-terra band. In terms of IC fabrication, Qorvo offers an extensive range of GaAs and GaN IC fabrication, whereas Wolfspeed offers GaN fabrication.

8.4.1 Transistor f_{max} and f_t – indicators of performance

The two key parameters that describe the transistor performance for high-frequency operation are f_{max} and f_t. The f_{max} of a transistor is the frequency where the power gain falls to unity, while f_t is the 'transition frequency' where its current gain falls to unity. A transistor generally will have different values of these two frequencies; nevertheless, they are similar in value, with f_{max} being higher. These are significant for circuit design because f_{max} is a valuable figure of merit for a power amplifier (PA) or phased array design. At the same time, f_t is a valuable metric for a switching circuit design such as a digital-to-analogue converter (DAC) or ADC.

As a 'rule of thumb', IC designers work to the maximum frequency of operation of $1/2\,f_{max}$ for either PA or phased array design and 1/3 of f_t for switching circuit design. However, working up to these values will present significant challenges in the initial design, resulting in lower yields and higher production costs. For higher yields, designers typically work to 0.4 of the f_{max} of the IC transistor process. Nevertheless, research prototypes can follow a more aggressive circuit design working to 0.5 f_{max} to show potential (the so-called 'hero results'). As IC processes improve, these can be transferred to production for higher yield with associated lower costs. Therefore, these two parameters represent excellent indicators of the ability of IC processes to meet system-on-chip (SoC) requirements in the sub-terra Hertz region.

Today, production-based ICs made with CMOS and SiGe are fast enough to operate up to frequencies close to 200 GHz with switching speeds of 100 Gbit/s. For example, Global Foundries' SiGe IC fabrication offers SiGe transistors with a f_{max} of 400 GHz and an f_t of 310 GHz. This effectively opens up 100 GHz from 95 GHz with speeds suitable for 6G. Nevertheless, on the frontier during the 6G lifetime, InP HEMTs [4] has achieved a f_{max} of circa 1.5 THz.

Figure 8.2 shows the status of the various semiconductor material systems and device performance. For f_{max} vs f_t low-cost, Si CMOS struggles to get beyond a f_{max} and f_t combination of 400 GHz, where higher device circuit parasitics and contact resistance offset device scaling for faster performance. This effectively limits its use to below 200 GHz frequency operation and 100 Gbit/s. Nevertheless, the 100–200 GHz could be the main frequency range for 6G initially explored if >100 Gbit/s applications are not required, as these would require channel bandwidths of at least 20 GHz. SiGe extends the frequency performance beyond that of Si with 720 GHz f_{max} and f_t of 505 GHz demonstrated [11].

Nevertheless, GaN HEMT transistors were developed over a decade ago with a f_{max} and f_t of >400 GHz. This is likely the leading candidate phased array for the 116–123 GHz unlicensed band GaN and the 100–200 GHz band for 6G because of its superior power handling capabilities, which exceeds SiGe by 12 dB. State of the art for peak output power, P_{sat}, is shown in Figure 8.3 for various semiconductor material systems up to circa 500 GHz with F_{max} for GaN, GaAs and SiGe.

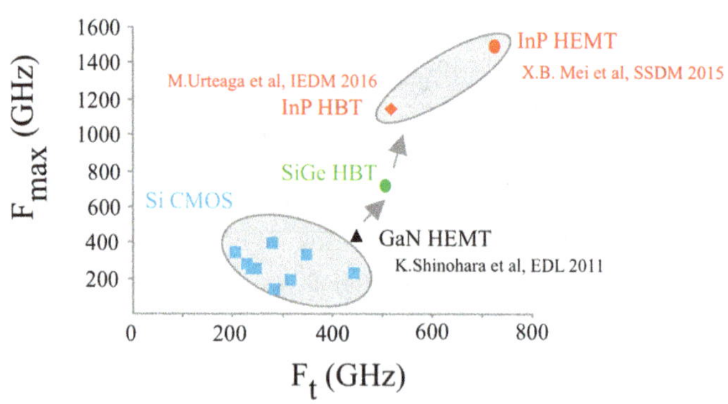

Figure 8.2 Status of the various semiconductor material systems and device performance (based on IMEC keynote, Scott McDermott at 5G Submit Dresden 2019)

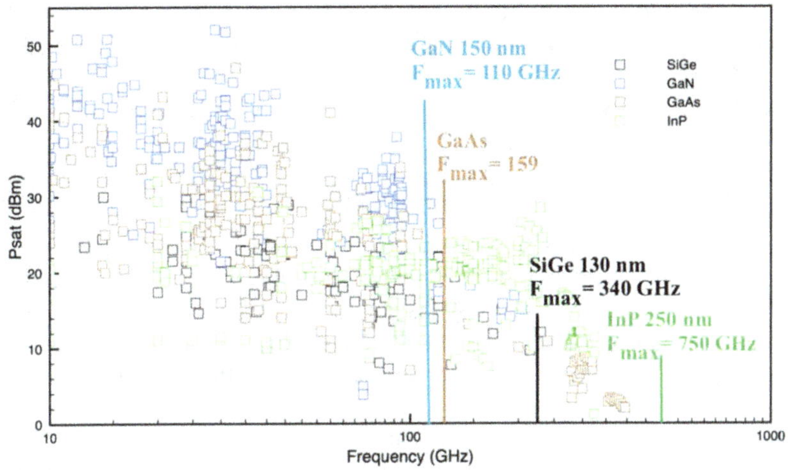

Figure 8.3 Power amplifier survey – Prof Hua Wang Georgia Tech et al. [13]

8.4.2 Challenges of transceiver design above 100 GHz

Operation above 100 GHz into the terahertz region (300 GHz–3 THz) will require new approaches to the transceiver front-end design due to challenges of transistor power efficiency and linearity coupled with the vastly increased number of antennas that will be possible in a small physical area [12]. At some point, the RF amplifiers will not be small enough to be physically located behind individual antenna elements. For example, at 240 GHz, the ½ wavelength dipole dimension on a substance with an ε_r of 4 is 312 μm while a single SiGe transistor node size today is 90 μm alone. While the node size will decrease somewhat in the future, you will see that at frequencies beyond 300 GHz, the amplifier size will exceed the size of the array element.

Furthermore, key transceiver performance metrics such as noise figure (NF), the linearity of the low-noise amplification, peak output power P_{sat} and power-added efficiency (PAE) all degrade rapidly with increasing operating frequency. A recent survey (2023) of the current literature from Prof. Wang's GEM Group [13] shows the product of P_{sat}, gain and PAE scales [14] as $1/f_0^2$, where f_0 is the operating frequency. This is the 'go-to' place for the latest results in mmWave Electronics.

Phase noise, which is the random fluctuation in the phase of a waveform that directly affects coherent systems, increases with frequency, creating increased noise levels for practical coherent demodulation. So, whereas today, in 5G, we are used to using higher order modulation schemes such as quadrature amplitude modulation (QAM) such as 256 and 1028 to increase data rates within a limited bandwidth, in mmWave and terahertz communications, this could drop to as low as say 64 (with 6 bits/Hz)!

Therefore, there are tremendous challenges in creating commercial transceivers at THz frequencies. However, global research is addressing the challenges. For example, the DARPA T-MUSIC programme [15] investigates SiGe HBT, CMOS/SOI and BiCMOS circuit integration to achieve PA threshold frequencies of up to 500–750 GHz. The T-MUSIC programme is a four-year effort that started in 2019. It seeks to develop wafer-scale technology for the United States on a Silicon CMOS-based foundry platform.

8.4.3 Performance and time scales of electronics

When discussing timescales for potential applications of transistor technology for 6G, it is essential to differentiate between what can be achieved with early prototypes for CMOS/SiGe/GaN and those that appear in commercial products with high yield and associated lower costs. Furthermore, the commercial products will require certainty over the regulatory frequency bands with an international agreement on bands of operation.

Today, one of the main focuses for integrated phased arrays is in the W-band (75–110 GHz) using SiGe with notable work at Nokia Bell labs under Shahriar Shahramian, Director of mmWave ASIC Research, Bell Laboratories, Nokia.

Nevertheless, significant challenges still need to be addressed, which will limit its immediate availability once the applications emerge. At the higher frequency of operation in this range, the SiGe transistor node size becomes compatible with the ½ wavelength dimensions of the antennas on the wafer. Many other considerations surround engineering the chip at these frequencies. Packaging is also a key element in the design at these frequencies, with packages effectively 'grown' with the IC to minimise interconnection loss.

Nevertheless, the authors believe 50–100 Gbit/s operation in the sub 150 GHz band would be possible depending on the channel bandwidths used. For example, 116–123 GHz could support up to 42 Gbit/s, assuming a conservative 6 bits/Hertz.

Extending the f_{max} and f_t of commercial SiGe fabrication to at least 750 GHz (one of the objectives of the DARPA T-MUSIC project) will enhance performance in SiGe and SOC fabrication. Global foundries also believe that their process could reach this performance level. What is learned in the next two to three years could be applied directly to the new process. Therefore, 95–300 GHz is the critical range of frequencies likely to be exploited between 2026 and 2028 on standard IC processes, depending on commercial demand. This effectively puts into play at least 200 GHz of additional spectrum for 6G, while until recently, we have concentrated on mainly the first 7 GHz of the EM spectrum (Figure 8.4). This is reminiscent of the early days of optical fibre research in the 1980s in which the author was involved.

8.4.4 'Hero' results on the way to 6G

In trying to gauge the capabilities of technology for 6G, one of the best merits is what is achieved in Lab prototypes, sometimes referred to as 'hero' results. While scaling and productionisation may take years, they inevitably point to future capabilities.

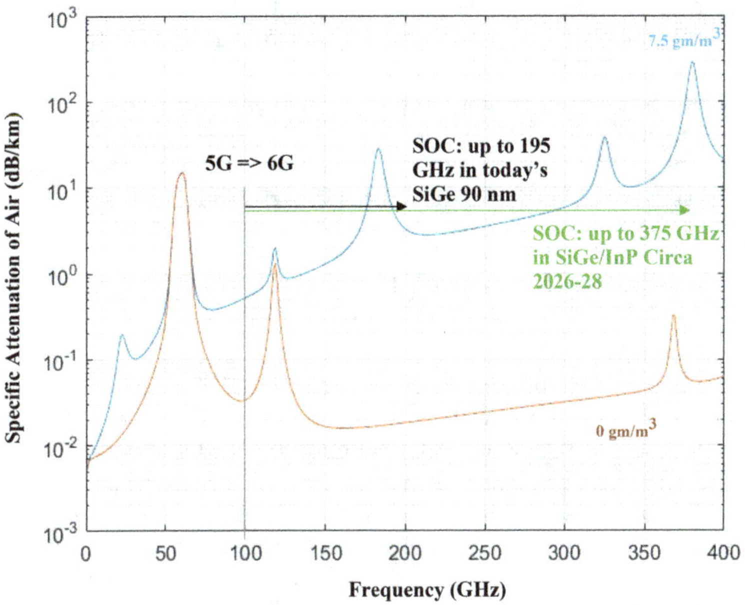

Figure 8.4 Range of operation for SiGe and InP

Cellular systems are advancing rapidly, potentially operating at 80 Gbit/s and beyond. IEEE Standard 802.15.3d has already achieved this speed at 300 GHz. The FCC has established a framework for innovation above 95 GHz, which is expected to encourage the development of prototypes for IC-based transceivers in the 100–200 GHz range. Additionally, semiconductor research is exploring transistor physics in the THz region, and two notable achievements in this field have been highlighted in their paper's abstract:

8.4.4.1 A fully integrated 384-element, 16-tile, W-band phased array with self-alignment and self-test – August 2019 – *IEEE Journal of Solid-State Circuits* pp (99):1–16 Shahriar Shahramian *et al.*

This article presents the design and implementation of a scalable phased-array system that operates at W-band frequency (75–110 GHz). The system is based on an integrated circuit transceiver chipset manufactured using the TowerJazz 0.18-μm SiGe BiCMOS technology. The chipset has an f_t/f_{max} of 240/270 GHz and comprises 24 phase-shifter elements, direct up- and down-converters, an analogue baseband, beam lookup memory and diagnostic circuits for performance monitoring. The chipset also integrates a phase-locked loop with a prime-ratio frequency multiplier. To produce a scalable phased array tile, they designed two organic printed circuit board (PCB) interposers with integrated antenna sub-arrays, which were co-assembled with the RFIC chipsets. The tiles are phase-aligned using a daisy-chained local oscillator (LO) synchronisation signal.

They present a statistical analysis of the effects of LO misalignment between tiles on beam patterns. They combined 16 tiles onto a carrier PCB to create a 384-element (256TX/128RX) phased-array system. The system provided a maximum saturated effective isotropic radiated power (EIRP) of 60 dBm (1 kW) at boresight for the 256 transmit elements. The wireless links were tested at 90.7 GHz using a 16-QAM constellation at a reduced EIRP of 52 dBm. The links produced data rates beyond 10 Gb/s for an equivalent link distance over 250 m.

8.4.4.2 An 80 Gb/s 300 GHz-band single-chip CMOS transceiver, S. Lee *et al.* 17–21 Feb. 2019: Conference: 2019 *IEEE International Solid-State Circuits Conference* (ISSCC)

IEEE Standard 802.15.3d, published in October 2017, defines a high-data-rate wireless physical layer that enables up to 100 Gb/s using the lower THz frequency range between 252 and 325 GHz (referred to as the '300 GHz band'). It stipulates that the 300 GHz band be channelised into 32 2.16 GHz-wide channels or a smaller number of wider channels whose bandwidths are all integer multiples of 2.16 GHz. Their paper presents a CMOS transceiver (TRX) chip targeted at channels 49–51 and 66 of 802.15.3d. The TRX was fabricated using a 40 nm CMOS process. Reports have been on solid-state transceivers (TRXs) operating in or near the 300 GHz band. Some of these were TX/RX or block-level chipsets, which can enjoy more flexibility in design and independent optimisation of TX and RX. They successfully achieved $\geq$ 64 Gb/s. On the other hand, single-chip TRXs did not always reveal achievable data rates, nor were they capable of supporting QAM. Nevertheless, the eventual development of full-featured single-chip TRXs is desirable, especially for applications requiring the deployment of many TRXs, as is envisioned implicitly by 802.15.3d. The single-chip QAM-capable CMOS TRX presented herein results from efforts in that direction.

8.5 Free space path loss at cm and mmWave frequencies

In the introduction, we explained that the perception is that the EM spectrum is seen as a scarce resource with low frequencies offering much better coverage than higher values. This has led to the massive concentration in the commercial exploitation of this resource for communications in frequencies from kHz to 6 GHz. This view stems from an interpretation of Frii's free space propagation law (FSPL) presented by the Danish-American radio engineer Harald T. Friis in 1946, described in its contemporary form with the formula:

$$P_r/P_t = G_r\, G_t(\lambda/4\pi d)^2 \tag{8.1}$$

or

$$P_r/P_t = G_r\, G_t(c/4\pi f d)^2 \tag{8.2}$$

where P_r = Received power
P_t = Transmitted power
λ = Wavelength ($c = f\lambda$)
d = Distance between Tx and Rx (large enough to ensure a plane wavefront)
G_r = Receiver antenna gain (with respect to an isotropic radiator)
G_t = Transmitter antenna gain
c = Speed of light in a vacuum
α = Loss exponent, where $\alpha = 2$ for Friis' Free space condition

In the design of a radio communications system, the choice of P_r, G_r, and G_t is such that P_r exceeds the minimum sensitivity of the system for its chosen modulation scheme. The right-hand side of (8.1) and (8.2) is effectively the transmission loss between the transmitter and receiver. The original analysis of FSPL assumes that the antenna gains of both the transmitter and receiver were frequency-independent, as are the transmitter power and receiver sensitivity. Prof. Ted Rappaport has pointed out that there is still a recurring misconception among researchers and radio engineers that the wireless channels at higher frequencies would experience more loss since they have been taught to only consider the free space path loss (FSPL) with omnidirectional antennas. Therefore, in (8.2), the loss from propagation increases at least quadratically with frequency and distance (where α the loss exponent = 2 for Friis' Free space). However, this consideration is incorrect if the transmitter and receiver use constant area phased arrays. Furthermore, there is a frequency dependence in the transmitter power that can be generated and the receiver sensitivity for phased arrays. So, in the following sections, we will explore all these elements to determine a correct formulation of Friis' equation in this case.

8.5.1 Phased arrays

The defining key technology for both cm and mmWave communications is phased arrays [16]. These are two-dimensional (2D) groupings of antennas in an array where the relative phases of the input signal to each antenna are varied so that the effective propagation pattern of the array is reinforced in a desired direction and suppressed in undesired directions. The systematic variation of the phases of each antenna element allows the beam to be moved vertically or horizontally over an angular range. This allows electronic beam steering in both directions and has applications in mobile systems and sensing, where the phased array can be used to scan an environment to create a three-dimensional (3D) spatial map.

> Interestingly, as Bodhisatwa Sadhu points out, Jagadis Chandra Bose first demonstrated mmWave transmission to the Royal Society in London in 1895, two years before Marconi founded his Wireless and Telegraph and Signal Company with an apparatus which is similar to a mmWave test setup found today in laboratories. While phased arrays were invented in 1905 by K. F. Braun, who won the Nobel Prize for Physics in 1909, his invention led to the

development of Radar, Smart antenna and MIMO. In particular, phased arrays were used to detect the Japanese attack on Pearl Harbour in WW2 using 36 antenna elements.

Different types of antenna elements can be used to make up the array. Those commonly used in phased arrays are either the dipole or the patch, see Figure 8.5. The dipole is a half-wavelength conductor fed at its centre, while the patch is usually a half-wavelength metallised square. The great advantage of phased arrays, as described in Rappaport *et al.*'s excellent book [17], is that rather than having to purchase a separate antenna with a lossy cable connection to a PCB which contains the rest of the transceiver an mmWave on-chip antenna may be directly etched in the on-chip metal during IC production process to form a system-on-a-chip (SoC). This has the potential to reduce the overall costs of mass production dramatically. Alternatively, the antenna for cm wavelengths may be fabricated in the packaging technology used to house the RF amplifier chip or perhaps integrated into the PCB used to house the transceiver, albeit at a significantly higher cost. For cm wave applications, the element sizes are proportionally larger than those for mmWave.

The short wavelengths at mmWave frequencies also allow both the transmit and receive antenna arrays of many antennas to fit within a small package or on a chip easily. Another factor that comes into play for the fabrication of phased arrays is the dielectric constant of the substance material. For example, at 140 GHz, a $\lambda/2$ dipole antenna is only 535 μm in length ($=\lambda/2\sqrt{\varepsilon_r}$) when implemented on a substrate with a relative permittivity (ε_r) of 4. In contrast, in a free space relative permittivity of 1, this dipole would be twice that size. The relative permittivity of 4 allows four times as many elements to be integrated into an IC in a given area compared with a material with a relative permittivity of 1. This dramatically helps with the realisation of massive phased arrays. While that might suggest finding dielectric materials with even higher relative permittivity values (ε_r), there are constraints on what can be used in an IC process and higher losses associated with high permittivity materials that mitigate against their use.

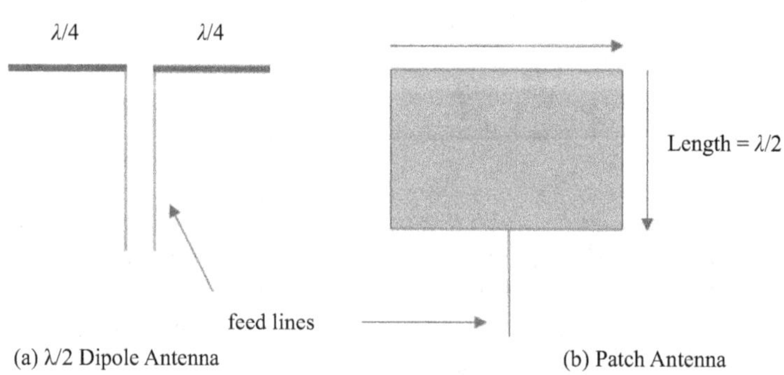

Figure 8.5 Typical antennas used in IC-based phased arrays

In a 2D phased array antenna with several regularly spaced antennas, as shown in Figure 8.6, the gain realised at the beam peak is equal to the number of elements times multiplied by the gain realised in the same direction when only one element is excited. This figure demonstrates that when using a constant area for the phased array and doubling the frequency of operation, the gain increases by four times.

$$\text{Gain of phased array} = \text{Array Gain } N\ (n \times n) \tag{8.3}$$

where $n \times n$: number of elements in the array

As is usual in engineering, things are not frequently that simple! In an excellent paper by Hannan [18], he explains that the active impedance of each antenna element in a phased array varies with its scan angle because of mutual coupling between the elements, as these are generally near each other (less than a wavelength). The associated mismatch causes part of the signal power driving each antenna element to be returned to its generator, thereby reducing the gain realised by each antenna element and that of the array. Equally, grating lobes can exist when the details are not closely spaced greater than $\lambda/2$. For example, the 3D radiation pattern of 4×4 and 8×8 rectangular arrays at 73 GHz using $\lambda/2$ spacing is shown in Figure 8.7(a) and (b). The strength of the signal, as identified by the colour scheme, is measured in dB, and the existence of various side lobes, albeit in low signal strength, is clearly shown. This is the typical radiation pattern frequency type found in 2D phased arrays. A $\lambda/2$ spacing is considered the optimum spacing for a phased array, particularly given a size restriction, which is the case in any IC.

8.5.2 mmWave massive phased arrays with equal area

Space has the highest premium in the design of handsets or even terminals. For mmWave phased arrays with equal areas, as the frequency of operation is doubled, the number of elements within the same area increases as a square law, as shown in Figure 8.6. In this case, the gain rises from 16 times to 64 times. This frequency-squared behaviour for the gain can be utilised in both the transmitter and receiver

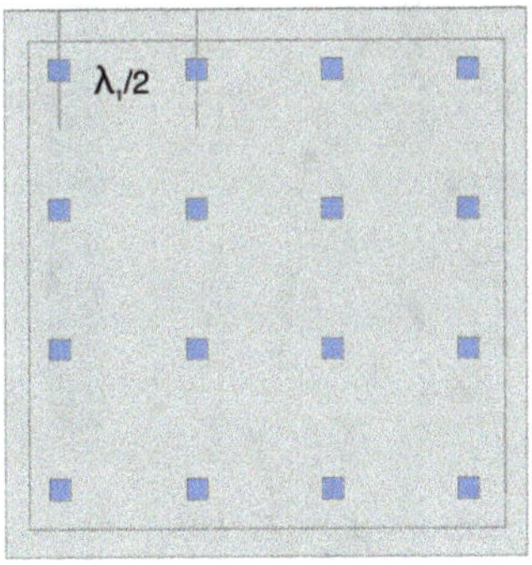

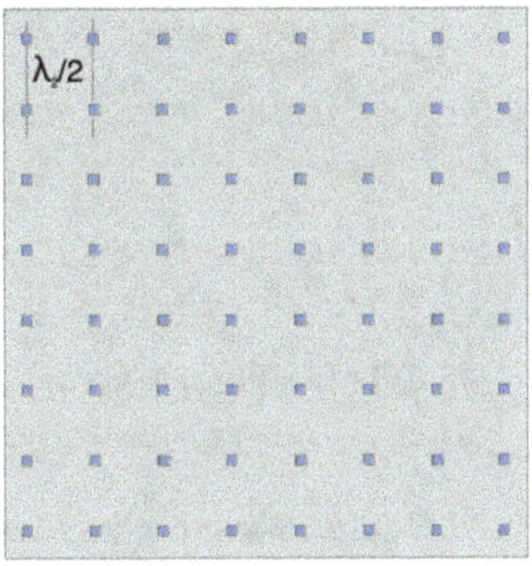

Figure 8.6 2D phased array with constant area at frequencies w_1=1 and w_2=2 w_1

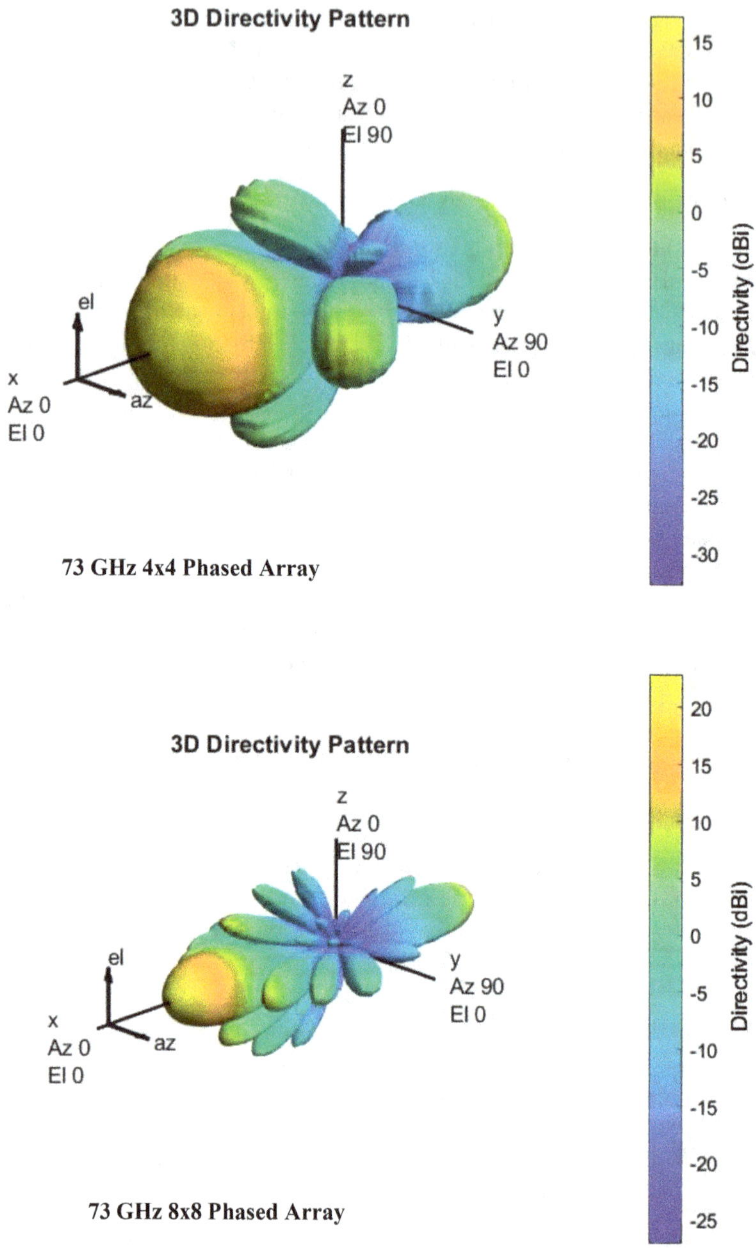

Figure 8.7 MATLAB antenna array visualisation at 73 GHz for 4 × 4 and 8 × 8 phased arrays

antennas. (Later, we will also consider the frequency dependence of the transmitter power and receiver sensitivity below.) Introducing this frequency behaviour into (8.2) completely changes the perception that it is better to operate at lower frequencies, as shown in (8.4)

$$P_r/P_t = G_r\ G_t(c/4\pi fd)^2 \Rightarrow G_r(f^2)\ G_t(f^2)(c/4\pi fd)^2 \\ \Rightarrow f^2 G_r\ G_t(c/4\pi fd)^2 \tag{8.4}$$

Whereas before, loss from propagation between the transmitter and receiver increases quadratically with frequency (8.2), now the transmission loss effectively decreases quadratically as the frequency increases so long as the antenna's physical sizes are kept constant. It is equivalent to moving f^2 from the denominator to the numerator. For example, moving from 3.5 to 35 GHz reduces the propagation loss by 100 times or 20 dB ($\Delta = 40$ dB better than (8.1)) with obviously higher values of improvements at higher mmWave frequencies. Of course, in both cases, the quadratic loss increases with the distance between the transmitter and receiver. However, you can think that the frequency square performance rises to compensate for the distance increase.

The results of improved coverage are seen in Figure 8.8. Here, we analyse the figuration presented in reference [19] for 28 and 140 GHz with phased arrays using an optimum wavelength spacing of $\lambda/2$. There are two use cases for these antennas: the lower set of curves representing the conventional view (omnidirectional) of Frii's FSPL as described in (8.2) with fixed gains in the antennas for the two frequencies where the loss increases quadratically with the frequency, And the top-performing group, where both the transmitter and receiver antenna array sizes

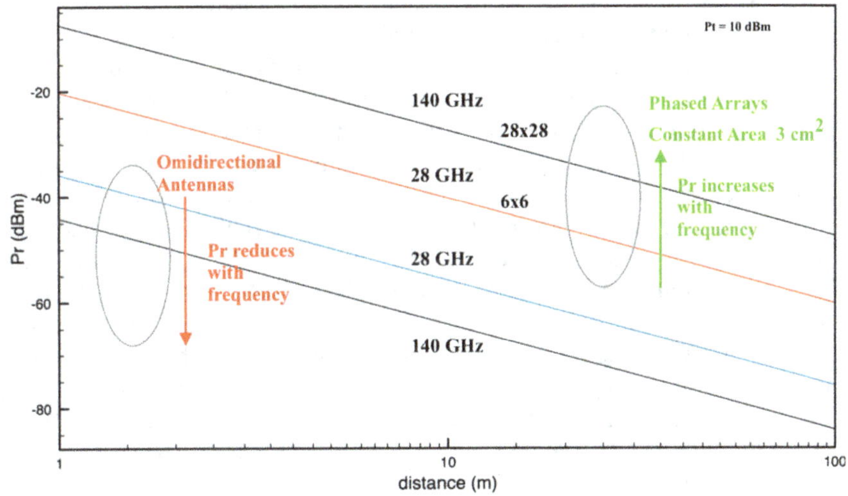

Figure 8.8 Comparison of the omnidirectional and constant area phased arrays antennas at 28 and 140 GHz

increase with frequency squared with the same phased array area of approximately 3 cm^2 in both cases.

Phased arrays are used in 5G and Wi-Fi systems today but offer even more significant potential in the mmWave spectrum for 6G and future Wi-Fi mmWave applications. The short wavelengths at mmWave frequencies also allow both the transmit and receive antenna arrays of many antennas to fit within a small package or on a chip easily.

8.5.3 Transmitter power dependence with frequency – Johnson's limit

While we have considered the surprising results of the improvement in Friis's FSPL with the frequency with constant-area phased arrays, we need to consider whether there is a frequency dependence also in the other elements of Frii's law – the transmitter power and receiver sensitivity. For example, does the ability to generate transmitter power levels fall as we move to higher frequencies, which would migrate against part of the benefits of higher frequency operation? How does receiver sensitivity influence this as well?

For the consideration of transmitter output power in the design of any electronic transmitter, the maximum output power is delivered as the output transistor is operated as close as possible to its device breakdown voltage across the output load. According to Johnson [20] from RCA, who derived the Johnson limit, the product of unity gain cut-off frequency f_t and breakdown voltage V_m of a transistor is approximately constant for a given semiconductor material, such that:

$$V_m f_t = E_m v_s / 2\pi \tag{8.5}$$

where E_m is the maximum electric field, and v_s is the saturated electron velocity. Table 8.2 describes these parameters for various semiconductor systems.

According to Wikipedia [21], semiconductor materials have different breakdown voltages, limiting their maximum power handling capabilities. Johnson's figure of merit measures a semiconductor material's suitability for high-frequency, high-output power operation. More specifically, it is the product of the charge carrier saturation velocity in the material and the electric breakdown voltage field under the same conditions. In Table 8.2, the results are normalised with silicon. From here, gallium nitride (GaN) is a fantastic technology for high-output powers in cm and mmWave transmitters, being 27 times better than silicon. So, this will be the semiconductor technology of choice for cm and mmWave transmitters. Given its advantage over other systems, such as indium phosphide (InP), which has a higher frequency of operation, it is likely to influence the choice of the lower frequency for operation unless there is an overriding reason for channel bandwidth. This would be a key candidate for the US unlicensed band at 116–123 GHz.

As a result, although advanced silicon (Si) process nodes can provide a higher frequency of operation (f_t/f_{max}), the breakdown voltage also drops,

Table 8.2 Johnson's figure of merit for GaN, SiC, GaAs, Si and InP [21]

Material	Saturation velocity $\times$ 10^5 m/s	$V_{Breakdown}$ (MV/cm)	Johnson figure of merit (Si = 1)
Gallium nitride (GaN)	2.5	3.3	27.5
Silicon carbide (SiC)	2.0	3.5	20
Gallium arsenide (GaAs)	1.5	0.4	2.7
Silicon (Si)	1.0	0.3	1.0
Indium phosphide (InP)	0.67	0.5	0.33

reducing the maximum voltage swing across the load and thus reducing the maximum output power delivered to the load – the power falls quadratically as a function of frequency. Other factors are described by Wang *et al.* in Chapter 19 [22].

Without the Johnson limit, the link budget (P_t/P_r) could improve remarkably as f^4. One of the key research areas is device physics, while a clever design circuit is another. In a private discussion with Prof. Hua Wang, Hua points out that it is true that for a single device using a given semiconductor material, assuming the critical device feature, say CMOS gate length, can be continuously varied, where its single device output power (P_t) degrades $\sim(1/f^2)$ due to Johnson's limit. However, he goes on to point out that if we look at the reported saturate output power P_t vs frequency in Figure 8.9, an exciting trend is that for operating frequency $f < (1/3) \cdot (f_{max}$ of the device), the P_t degrades a bit slower than $1/f^2$. This is because the active devices can provide sufficient gain in this frequency range so that the designers can realise more complex PA/transmitter designs. For example, with on-chip power combining and device stacking, resulting in P_t boosting, see below for more detail. Moreover, amazing device engineering by Foundries is trying to keep the device breakdown voltage while decreasing feature sizes and increasing the speed. Both aspects help reduce P_t degradation over frequency in practical designs.

On the other hand, for $f \sim (1/3) \cdot f_{max}$ to $(1/2) \cdot f_{max}$, the P_t degrades much faster. This is because active devices cannot support sufficient gain anymore, and one has to rely on nonlinear circuits and harmonics for power/signal generation. (However, nonlinear signal power/signal generation cannot support linear amplification of complex modulations and will lose spectral efficiency, so this is probably not a good fit for future 6G communications.) Therefore, it is critical to have faster and more powerful devices. This is the focus of the DARPA T-MUSIC programme [23]. Moreover, operating below device $(1/3) \cdot f_{max}$ will result in sufficient power gain and even more graceful degradation of P_t.

To support operation up to 400 GHz, we would need devices with a f_{max} of 1 THz

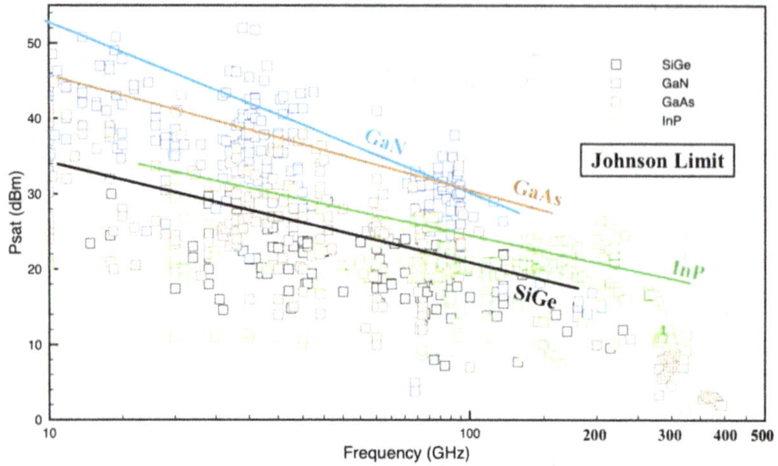

Figure 8.9 PA surrey from Georgia-Tech Electronics and Micro-systems (GEM) [13]

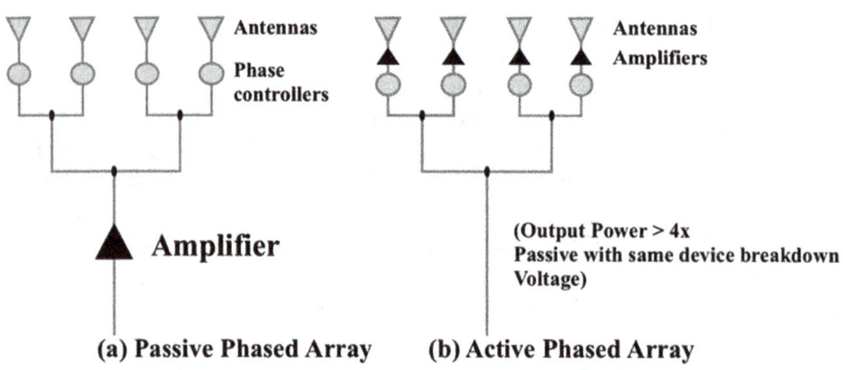

Figure 8.10 Active and passive array configurations: (a) Passive phased array and (b) Active phased array

Georgia-Tech Electronics and Micro-systems (GEM) Lab, under the leadership of Prof. Hua Wang, publishes regular reviews [13] of the reported power amplifiers, with over 2000 references, with carrier frequencies of between 500 MHz and 1.5 THz implemented in commercially available technologies, including CMOS, SiGe, LDMOS, GaN InP and GaAs. This is an outstanding source of information. They aim to help the scientific and engineering communities identify technology trends and other aspects of PA design, as shown in Figure 8.9, based on their data set.

Consider the diagrams in Figure 8.10, which show two typical configurations for driving phased arrays. Namely, a passive array configuration is in (a), and an active configuration is in (b). For the same EIRP, the passive device in (a) is limited by the

Johnson Limit applied to the array's input amplifier output. In (b), where each separate antenna is driven by its amplifier, its output power can be 1/*N* for the single drive amplifier in (a) at the expense of the transistor count. This could be used to overcome the Johnson limit and has other benefits, such as improving the linearity of the array. The design of phased arrays has recently been covered in reference [24].

8.5.4 Receiver sensitivity improvement in a phased array

Finally, to complete our picture of the performance of phased array wireless systems for Friis's FSPL analysis with the frequency, we now consider how a receiver's performance is influenced by frequency, taking advantage of the possible square law increase in array size for a constant area. An excellent analysis of the noise improvement performance of phased array receivers was published by Kim *et al.* [25]. They showed that

$$SNR_{0,eff} = N \cdot SNR_0, \tag{8.6}$$

SNR_{oEff} is the effective output signal-to-noise ratio of the phased array, N is the array size with an f^2 dependence for a constant area phased array, and SNR_o is the signal-to-noise ratio of a single array element.

The phased array receiver's SNR improves with f^2 as the frequency increases because of the increase in N. This, of course, has a direct impact on the radio link budget, and this improvement is in addition to the rise in the gain of the array.

For a given Noise Factor (F), an array receiver improves the sensitivity by 10 $\text{Log}_{10}(N)$ decibels compared to a single-path receiver. This result may not be surprising because the array receiver collects an N-fold signal power, which is added coherently. In contrast, the noise in each array element path is mutually uncorrelated and combined incoherently. For instance, an eight-element phased array can improve the receiver sensitivity by 9 dB.

8.5.5 Conclusions to Frii's FSPL law for constant area devices

Put simply:

$$P_r(f)/P_t(f) = f^2\, G_r\, G_t(c/4\pi d)^2 \tag{8.7}$$

While the transmitter power may fall quadratically with frequency due to Johnson's limit, this is directly compensated by improving the receiver's sensitivity with frequency. This gives us exciting possibilities in designing future wireless systems with smartphones and other terminal devices.

8.6 Applications of mmWave and THz

With 5G exploring during its lifetime the use of the lower part of the mmWave band from 30 to 100 GHz with assignments in place from regulators such as the FCC [26] for 70/80/90 GHz. This leaves 6G open to use the bands above 100 GHz with access to over 275 GHz of spectrum up to the water vapour absorption peak at 380 GHz for entirely new ranges of applications. In practical terms, there is plenty of wireless spectrum!

One advantage of using extremely high frequencies is that the shorter wavelengths allow for precise positioning systems with millimetre accuracy, as explained in Chapter 14. Another benefit is that the wide channel bandwidths associated with such high frequencies can support operations at 100 Gbit/s or higher. However, achieving speeds of 1 Tbit/s would require a channel of at least 150 GHz. Currently, we do not have any wireless applications that require such high speeds.

Today, commercial foundries are focusing on the needs of 5G, and their mmWave uses up to circa 100 GHz, with GaN leading the charge for very high-output powers in power amplifiers and phased arrays. At the same time, CMOS can still reach up to 200 GHz (0.4 $\times$ F_{max} of circa 500 GHz). The exploitation of the spectrum can emerge in stages based on the emerging performance of IC technologies, with an evident focus on 100–200 GHz in the first phase, where both high-performance GaN and low-cost silicon are available.

GaAs and SiGe will take the use beyond these performance levels, and other semiconductor technologies, such as InP, could take us beyond this towards operation in the 200–400 GHz region.

What are the potential applications that may demand these levels of performance? Prof Ted Rappaport *et al.* published an excellent paper [24] on applications in this space. At the same time, Prof Hua Wang, GTech and Ned Cahoon with Anirban Bandyopadhyay from Global Foundries presented an excellent in-depth analysis of the associated electronics and state of play [27].

For communications, while the ultra-high data rates facilitated by mmWave and THz wireless local area and cellular networks will enable super-fast download speeds for computer communications, there are nevertheless applications for mmWave and terahertz in an enormous wide applications fields of spectroscopy gas sensing, material research, hydration monitoring, security scanning, high-definition XR/VR, general entertainment uses, ultra-high speed on-chip communications, autonomous vehicles, robotic controls, and many more! The significance of this broad applications domain is that it will encourage the most comprehensive possible research to address their technical challenges while delivering increased scale and eventually lower costs and accelerating the time scales for the adoption into the wireless communications domain.

Some notable ideas in the following domains include:

8.6.1 In communications

- ***Small cells:*** both in 5G and 6G where mmWave frequencies could provide speeds of 100s of Gbit/s.
- ***Information showers:*** here, in a typical scenario, a radio with a direct line of sight from above floods a small area below where a user can download or upload at extremely high speeds – 100s of Gbit/s. These can be positioned in

places like the old red telephone boxes in the UK, which is increasingly protected as a listed site. This is much more like a WLAN application than a cellular small cell as there is no handover to other cells.

- ***Fixed wireless access:*** while the economics of FWA has always been challenging in developed countries as invariably these compete with existing fixed networks as overbuilds, one of the most significant technical issues has been the restricted range of spectrum they can access, which limits their ability to compete with cable or FTTH which can offer 10 Gbit/s today. Using mmWave or THz later could solve this bandwidth problem with other uses in backhaul for 5G/6G deployments.
- ***Real-time Holographic communications:*** This use case requires potential data speeds up to several Tbps. It is often cited as crucial in developing 6G technology, highlighting the need for extensive spectrum resources. In the final chapter, we discussed the progress made towards holographic communications, but we need to be more sceptical about its availability in 6G and its practical use. Moreover, there is an assumption that people genuinely require it. Despite 3D cinema technology being available for over a decade, with impressive recent advancements, very few films are still released in cinemas, and 3D TV is no longer available.

8.6.2 In sensing and imaging

An enormous range of sensing applications benefits from the properties of mmWave and THz radios. These include millimetre to sub-millimetre resolution accuracies, extremely wide channel bandwidths, and extremely high directional antennas implemented in small areas with the ability to steer the beam to scan the environment for 3D spatial mapping. We have seen earlier how relatively little radio transmission in the GHz to 1 THz is affected by water vapour compared with light or infra-red in fog, so this technology can provide sight in the dark or smoke for rescue services. This could form some revolutionary aspects for 6G outside the traditional communications use cases.

- For example, a future device would be worn on the user's head to scan an internal environment to form a 3D map in real time. This would be unaffected by darkness or smoke, with applications for blind people and see-in-the-dark applications. Such fast and accurate scanning systems would also have commercial applications and potential gaming scenarios.
- In the high mmWave or THz bands, imaging applications could include gesture detection, spectrometers for detecting food allergies, explosive device detection at airports,
- High-definition video radars operating with a few hundred GHz could provide HD TV-like images and complement lower-frequency radars.
- THz radio could augment human and computer vision to see around corners. As many surfaces appear as mirrors at these frequencies, such systems can see around corners or behind walls with sufficient scattering reflections. This could be very useful in rescue or military applications.

8.6.3 In precise measurement

- In a 'future 6G iPhone', a combination of imaging and communications for a centimetre or even millimetre location is an exciting, revolutionary feature. In a gaming scenario, a team leader could explore an environment while team members in other locations wearing VR headsets could form a virtual team. The ability to resolve to a cm level would take such a revolutionary concept to a new level.

8.6.4 In human cognition – 100 Tbit/s

Human cognition is one of the most intriguing applications, requiring 100 Tbit/s [24,28]. Yes, we know we do not have enough spectrum for this. Nevertheless, this application can calculate the human brain's speed and estimate its storage capacity.

The human brain has 1011 neurons, each connected to about 1000 others. Each neutron can fire at approximately 200 per second, resulting in:

$$\begin{aligned}\textit{Computation speed of a Human Brain} &= 20 \times 10^{15} \text{ floating} \\ &\quad - \text{point operations per second}(\textit{flops}) \\ &= 20{,}000\text{Tbit/s}\end{aligned}$$

For comparison, the Apple M2 SOC processor has a theoretical maximum of 3.6 TFLOPs, which is over three orders of magnitude slower than a human brain. According to Moore's Law, assuming no device scaling problems (yes, there will be), it is at least 25 years behind us.

Each neuron has to write access to 1,000 bytes. Thus, a memory size can be computed as follows.

$$\begin{aligned}\textit{Storage of a Human Brain} &= 10^{11} \text{ neurons} \\ &\quad \times 10^{3} \text{ bytes per neuron} = 10^{14} \text{ bytes} \\ &= 100 \text{ TB}\end{aligned}$$

In References [24,28], they assumed that 6G wireless generation has RF channels up to 10 GHz for each user. By taking that, each user can exploit 10 bits/symbol modulation methods (optimistic) and a 1,000-time increase in channel capacity. In addition to using massive MIMO, data rates are expected to be up to 100 Tbit/s, providing only 0.5% of real-time human computational power. Perhaps 7G or 8G!

8.7 Performance of cm and mmWave in real environments

Free space path loss does not always hold in practice because of interference caused by multiple ray paths from the transmitter combining to cause cancellation at some distant point. At that point, the loss exponent associated with the distance d will

Table 8.3 Beamwidth narrows as array size increases

Frequency (GHz)	Array size (n)	Antenna effective area Ae (cm^2)	Beam width (°)
28	3 × 3	2.9	37
73	8 × 8	2.9	15
140	16 × 16	2.9	7.5
	64 × 64	47	1.8

increase from the free space value of 2 to higher values, such as typically 3–5 in UHF cellular systems, depending on the environment. However, it is essential to realise at cm and mmWave, where massive phased arrays are used, the transmitter beam width is much narrower than in low-band, mid-band, and high-band frequencies found in those cellular systems, as shown in Table 8.3. This allows increasing like 'pencil beam' transmission where the means of transmission is via line-of-sight (LOS) or coupled with a reflection from an object such as a metallised window in an office building in a city. This suits application use cases where the LOS view is likely dominant and ultimately the limiting factor in the system range.

The transmitter beam width narrows as the array size increases within a constant area as the frequency increases. The rule of thumb for the beam width of an antenna array approximates 60°/(D/λ), where D is the electrical length.

New York University's Prof. Ted Rappaport's group has done a lot of work studying the propagation of mmWave in this area. For example, in their measurements of 60 GHz indoor wireless channels with a clear LOS path, they have shown varying α values (distance loss exponent) in different LOS environments, including open space, hallways and laboratories [17]:

- For open space scenarios of limited distance where atmospheric absorption has minimal impact and where there are few reflectors, they found $\alpha \approx 2.0$, which is the free space value.
- In hallways, where a waveguiding effect can form that $1.2 \leq \alpha \leq 1.32$! The system acts similarly to a FWA system, which is a perfect use case compared with FWA, which does not use mmWave. For a laboratory environment, where both constructive and destructive reflections were present, the estimated value is $1.71 \leq \alpha \leq 2.71$. This would be a good predictor for an indoor WLAN application.
- In outdoor LOS measurements for mmWave cellular at 28, 38, 60, and 73 GHz, α is between 1.8 and 2.2 but changed to about 4 or 5 in non-line-of-sight (NLOS) conditions. This bodes well for micro-cellular or WLAN applications in open spaces.

As concluded in Ref. [17], Section 3.7.1, they found little difference between the omnidirectional channel path loss at mmWave frequencies and today's UHF/microwave channels. This is a truly amazing result. In particular, the FSPL would provide an adequate model of the significant path losses associated with LOS conditions, if not slightly generous. The FPSL is too generous to represent reality without a LOS signal path, but this is equally true for UHF/microwave channels.

8.8 Material penetration losses at mmWave

Earlier, we discussed the suitability of Friis' free space path loss (FSPL) for the propagation of signals between the transmitter and the receiver in open spaces. However, in many cases, the user may be indoors while the transmitter is outside on a pole or tower. In such a scenario, Low-band frequencies, below 1 GHz, are advantageous as they are suitable for long-range transmission with relatively low transmission losses from outside to indoor spaces through building materials, compared to microwave frequencies. Nonetheless, the actual materials used in the construction of buildings play a crucial role in determining the extent of this suitability for indoor reception.

In many cellular systems that use Mid-band frequencies, approximately 2.5 GHz for 4G and 3.5 GHz for 5G, the rule of thumb is that indoor coverage requires between 15 and 20 dB of link budget for in-building penetration. While indoor coverage for mmWave frequencies may require a higher link budget allowance, the ability to realise massive EIRP with massive phased arrays presents some opportunities for indoor coverage. To quantify this opportunity, several references have measured material loss at various mmWave frequencies. An excellent detailed summary can be found in the reference [24]. They found, for example.

- Typical attenuations for transparent glass thicknesses from 3 mm to 1.2 cm were approximately 1 dB/mm from 60 to 140 GHz.
- For a steel door of 5.3 cm thickness at 73 GHz, approximately 50 dB!
- For a strongly reinforced uniform concrete wall of 35 cm, 22 dB attenuation at 1–4 GHz rose rapidly to 64 dB at 9 GHz. In comparison, a cement tile of 2.5 cm thickness was 39 dB at 100 GHz.
- For a uniform plasterboard wall of 12 cm thickness, typical attenuations were 10 dB at 5 GHz, rising to 11–18 dB at 15–18 GHz, respectively.

8.8.1 Low-emissivity glass windows

As shown above, transmission through windows may be attractive in some scenarios, such as for FWA, as transparent glass has relativity low transmission losses. However, modern windows include metal-containing coatings to reduce buildings' energy consumption. These coatings strongly attenuate the microwaves used for mobile communications, which at sub-terahertz frequencies could be more than 40 dB, making an ideal Faraday Cage. There has been recent work to mitigate this [29]. Laser ablation scribes a line pattern on the low-emissivity coating. The microwave attenuation of the initial coating ranges between –25 and –30 dB between 850 MHz and 3 GHz. The optimised patterning reduces it down to -1.2 ± 0.6 dB. The fraction of the ablated area is relatively low, allowing it to maintain its thermal qualities. This will not address the installed base of low E-glass, which will persist until this is introduced. Future work is required to develop laser ablation techniques applied across frequencies from mid-band to sub-terahertz.

8.8.2 Summary of material penetration losses at mmWave

The use of mmWave faces a much more hostile environment than UHF or microwave frequencies for ubiquitous indoor coverage, and an additional link budget allowance of over 30 dB may be required, which would, in practice, rule it out in most cases for indoor range from local poles or towers located outside. However, there may be opportunities to use the narrow beam from the array targeted at a clear glass window where the receiver is mounted directly inside. This issue here would be E-Glass, which is used to keep thermal energy inside buildings, and this type of glass has attenuations of over 30 dB alone. Even laser ablation may be too late as it only applies to new buildings and will likely incur a higher cost. For WLAN using mmWave, the typical losses through concrete walls would limit uses within a room.

8.9 Modelling the performance of mmWave phased array-based communications system.

To provide a better insight into the possible application's performance of mmWave phased array-based communications systems, a detailed Excel-based model was created by the author, described in Appendix A, which calculates the wireless link budget for various parameters of the wireless systems with the critical elements of the phased array. This brings together both the device physics and the system performance. This model has been checked with published experimental results and will provide a realistic performance prediction level. Nevertheless, the results should be interpreted as indicative.

In summary, for the wireless parameters, the user can specify the frequency of operation, the system speed, transmitter power and the distance between the transmitter and receiver. The minimum SNR of the receiver is calculated using an energy per bit to noise power spectral density ratio of 18.8 dB (suitable for 64 QAM and 6 bits/Hz), which is ideal for a range of modulation systems and, in many cases, exceeds the required ratio for a BER of 10^{-6}. However, it is simple to modify this for other modulation systems. 64 QAM was chosen for transistor linearity concerns as mmWave systems will not achieve the same level of performance as those mobile/fixed systems below 30 GHz. This supports the recognised requirement by the mmWave Coalition that channels of 20 GHz will be required for 100 Gbit/s operation. The device structure of the phased array is described by the maximum aperture length using the wavelength of operation, the relative permittivity of the substrate material, assuming a ½ wavelength dipole element and similar spacing. From these parameters, the aperture areas A_e of the transmitter and receiver can be calculated, determining the gains directly. Here, we assume that with ½ wavelength spacing, the physical area of the array is approximately equal to the aperture areas A_e. However, the emission efficiency of the antennas is assumed to be 100% for simplicity. The merit of this approach is that published experimental results can be analysed to reveal the likely device structures. And finally, like any model, there is always more work to improve and refine.

8.10 Performance analysis of typical applications of mmWave

Let us discuss some of the leading use cases for mmWave in communications to gain insight into its realistic performance. This will help highlight some of the key advantages and challenges of moving to higher frequencies. The aim is also to show how such systems could be realised before the launch of 6G circa 2030 while painting a roadmap for how higher frequencies can further improve performance when other IC technologies become available in commercial foundries. It will also help highlight the precepted value of spectrum when 100s of GHz becomes available with superior performance to lower frequencies.

8.10.1 Information shower/petrol station in FCC band: 116–123 GHz

One of the authors proposed the concept of an information shower for infra-red communications in places to augment the original 3G networks. The simple idea then was to provide orders of magnitude higher downloads and uploads than on the cellular network using the optical spectrum to laptop computers in places people regularly frequent, such as petrol filling stations, coffee shops, etc. Since then, others have proposed a similar idea for ultra-short-range wireless technologies providing extreme data rates. This typical use case can assess the likely performance of short-distance systems for mm wireless communication, such as WLANs for VR Goggles or possibly holographic images.

Figure 8.11 presents the wireless link budget for the mmWave system using the US FCC band of 116–123 GHz. Here, the same array size as the 60 GHz WiGig

Summary Link Budget Information shower using the unlicensed band 116-123 GHz		
Frequency (GHz)	120	**Output Powers**
Distance (m)	5	GaN @110GHz (imec): 28 dBm
LOS Path Loss dB	88	
Tx Power (dBm)	0	SiGe @114GHz: 17 dBm
Tx Antenna Gain (dB) = 10Log N	24	
Beam width (degrees)	13.3	CMOS @ 60GHz: 19 dBm
Rx Antenna Gain (dB) = 10 Log N	24	
Tx EIRP (dBm)	24	
Receiver bandwidth (GHz)	7	
Bit rate (Gbit/s) 64 QAM (5 bits/Hz)	35	
Received power (dBm)	-40	
Receiver sensitivity (dBm)	-60	NF=5 dB
Noise improvement via Array	12	
Margin (dB)	**20**	
Array Physical area Est (cm^2)	1.3	
Tx Aer (cm^2)	1.3	
Rx Aer (cm^2)	1.3	
Total Tx elements N1d => SQRT(Nt)	256	
Total Rx elements N1d => SQRT(Nt)	256	
dielectric constant	**4**	
λ/2 dipole and spacing (μm)	625	

Figure 8.11 Link budget of the information shower

(802.11 ad) system of 2015 of 16 × 16 was used, which could be implemented in CMOS, GaAs, GaN or SiGe. The effective area A_e of the array is 1.3 cm^2 with a $\lambda/2$ dipole spacing. As the A_e is less than or equal to the physical area with a $\lambda/2$ spacing, the physical area is estimated to be close to A_e. However, in practice, the size of the IC would be determined by the space requirement of all the associated system electronics. For example, the WiGig system, at 60 GHz, also used a similar 7 GHz channel, measured 41.6 mm by 41.6 mm using the 256-element array. A calculation of the A_e @ 60 GHz here would yield only 5 cm^2 with $\lambda/2$ spacing, which is one-quarter of the actual size. Nevertheless, even considering the area required for electronics from 2015, this demonstrated that such devices can fit into today's smartphone.

GaN would be preferred for high-output power applications as it has a reported output maximum power of 28 dBm at 114 GHz, while SiGe has a reported maximum output power of 17 dBm. However, CMOS would be preferred here due to its lower cost and an output power requirement of only one mW, which is desirable for a smartphone. The proposed system uses 64 QAM with a theoretical 6-bit/Hz while 5-bit/Hz would reduce its speed due to transistor linearity issues. This limits the system's speed to 35 Gbit/s rather than the theoretical speed of 42 Gbit/s. A future 100 Gbit/s system would require a channel bandwidth of 20 GHz, arguing for much wider channel bandwidths than regulators currently propose in the 100–200 GHz region.

The overall system for 35 Gbit/s has a very healthy system margin of 20 dB, while the path distance is assumed at 5 m line of sight (LOS). There would be no significant radio wave absorption at this frequency and length. Such a system could easily be scaled over 100 Gbit/s with a wider channel bandwidth of 20 GHz and using the same device parameters while achieving a system margin of 15 dB. Even considering any additional impairments, these levels of system margin demonstrate the excellent potential for an information shower in the 116–123 GHz band. It would be future applications for communications, such as VR Goggles, for example.

8.10.2 Mobile and fixed wireless communications – comparison of cm and mmWave

The 5G technology has followed the traditional 'spectrum staircase' method. New applications require more bandwidth or a different frequency band, and the industry/regulators move to the next available band. For cellular and FWA, the current frequencies used in 5G are mainly in the cm Wave range, around 28 and 39 GHz. mmWave is also available at 70/80/90 GHz. This method assumes that spectrum is a scarce resource, and therefore, it usually achieves high valuations in spectrum auctions. For instance, in May 2017, Verizon bought Straight Path Communication for about \$3.1bn, winning a bidding war with AT&T. Straight path communication owned valuable 28 and 39 GHz spectrum (222 billion MHz-POPs – megahertz per population) and was considered necessary for the industry.

When considering whether to use mmWave spectrum for cellular or FWA, choosing between frequencies like 28 GHz or 39 GHz and those over 100 GHz depends on the deployment scenarios and propagation considerations. Figures 8.12 and 8.13 show two deployment scenarios being considered today for 28 and 39 GHz, as well as their potential use above 100 GHz.

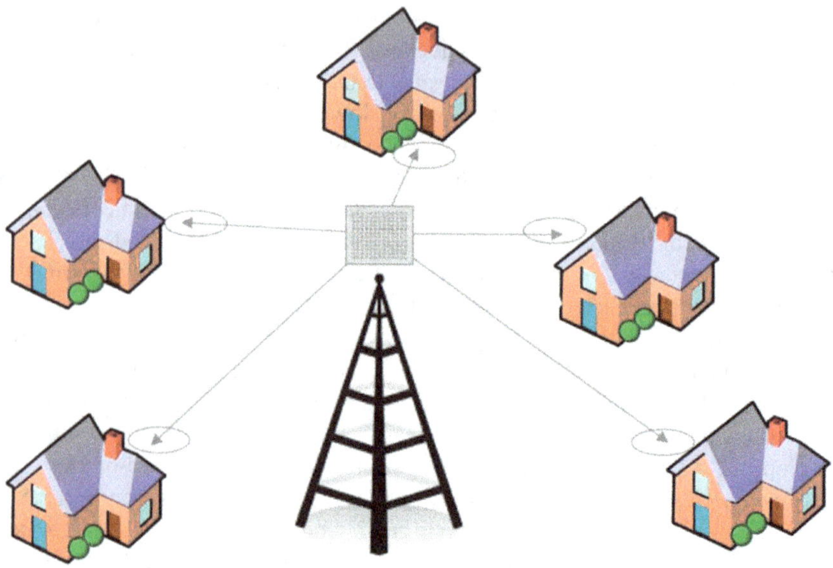

(a) Fixed Wireless Access in Suburaban enviroments

Figure 8.12 FWA in suburban environments

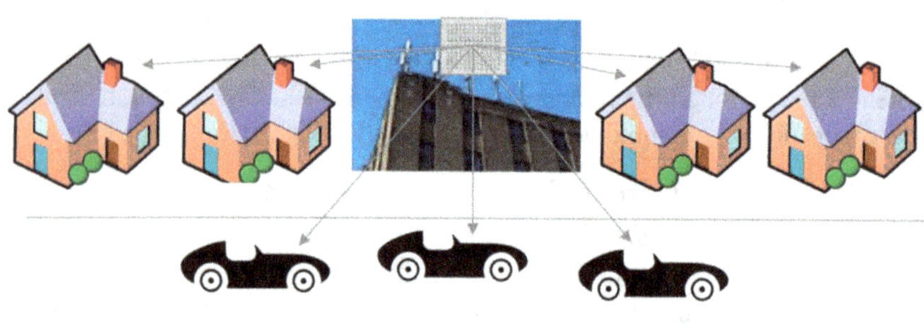

(b) Fixed and Cellular deployment in Dense urban

Figure 8.13 Deployment options for cm and mmWave in FWA and cellular

In Figure 8.12, for the FWA case, the base station is designed to deliver a high bandwidth service to homes in a suburban setting and mounted on a utility pole or tower. This deployment must cover as large an area as possible for the best business case. And, within that, the best case is when there are outdoor wireless communications. This avoids the issue of penetration through walls and into the home, which is not usually feasible for microwave links as it demands an additional 30–40 dB of link loss, as demonstrated by the material losses illustrated in Section 8.8. The transmitter EIRP powers for FWA are usually in the range of 60–70 dBm [30].

In Figure 8.13, for a typical dense urban scenario, the base station is mounted at a lower height to the ground on a building rooftop, if available or on a façade. In this case, the transmitter power may not be as high as in the first case.

In both cases, the transmitter must steer its wireless beam to cover the deployment areas. Finding a single case broadly representing a whole area is not easy. That is why we examine reference [31], published by one of the authors. It has modelled a 1 km^2 area of London to make a more meaningful comparison with other wireless technologies.

In assessing the value of spectrum for wireless applications such as cellular and FWA between the current 5G frequencies mainly based around cm Wave (in the 28 and 39 GHz range) and future mmWave (>95 GHz), it is essential to consider all the aspects. In particular, the mmWave limitations include transmitter power and its advantages in increased antenna gain with reduced wavelength. Such a comparison might say something meaningful about the value of the spectrum in these two cases, as we know how much has been paid in the 5G case. With 275 GHz of spectrum available for mmWave between 100 and 375 GHz, it would be reasonable to assume much lower spectrum prices. Thomas Cameron from analog devices [30] has published an excellent review examining the performance at 28 GHz (cmWave). Table 8.4 compares that with an mmWave system operating in the US FCC band of 116–123 GHz.

The link is assumed to be 200 m, requiring an NLOS path loss of circa 135 dB compared with an LOS of 107 dB. In the 28 GHz case, 256 elements were assumed in

Table 8.4 5G base station link budget and the two FWA and cellular use cases based on a comparison with [30] @ 28 GHz and 116–123 GHz

Link budget 200 m link @ 28 GHz 800 MHz bandwidth	**Downlink (base station)**	**Uplink (CPE)**	**Link budget 200 m Link @ 116 GHz 800 MHz bandwidth**	**Uplink (CPE)**
Antenna element count	256 (23.4 cm^2)	64 (5.4 cm^2)	4096 (22.8 cm^2)	1028 (5.4 cm^2)
Total conducted PA power (dBm)	33	19	28	
Antenna gain (dB)	27	21	39	36
Tx EIRP (dBm)	60	40	67	
Beam width (degrees)	13	26	3	7
Path loss (dB)	135 (NLOS) =>107 (LOS)	135	148 (NLOS) =>120 (LOS)	
Received power (dBm)	−74	−94	−52	
Thermal noise floor (dBm)	−88	−88	−88	−88
Rx noise figure (dB)	5	5	5	5
P_{min} (post array)	−65		−71	
Array noise improvement (dB)	9		15	
Margin (dB)	12 (NLOS) 42 (LOS)		27 (NLOS) 55 (LOS)	

the base station and 64 elements in the CPE, while for the 116–123 GHz case, for the same area size of the array, this allows 4096 and 1028 elements, respectively. This increase has a profound impact on the system budget. However, before we consider that, let us discuss the difference between NLOS and LOS. An excellent review of the propagation models for 5G mmWave from NYU is available online. Rappaport *et al.* [32] compared the breakpoints between the different 5G models for LOS and NLOS for 28 GHz. It points out that breakpoints are controversial as they have yet to be reported in the measurement. In their Figure 5, the breakpoint from one of the models ranges from 60 to over 500 m for a base station height of 4 m and a terminal height of 1.5 m, which increases to 170 m and above 1000 m for a base station height of 10 m. One of the advantages of the 116 GHz system over the 28 GHz system is that the beam width has reduced from circa 13° to only 3°, which should move these breakpoints further out. This bodes well for a LOS model for propagation for 116 GHz for a 200 m range; however, blockage (buildings, foliage, people, etc.) is another major factor to consider.

Considering any wireless system, it is essential to establish a suitable fade margin to deal with, e.g. the effects of rain [33]. To produce a conservative fade margin of 100 mm/h for tropical rain, a rain attenuation of 30 dB/km fade margin is assumed. At the same time, this figure falls to circa 3 dB per km for light rainfall. Even a figure of 6 dB for the 200 m system does not considerably eat into the predicted system margins, particularly for the 116 GHz.

Comparing the link budgets for 28 GHz and 116–123 GHz reveals some remarkable results. Meanwhile, a GaN implementation at 116 GHz is limited to circa 28 dBm PA output power cf. 33 dBm at 28 GHz, the EIRP is significantly higher, 7 dB, due to increased array gain. Even given the higher propagation losses in the LOS and NLOS cases at 116–123 GHz, this system delivers a significant increase in the overall margin of 13–15 dB and is the clear winner. Furthermore, while the 28 GHz can support an 800 MHz channel, the 116–123 GHz system can 7 GHz. This makes using 100 GHz and above much more competitive for FWA with DOCSIS 4.0 and FTTH fixed networks, which support 10 Gbit/s.

Nevertheless, it is still illuminating to compare the use of mmWave with other 5G wireless technologies in a scenario of 1 km^2 deployment in London by one of the authors, Wisely *et al.* [31], which considers the capacity and costs of 5G networks in dense urban deployments. In their work, a techno-economic analysis is carried out of a 5G enhanced mobile broadband scenario in dense urban areas using radio capacity modelling of probable 5G technologies within a 1 km^2 grid representing central London. Different density networks were modelled at 700 MHz (macro network), 3.5 GHz (micro network) and 24–27.5 GHz (hot spots) – together with 802.11ac access points. They use published data on network costs of various deployment options to evaluate capacity, headline rate and capital expenditure/operating expense. It has been shown that reaching headline rates of 64–100 Mbit/s everywhere is possible with several different technology options; However, massive capacity increases (over 100 Gbit/s/km) can only be realistically achieved with a millimetre wave (outdoor) and 802.11ac (internally). However, the cost of deploying such capacity will be several times that of LTE – the authors estimate a 4–5 times increase in costs for a 100 Mbit/s everywhere network with a $\times 100$ capacity increase over existing LTE networks.

The key results are summarised here:

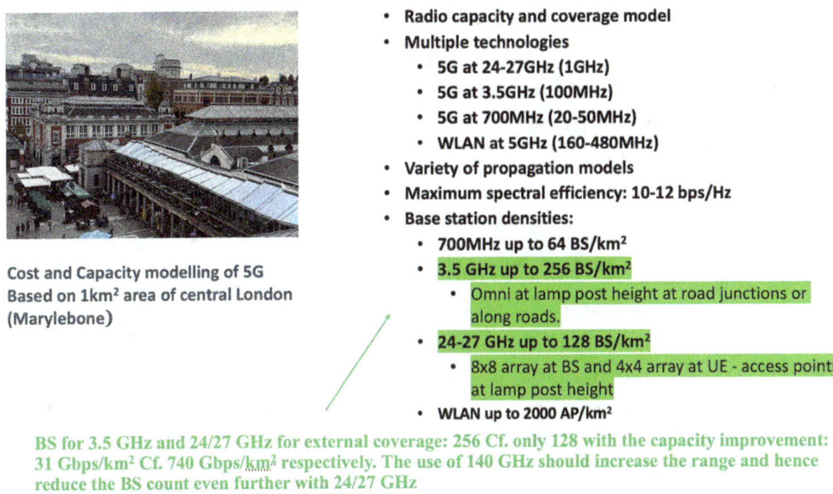

Figure 8.14 Costs and capacity modelling 5G in central London, UK [31]

In Figure 8.14, you can see that in the green comparison of 3.5 GHz with 24–28 GHz, the base station count is half of 3.5 GHz while capacity has increased from 21 to 740 Gbit/s per km^2. Moving to higher frequencies above 100 GHz would have further performance advantages, as we have just described while allowing access to much greater bandwidths. Of course, this capacity level would be questionable, with little or no coverage from outside to inside.

Consequently, with 275 GHz of spectrum available above 100–375 GHz, just below the water absorption peak at 380 GHz, one would expect much lower valuations associated with a spectrum that Verizon paid.

8.10.3 The world's largest antenna array (phased array) in 2035!

Speculating what the largest phased array could be in the next 10–20 years and what performance level could be achieved is fun. As we previously discussed, the limiting factor would be the space required for the size of the electronics rather than the array element. Here, we consider two frequencies of operation to also contrast the performance of two semiconductor materials in terms of output power GaN at 140 GHz and SiGe at 300 GHz.

8.10.3.1 SiGe case @ 300 GHz: 4 million element array

If a 30 cm diameter SiGe Wafer were allocated to a single array at 300 GHz, this would allow the realisation of a 2000 × 2000 element array with an array gain of

66 dB. While assuming the case of a single element as an isotopic radiator and an output power at this frequency for SiGe of circa 15 dBm, the EIRP would be 81 dBm! The associated beam width would only be 0.1°.

The $\lambda/2$ dipole would be 250 μm with a dielectric constant of 4. This spacing would allow perhaps a simple amplifier of two transistors with coupling elements.

To put this type of device into context, a steel door with a thickness of 5 cm has an attenuation loss of 50 dB, so this would transmit through this door with a significant margin, neglecting surface reflection. Meanwhile, its use on the transmitter of the 200 m system in Table 8.4 would increase the NLOS from 27 to over 40 dB. This could give FWA indoor coverage at 300 GHz!

8.10.3.2 GaN case @ 140 GHz: 1 million element array

Again, using the same wafer size of 30 cm diameter would allow an array size of 1000 × 1000 elements with an array gain of 60 dB. Assuming a single element as an isotopic radiator and an output power at this frequency for GaN of 25 dBm, the EIRP would increase to 85 dBm! The associated beam width would only be 0.2°.

The $\lambda/2$ dipole would now be 536 μm with a dielectric constant of 4. This spacing would allow much more space for the electronics. Similarly, the previous use cases that were previously considered would be further improved.

8.10.3.3 What does this mean?

While both cases may seem difficult to realise today, they assume a radical change in telecommunications and other applications for massive, phased arrays in the mmWave bands. It demonstrates the value of GaN at the lower mmWave bands for many future applications in the 100–200 GHz range where high-output power is required. Improvements in GaN HEMT technology should be able to deliver this for the launch of 6G, while today, it could meet the demands of the 116–123 GHz band.

8.11 ADC and DAC considerations at mmWave – the energy challenge

In Prof. Hua Wang *et al.*'s book [34, Chapter 19], he considers the analogue digital conversion (ADC) and digital analogue conversion (DAC) aspects at mmWave frequencies. These larger bandwidths will challenge the power requirements with conventional methods between the analogue and digital domains in both the receiver and transmitter.

The signal-to-noise and distortion ratio SNDR-based Schreier figure of merit (FoM) is the widely accepted metric for ADC, which is defined by

$$\text{FoM} = \text{SNDR} + 10\text{Log}_{10}(f_s/2/P) \tag{8.8}$$

With SNDR in dB, power consumption in Watts, and the Nyquist sampling frequency f_s in Hz (=2× bandwidth for most converters).

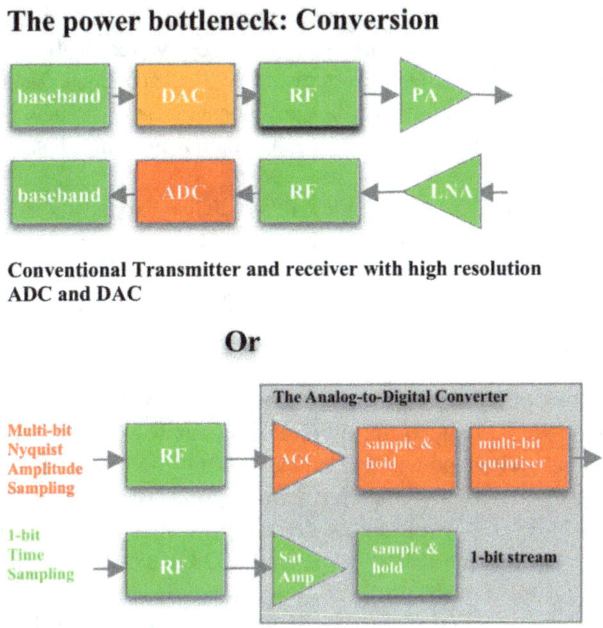

Figure 8.15 Transmitter and receiver chains with 1-bit resolution ADC and DAC [35] cf. conventional

In a conversion with Prof. Gerhard Fettweis, he pointed out that at 1 Tbits/s with a 100 mW link budget, the PA would require 1 W while the DAC/ADC would require 10 W with a 10–12-bit resolution. Each bit doubles the power consumption. To address this problem, see Figure 8.15, work [35] at Prof. Gerhard Fettweis 5G/6G Labs has focused on using the time domain with one-bit resolution and massive oversampling. (This technique is also used in high-resolution audio systems [24 bits] to address the problem that it is not possible to achieve the dynamic range with a conventional 24 bit DAC because the lower bits are lost in the noise). With this technique, they have demonstrated 100 Gbit/s transmission at 300 GHz with a power consumption of 40 pJ/bit.

8.12 Summary and conclusions

Over the past decade, the perception of frequencies beyond 100 GHz has changed from being considered a forbidden zone to becoming an exciting new frontier for communication and other applications. This is due to the remarkable advancements in IC electronics over the past two decades, combined with the pioneering work of Professor Ted Rappaport at New York University on cellular applications of cm and mmWave, and their use in phased arrays. With equal-area phase arrays, link budget is now recognised as increasing quadratically with frequency, which challenges the widely held belief that operating wireless systems at higher frequencies is much worse. This also challenges the traditional method of allocating bands to new generations of mobile devices based on the nearest highest neighbour and their historical spectrum valuations.

During its lifetime, 5G technology will gradually expand its frequency spectrum towards 100 GHz. This will pave the way initially for the next generation of telecommunication technology, 6G, to explore the electromagnetic spectrum's 100–200 GHz range. Later, 6G may use 200–400 GHz frequencies, with 300 GHz being a likely possibility.

While almost 3 THz awaits, this will likely be outside the scope of 6G. The present capabilities of IC foundries will allow the first phase of 6G to tap into the 100–200 GHz of the EM spectrum, utilising low-cost CMOS or SiGe; GaN here allows exceptional output power levels, making an enormous difference in link margins in critical wireless systems. Furthermore, improvements in device physics extending F_{max} to 1 THz would allow coverage of 200–400 GHz in the second phase of 6G. However, even just an additional 100 GHz to 5G use will provide as much spectrum used over the last 120 years for 6G. Nevertheless, many other potential non-telecoms applications await to tap into this region beyond cellular and Wi-Fi. This vast universe of ideas and innovation could accelerate the adoption of mmWave technology and lower its cost. Here, 6G is likely to differentiate itself from 5G by expanding into these applications beyond communications; these include sensing imaging and spectrology, creating revolutionary use cases.

Furthermore, the massive improvements in electronic performance and size also allow the integration of the whole system onto a chip, including antennas. This will dramatically reduce costs for a base station for mmWave. Using a SOC implementation, operating at as high a frequency as possible within the frequency range of a given IC technology will allow a higher link budget.

While wireless regulation had historically ended at 275 GHz, the World Radio Conference in October 2019 extended this to 3 THz. Spectrum regulators are identifying bands for potential 6G use focused on the 100–200 GHz band. Here, they have indicated channel sizes of up to 7 GHz. The issue is that to realise 100 Gbit/s speeds, it is necessary to have channels of at least 20 GHz due to transistor device linearity issues. This raises the issue of spectrum sharing here. Why auction channels of 20 GHz in the valuable 100–200 GHz range when it is so easy to share its use? We discuss a scheme in Chapter 7 using weak signal propagation reporting; however, the mmWave energy is confined to small cells with pencil beams in this range. Therefore, these sites can be registered, and the spectrum sharing is relatively straightforward.

However, there is no practical limit to the availably of the spectrum now! Of course, engineering challenges still exist to overcome, but mmWave is waiting for innovation in the applications space to drive its take-up. These applications may not be those generally associated with a mobile generation- a true revolution.

Appendix A System calculations of a phased array-based communications

This appendix summarises the equations to help create a spreadsheet calculator for the performance of phased arrays in telecommunications and other applications.

A.1 Directivity of phased array

N is the "array gain" ($n_x \cdot n_y$), where n_x and n_y are the element counts in the x and y directions of the array. This is not the overall gain, called the array's directivity, as this is determined by N multiplied by the gain of each element.

The ratio of effective aperture to directivity (D) is a universal constant $\lambda^2/4\pi$.

$$D = A_e \cdot 4\pi/\lambda^2 \tag{A.1}$$

The effective area (A_e) is always less than or equal to the physical area (PA).

The PA is $N \cdot dx \cdot dy$, where dy and dx are the spacing between elements in the x and y dimensions.

Thus $A_e = N \cdot dx \cdot dy \cdot \eta$

where η is a factor called the "illumination efficiency" that accounts for tapering of the array and spacing of the array.

A.2 Effective isotropic radiated power (EIRP)

$$\text{EIRP} = P_{out} \cdot D$$

A.3 Path loss (dB)

$$\text{Path Loss} = 20 \cdot \text{Log}_{10}(\lambda/4\pi d)$$

A.4 Received power into the receiver

This considers the output power, the transmitter array Directivity, the similar Directivity in the receiver array and the path loss – the reference [13] gives the latest reported output powers for power amplifiers in different semiconductor material systems at various frequencies.

A.5 Minimum receiver power for a given QAM and BER

This reference, Gaussian Waves, is an excellent review of the performance comparison of digital modulation techniques, which we have used for phased arrays [36]. It contains the MathLab script for the theoretical BER (given SNR per bit–E_b/N_0) for various linear modulations shown here. Note that the E_b/N_0 values used in that table are linear, not dB.

Table A.1 Capacity of various modulation schemes; their efficiency and channel bandwidth (for other schemes from figure see [36])

Modulation	Required E_b/N_0 (dB)	Max n (log2(*M*)) (bits/s/Hz)	*C/N* (dB)	Min channel BW for ISI free signalling
4 QAM	10.6	2	13.6	0.5 Rb
16 QAM	14.5	4	20.5	0.25 Rb
64 QAM	18.8	6	26.6	0.17 Rb

E_b/N_0 vs BER for various digital modulation schemes based on MathLab code is plotted in [36].

For a given Noise Figure, an *N*-array receiver improves the sensitivity by 10 log(*N*): Xiang Guan *et al.*: *IEEE Journal of Solid-State Circuits* Vol 39. No. 12 (2004). However, depending on the detailed configuration of the receiver gains, the improvement can range from *n* to *N*. Here, we take the conservative value of 10 $\text{Log}_{10}(n)$

The minimum required signal-to-noise level SNR_{omin} is given by

$$SNR_{omin} = E_b/N_0 + R/B$$

where *R*/*B* is the bits/Hz, for example, using the Table A.1, which has E_b/N_0 for a BER of 10^{-6}, the *R*/*B* for 64 QAM is a maximum of 6 bits/Hz.

Therefore, the minimum received power at the input to the receiver post-phased array is

$$P_{\min} = -174\ \text{dBM/Hz} + 10\ \text{Log}_{10}(B_{rx}) + NF_{rx} + 10\text{Log}_{10}(n)$$

A.6 System margin (dB)

This is simply the difference between the received power and $P_{\min}$.

A.7 System speed

System speed is *R*/*B* · bandwidth.

References

[1] Ericsson, "5G to account for 75% of mobile data traffic in 2029," [Online]. Available: https://www.ericsson.com/en/reports-and-papers/mobility-report/dataforecasts/mobile-traffic-forecast.

[2] IWWC, "IWWC. World Wireless Conference 2019," [Online]. 2019. Available: https://www.itu.int/en/ITU-R/conferences/wrc/2019/Pages/default.aspx.

[3] FCC, "FCC 19-19.," [Online]. 2019. Available: https://docs.fcc.gov/public/attachments/FCC-19-19A1.pdf.

[4] K. Sengupta, T. Nagatsuma, and D. & Mittleman, "Terahertz integrated electronic and hybrid electronic–photonic systems," *Nature Electronics*. 2018; 1: 622–635. https://www.brown.edu/research/labs/mittleman/sites/brown.edu.research.labs.mittleman/files/uploads/Sengupta_Review.pdf.

[5] ITU. [Online]. 2013. Available: https://www.itu.int/dms_pubrec/itu-r/rec/p/R-REC-P.676-10-201309-S!!PDF-E.pdf [accessed November 2013].

[6] Nokia. The Golden Bands and the Future of 6G. [Online]. 2024 [cited 15 February 2024]. Available from: https://www.nokia.com/about-us/news-room/ articles/the-golden-bands-and-the-future-of-6g-0/.

[7] T. S. Rappaport, S. Sun, R. Mayzus, *et al.*, "Millimeter wave mobile communications for 5G cellular: It will work!" *IEEE Access*, 2013; 1: 335–349, doi:10.1109/ACCESS.2013.2260813.

[8] mm Wave Coalition. [Online]. 2024. Available: https://mmwavecoalition.org.

[9] TowerJazz. [Online]. Available: https://towersemi.com/wp-content/uploads/2018/08/TowerJazz_PR_20150514.pdf.

[10] Z. Hu, M. Kaynak, and R. Han, "High power radiation at 1 THz in silicon: a fully scalable array using a multi-functional radiating mesh structure," *IEEE Journal of Solid-Sate Circuits*. 2018; 53(5): 1313–1327.

[11] B. Heinemann, H. Rücker, R. Barth, et al., "SiGe HBT with ft/ fmax of 505 GHz/ 720 GHz," *IEEE Int. Electron Devices Meeting (IEDM)*, December 2016.

[12] T. S. Rappaport, "6G and beyond: Terahertz communications and sensing," 2019 *Brooklyn 5G Summit Keynote*, April 2019.

[13] H. Wang, K. Choi, B. Abdelaziz, *et al.* "PA survey." [Online]. 2023. Available: https://ideas.ethz.ch/Surveys/pa-survey.html.

[14] H. Wang, "Short course: MIMO beam forming circuit and algorithm of 5G," *Symposia on VLSI Technology and Circuits 2019*, 2019.

[15] D. T.-M. Programme. [Online]. Available: https://www.darpa.mil/program/technologies-for-mixed-mode-ultra-scaled-integrated-circuits.

[16] B. Sadhu, "Silicon based mm-Wave Phased arrays," *IEEE Microwave Theory and Technology Society*. Available from: https://www.youtube.com/watch?v=EsHZg4tk2YY , June 2022.

[17] T. Rappaport, R. Heath, R. Daniels and J. Murdock, *Millimeter Wave Wireless Communications*, 1st ed., Prentice Hall Communications, 2014.

[18] P. Hannon, " The element-gain paradox for a phased array antenna," *IEEE Transactions on Antennas and Propagation*. 1964; 12(4): 423–433.

[19] Y. Xing and T. S. Rappaport, "Propagation measurement system and approach at 140 GHz–moving to 6G and above 100 GHz," *IEEE Global Communications Conference (GLOBECOM)*, Abu Dhabi, UAE, pp. 1–6, 2018.

[20] E. Johnston, "Physical limitations on frequency and power parameters of transistors," *RCA Review*, 1965; 26: 163–177.

[21] Wikipedia. [Online]. Available: https://en.wikipedia.org/wiki/Johnsons_figure_of_merit.

[22] H. Wang and S. Kaushik (eds.) *RF and mm-Wave Power Generation in Silicon*, 1st ed. Amsterdam: Elsevier; 2015.

[23] DARPA. [Online]. Available: https://grantbulletin.research.uiowa.edu/darpa-technologies-mixed-mode-ultra-scaled-integrated-circuits-t-music.

[24] T. S. Rappaport, Y. Xing, O. Kanhere, *et al.*, "Wireless communications and applications above 100 GHz: Opportunities and challenges 6G," *IEEE Access*, 2019; 7: 78729–78757.

[25] J. Kim, J. Jeong and S. Jeon, "Improvement of noise performance in phased-array receivers," *ETRI Journal*, 2011; 33(2): 176–183.

[26] FCC, "Millimetre Wave 70/80/90 GHz Service rules part 47 C.F.R, part 101," [Online]. Available: https://www.fcc.gov/wireless/bureau-divisions/broadband-division/microwave-services/millimeter-wave-708090-ghz-service-0.

[27] H. Wang, N. Cahoon and A. Bandyopadhyay, "Sub-THz wireless communication & sensing – a perspective on device, circuit, and system," [Online]. 2019. Available: https://mmwavecoalition.org/wp-content/uploads/2019/05/Sub-THz_slides_GT_GFUS_final_05_09_19.pdf [accessed 5 September 2019].

[28] H. Moravec, "When will computer hardware match the human brain," *Journal of Evolution and Technology*. 1998; 1(1): 10.

[29] O. Bouvard, M. Lanini, L. Burnier, *et al.*, "Structured transparent low emissivity coatings with high microwave transmission," *Applied Physics A – Material Science and Processing*. 2017; 123: 66.

[30] T. Cameron, "Bits to beams: RF technology evolution for 5G millimeter wave radios," [Online]. 2018. Available: https://www.analog.com/en/signals/thought-leadership/bits-to-beams-rf-tech-evolution-5g-milli-wave-radios.html.

[31] D. Wisely, N. Wang and R. Tafazolli, "Capacity and costs for 5G networks in dense urban areas," *IET Communications*, 2018; 12(19): 2502–2510. Available: https://doi.org/10.1049/iet-com.2018.5505.

[32] T. Rappaport, Y. Xing, G. MacCarthy, A. Molisch, E. Melios and J. Zhang. Overview of millimeter wave communications for fifth-generation (5G) wireless networks—with a focus on propagation models. *IEEE Transactions on Antennas and Propagation*. 2017; 65(12): 6213–6230. Available: https://doi.org/10.1109/TAP.2017.2734243.

[33] Niu Y, Li Y, Jin D, et al. A survey of millimeter wave communications (mmWave) for 5G: opportunities and challenges. *Wireless Networks*. 2015; 21: 2657–2676.

[34] H. Wang and K. Sengupta (eds.), RF and mm-Wave Power Generation in Silicon, 1 ed., Academic Press, 2015.

[35] G. Fettweis and L. Landau, "Information Rates Employing 1-bit Quantization and Oversampling at the Receiver in Proceedings of IEEE International

Workshop on Signal Processing Advances in Wireless Communications (SPAWC 201," *Proceedings of IEEE International Workshop on Signal Processing Advances in Wireless Communications (SPAWC 2014),* https://www.vodafone-chair.org/pbls/legacy/l-landau/Information_Rates_Employing_1-bit_Quantization_and_Oversampling_at_the_Receiver.pdf, 2014.
[36] Mathuranathan, "Performance comparison of Digital Modulation techniques," [Online]. Available: https://www.gaussianwaves.com/2010/04/performance-comparison-of-digital-modulation-techniques-2/ [accessed 14 April 2010].

Chapter 9
6G convergence

Isolation is a blind alley.... Nothing on the planet grows except by convergence.

– *Pierre Teilhard de Chardin*

9.1 Introduction – why convergence?

For more than two decades, people have discussed the concept of convergence in the telecommunications industry, specifically the convergence of fixed and mobile networks. However, apart from mergers and acquisitions between fixed and mobile operators, there has been little to no progress. However, with the advent of 6G technology, we may finally see a new form of convergence between fixed and mobile networks.

Originally, fixed-mobile convergence (FMC) was seen as removing the differences between fixed and mobile networks, so it effectively became one network. In 2004, the Fixed-Mobile Convergence Alliance (FMCA), a group of carriers including British Telecommunications (BT), Brazil Telecom, Korea Telecom, NTT, Rogers Wireless and Swisscom, defined this as

Fixed Mobile Convergence is a transition point in the telecommunications industry that will finally remove the distinctions between fixed and mobile networks, providing a superior experience to customers by creating seamless services using a combination of fixed broadband and local access wireless technologies to meet their needs in homes, offices, other buildings and on the go.

One of the first FMC products was the BT Fusion phone, which used a combination of Wi-Fi, focusing on use indoors and in hot spots from its broadband network, with mobile coverage via its partnership with Vodafone in a mobile virtual network operator arrangement. Here, seamless service refers to seamless handover between the Wi-Fi and mobile networks with no break in service for voice calls. As discussed in our book, this failed due to limited handset support and was withdrawn in 2009.

Since the emergence of the iPhone 3G in 2008, followed by other smartphones, people have obtained their applications from App stores, bypassing fixed and mobile carriers. These apps were used on Wi-Fi and mobile networks, where service continuity was unimportant. Nevertheless, Apple does offer this feature with its apps with 'Handoff'. With 'Handoff', we can start working on one device, switch to another nearby device on a different network and pick up where you left off. Therefore, we have achieved FMC from a service or applications perspective. Therefore, why would this change with 6G?

Chapter 6 examines the core focus of 6G, which is to merge the physical, biological and digital worlds into a multi-sensory experience for society and industry. This integration could create unprecedented economic opportunities while presenting unique societal challenges. These three worlds typically operate in separate domains, and the central theme of 6G is convergence across multiple levels of the architecture, not only at the access layer. Let us take a moment to review the six main elements of 6G technology that show 6G convergence, which is about new services that are much more than just best efforts IP, and explore the argument for broader context in each one supporting convergence.

1. Connecting intelligence:
   In the future, 6G technology will be important in deploying intelligence throughout society. It will provide a framework for supporting and enhancing artificial intelligence (AI) and machine learning technologies, enabling real-time trustworthy control. This will help improve efficiency and service experience, focusing on including human input in the process.

   Yet 6G will not be the only source of intelligence, and it needs to recognise that hyperscalers are set to offer many new services on their cloud and AI infrastructures or in partnerships with carriers. This would be a significant change from 5G.
2. Network of networks:
   The 6G network will combine various resources, such as communication, data and AI processing, to connect on different access levels, from within the body to indoor locations, data centres and wide area networks.

   Yet this implies that the resources will exist on different networks or sub-networks, so a form of network convergence is needed for this to be fully effective.
3. Global service coverage:
   While the prime aim within 6G is to address the digital equity/digital divide with an emphasis on satellite technologies, this does involve integration with a completely different network, and satellites are a new form of Network of Networks described in Step 2.

   As discussed in Chapter 10, satellites are essential in extending coverage outside the usual cellular network. They are designed for economic coverage for the new industrial use cases, including drones and robots across vast areas of land and oceans.
4. Trust and security:
   The goal of 6G is to maintain the privacy and security of end-to-end communications while ensuring data privacy, operational resilience and security. This will help build trust in wireless networks and their applications for consumers and enterprises.

Yet, why is this limited to cellular or wireless systems to be fully effective? This will be particularly important for a 'network of networks', as implied in Step 2 or with private 5G/6G networks connected to public networks. Many 6G projects and visions support this concept, yet it is not normally articulated. This, again, is another element supporting a form of convergence, in this case in trust and security.

5. Sustainability:
   Including 'zero energy' devices for increased Internet of Things (IoT) roles: However, most IoT devices will not be cellular-based. Cellular has less than 26% (2022) of the IoT market and is projected to fall to 19% by 2028 [1], with access technologies bridging fixed and mobile domains. So, for 6G to achieve its aims, it needs to potentially bring all forms of IoT within its reach, including networks of networks, supported by trust and security and connecting intelligence.
6. Extreme experience:
   In the future, 6G systems can access spectrum above 100 GHz, allowing for extreme data rates greater than 100 Gbit/s. Additionally, these systems will provide sensing and positioning accurate to the cm level.

   For 6G to be an effective bridge between worlds, it must share network sensing capabilities and combine them with other wireless sensing technologies on different access networks. This will often be done within.

   While 6G will offer extreme speeds, Wi-Fi is likely to be used for consumer applications in homes with fixed broadband networks. Many extended reality applications work well on speeds of less than 100 Mbit/s and latencies of 5–10 ms. Wi-Fi is expected to support Gbit/s speeds and these low latencies. Wi-Fi allows for indoor and cellular handover to outdoor use for many extended reality use cases. For converged operators, the use of Wi-Fi as a trusted technology offers many cost advantages over small or femtocells and much better coverage than the proposed 8 GHz bands.

6G is expected to build on the foundation laid by 5G. Still, it is essential to note that cellular technology alone cannot wholly transform society and industry. A fixed broadband network, which includes cable and fibre-to-the-home (FTTH), is also crucial for delivering high-speed access that drives the overall service experience. This is where wireless technologies such as Wi-Fi and cellular small or femtocells come into play. Other wireless IoT technologies are also highly optimised for many new use cases. The Core and Multi-access Edge Computing (MEC) technologies that support many aspects of these new use cases are also required for devices connected to fixed networks, directly or indirectly. Wi-Fi, which we discuss in detail in Chapter 3 as an access technology, is expected to support many 5G use cases over the next decade. This could make 6G a true network of networks, which is essential for MNOs and hyperscalers.

Caroline Gabriel, Analysys Mason, has also argued that network convergence is essential if 6G is to address the shortcomings of 5G [2]:

> *5G is too mobile-centric to deliver on all its promises; a new 6G converged architecture is essential for true digital transformation.*

The telecoms, media and technology (TMT) industry needs to make progress with 6G because the true digital transformation will not be enabled by a siloed mobile network, but rather by one that is built from the outset around the convergence of various access technologies, and of networks and cloud.

For these reasons, 6G will be a network of networks, mixing fixed and mobile telecom networks with cloud AI providers. It has the challenge that consolidated networks increase the risk of security breaches, requiring robust measures to protect sensitive data, as discussed in Chapter 12 with privacy-enhancing technology (PET).

We see Wi-Fi in trusted or non-trusted forms as a true partner to enable 6G services inside buildings for domestic and industrial use cases. Industry sites may prefer small cells or femtocells to cover larger areas with higher performance levels. Wi-Fi has the advantage in homes of coverage from a single Access Point supporting the new use cases of extended reality (XR).

6G must unify various architectures to promote greater mix and match with plug-and-play. A common core for fixed and mobile and MEC for low latencies will add value to potential partnerships with other Cloud and AI vendors. Nevertheless, carriers will also offer cloud and AI environments to support their applications and services.

This chapter supports the argument for 6G convergence with an overview of its technologies. It describes the challenges facing 6G convergence, Section 9.7.3 proposes some architectural principles, and Section 9.7.4 gives an illustrative example. Section 9.3 looks at existing technologies and fixed-mobile and Wi-Fi convergence standards and why they cannot be considered candidates for 6G convergence. Understanding the complexities and limitations of these existing technologies and standards places the 6G convergence challenges in context. Then, in the context of convergence, Section 9.4 explores inside-out networks for indoors, and Section 9.5 inside-out networks for outdoor coverage. We question if cellular networks' indoor and outdoor coverage should remain converged on the same macro-tower infrastructure. There are performance, cost and energy-saving benefits for MNOs and customers with an architecture where indoor access points serve indoor wireless users. Wi-Fi may be the technology of choice. Intrinsically, it is best suited to the consumer's home. Then, Section 9.6 describes neutral hosting, which allows the convergence of many MNOs on shared infrastructure in small businesses, hospitals, schools, shopping malls, stadia, ports and airports to deliver 6G services. The conclusion summarises and contrasts the evolutionary and revolutionary approaches to 6G convergence.

9.2 Commercial strategy

The success of 6G technology hinges on its ability to generate new revenue streams. Telecom operators recently investing in 5G and FTTx (fibre to the home/

premise/cabinet) roll-outs may have a weak business case to invest in another network generation like 6G. Remember, many large carriers have fixed and mobile networks achieved through mergers and acquisitions. However, in the past decade, the growth of the Technology, Media, and Telecommunications (TMT) sector has been primarily driven by Internet and cloud service providers. These players are willing to invest in new infrastructure only if it increases revenue, as Caroline Gabriel, Analysys Mason, points out:

> *The convergence of 6G technology will create a new infrastructure that enables further digital transformation and generates additional revenue streams. The shortcomings of 5G technology will likely expedite the need for a new converged infrastructure under the 6G banner.*

As society and industry become increasingly digitalised, the need for 6G convergence becomes more apparent. However, the architecture for these products should be fundamentally different. Future applications will require more than a simple IP connection. They need MEC, Cloud and AI/machine learning (ML), with interworking with hyperscalers. While many organisations may see the Internet as good enough for a network of networks, 6G convergence goes beyond basic network connectivity for new use cases. The value of 6G convergence built on private or trusted networks challenges the hyperscalers' and governments' perceptions that best-efforts Internet service is sufficient to deliver broadband services and the digitisation of society.

That is not to say that all 6G use cases/applications need converged networks. The OTT cloud and AI support lower-value applications that use best-effort IP and exist alongside the higher-value use cases requiring some or all of the six themes described in Section 9.1. Device manufacturers such as Apple prefer this route for their devices and operating systems while supporting applications that use the 6G network suite of assets.

Hyperscalers, in partnership with carriers, need to support more open multi-cloud and MEC standards. Similarly, MNOs want to protect their investment in infrastructure and spectrum and will be reluctant to take any actions that devalue that investment in the short term, even if they create value in the long term.

Therefore, the development of 6G technology requires a commercial revolution and changes in mindset. All stakeholders must discuss how openness and convergence create value for everyone involved in 6G technology. From a technical standpoint, the convergence of 6G must be designed to allow for tussles [3] between stakeholders. Figure 9.1 shows the 6G convergence value chain. As they do today, the customer pays the application providers, hyperscalers, networks and user equipment vendors directly. 6G convergence supports the value propositions or services of trust (Chapter 12), cloud (Chapter 11), sensing (Chapter 14) and intelligence (Chapter 13). The platform and infrastructure providers have commercial relationships that may involve payments but importantly the common interfaces and abstractions, as indicated by the use of Lego bricks, allow them to be

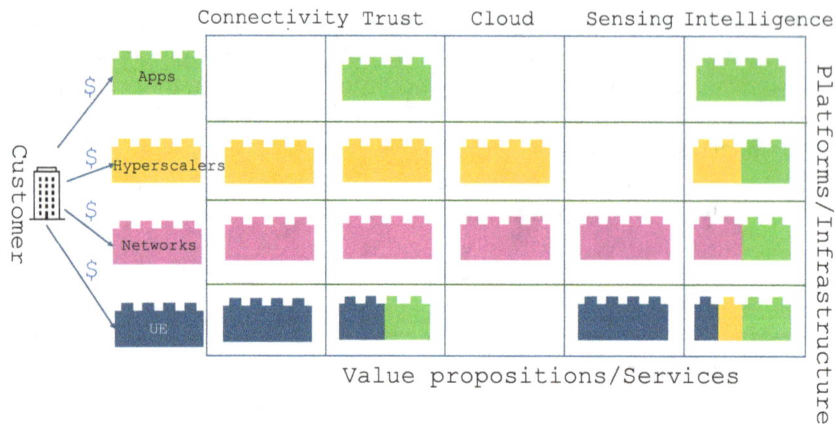

Figure 9.1 6G convergence value chain

connected to offer services. The Lego brick's colour indicates the component's provider, and the app providers can supply AI software to run on any platform. Today's networks provide connectivity across the platforms, and trust is not integrated except in some private networks. MEC and hyperscaler cloud interconnect is limited today due to a lack of MEC deployments, while connected sensing and intelligence are new services.

9.3 Fixed-mobile and Wi-Fi convergence standards

Let us be clear: Existing FMC standards do not describe how a network operator with fixed and mobile networks can converge their infrastructure to reduce costs. The 3GPP, Broadband Forum (BBF) and CableLab FMC standards only describe how UE and other devices can access the 5G core services via non-3GPP accesses. While the IETF FMC standards treat the cellular network as just another IP link.

This section examines the evolution of fixed mobile and Wi-Fi convergence standards to discover what revolutionary 6G convergence should address if Wi-Fi is to offer an equivalent service to 5G, i.e. make 6G access agnostic or use Wi-Fi to drive an 'inside-out' strategy. It describes the existing standards for the mobile core to offer 5G services to fixed and Wi-Fi networks. It examines the issues of the current 5G converged services architecture and whether Wi-Fi can deliver the same equivalent services as 3GPP radio using a licensed spectrum. It identifies issues with standards-based convergence using a complex mix of functions and treating Wi-Fi devices differently according to their ownership and access network.

We cannot avoid 5G and fixed networks using different protocol stacks for the access network or 'stratum' where protocols must reflect the medium's physical properties, e.g. radio, copper line and fibre. Today's convergence assumes all networks carry the Internet protocol (IP), which, from the customers' point of view, provides service continuity and interworking, except for QoS. Because the protocol stacks are different below the IP layer, the network operators use different network equipment with different management and signalling systems, and typically, these different networks locate their equipment in different locations. A converged fixed-mobile core should offer customers better service continuity with the same QoS and IP addresses between networks. Today's FMC standards use gateways between different networks to give customers some service continuity, so they do not help network operators with fixed and mobile networks converge their network infrastructures.

5G, with a standalone 5G core, can offer services by multiple non-3GPP access methods, but it is a complex matrix of

1. Wi-Fi, wired broadband and wired cable access networks.
2. 3GPP, BBF and CableLabs procedures.
3. Trusted and untrusted access networks.
4. 5G capable UE, non-5G capable WLAN and non-5G capable end systems.
5. Wi-Fi and 5G, Wi-Fi only and wired LAN only devices.

However, not all combinations are defined or supported by standards.

The access types standardised so far are Wi-Fi, broadband wireline and cable wireline. 3GPP categorises access networks as trusted or untrusted. Figures 9.2–9.8

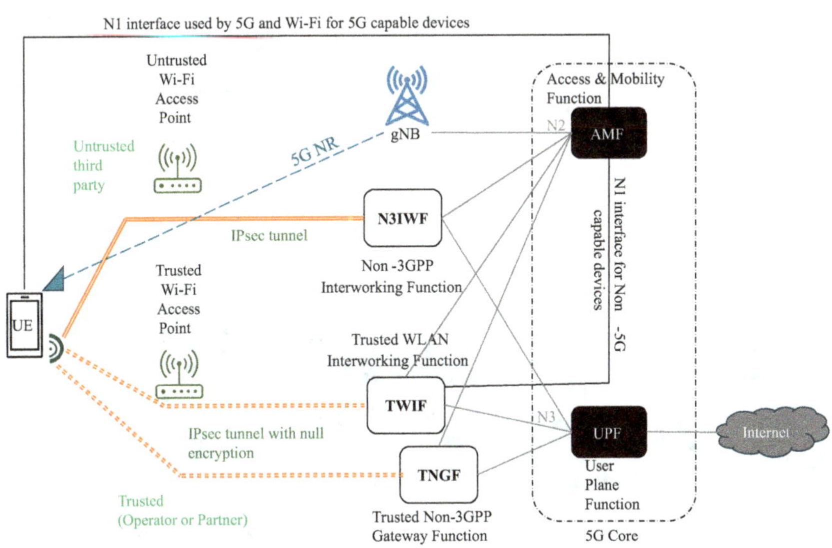

Figure 9.2 Wi-Fi connection to 5G

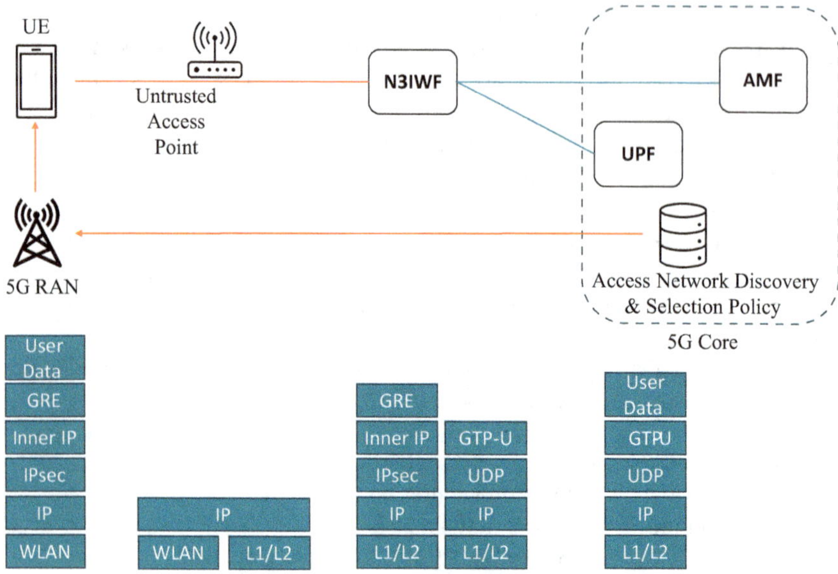

Figure 9.3 5G capable UE using untrusted Wi-Fi access point

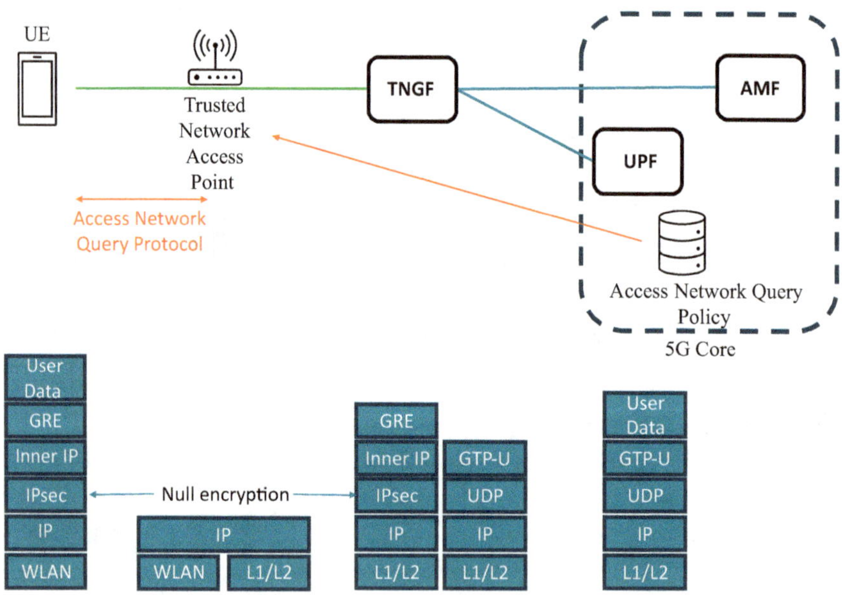

Figure 9.4 Trusted Wi-Fi access point with 5G capable UE

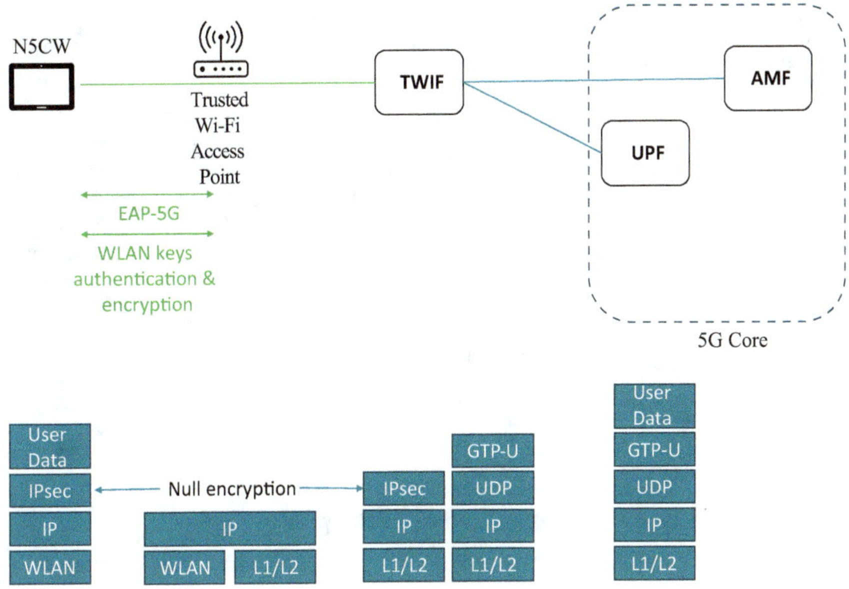

Figure 9.5 Trusted Wi-Fi access point with non-5G capable Wi-Fi device

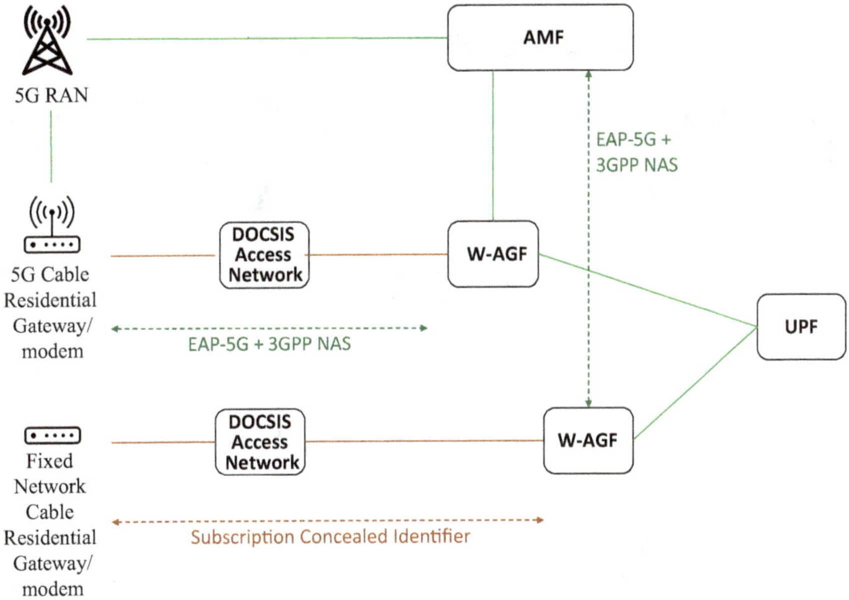

Figure 9.6 CableLabs FMC architecture

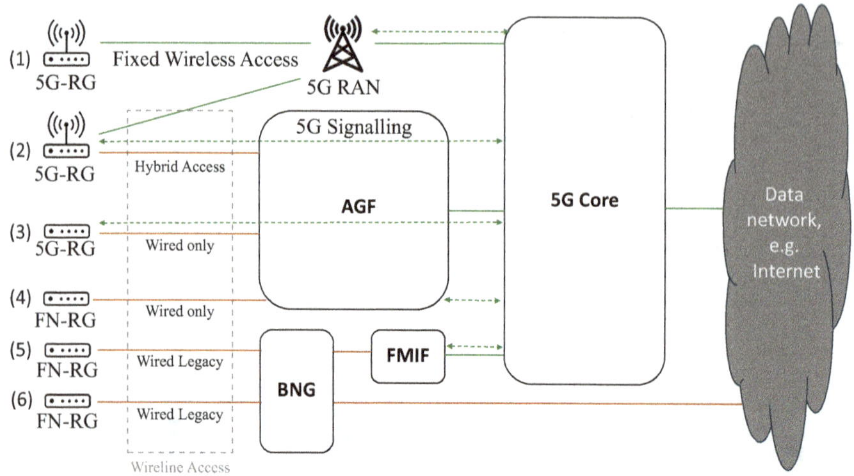

Figure 9.7 BBF FMC options

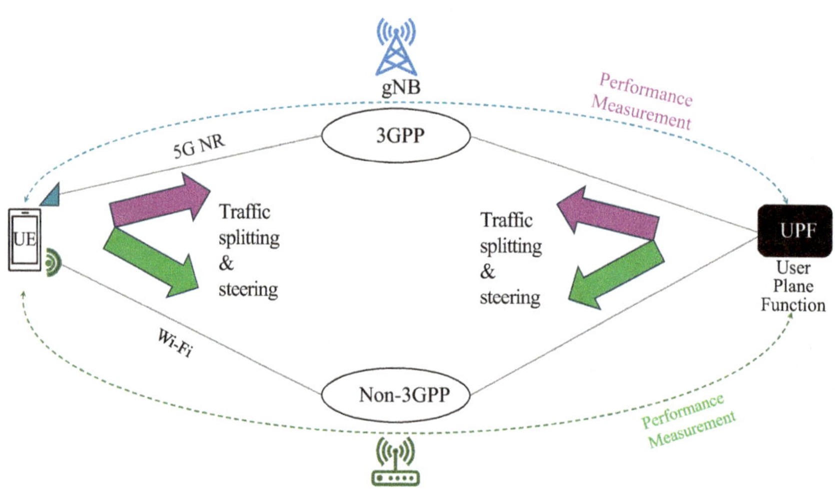

Figure 9.8 Access traffic steering-switching-splitting

describe the methods and differences between these for completeness. The 3GPP does not define trust but often assumes that a trusted public access network has a well-defined commercial relationship with the MNO offering the public 5G services. The broadband wireline standards are the BBF's responsibility, while the cable wireline standards are the responsibility of CableLabs. Box 9.1 and Figure 9.2 show some of the key components and terminology.

Box 9.1: 3GPP 'Non-3GPP' access terminology and components

- **Non-3GPP**: An access technology that is not standardised by the 3GPP, but the UE may still implement 5G non-access stratum capabilities.
- **Non-5G capable over WLAN (N5CW)**: A Wi-Fi device with no 5G capabilities that uses WLAN keys for authentication and encryption.
- **Non-3GPP interworking function (N3IWF)**: Gateway between untrusted access networks (usually Wi-Fi) and the 5G core.
- **Trusted network access point (TNAP)**: Connects 5G capable UEs to the trusted access network using non-3GPP wireless or wired technology.
- **Trusted non-3GPP gateway function (TNGF)**: Connects TNAPs to the 5G core.
- **Trusted WLAN access point (TWAP):** Connects N5CWs to the trusted access network.
- **Trusted WLAN interworking function** (TWIF) – Connects TWAPs to the 5G core.
- **Wireline 5G broadband access network (W-5GBAN)**: A broadband access network using BBF specs.
- **Wireline 5G cable access network (W-5GCAN):** A cable network using CableLabs specs.
- **5G residential gateway (5G-RG):** Acts like a fully 5G capable UE to connect 5G UEs or non-3GPP devices on a residential Wi-Fi/LAN to the 5G core.
- **Fixed network residential gateway (FN-RG)**: That is not 5G capable.
- **End system (ES):** Non-5G capable device on a LAN connected via FN-RG/5G-RF to a fixed wired network.
- **Wired access gateway function (W-AGF):** Connects wired access networks to 5G cores.

For private 5G networks, the organisations purchasing and using the private 5G networks must define how and which networks and devices they trust. The gateway functions defined for public FMC may or may not be appropriate for private 5G networks.

Table 9.1 shows the main gateway functions for interworking the combinations of trusted/untrusted network access types and 5G capable user equipment and not 5G capable (N5CW/ES) devices. Figure 9.2 shows the trusted TNGF, the N3IWF and the TWIF. Figure 9.6 shows the W-AGF, and Figure 9.7 shows the access gateway function.

Figure 9.3 shows how a 5G capable UE, e.g. a smartphone, can use an untrusted Wi-Fi access point and an untrusted network to access 5G core services. The non-3GPP interworking function (N3IWF) provides the gateway between untrusted networks and the 5G core. The UE and non-3GPP interworking function set up a secure IPsec tunnel over which 5G protocols run. Figure 9.3 also shows the

Table 9.1 Fixed-mobile interworking functions

Trust	Access type	5G capable UE	N5CW/ES
Trusted	Wi-Fi	TNGF	TWIF
	Broadband	AGF	AGF
	Cable	W-AGF	W-AGF
Untrusted	Wi-Fi Broadband	N3IWF N3IWF	
	Cable	N3IWF	

protocol stacks involved. The UE must download an 'access network discovery and selection policy' from the 5G core before using an untrusted access point. 5G does not specify how the UE authenticates to the untrusted Wi-Fi- access point.

Figure 9.4 shows how a 5G capable UE can use a TNAP to use a 5G core via the trusted network gateway function (TNGF). Trusted network access works similarly to untrusted network access except that the IPsec tunnel can use null encryption to save processing overhead, as the MNO trusts the access point and network. The UE uses the Access Network Query Protocol, as also used by Hotspot 2.0 [4] and Passpoint [5], to select a suitable trusted Wi-Fi access point.

Figure 9.5 shows how a non-5G capable Wi-Fi device can use the 5G core services via a trusted Wi-Fi access point and the trusted WLAN interworking function. The non-5G capable Wi-Fi device still uses the EAP-5G authentication protocol and SIM/eSIM identities, and 5G does not define how non-5G SIM authentication would work. Security depends on the WLAN keys for authentication and encryption, determined by the 5G procedures operating on the trusted Wi-Fi access point. The protocol stack for the non-5G capable Wi-Fi device and how it selects a trusted access point is not explicit, but the protocol stack should be more straightforward than for 5G capable UE.

Figure 9.6 shows the CableLabs architecture for interworking cable residential gateways with a 5G core. In the case of a 5G capable 5G Cable Residential Gateway, the standard 5G authentication (EAP-5G) and signalling protocols (NAS) are used. For the Fixed Network Cable Residential Gateway, the W-AGF implements the 5G authentication and signalling protocols.

Figure 9.7 shows the BBF's options for interworking broadband services (typically DSL and fibre-based) with the 5G Core. For Fixed Wireless Access (1), the 5G-RG connects to a 5G RAN as per a UE. For Hybrid Access (2), the 5G-RG connects via the 5G RAN and the Access Gateway Function (AGF). A 5G-RG (3) can connect via a wired-only option. In cases (1)–(3), the 5G-RG and the 5G Core perform 5G signalling as per a UE. A Fixed-Network Residential Gateway (FN-RG) (4) does not have 5G capabilities, so the AGF proxies the 5G signalling for the Fixed-Network Residential Gateway. A legacy Fixed-Network Residential Gateway (5) can connect via a Broadband Network Gateway, wherein a Fixed Mobile Interworking Function

(FMIF) must proxy the 5G signalling to the 5G core. Case (6) is the classic legacy connection to the Data Network or Internet via a Broadband Network Gateway. The 3GPP considers cases (5) and (6) out of their scope.

Connections via 5G and Wi-Fi simultaneously from the same UE are possible using the 3GPP Release 16 Access Traffic Steering-Switching-Splitting (ATSSS) standard. ATSSS can steer traffic onto the best available network based on cost, speed or latency, as shown in Figure 9.8. It can switch traffic seamlessly between 5G and Wi-Fi networks. It can also split traffic over 5G and Wi-Fi according to policies.

ATSSS achieves this using multipath TCP (MTCP) as defined by the IETF's RFC 8684. ATSSS requires performance measurement functions (PMFs) to be embedded in the UE and UPF to measure the performance of the separate paths. The UE and UPF require ATSSS lower layer functions. While the UE must implement MPTCP, and the UPF has an MPTCP proxy, the 5G core network must include ATSSS capabilities.

ATSSS chooses a network as an active-standby for rapid service restoration in case of a network fault. It can choose paths for specific traffic according to policies based on latency, load balancing or importance.

My experience with load-balancing traffic or choosing the best path over networks with very different and potentially rapidly changing characteristics has shown it to be complex and possibly intractable. The traffic generated by applications can change quicker than the periods measuring and averaging performance, hence basing routing decisions on out-of-date information. Traffic load influences active probe performance measurements, which can cause traffic to alternate between paths, making bandwidth estimation algorithms, e.g. in TCP or QUIC, inaccurate, leading to under or overutilisation and possible packet reordering. AI might provide a solution to optimise ATSSS load balancing as it is a multi-factor problem with significant amounts of measurements to be analysed.

ATSSS could be helpful for FMC, but only time will tell if ATSSS is a chimaera.[1]

In conclusion, the plethora of FMC options appears complex. However, the number of options is driven by the combinations of 5G capable UE and non-5G capable end systems, trusted or untrusted Wi-Fi access points, usually determined by who supplies and manages the Wi-Fi access point, cable or broadband access lines. No one convergence gateway is better than any other; the customers' specific equipment and network operators drive the choice. This is why we recommend a generic convergence gateway for 6G.

For an in-depth tutorial on non-3GPP accesses in 5G systems, see Lemes *et al.* [6].

9.3.1 FMC 3GPP issues

Despite the solid work of the 3GPP, CableLabs and BBF, there is little real-world deployment experience with their convergence procedures and protocols. For 6G, there are the following issues that research and development should address if

[1]A thing that is wished for but is impossible to achieve.

Wi-Fi is to offer an equivalent service to 5G, i.e. make 6G access agnostic or use Wi-Fi to drive an 'inside-out' strategy.

- Foremost, the FMC standards do not describe how a network operator with fixed and mobile networks can converge their infrastructure to reduce costs.
- In the existing standards, it is unclear how an MNO or a private enterprise establishes trust with 'trusted' non-3GPP networks. Many assume that MNOs can only trust access points and networks they own or have a close commercial relationship with. Even then, it is not clear that non-3GPP access networks would be as secure as 5G networks by default. No MNO can own all enterprise and residential Wi-Fi access points, so a global alliance or consortium, which can agree on security, service and trust standards, is required to enable global roaming between Wi-Fi access points.
- There are missing details on how non-5G capable devices connect, especially devices without 3GPP credentials, e.g. no SIM. 3GPP considers the case of non-5G capable Wi-Fi devices using an untrusted Wi-Fi access point outside their scope.
- The current 3GPP, BBF and CableLabs FMC methods are varied and complex, increasing attack vectors and the probability of bugs. These methods may not be suitable for low-power IoT devices. The complexity makes integration for enterprises using Private 5G networks difficult.
- Untrusted Wi-Fi access requires the UE to have an 'access network discovery and selection policy', which either requires a 5G NR connection or factory install on the UE, which may not be feasible for IoT devices nor support an 'inside-out' strategy.
- Establishing IPsec tunnels adds latency to setting up a connection; Lemes *et al.* [6] measured initial registration and authorisation times of ~1 s and session establishment times of ~0.2 s. Designers should consider IPsec tunnel set-up latency in the design of 5G solutions, i.e. do not design assuming low latency and frequent IPsec registration or session establishment.
- Adding another layer of gateways increases the number of network functions and equipment, which is not the right direction for cost and resource optimisation.
- Fixed, Wi-Fi and 5G networks use different QoS mechanisms; therefore, QoS is mapped approximately between networks, achieved with a significant amount of provisioning orchestration. Wi-Fi is also subject to interference from legacy Wi-Fi devices, which may prevent Wi-Fi from delivering the required QoS.
- 5G does not address private 5G networks using Wi-Fi access points managed by an enterprise or sometimes outsourced to a cloud-based management system, e.g. Cisco Meraki, as part of a 5G access agnostic service.

The existing FMC standards have many issues if they were to be used for the more complete 6G convergence vision. Whether 6G convergence can reuse any parts of existing FMC standards or take a clean-slate approach is for further study. However, 6G convergence standards must be complete, address how trust with non-3GPP components is established, address private networks, and have a more consistent and generic approach to convergence gateways. The architecture principles Section 9.7.3 and the architecture example in Section 9.7.4 highlight how this is achieved. A complete and more flexible 6G convergence solution would enable alternative ways to build mobile networks, which could deliver significant cost and energy savings. One such alternative is 'inside-out' networks, the subject of Sections 9.4 and 9.5.

9.4 Inside-out networks for indoors

The mode of serving customers' mobiles via a macrocell-type architecture has changed very little since 1G when all the traffic was voice in outside locations, i.e. cars. Yet today, 70–80% [7–10] of traffic on mobile networks is indoor data, and we are still trying to serve this indoor data with the same approach, which is wasteful in terms of both energy consumption and radio resources. If 6G is about reducing energy consumption, then indoor base stations could make a significant difference. There is an extra radio loss to the indoor user from a macro tower of approximately 15–30 dB, depending on the building construction and the number of intervening walls. High-order MIMO arrays and smaller cells overcome the losses at higher frequencies, with a significant loss for 5G at 3.5 GHz. This radio loss directly translates into reduced capacity, coverage, higher energy consumption and increased costs. Table 9.2 shows the power budget for serving an indoor user from a remote macrocell and a femtocell located in the same building, which could be an office. It illustrates the great advantage of placing base stations indoors, close to the users. The femtocell's much better signal-to-interference and noise ratio (SINR) can offer much higher data rates despite the transmitter power being 400 times lower, clearly demonstrating the difficulty of using macrocells to reach the building and then penetrate the outside wall from a distance.

If performance can be increased greatly by using internal base stations, then the improvement in energy efficiency is even more dramatic. The EARTH project [https://earth-project.eu] estimated the power used by macro and femtocells a:

$$\text{Power Macro} = 118.7\ \text{W} + 5.32 \times T_{\text{max}} = 331.5\ \text{W/sector (using } T_{\text{max}} = 40\ \text{W)}$$
$$\text{Power Femto} = 4.8\ \text{W} + 7.5 \times T_{\text{max}} = 3.1\ \text{W (using } T_{\text{max}} = 100\ \text{mWmax)}$$

Yunas *et al.* [11] modelled a Manhattan-style grid of a typical North American city and compared the energy costs of using macrocells or femtocells to cover a 1 km^2 area from the inside-out (a topic we come back to later in this chapter). They considered 120 macrocells/km^2 and found that 3125 femtocells/km^2 could offer 400 times more indoor and 12 times more outdoor capacity than a macro network for the same spectrum usage. For a given amount of energy, the total bits transferred was 600 times more for the femtocells when serving indoor traffic and 20 times more for outdoor traffic (compared to a macrocell). If 80% of the traffic is indoors, then the overall energy gain of using this 'inside-out' network is 484 times.

First, inside-out for indoors could also provide a much-improved service for users. It is here that 70–80% of mobile data is used. The aim is not to build a mobile network using only indoor femtocells or Wi-Fi access points but to support most indoor traffic on either femtocells or Wi-Fi access points with broadband connection to the mobile core for 6G services. This inside-out network can take advantage

Table 9.2 Typical link budgets for a macrocell and an indoor femtocell using 100 MHz (3.5 GHz)

	Macrocell	**Femtocell**	**Notes**
Transmit power	46 dBm (40 W)	20 dBm (100 mW)	
Cable loss	2 dB	0 dB	
Antenna gain	18 dB	+2.2 dB	Assuming macro 3 sector down tilt antenna
Propagation loss	−132.3 dB	−57.3 dB	WINNER II D112 v1.2 model. Femto 5m from terminal NLOS. 10m Macro height and 3m receiver using 3DUMi TR36.873 model at 400m
Shadowing loss	−6 dB	0 dB	Modifies macro loss for large-scale fading due to buildings/vehicles/trees, etc.
External wall loss	−15 dB	0 dB	10–20 dB/wall – assuming 1 or 0 walls
Internal wall loss	−4 dB	−4 dB	3–5 dB/wall
Internal floor loss	+1.5 dB	−4 dB	4 dB/floor loss for femtocells (WINNER II D112 v1.2 model) User upstairs 1.5 dB/floor gain for macrocells
Interference margin	−5 dB	0 dB	Accounts for interference from neighbouring cells with single-frequency reuse
MIMO gain	+12 dB	+6 dB	It can be used to increase power or throughput
Received power	−86.8 dBm	−35 dBm	
Receiver sensitivity	−88 dBm	−88 dBm	For 100 MHz channel
SINR	+1.2 dB	+53 dB	SINR ratio
Data rate	100 Mbit/s	700 Mbit/s+	Macro SINR supports QPSK at 1/2 rate coding. Femto supports 256QAM at 7/8 coding

of the MEC, digital twins and 6G AI. This then allows the macrocells to carry the outdoor traffic. Second, there could be serendipitous coverage of other locations from the AP as some of the signal penetrates walls or windows to the outside. This is the experience of Community Wi-Fi, for example. It further reduces the traffic served by the macrocell. The significance of the inside-out approach grows exponentially with the number of APs. Especially if these are open to other converged operators to use where neutral hosting plays a key role.

This section also explores how indoor APs could supplement macrocell coverage for some specific scenarios. Shipping indoor Internet traffic, which comprises most of the traffic on cellular networks, onto indoor access points; hence, the fixed broadband infrastructure also converges future investments in fixed infrastructure growth for network operators that own both fixed and mobile networks.

One of the additional benefits of 6G convergence is that it opens the possibility of changing the cellular delivery architecture from mainly a macrocell first in an outside-in approach to mainly an inside-out approach using Wi-Fi access points and small or femtocells. This migration can happen over time, driven by MNO's need to increase capacity and the lifespan of customer premises equipment. It allows 6G to reduce energy consumption and gives the converged operators' customers a better indoor experience.

9.5 The inside-out network for outdoors

So far, we have looked at moving the base stations inside buildings and seen how this results in a very significant uplift in energy efficiency and radio performance. The next step is to consider if these indoor base stations might also be able to serve, as a bonus, outdoor users or those in nearby buildings by 'leaking' radio signals over a wider area. The classic example of this 'inside-out' network is Wi-Fi. Broadband providers, such as BT in the UK, now include multiple Wi-Fi identifiers on their home and business Wi-Fi hubs that all offer connection to their broadband customers. This system was designed to offer a community Wi-Fi service so visitors to homes, offices or shops can use the Wi-Fi. As a bonus, you can legally and legitimately connect to the Wi-Fi from the flat below, next door or outside if nearby, and BT provides the Wi-Fi. This sounds very useful, but in practice, the range of Wi-Fi in the 5–7 GHz band (Chapter 3) is quite limited, and even though you might manage to connect, the throughput might be very low at the edge of the coverage range.

Kawade and Nekovee [12] tried to model how far Wi-Fi would go towards creating an inside-out network. They looked at a 1 km^2 residential area in a typical European Union city with 5000 dwellings. They compared the use of Wi-Fi over 2.4 GHz, 5 GHz and TV White Spaces (TVWS) spectrum (there is a lot more on TVWS in Chapter 7). In many countries, TV is still broadcast (digitally) over channels in the 400–900 MHz range. Crucially, it is possible to find many unused channels at any given location, and, provided they are used at low power, they can be put to a secondary use.

TVWS consults a location database in the same way as Wi-Fi 6E, described in Chapter 3, or CBRS, described in Chapter 7. The database gives the channels and maximum power combination for a given location. The modelling considers the UK TV broadcast band 470–862 MHz. Kawade *et al.* [13] describe the concept in greater detail and provide further information about modelling an inside-out network. The available TVWS spectrum in the UK offered a location-dependent 50–150 MHz in the greater London area with a maximum transmit power of 100 mW (similar to the Wi-Fi maximum power at 2.4 GHz). Table 9.3 shows the key coverage result. The coverage is the fraction of the 1 km^2 area that could achieve at least the data rate shown. The coverage was low when the deployment density (fraction of dwellings with an access point) was low. It grew to the maximum coverage value at the deployment density shown in the table and then declined due to interference as the deployment density increased.

Table 9.3 Coverage results of Kawade et al. [13] modelling Wi-Fi technology over different spectrum ranges with different power regulations

	TVWS (400–900 MHz)	2.4 GHz	5 GHz
Maximum achievable coverage	95%	60%	55%
Deployment density for max coverage	12%	45%	>50%
Normalised data rates	9 Mbit/s	14 Mbit/s	17 Mbit/s

The result showed the limitations of using Wi-Fi for universal coverage. Sixty per cent coverage was obtainable at 2.4 GHz when 45% of buildings/houses/flats were sharing it. The 5 GHz results were less favourable due to the lower propagation range at these frequencies. A significant result, however, was that 95% of coverage could be obtained if only 12% of dwellings shared Wi-Fi using the TVWS spectrum. Given that the transmit power was the same for 2.4 GHz Wi-Fi and TVWS, the lower frequency propagation was the primary driver. TVWS could offer much higher power limits in some areas, meaning that 95% coverage requires less than 12% of buildings with base stations, and much higher data rates would be possible.

TVWS has not progressed mainly because it offered nothing that Wi-Fi and 4G/5G do not already provide. It was also a victim of the success of 4G, coupled with the cost of the database, lack of development of new chips and a dearth of worldwide standards and frequencies. However, new bands are available for 6G, such as the 6–7 GHz band (Chapter 3) and CBRS (Chapter 7). These could also offer higher powers, compensating for the poorer propagation at higher frequencies. US cable operators plan to use CBRS spectrum for an inside-out strategy to build urban/semi-urban networks. According to Sharma [14], Comcast is building a CBRS 5G network:

> *Samsung and Comcast announced they are collaborating to enhance 5G connectivity for Xfinity Mobile and Comcast Business Mobile customers in Comcast service areas. Samsung will supply 5G RAN solutions to Comcast, helping the company cost-effectively enhance 5G cellular connectivity to its consumer and business customers using CBRS and 600MHz spectrum within Comcast service areas. As demand for reliable wireless Internet access increases, Samsung's 5G RAN solutions and Comcast's mid-band (CBRS) and low-band (600MHz) spectrum will enable Comcast to supplement its existing Xfinity Wi-Fi network and cellular network partnership with additional 5G coverage in certain high-traffic areas within its service areas.*

Interestingly, Comcast is adding a 600 MHz layer, presumably to carry the control plane and offer 100% coverage, albeit with limited bandwidth. This is the sort of layered approach that 6G will probably adopt.

Yunas *et al.* [11] further studied an 'inside-out' network. These authors modelled a Manhattan-style 1 km^2 grid and compared a macro solution with 3000 indoor femtocells. They found substantial cost savings, performance improvements and (as already noted) much lower energy consumption. They also showed that the femtocells alone could offer excellent outdoor coverage even with the loss of penetrating the external walls of the buildings. Qualcomm presented a similar proposal as neighbourhood small cells [15].

It is possible that 6G could utilise an inside-out network build. This might use higher powers than Wi-Fi with improved AI interference management. The potential performance gains and energy savings over a macro build make this a potentially attractive option for 6G. Chapter 16 presents this idea in the broader analysis of a revolutionary 6G vision.

9.6 6G neutral hosting

Industry 4.0 is much more than large factories, construction, agriculture and mining. 6G services must also be delivered to places such as small businesses, hospitals, schools, shopping malls, stadia, ports and airports to achieve the broader digitisation of industry and society. This type of location has two key characteristics. First, one entity owns and controls how and when communications infrastructure is installed; second, they are shared spaces.

Meanwhile, homes and small offices can use Wi-Fi via the interworking gateways and convergence functions described in Section 9.3, but more significant Industry 4.0 sites require small cells or femtocells. Neutral hosting delivers the performance and energy savings needed for 6G applications in these larger shared spaces. To some extent, this is already happening today at the University of Michigan (with a massive Wi-Fi network [16]) and Caribe Royale Resort near Disneyland Orlando (private 5G using CBRS spectrum [17]).

Since the end users of these networks are typically customers of multiple MNOs, the networks in these locations are usually configured to connect to multiple MNO networks. This is the origin of the term neutral hosting. The example above of the Caribe Royal Resort is a typical set-up in that they have arranged to install a private 5G network and then signed deals with MNOs to allow visitors to roam onto the private network.

Two cellular standards have been around since LTE that support this sharing (Figure 9.9). Multi-operator core network (MOCN) shares the RAN (including base stations, spectrum and carriers), but each operator has its own core network. T-Mobile used it after it bought Sprint to help integrate the two networks quickly. 5G private networks are trialling MOCN to add mobile connectivity. Comcast Business and Ballast Networks are reported to have built a private wireless network in the Sound Hotel Seattle Belltown and are testing an MOCN connection between the hotel's private network and that of a mobile network operator (MNO). This is likely the same technology used at the Caribe Royale Resort, as noted above. Light Reading, however, [18] notes that there is reluctance for MNOs to offer services over networks managed and owned by third parties and that they would also have to negotiate with (potentially) thousands of private network owners.

Multi-operator radio access network (MORAN) provides a slightly different solution where the RAN is completely shared (including the base-band unit, RF

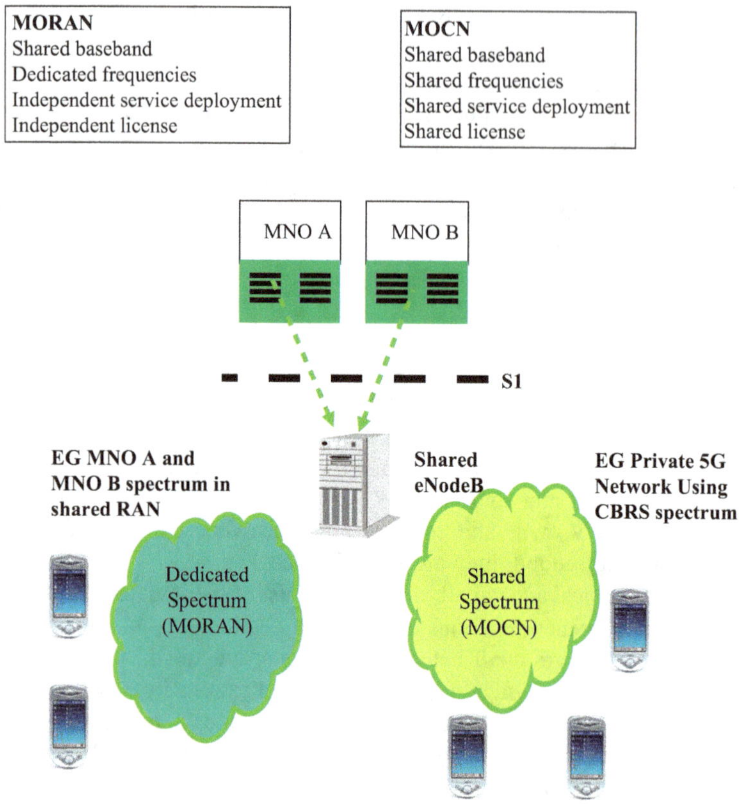

Figure 9.9 MORAN and MOCN network sharing standards

elements and antennae), but each MNO has its own spectrum and carriers. SoftBank Corp. and KDDI use MORAN to deploy a shared RAN in Japan.

Swedish real estate management company Proptivity has deployed an Ericsson indoor neutral host solution in a Stockholm shopping mall [19]. Three (3 Sweden) have signed up to use the system. It uses small radio heads (called dots by Ercisson) connected to base-band units with CAT 6 cable for power and data (10 GigE – 10 Gbit/s up to 180 ft). It is unclear if the solution uses MORAN, but at least it uses the same principles.

Much research has been published on neutral hosting in 5G and using private networks for roaming. Kibria *et al.* [20] describe a neutral host micro-operator using a shared spectrum. The neutral host micro-operator is a closed area such as an airport, mall, port or campus. These micro-operators create a 5G network using spectrum (licensed or shared), enter agreements with multiple MNOs and run their own services through a mini-core. Ahokangas *et al.* [21] analyse the business models of these micro-operators. One example described is a hospital that uses a private network to connect sensors, manage patient records (security and reliability), test new devices and monitor patient broadband and data collection/analysis. Bajracharya *et al.* [22] present a survey of neutral host technology. Project 5G ESSENCE [23] has developed a framework for Edge cloud computing and small cells as a service.

Another EU project called 5GCity [24] also explored the concept of neutral hosts putting infrastructure across a city (with small-scale trials in Barcelona, Bristol and Belgium). The project architecture has a flexible business model, but a third party (the city authority) installs small cells on street furniture or owned/leased buildings. 5GCity distributed a substantial MEC (multi-access edge computing) platform around the urban area. The project explores several spectrum options, including using MNO spectrum and unlicensed or shared spectrum. The trials also include Wi-Fi networks. The project describes two test applications: First, CCTV monitors the city, and the AI detects illegal rubbish dumping. Second, the application of TV or film crews uploading unprocessed footage to a central (or distributed) editing/control unit. While not specifically targeting indoor coverage, the techno-economic analysis associated with the project does show that large-scale neutral hosting can reduce CAPEX and OPEX costs for all players. A further project, SEASAME [25], developed a framework for a neutral hosting platform and introduced the concept of the Light Data Centre (Light DC) and cloud-enabled small cells (Figure 9.10).

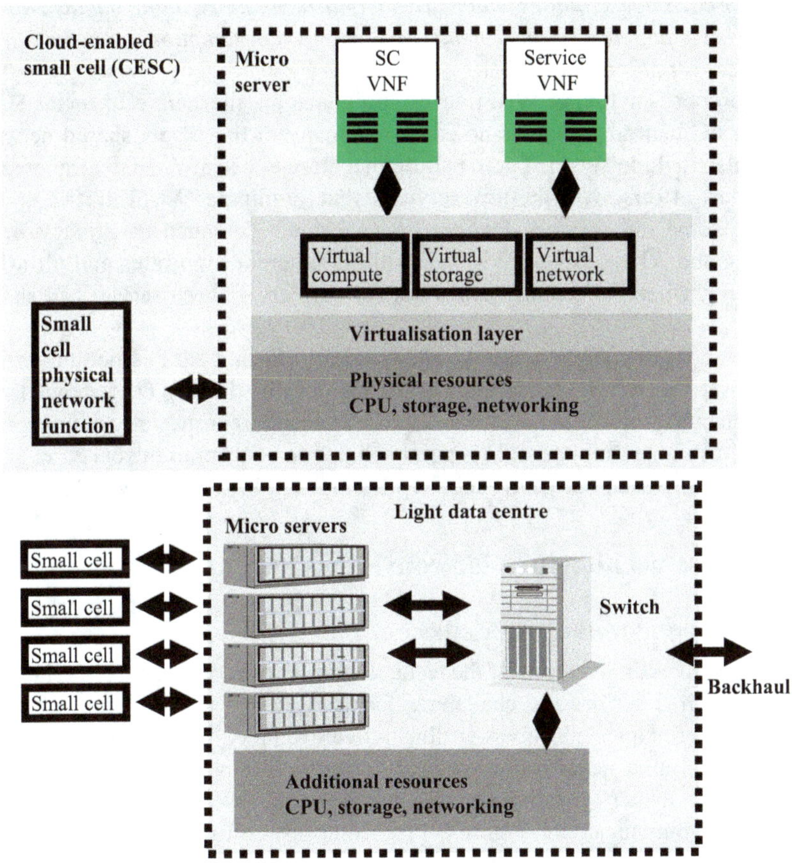

Figure 9.10 Neutral host micro-operator concept (from EU Project SESAME)

The most apparent barrier to neutral hosting is spectrum. Neutral hosts can use spectrum licensed to MNOs, but that puts severe constraints on using that spectrum. Another barrier to neutral hosts is the number and variety of them. The solution is for MNOs to agree on a common set of neutral host standards, including service definitions, for 6G. These would include guaranteed QoS levels and traffic-neutral connections. This agreement should ensure that MNOs can sign up for neutral hosting with much greater certainty that their customers can receive the same services and QoS that they themselves offer with appropriate reconciliation mechanisms/remedies. Ideally, new standards would also introduce more flexibility with the equipment splits, taking advantage of OpenRAN and allowing a higher level of diagnostics. This might be dubbed MOCN 2.0. There would also need to be proper interoperability standards.

One possible start on these standards is the Joint Operator Technical Specifications for Neutral Host In-Building (JOTS NHIB). This was agreed between the UK mobile operators in 2021 and sets out technical requirements for shared in-building solutions using small-cell technology [26]. Vodafone stated: '*The JOTS forum offers great opportunities for the sector. We can work together to find indoor coverage solutions which are not only technically viable but also offer a reduction in equipment requirements and energy consumption alongside some other great cost efficiencies*' [27].

We propose including Wi-Fi access points as an alternative to using 3GPP RANs for 6G neutral hosts. As noted above, many of these new shared network projects also include Wi-Fi. There is little incentive to offer Wi-Fi neutral hosting for the best-efforts over-the-top services that dominate Wi-Fi traffic today. However, some 6G services involve network functions such as privacy, local computing and AI resources. Wi-Fi 8, with its inherent high rates and ultra-low latency, might be an alternative solution for delivering these services in shared spaces.

We propose that Wi-Fi Neutral hosts could use Hotspot 2.0/Passpoint to offer a 6G Wi-Fi service. R&D is, however, required to develop this to a 6G femtocell type solution that integrates into the 3GPP 6G core architecture, especially defining how an MNO trusts a neutral host access point solution to connect to the 6G core. 3GPP MOCN could be used but would require development to work for Wi-Fi.

9.7 Future architecture for converged networks

9.7.1 Introduction

This section gives an overview of the technologies for 6G convergence, as shown in Figure 9.11. It describes the challenges facing 6G convergence, proposes some architectural principles and gives an illustrative example. The physical categories cover personal area networks where, for example, bio-medical devices communicate via the owner's nearby smartphone to global coverage via satellites for emergency communications. Figure 9.11 encompasses all the use cases described in this book. The 'network of networks' brings together a plethora of network

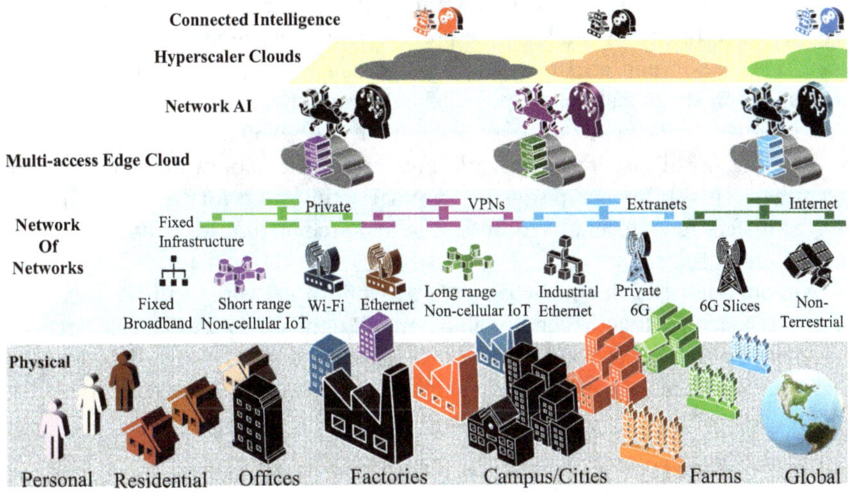

Figure 9.11 Scope of 6G convergence

technologies including, but not limited to, fixed broadband, short-range non-cellular IoT, Wi-Fi, wired Ethernet, long-range non-cellular IoT, industrial wired Ethernet, private 6G networks, 6G network slices on a public mobile network and non-terrestrial networks. We include fixed infrastructure partitioned into private, virtual private networks (VPN), extranets and the Internet. This plethora of network technologies reflects the specialised nature of network technologies and the recognition that a single cellular network cannot meet all the requirements of society and Industry 4.0.

The MEC, whether public or private, on customer or service provider premises, is essential to provide low latency services and optimise content delivery and data processing.

The 6G network will use network AI to optimise its infrastructure and services and offer AI services to customers.

It is inevitable that 6G networks and services will cooperate and interwork with the hyperscalers' clouds.

Connected intelligence is a nascent concept. The aim of connected intelligence is cooperation between diverse organisations' AIs. This AI cooperation allows organisations to work together to improve their overall efficiency and individual efficiencies. This would apply to health and local government services as much as to industrial production. Chapter 5 presents an example of connected intelligence when MNOs' network configuration and optimisation AIs communicate with Industry 4.0 organisations' operational technology AIs to optimise communication services, enabling supply chain optimisation. Chapter 13 presents another example of connected intelligence for accelerating federated learning by performing in-network computation, which could accelerate the creation of global AI models in hyperscaler clouds.

The Internet or IP alone is not the 'network of networks' for 6G convergence. The Internet is built on the premise that everything is permitted everywhere. The end-to-end principle [28] is important but often abused for practical reasons, which states that the endpoints should look after themselves. Industrial or private networks are built on the principle that only explicitly permitted devices and activities are allowed, and networks are partitioned or firewalled into different trust zones. This difference means the optimal architecture for delivering Internet services differs fundamentally from the optimal architecture for delivering a digitalised society and industry.

Although most of the traffic on a 6G network of networks will be IP packets, 6G convergence must go beyond or above the IP network layer. MEC, hyperscaler and AI interworking need to be addressed. There needs to be a trust architecture that works across many network technologies and devices. IP alone does not deliver that. In the words of Professor Fettweis [29]:

> *Trustworthiness for 6G is key. It comprises: Privacy, Security, Integrity, Resilience, Reliability, Availability, Accountability, Authenticity, Device independence.... A new layer in network operations should be included, as trustworthiness will be a process not only for designing systems but must also be guaranteed during services. This is a grand challenge for electrical and computer engineering.*

5G considers trust to decrease further away from the core network functions, as shown in Figure 9.12. For 6G convergence, a more sophisticated trust model is required, especially if the 6G core is distributed and other networks have different trusted authentication methods. Section 9.3.1 describes the issues with existing standards for 5G convergence.

Privacy-enhancing technologies, MEC interconnect, multi-cloud, sensing, federated learning, in-network computing and digital twins are emerging

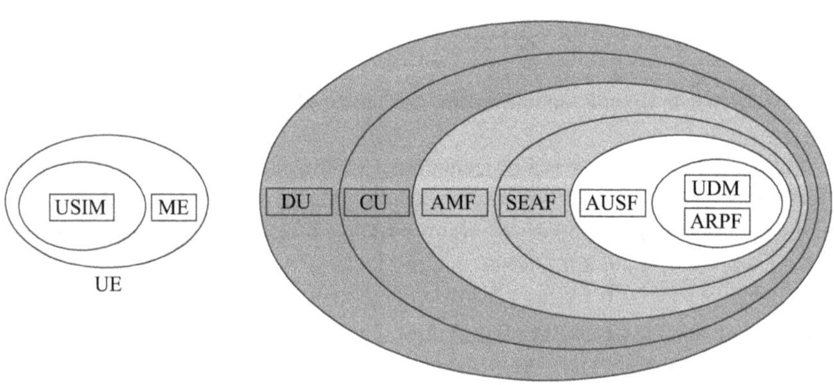

UDM=Unified Data Mgt.. ARPF=Authentication Repository Processing Fn. AUSF=Authentication Server Fn. SEAF=Security Anchor Fn. AMF=Access control & Mobility Fn. CU=RAN Centralised Unit. DU=RAN Distributed Unit. ME=Mobile Equipment USIM=Universal Subscriber Identity Module. UE=User Equipment

Figure 9.12 5G trust model

technologies that play a role in 6G convergence. Many of these new technologies are nascent and evolving, and some are not yet widely researched. For example:

- Digital twins as a service (DTaaS) is valuable, but a digital twin is not an off-the-shelf software package or application. Digital twins require significant integration work across an organisation's IT and operational technology footprints. A 6G convergence initiative should provide the abstractions necessary to automate the integration work required to create a DTaaS solution. A DTaaS solution would allow smaller organisations to exploit digital twins more quickly to improve efficiency while speeding up the development of digital twins in larger organisations.
- Connected intelligence is a nascent concept, let alone a well-developed technology. Semantic communications, as seen in Chapter 13, may be an appropriate approach to connecting multiple AIs to optimise the effectiveness of global tasks.

6G convergence crosses the areas of interest and responsibilities of many Standards Development Organisations (SDO). SDOs are well versed in liaising to create specific standards, but all involved need to buy into the revolutionary 6G convergence vision. Geopolitical tensions may also prevent achieving the global standards required for 6G convergence.

9.7.2 6G converged technology survey

This section briefly examines existing technologies that may be candidates for 6G convergence. The Internet is the most ubiquitous 'Network of Networks' built upon the 'IP over Everything'[2] principle. It works well most of the time. Most 4G and 5G traffic is from consumers connecting to the Internet. However, the Internet has some significant issues as the 6G 'Network of Networks' for Industry 4.0 applications:

- It is a best-effort service, so there are no service level guarantees or QoS, especially across multiple peer Internet service providers.
- It cannot differentiate between applications.
- It is not secure, especially for low-power IoT devices that cannot defend themselves.

Nevertheless, some multinational corporations have built their corporate networks over the Internet using encrypted tunnels. A tunnel encapsulates the corporate traffic to transport it securely across the Internet. Software-defined wide area networks (SDWAN) is a technology *du-jour* for automating the configuration and operation of secure networks over the Internet. Although the corporate traffic is encrypted, the corporate Internet gateways are subject to denial of service or hacking attacks from anywhere on the Internet. It can perform well most of the time

[2]Vint Cerf's joke was to hand out 'IP over Everything' T-shirts.

but has the fundamental weakness of using the Internet. For these reasons, it does not meet the challenges outlined by Professor Fettweis.

Fixed virtual private networks, typically using MPLS [30] networks operated by fixed network operators, address many of the concerns of using the Internet. Corporations on the same network are isolated, typically using MPLS. Service providers implement QoS and Service Level Agreements (SLA). Multiple fixed network operators can peer, and the peerings can support QoS and SLAs. VPN operators often support using cellular networks as a backup access mechanism in the case of the failure of fixed access connections. MPLS can be used to build extranets, a requirement for Industry 4.0. The main issue with fixed VPNs, or MPLS, as a 6G 'Network of Networks' is that they are statically configured and are not dynamic so that a cellular network user can roam to other MNOs.

Section 9.3.1 explains how gateways connect non-3GPP access mechanisms to a 3GPP core network. The main issues with FMC gateways are the complexity created by the variety of gateways and the split of standards across 3GPP, BBF and CableLabs. It is not clear what makes a gateway trusted. The standards assume 5G signalling capabilities in the user equipment, which may not always be accurate, especially for IoT devices. Non-3GPP access gateways are not widely in operation so there is little operational experience to call upon for 6G.

6G convergence goes beyond the 'Network of networks' concept, and cloud interconnect and interworking must be considered. Chapter 11 describes the work of 3GPP, GSMA Operator Platform and ETSI MEC to address the interworking of mobile edge computing. Figure 9.13 shows the complexity of aligning these standards, which are still in development and not widely implemented.

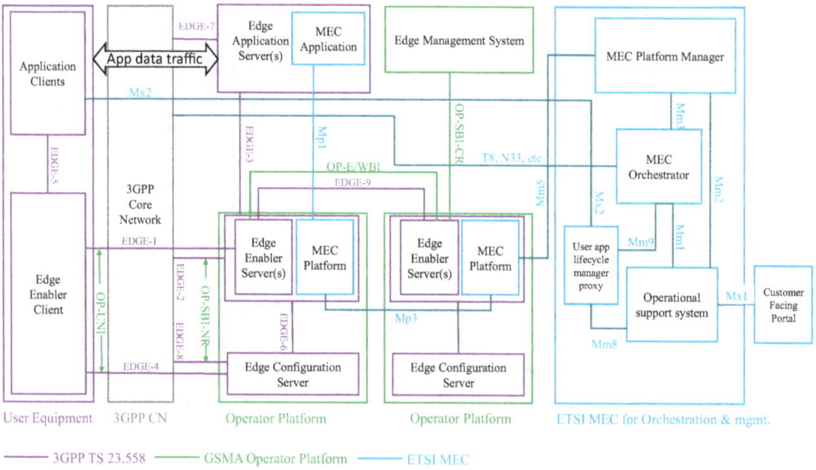

Figure 9.13 3GPP TS 23.558, ETSI MEC and GSMA OP specifications compared

Multi-cloud operation using Infrastructure as a Service (IaaS), as seen in Chapter 11, from multiple hyperscalers, is a more mature technology. Corporations have adopted a multi-cloud approach for resilience and commercial leverage. Multi-cloud technology depends on another layer of cloud abstraction, often provided by a third party, such as a cloud broker. Some open-source cloud management systems, such as Kubernetes, can manage multi-cloud IaaS aspects. The Distributed Management Task Force (DMTF) [31] has created a standardised Cloud Infrastructure Management Interface ISO/IEC 19831:2015 [32] that, in theory, is a standard interface to Infrastructure as a Service, but it is not widely implemented.

MEC and hyperscaler clouds will inevitably have to cooperate and work together, but it is unclear how the interfaces and abstractions can be standardised across all MNOs and hyperscalers.

Cellular networks alone cannot do everything; for example, they are losing IoT market share, often do not work inside buildings, cannot replace industrial Ethernet and only cover a minority of the globe. Yet they are the only grouping via 3GPP with a vision for the digitalisation of society and industry, and they have links to other standards organisations as described in Chapter 6.

9.7.3 Architecture principles

Creating a 6G convergence architecture is significant even before considering the work required to achieve a global agreement on standards. In this section, we propose some architectural principles for consideration.

For Internet services, reducing the amount of times the traffic is touched or processed by network equipment reduces the cost per bit. The Internet starts with the premise that everything is permitted everywhere and there should be no 'middleboxes'. Therefore, we propose that 5G core functions associated with policy control and implementation, e.g. PCF and some UPF capabilities, are not required for 6G Internet services. If all Internet service tariffs are flat rate, this simplifies billing and network metering functions. Internet services are best efforts; therefore, QoS policy and implementation are not required.

However, non-Internet or VPN or converged services for the secure digitisation of society and industry require sophisticated policy control and implementation and per-application QoS with SLA. All non-Internet traffic is subject to strict enforcement of connectivity and QoS policies.

The bifurcation of Internet and VPN traffic seems anti-convergent, but recognition of their fundamental differences is required to optimise both architectures. A compromised convergence adds excess cost to delivering Internet traffic and limits the value-added services 6G operators can provide to digitise society and industry.

Infrastructure providers/functions and authentication providers/functions should be separated. Multiple and distributed methods of authentication should be allowed. However, the authentication methods must be approved as secure and trustworthy.

The 6G core could be distributed, stateless, or 'core-less'. Convergence is a factor in re-engineering the 5G core alongside the need to make it more robust, scalable, and efficient.

All non-Internet traffic should use the zero-trust architecture. The 'zero-trust architecture' is a security framework used by many enterprises operating on the principle that every device, user and application must be authenticated for a particular purpose regardless of whether they are inside or outside of the enterprise's network. The zero-trust architecture deprecates the assumption that everything inside the enterprise's network can be trusted. It adds another layer of defence in case attackers breach corporate firewalls and protects against insider attacks.

All non-Internet traffic handling infrastructure, which includes software and user equipment, should be attested. Any unauthorised changes should be detected and isolated. The compromise with this principle is fragility; for instance, if a device runs version 1.11.7 software instead of version 1.11.8, the consequence of isolating it would be greater than allowing it to continue in service. Some fuzziness needs to be allowed in the attestation.

Common abstractions are required for networks, QoS, in-network computing, gateways, cloud infrastructure, common cloud and AI services. Common abstractions mean there is a common language for APIs, configuration and management. Many common abstractions are available, e.g. from the TMForum, IETF, and DMTF. The challenge is to agree on the many common abstractions to adopt.

Layered networks [33] should be recognised. Not to be confused with the old OSI 7 layered model, a 'layer network' represents the collection of access points of the same type that can transfer information, e.g. all the devices connected to the same optical network, the same Wi-Fi access point, the same MNO, a 5G network. It is not all built on the Internet, nor will any organisation own the infrastructure from the connected AIs down to the ducts in the ground. Mutual authentication is required between the layered networks. Mutual authentication is also required between the user equipment and the network the user equipment connects to. For example, this prevents a Wi-Fi hub from being plugged into the wrong Ethernet or an unauthorised Wi-Fi hub plugged into the organisation's private network.

9.7.4 Architecture example

This section outlines our example of a revolutionary approach to 6G convergence focused on building trustworthiness. This example architecture could be used for applications where trust is essential, such as connecting AIs automatically optimising production lines across organisations.

Figure 9.14 illustrates an example of a non-Internet 6G convergence solution. Trust in non-3GPP networks, clouds and applications is the key[3] to 6G convergence. This example shares much with the 3GPP architecture, using mutual authentication between the user equipment and the network and with a centralised global authentication server. The main difference is that the example separates the authentication

[3]No pun intended.

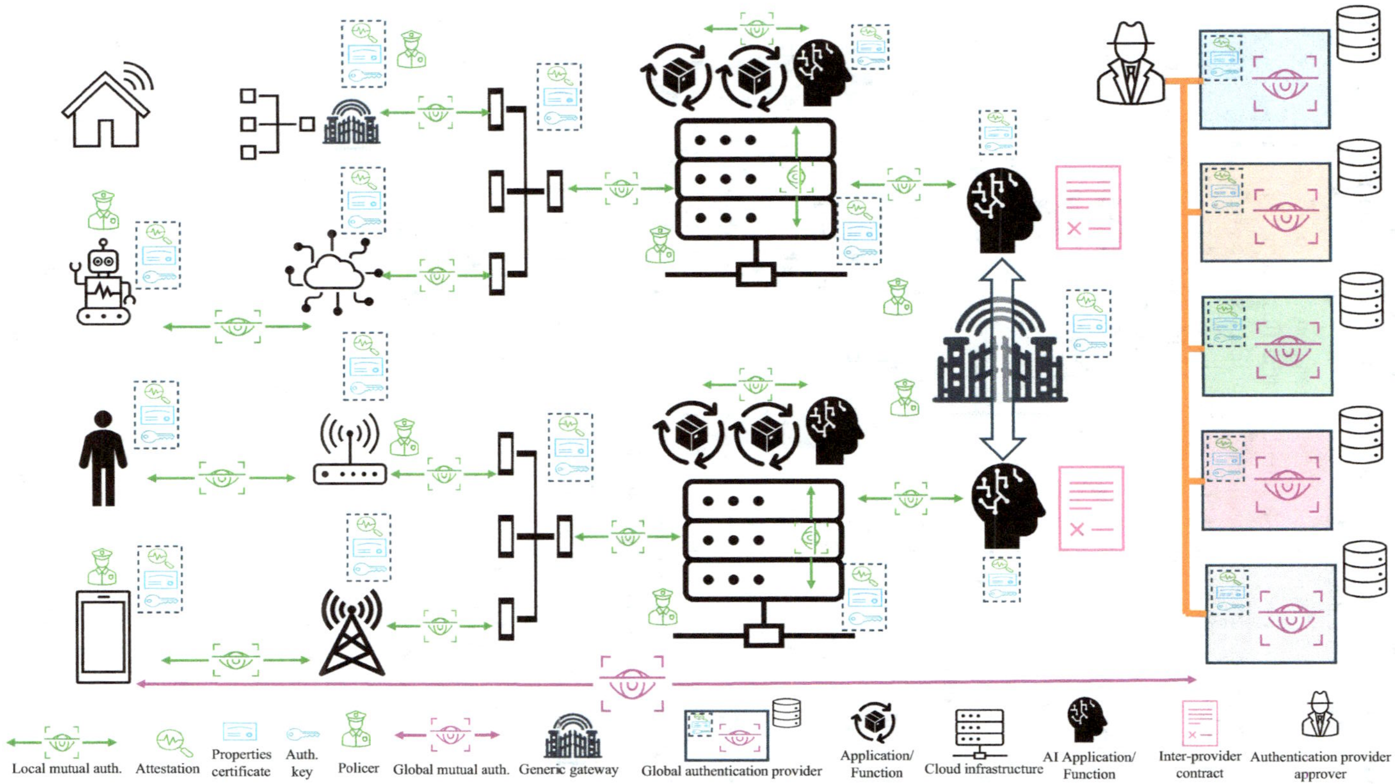

Figure 9.14 Example 6G convergence non-Internet solution

provider from the infrastructure provider, which is a step further than separating the security anchor function (SEAF) and the access management function in 3GPP. In this example, multiple authentication providers may use different authentication methods, although there is a need for a process to approve authentication methods as secure and trustworthy. The other difference is that this example assumes UE does not need a tamper-proof SIM; attestation of the UE hardware and software, supplemented with trusted execution environments where necessary, provides tamper-proofness. Mutual authentication extends to network and cloud infrastructure. A secure generic gateway is used when an end device cannot be attested or run cryptographic authentication methods, e.g. a simple IoT device. The gateway is generic because it conforms to a new common gateway abstraction even though its network interfaces may be specialised. Each capable UE and network device has a digital certificate of its properties, including details about device ownership, location (if fixed), vendor, installer, physical connection identifiers or network supplier. System-wide audits and attestation require this information.

QoS and connectivity require traffic policing. Policing is pushed as close to the edge as possible for scalability. Fully attested UE can be trusted to police themselves according to policies delegated to them. The first network device or gateway the unattestable UE connects through performs the policing. In the other direction, all traffic coming into the network from peers, other network domains or providers, is also policed.

Contract negotiation between network providers is automated using AI-powered multi-agent reinforcement learning (MARL). This allows a much finer granularity more dynamic optimisation of contracts and services than human-negotiated contracts. The contract negotiating AIs may be given requirements by AIs proxying for application requirements.

Semantic communication methods may optimise overall communication and improve task effectiveness.

Much of the fundamental technology to build this architecture exists today. The bulk of the development required is agreeing standards and common abstractions. Dynamic attestation, MARL for network contract negotiations and semantic communications to optimise tasks across organisations requires further research and development.

In this architecture example, a significant issue is the unlicensed spectrum used by Wi-Fi, and many wireless IoT protocol networks prevent the delivery of guaranteed QoS due to interference from other unlicensed users.

9.8 Conclusion

6G convergence is not about the old FMC paradigm intended to give seamless services to customers using cellular and Wi-Fi networks, which is no longer required because UE handles this now. 6G convergence is about breaking out of the mobile stove pipe and recognising that 6G is not the only source of intelligence, cloud and communications services. 6G must provide global service coverage, be trustworthy and sustainable and offer extreme experience across a network of networks.

Future applications will require more than a simple IP connection. They need MEC, Cloud and AI/ML, with interworking with hyperscalers. 6G convergence requires a new mindset. All stakeholders must discuss how openness and convergence create value for everyone involved in 6G technology. From a technical standpoint, the convergence of 6G must be designed to allow for tussles [3] between stakeholders.

6G convergence has a broad scope, and existing Internet, IP, VPN, SDWAN, networks, cloud interworking and FMCS are insufficient. We have proposed architecture principles to establish trustworthiness and standard abstractions, of interfaces and services, that are essential for 6G convergence.

Wi-Fi and small cells or femtocells could be used to build inside-out networks to support indoor customers primarily, but as a bonus, give some outdoor coverage that is more energy and cost-efficient than today's model of outside-in networks using macro towers. Studies have shown that inside-out networks cover a surprisingly large area with as little as one in ten properties sharing small cells. This creates an unprecedented opportunity for new players and new service offerings for existing players in the 6G value chain.

Many of the likely indoor locations, such as offices, shopping malls, sports stadiums, etc., are owned and controlled by various non-telco organisations. Neutral hosting is the most promising solution for these organisations to leverage their estate while offering 6G services to users from several 6G providers or MNOs. There is already significant activity in 5G, typified by 5G private network owners signing agreements with MNOs. However, current standards need development to allow cellular and Wi-Fi neutral hosting to deliver 6G services across various locations.

Existing FMC standards have many issues; evolving these for 6G would not only be insufficient, but it would prolong the life of a dated and failed FMC paradigm. Instead, we recommend a revolutionary 6G convergence approach where MNOs must adopt a 6G convergence or network of networks approach to digitalise society and industry fully and avoid being mobile siloed. It makes excellent commercial sense for MNOs.

Revolutionary 6G convergence is not the Internet nor IP alone; it must go beyond and above the IP layer to address trustworthiness, interconnecting and interworking clouds, intelligence and digital twin services.

Trustworthiness of 3GPP and non-3GPP networks is key. Networks using different authentication methods need to be trusted. We propose a trust architecture that includes the attestation of software, hardware, network components and service configuration. Trusted execution environments on UE and in clouds may enable policy policing to be delegated to the UE and removed from the network, reducing the cost of network equipment.

A complete set of standard abstractions of network services, compute infrastructure, and cloud and AI services is required to integrate diverse services across many providers. Many of these abstractions have been standardised, for example, in the TMForum, DTMF and ITU, but 6G convergence requires the adoption of these across multiple types of service providers.

Creating a 6G convergence architecture is significant even before considering the work required to achieve global agreements. However, the rewards for mobile network operators more than justifies the work required.

References

[1] News R. Total IoT connections to surge, but cellular IoT to lose share with 2G/3G switch-off. [Online]. 2023 [cited 2023 June 22]. Available from: https://www.rcrwireless.com/20230622/internet-of-things-4/total-iot-connections-to-surge-but-cellular-iot-to-lose-share-with-2g-3g-switch-off.

[2] Mason A. Network convergence will be essential if 6G is to address the shortcomings of 5G. [Online]. 2022 [cited 2022 July 22]. Available from: https://www.analysysmason.com/research/content/articles/6g-network-convergence-rma08/.

[3] Clark D, Wroclawski J, Sollins K, and Braden R. Tussle in cyberspace: defining tomorrow's internet. In *2002 Conference on Applications, Technologies, Architectures, and Protocols for Computer Communications*; 2002. pp. 347–56.

[4] Bigelow SJ. https://www.techtarget.com. [Online]. 2013 [cited 2024 Mar 27]. Available from: https://www.techtarget.com/whatis/definition/Hot-Spot-20-HS-20.

[5] Wi-Fi alliance. https://www.wi-fi.org/. [Online]. 2024 [cited 2024 Mar 27]. Available from: https://www.wi-fi.org/discover-wi-fi/passpoint.

[6] Lemes M, Alberti A, Both C, Júnior A, and Cardoso K. A tutorial on trusted and untrusted non-3GPP accesses in 5G systems—First steps toward a unified communications infrastructure. *IEEE Access*. 2022; 10: 116662–85.

[7] Bajracharya R, Shrestha R, Jung H, and Shin H. Neutral host technology: The future of mobile network operators. *IEEE Access*. 2022; 10: 99221–34.

[8] Cisco. https://www.cisco.com/. [Online]. 2020 [cited 2024 Mar 27]. Available from: https://www.cisco.com/c/en/us/solutions/collateral/executive-perspectives/annual-internet-report/white-paper-c11-741490.html.

[9] Carson S and Davis S. https://www.ericsson.com/. [Online]. 2023 [cited 2024 Mar 27]. Available from: https://www.ericsson.com/49dd9d/assets/local/reports-papers/mobility-report/documents/2023/ericsson-mobility-report-june-2023.pdf.

[10] 5G Americas. https://www.5gamericas.org/. [Online]. 2016 [cited 2024 Mar 27]. Available from: https://www.5gamericas.org/wp-content/uploads/2019/07/SCF191_Multi-operator_neutral_host_small_cells.pdf.

[11] Yunas S, Niemelä J, Valkama M, and Isotalo T. Techno-economical analysis and comparison of legacy and ultra-dense small cell networks. In *39th Annual IEEE Conference on Local Computer Networks Workshops*; 2014. pp. 768–76.

[12] Kawade S and Nekovee M. Broadband wireless delivery using an inside-out TV white space network architecture. In *2011 IEEE Global Telecommunications Conference (GLOBECOM 2011)*; 2011. pp. 1–6.

[13] Kawade S, Abhayawardhana V, Hodgkinson T, and Wisely D. Cluster based channel allocation for Public WLANs. In *IEEE 65th Vehicular Technology Conference (VTC2007)*; 2007. pp. 2879–83.

[14] Sharma R. https://www.thefastmode.com/. [Online]. 2022 [cited 2024 Mar 27]. Available from: https://www.thefastmode.com/technology-solutions/27689-samsung-powers-comcast-s-5g-network-using-cbrs-600mhz-spectrum.

[15] Sangam P. https://www.qualcomm.com. [Online]. 2013 [cited 2024 Mar 27]. Available from: https://www.qualcomm.com/news/onq/2013/02/getting-opportunistic-small-cells.

[16] Gottschalk A. https://michigan.it.umich.edu/. [Online]. 2022 [cited 2024 Mar 27]. Available from: https://michigan.it.umich.edu/news/2022/06/28/u-m-first-in-nation-to-offer-wifi-6e-universitywide-accomplishment-is-an-investment-in-the-future/.

[17] DeGrasse M. https://www.fiercewireless.com/. [Online]. 2021 [cited 2024 Mar 27]. Available from: https://www.fierce-network.com/private-wireless/caribe-royale-deploys-cbrs-neutral-host-network.

[18] Dano M. https://www.lightreading.com/. [Online]. 2023 [cited 2024 Mar 27]. Available from: https://www.lightreading.com/private-networks/can-mocn-save-private-wireless-.

[19] Ericsson. https://www.ericsson.com/. [Online]. 2023 [cited 2024 Mar 27]. Available from: https://www.ericsson.com/en/press-releases/2023/6/ericsson-and-neutral-host-proptivity-deploy-shared-indoor-5g-ran.

[20] Kibria M, Villardi G, Nguyen K, Liao W, Ishizu K, and Kojima F. Shared spectrum access communications: A neutral host micro operator approach. *IEEE Journal on Selected Areas in Communications*. 2017; 35(8): 1741–53.

[21] Ahokangas P, Moqaddamerad S, Matinmikko M, *et al.* Future micro operators business models in 5G. *The Business & Management Review*. 2016; 7(5): 143.

[22] Bajracharya R, Shrestha R, Jung H, and Shin H. Neutral host technology: The future of mobile network operators. *IEEE Access*. 2022; 10: 99221–34.

[23] 5G Essence. https://www.5g-essence-h2020.eu/. [Online]. 2024 [cited 2024 Mar 27]. Available from: https://www.5g-essence-h2020.eu/.

[24] 5g City. https://www.5gcity.eu/. [Online]. 2020 [cited 2024 Mar 27]. Available from: https://www.5gcity.eu/.

[25] European Commission. https://cordis.europa.eu. [Online]. 2017 [cited 2024 Mar 27]. Available from: https://cordis.europa.eu/project/id/671596.

[26] Holyoak S. https://www.linkedin.com/. [Online]. 2021 [cited 2024 Mar 27]. Available from: https://www.linkedin.com/pulse/new-jots-nhib-specification-good-news-indoor-wireless-stuart-holyoak/.

[27] Mobile UK. https://www.mobileuk.org. [Online]. [cited 2024 Mar 27]. Available from: https://www.mobileuk.org/jots.

[28] Saltzer J, Reed D, and Clark D. End-to-end arguments in system design. *ACM Transactions on Computer Systems*. 1984; 2(4): 277–88.

[29] Fettweis G and Boche H. On 6G and trustworthiness. *Communications of the ACM*. 2022; 65(4): 48–9.

[30] Wikipedia. https://en.wikipedia.org. [Online]. 2024 [cited 2024 Mar 25]. Available from: https://en.wikipedia.org/wiki/Multiprotocol_Label_Switching.

[31] DMTF. https://www.dmtf.org. [Online]. 2024 [cited 2024 Mar 25]. Available from: https://www.dmtf.org/about.

[32] ISO. https://www.iso.org. [Online]. 2021 [cited 2024 Mar 25]. Available from: https://www.iso.org/standard/66296.html.

[33] ITU-T. https://www.itu.int. [Online]. 2001 [cited 2024 Mar h25]. Available from: https://www.itu.int/rec/T-REC-G.805-200003-I/en.

Chapter 10
Non-terrestrial networks (NTNs)

Heavier-than-air flying machines are impossible.

– Lord William Thomson Kelvin, British mathematician and physicist, president of the British Royal Society, 1895

10.1 Introduction

In this chapter, we move away from terrestrial networks (TNs) covering cellular and Wi-Fi to look at non-terrestrial networks (NTNs). NTNs include satellites, drones and high-altitude platforms (aircraft and balloons). There has been much debate about the possible role of satellites in 5G and 6G [1–4], and the first part of the chapter examines the different types of satellites based on their orbital height and associated coverage, latency and capacity. There has been a significant investment in low Earth orbit (LEO) satellites in recent years, typified by Starlink [5]. These satellites potentially offer a substantial uplift in capacity due to their high numbers and lower latency than higher geostationary satellites. To try and answer critical questions about which applications these satellite constellations will be suitable for, we analyse the coverage, capacity and cost of these LEO systems.

The chapter then examines drones, or uncrewed aerial vehicles (UAV), and high altitude platforms (HAPs). These have also been suggested for various applications, including temporary coverage boosts and rural connectivity. The following section examines how NTNs might connect to cellular networks and describes the evolving standards. The application section explores possible use cases that NTNs might enable in 5G and 6G. There are many challenges for NTNs, and the final section examines these, including the Kessler syndrome, which might pose an existential threat to satellite constellations.

10.2 Satellites

Most satellite books and articles start with a nod to Arthur C. Clarke, who published a paper in October 1945 in the *Wireless World* journal entitled 'Extra-Terrestrial Relays – Can Rocket Stations Give Worldwide Radio Coverage?' This paper is widely considered to have foreshadowed all modern satellite systems.

Sputnik was the very first satellite launched in 1957 [6]. Later, the first successful geostationary (GEO) satellite was Syncom 2, launched in 1963. GEO satellites orbit 35,785 km above the Earth and appear stationary in the sky. This is a consequence of Newton's laws of motion, and it is only at this orbital distance that the satellite's rotation is synchronised with the Earth's. GEO satellites have the advantage that terrestrial users can erect small, cheap, stationary antennas that only require initial alignment. They are frequently used for satellite TV. Moreover, each satellite has a very sizeable footprint on the Earth that does not change, meaning that no inter-satellite links (ISLs) are necessary, unlike LEO. Table 10.1 compares GEO and LEO satellites with TNs.

The advantage of geostationary satellites is their coverage. For example, the global positioning system (GPS) requires only about 32 satellites to cover the surface of the Earth, while a single satellite can cover the whole UK for broadcast TV. The major drawbacks with GEO satellites, in the context of 6G at least, are their limited capacity and high latency for communication services. The small number of these satellites limits the bandwidth per square metre on the Earth's surface, while the latency is simply the time of flight for GEO systems to send a radio signal from the Earth to and from the satellite: 480 ms at the equator and greater at higher latitudes. These latencies rule out interactive services such as voice or two-way video with GEO satellites. However, such satellites are ideal for broadcast TV, weather and navigation, for which they are extensively used. Fixing the issues for 6G requires using lower-orbit satellites where the coverage area is proportionally smaller than geostationary, allowing significantly higher capacity for the same spectrum use. At the same time, the lower orbits reduce the latency significantly. Medium Earth orbit (MEO) satellites were introduced in 1962/63 with Telstar 1 and 2, respectively. Telstar 1 was the first TV satellite relay – with pictures sent across the Atlantic to Goonhilly in Cornwall. MEOs have an orbit between 2000 and 35,785 km above the Earth. However, MEOs (and LEOs) require earth stations to track, align to and hand over between the satellites. MEO satellites are launched and aligned in constellations to provide global coverage. In the 1990s, satellite operators like Iridium and Globalstar offered communication services (such as voice) from space. However, these services struggled commercially [7].

Table 10.1 Simple comparison of TN and NTNs

	TN	NTN
Coverage on Earth (diameter)	0.1–100 km	100–1500 km (LEO) 3500 km (GEO)
Propagation delay	0.67 ms (1 km)	2–20 ms (LEO) 540 ms (GEO)
Path loss	140 dB (100 km cell 2 GHz)	190 dB (GEO at 2 GHz)
Doppler shift	1 kHz (high-speed train 2 GHz)	Max 48 kHz (LEO at 600 km GHz)
Handovers	On UE moving cells	Regular handover for static users with LEO platforms. No handover is needed with GEO

There are no satellites between the LEO and MEO orbital distances as the radiation from the Van Allen belt destroys electronic components.

A new development is the emergence of large-scale LEO constellations. However, LEOs themselves are not new. The downside of LEO constellations is that more satellites are needed to give global coverage and that the tracking, alignment and handover are more complicated. They also have a very short lifespan of 7–10 years [8] due to orbit decay and need to be replaced after this time. GEO, on the other hand, can survive for 100 years theoretically. For LEO, this dramatically adds to the cost of running the constellation. Imagine replacing the complete radio access network (RAN) in a mobile network every 7–10 years, including replacing the sites and towers. Table 10.2 shows some of the (numerous) commercial satellite systems currently in operation and the varied services they provide. It contrasts the different applications across GEO, MEO and LEO systems.

10.2.1 Satellite spectrum bands

The various satellite systems use low-band, mid-band and high-band spectra up to 40 GHz. The Satellite spectrum bands are designated as follows:

- L – 1–2 GHz,
- S – 2–4 GHz,
- C – 4–8 GHz,
- X – 8–12 GHz,
- Ku – 12–18 GHz, and
- Ka – 26–40 GHz.

Operating worldwide requires a globally, or at least regionally, designated spectrum covering the countries in which they operate. This is difficult to obtain in

Table 10.2 Example GEO, MEO and LEO systems – showing the range of current applications (table shows actual satellite numbers in orbit as of 2024)

	Constellation	Orbital height	Number of satellites	Frequency bands	Applications
GEO	Intelsat	35,786 km	52	C, Ku, Ka	Broadcasting, data and mobile comms
GEO	Eutelsat	35,786 km	36	C, Ku, Ka	Video and data services
MEO	GPS	20,180 km	31	L	Positioning and timing
MEO	GLONASS	19,130 km	24	L	Positioning and timing
MEO	Galileo	23,220 km	30	L	Positioning and timing
LEO	Globalstar	1414 km	48	S L	Voice, data and iPhone emergency services
LEO	Orbcomm	720 and 750 km	54	VHF	IoT and M2M services
LEO	OneWeb	1200 km	648	Ku	Broadband
LEO	Starlink	540–570 km	6000	Ku/Ka	Broadband

the lower bands, with intense competition from many other users (see Chapter 7 for more details). Most of the L band is used for non-satellite applications, and only a very narrow part of it is available for location and timing satellite services, such as GPS. All GPS satellites broadcast on at least two carrier frequencies: L1, 1575.42 MHz, and L2, 1227.6 MHz. The newer satellites also broadcast on L5 at 1176 MHz, but these are extremely narrowband transmissions. Another example of spectrum usage is Globalstar, which uses the L band (1610–1618.725 MHz) for their handset to satellite uplink and the S band (2483.5–2500 MHz) for the downlink.

Many satellites use the higher frequency Ku and Ka bands, with more spectrum available for satellite services supporting higher bandwidth services. However, these spectrum bands might also become targets for cellular use. It is essential to recognise that Ku and Ka bands do not penetrate indoors and offer no service there. Even the lower frequency satellite bands do not generally provide indoor service due to the much more significant propagation loss for satellite systems than a macro network (Table 10.1).

10.2.2 LEO satellite analysis

However, the most recent publicity around satellites has been around the launch, or planned launch, of four prominent LEO constellations. With up to 40–50,000 satellites in the planning stage for the 6G time frame, this is a significant expansion compared to the (typically) tens of satellites in previous constellations (Table 10.2). Starlink is the most advanced, with over 6000 satellites launched in Q1 2024.

LEO satellites are not new, with the first launch in 1957. During the 1980s, they were considered a way of providing narrowband services, but construction and launch costs proved prohibitive. Launching costs have fallen recently, with the Falcon X reusable launcher being a prime example. Together with the development of smaller satellites, this has given rise to new momentum for LEO constellations. Apple iPhones have recently (from iPhone 14) added an emergency SOS service using the Globestar LEO network. This service is described as

> *Emergency SOS via satellite can help you connect with the emergency services under exceptional circumstances when no other means of reaching the emergency services are available. If you call or text emergency services and can't connect because you're somewhere with no mobile and Wi-Fi coverage, your iPhone tries to connect you via satellite . . . .*
>
> *To connect to a satellite, you must be outside with a clear view of the sky and horizon. When you use a satellite connection, the experience differs from sending or receiving a message via mobile [9].*

The service is free for two years after activation, but there is no indication of the pricing in the future. The fact that a satellite transceiver has been integrated into a smartphone at a relatively low cost shows the possibility of using satellites as part of an extended 6G system. It also shows that such services could be supplied by a smartphone vendor rather than an MNO as a virtual operator. They can partner with a single LEO operator for a global service with the LEO satellite band. This avoids many issues associated with 6G cellular use, where spectrum use is fragmented across operators and countries. These LEO satellite services could include text, voice and broadband access.

LEO satellites have several significant disadvantages when compared to GEO and MEO satellites. First and foremost is the short, 7–10-year lifetime. Another problem is that the lower orbit means many more satellites are needed in a constellation to achieve global coverage. Unless there are links between the satellites, as described later, the user and a ground station must be within the satellite footprint (field of view). The relative motion of the satellites gives rise to a significant Doppler shift that has to be mitigated and does reduce radio performance. Unlike geostationary satellites, there are also numerous handovers to be managed as the satellites appear to sweep across the sky to the terrestrial user, travelling at an orbital speed of around 25,000 km/h. It takes 12–20 min for an LEO satellite to go from horizon to horizon, which determines the frequency of handovers between satellites, impacting signalling performance. The upsides of LEO constellations are the much lower latency, simply because they are much lower in orbit and the significant increase in capacity due to the smaller footprints and higher numbers in the significant planned constellations. This section quantifies how much more capacity is added, the latency values, and estimates the capacity.

Four significant systems are planned or have begun launch (summarised in Table 10.3). There are some notable differences between the architectures of these systems. Still, they all connect to a user and a ground station that further links to a terrestrial network for backhaul. One of the difficulties that LEOs face is the need always to have a ground station in their 'field of view'. However, these ground stations have much larger antennae than user equipment, typically 3 m in diameter, and use different frequencies to communicate with the satellite to provide the backhaul connection for all the users. One way to reduce the number of ground stations is to add inter-satellite links (ISL) – using either optical or radio communication systems. That allows satellites to create a mesh network (Figure 10.1). Starlink uses optical ISL links to transmit 42 million GB of data daily in 2024 [10].

The early satellites in a new constellation are launched in polar orbits to achieve global coverage. When this is reached, subsequent satellites are aligned on an inclined orbit to cover the more populated areas of the Earth. This ability to direct satellites to more populated or higher user densities is one advantage of LEO systems. MEO systems such as Globalstar and Iridium have communication latencies of around 40 ms, and a recent study [11] showed that Starlink has a latency that varied (due to changing orbital positions and user location) between 1.8 and 22.8 ms in the Metro Vancouver area.

10.2.2.1 LEO capacity

Various factors determine the capacity of a constellation:

- Atmospheric impacts: The frequencies used are subject to absorption, scattering and diffraction by clouds, molecules, dust and rain.
- Operating frequencies: The frequency dictates the fade model and spot size.

Table 10.3 Comparison of the four large constellation LEO systems

	Telecast	**OneWeb**	**Starlink**	**Amazon**
Current status	Testing is taking place	In operation	Starlink constellation is up and running	Kuiper Phase 1 FFC approved
Operational date	Testing is taking place	In operation but limited coverage	2 million subscribers in 2023	Expected launches in 2024 and testing later that year
Satellites in orbit Q1 2024	198	544	6000	0
Total satellites planned	1671	6372	12,000 (Possible extension to 32,000)	3236
Altitude	1015–1325 km	1200 km	540–570 km	590–630 km
DownLink Satellite to user	17.8–20.2 GHz	10.7–12.7 GHz	10.7–12.7 GHz	17.7–20.2 GHz
DownLink Satellite to gateway	17.8–20.2 GHz	17.8–19.3 GHz	17.8–19.3 GHz	17.7–20.2 GHz
UpLink User to satellite	27.5–30 GHz	14–14.5 GHz	14–14.5 GHz	28.35–30 GHz
UpLink Gateway to satellite	27.5–30 GHz	27.5–30 GHz	27.5–30 GHz	27.5–30 GHz

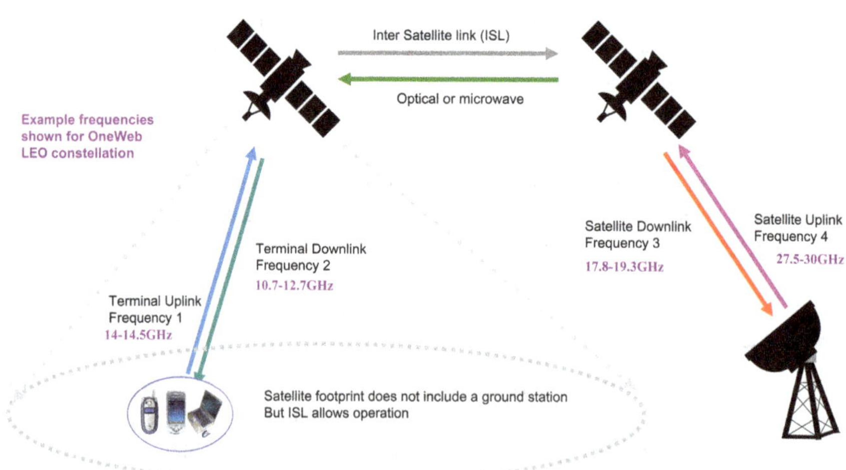

Figure 10.1 Inter-satellite links (and up and downlinks)

- Link budget: Just like a terrestrial network, the link budget has a trade-off that can be used to serve fewer users at higher data rates or more significant numbers at lower rates.
- Satellite orbital configuration: The number of satellites in different orbital inclinations – polar orbits for global coverage and more inclined orbits to cover populated areas.
- ISL: These increase capacity by removing the need for both the user and a ground station to be in the field of view.
- Ground stations: The more numerous the ground stations, the higher the capacity (and cost) and the more chance a satellite connects to a ground station.

Pachler *et al.* [12] have produced a comprehensive model of the four major systems listed in Table 10.2 – using actual and planned orbits with realistic weather models and user location spreads. The modelling was done for two phases of satellite launch, and the number of satellites in each phase is shown in the table. The results are only indicative, and, as noted above, Starlink has already launched more than the phase two modelling numbers. Pachler's findings are summarised in Table 10.4.

Li *et al.* [13] suggest that 28 Tbit/s would allow a constellation to support 1.86 million users with a service rate of 100 Mbit/s. There is also the issue that LEO satellites using the Ka and Ku bands have a minimum footprint of around 350 km. Wales in the UK, for example, might be covered by a single satellite. Table 10.4 shows that, for OneWeb, each satellite has a capacity of about 1.4 Gbit/s. That would allow 114 users to access 100 Mbit/s at the same time. Because users never actually use capacity simultaneously that might support a thousand users overall.

Table 10.4 Capacity of LEO satellite systems – from Pachler [12]

	Telecast	**OneWeb**	**Space X**	**Amazon**
Total constellation capacity for initial deployment No ISL	5 Tbit/s	1 Tbit/s	7.5 Tbit/s	6 Tbit/s
Number of satellites in Phase 1	298	716	1584	578
Total constellation capacity for initial deployment With ISL	6 Tbit/s	1 Tbit/s (ISL in use for phase 1)	7.5 Tbit/s (ISL in use for phase 1)	10 Tbit/s
Number of satellites in phase 2	1671	6372	1408	3236
Total Constellation capacity for phase 2 deployment No ISL	18 Tbit/s	20 Tbit/s	18 Tbit/s	28 Tbit/s
Total constellation capacity for phase 2 With ISL	23 Tbit/s	30 Tbit/s	28 Tbit/s	48 Tbit/s

Still, the population of Wales is 3 million, which gives some idea of what satellites alone can and cannot achieve in supplying broadband to rural areas.

10.2.2.2 LEO coverage

A close look at Table 10.3 will show all proposed LEO constellations communicating with users using frequencies between 14 and 30 GHz. None of these frequencies can penetrate indoors. Even systems that use lower frequencies do not generally work indoors because the link budget for satellites (all the losses between the transmitter and receiver) is at least 30 dB worse than for a typical macro radio link budget (Table 10.1).

10.2.2.3 LEO cost

One crucial factor relating to the role of satellites in 6G is the relative costs between satellite and earth-based services. There is little doubt that the new LEO constellations offer a massive increase in total capacity compared to previous generations of satellites. However, when discussing possible 6G applications that involve a satellite element, it is essential to have a good idea of the estimated cost.

Several factors determine the cost of satellites – some of which have the potential to reduce over time and some that look more likely to increase:

- Satellite replacement costs: LEO satellites typically have a lifetime of 7–10 years and a considerable ongoing cost for their replacement.
- Satellite manufacturing costs – The expense of the satellites themselves.
- Launch costs – Different systems use different launch platforms, and the number of satellites launched from a single rocket can be substantial, with a Space X rocket having a capacity of 60.
- Gateway costs – Gateways need to be provided for both coverage and capacity.

Li *et al.* [13] have created a techno-economic model for LEO constellations that offers excellent insight into the likely range of the satellite price premium over TNs. They present results for 1775 satellites orbiting at 550 km. In their model, they arrange for a subset of the satellites to provide global coverage and then add satellites to serve more populated areas. They assume users are offered a 100 Mbit/s service but that the total capacity is 50 times less than if all users were accessing the service at the same time (which they do not – this is called the overbooking factor and is discussed in the section on 5G techno-economic modelling in Chapter 2). Table 10.5 shows the results of their modelling.

This modelling is close to the number of satellites in the major LEO systems, as shown in Tables 10.2 and 10.3. They suggest a monthly cost of 100 Mbit/s service of around €44. It is important to note that this is just the cost of the satellite element in their model and would not include other operating costs that would need to be added to get the end-user price. These extra costs might include customer acquisition and margin, R&D, spectrum and regulatory issues. The equivalent cost for an urban 5G macro network is around €4–€7 (see Chapter 2 for the details).

This leads to an indicative satellite cost premium of around 6–11 times the typical terrestrial cost. However, there is scope for lowering these costs, including tuning the RF parameters (as related in [13]) and employing reusable and higher-capacity launchers.

Table 10.5 Techno-economic modelling of LEO constellations from Li et al. [13]

	LEO 550-a	**LEO 550-b**
FoV radius	940 km	940 km
Total number of satellites	1775	1775
Average throughput per satellite	10 Gbit/s	14 Gbit/s
Throughput (constellation)	17.75 Tbit/s	28.85 Tbit/s
Effective throughput	2.7 Tbit/s	3.7 Tbit/s
Total number of users	1,331,250	1,868,750
Launch cost (all satellites)	€1.6 billion	€1.6 billion
Manufacture cost (all satellites)	€1.3 billion	€1.3 billion
Number of gateways	222	311
Total cost Earth segment	€289 M	€404 M
Yearly cost (constellation)	€630 M	€636 M
Cost per year user terminal (per user)	€523	€340
Cost per month per customer	€43.6	€28.4
Cost per Mbit/s	€0.44	€0.28

As a further quick reference, highlighting that cost is not the same as price: 'Starlink costs anywhere from $90 per month for Starlink Standard with standard data in low-capacity areas to $5,000 per month if you want the Starlink Mobility 5TB of Priority Mobile data. Starlink also requires upfront equipment fees ranging from $599 at the lowest to $2,500 at the highest' [14].

One area where satellites will have an advantage is that they generate electricity from solar panels, which is free. This is a significant cost for TNs: Ericsson estimates the global energy cost of running mobile networks will exceed $25 billion annually. GSMA notes that energy consumption typically accounts for 15–40% of telcos' network Opex [15].

In summary, LEO systems offer much lower capacity when compared to TNs but can serve areas where cellular coverage is uneconomic. They can provide similar speeds and latency but have much higher operating costs. This will influence their use cases and limit them to areas with no mobile coverage. They do compete directly with mobile.

10.3 Non-satellite platforms – drones, aircraft and balloons

This category generally covers UAVs (basically drones) and HAP (basically balloons/zeppelins/aircraft). HAPs have been suggested in 5G to cover rural areas that do not get good terrestrial or mobile Internet access. British Telecommunications

(BT) and Stratospheric Platforms recently announced a trial of 5G coverage from an aircraft [16] – although initial tests appear to be from a fixed tower. It is claimed, ‘The technology is capable of delivering up to 150 Mbit/s across areas as big as 15,000 km^2 through 500 individually steerable beams’. BT intends to ‘connect unserved rural areas and enable exciting new use cases for private users’ [16]. A previous test in the Middle East involved an aircraft at 45,000 ft. In another test: ‘SoftBank and the government of Rwanda conducted a 5G communications trial from a high-altitude pseudo-satellite or platform station (HAPS), an uncrewed aerial vehicle (UAV) prototype. For the test, the pair used SoftBank’s proprietary 5G communications payload in the stratosphere, which ranges from four to 12 miles above the Earth’s surface to around 31 miles’.

There are proposals for drones to connect to a 5G network at a lower altitude. This is a different architecture from HAPS and satellites in that the drones do not (generally) provide service for other users but are the end terminal. The GSMA says, ‘... 5G drones will help deliver digital twins. Autonomous drones could save infrastructure owners tens of billions of dollars a year’ [17].

A 5G/6G connection to a drone could allow remote piloting and integration with TNs (e.g. in security or inspection). It would also allow the proper enforcement of flight rules (e.g. no-fly zones at airports and prisons). Some proposals for larger drones offer a relay for 6G services to end users. These might temporarily provide enhanced capacity, such as at concerts and sporting events. Gharib [18] extensively details the operation of drones over existing networks and describes many possible 6G applications.

10.4 3GPP standards for NTN configurations

As of Q1 2024, satellite voice and broadband services require a specialised handset or dish. The protocol used between this user equipment and the satellite is proprietary. It is expected that some user equipment will begin to use 3GPP standards as 5G support is added in the next few years. 5G standards have been extended to allow satellite operation from Release 17 onwards (Figure 10.2). This section looks at some of the architectural options that have been proposed.

In Figure 10.2, the satellite (HAP or drone) is a transparent relay for the 5G service. Other than possibly using a new transceiver and frequency band, the terminal would not require new protocols or specific software. Starlink is also equipping some satellites to receive LTE signals from handsets. The satellites are equipped with a large antenna for LTE service to receive weak signals from a standard LTE handset at the LTE frequencies, mid-band. The LTE signal is then transparently relayed to a terrestrial eNodeB. The signal is sufficiently weak that it will not provide service indoors. The satellite connection is only used when there is no terrestrial coverage, so there is no need for frequency planning between terrestrial and non-terrestrial transmissions, and the service is planned to be country-specific since LTE frequencies differ in different countries and regions. This is possible with LEO satellites because their narrow beams and small footprints allow

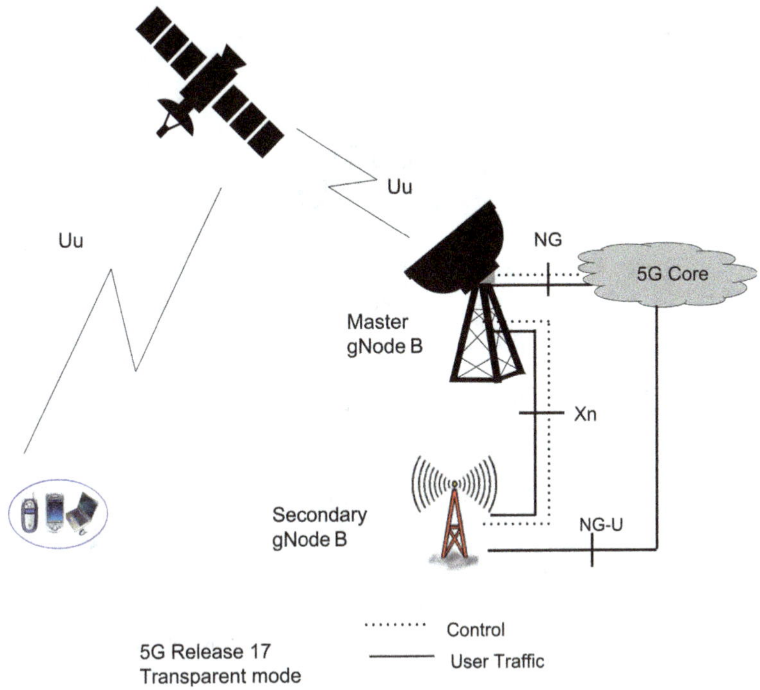

Figure 10.2 5G NTN transparent connect as introduced in Release 17

some geographic selectivity. Indeed, they currently do not provide service in countries such as Iran and North Korea. The MNO that owns the LTE spectrum must also reach an agreement with the satellite provider to use it in relevant regions since they are the only licensed users of that spectrum. There is a very different economic model in which a handset manufacturer buys capacity directly from a constellation and then includes radio transceivers for the appropriate frequency in each handset. These issues would be significantly simplified if the spectrum used were globally available. We term such a manufacturer as a Global Virtual Satellite Operator (GVSO), and this concept is further discussed in Chapter 16.

One announced LTE service is by T-Mobile US:

> *Initially, T-Mobile US will be able to offer text messaging via Direct to Cell with voice and data to follow 'in the coming years.' [19]*

5G advanced will add new NTN capabilities and enhance the overall performance, but 6G is expected to allow some elements of the RAN to be located within the NTN node. Figure 10.3 shows the entire gNode B located onboard a satellite, but there are also proposals to split the gNode B, as can happen in 5G (see Chapter 2 for details). This is known as a regenerative mode and offers much more flexibility and capabilities to the network. These would include moving from a three-frequency repeat pattern for the footprint coverage to a fractional reuse system, as is used in 5G and increases efficiency by up to three times. It would also

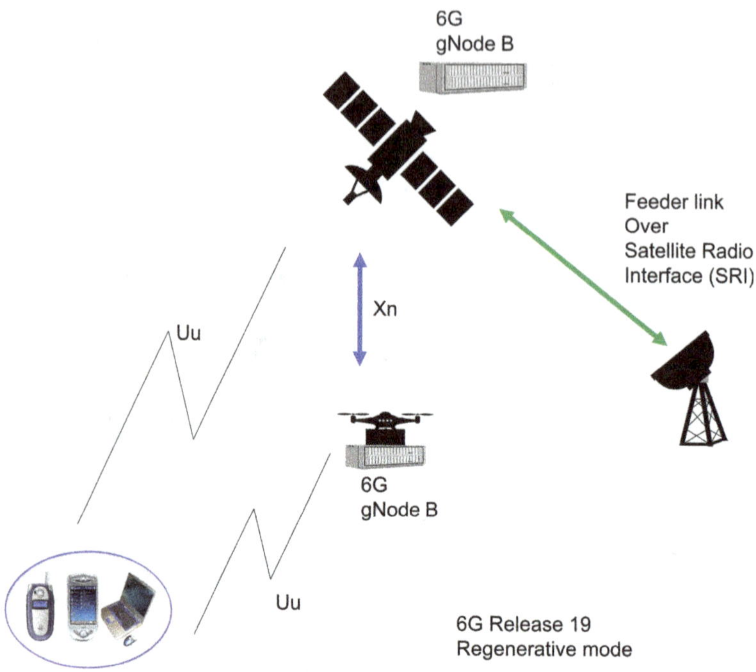

Figure 10.3 NTNs in regenerative mode – 6G

improve channel estimation, making it possible to improve Doppler compensation and higher modulation rates. Mobile edge computing (MEC) capabilities may also be located on NTN terminals, drastically reducing latency for some services, improving security and significantly increasing the system's overall efficiency.

Figures 10.4–10.8 show the architectural options that 6G NTNs can unlock. There are four options based on whether the NTN is a user, a relay for end-users, a backhaul link or a base station. These different configurations make NTNs very flexible. Figure 10.5 shows the NTN system's various elements combined to offer users 5/6G services directly.

Figure 10.5 shows the NTN being used as backhaul for an isolated segment of a 5/6G network. This could be a remote factory or campus with poor cellular coverage. It might also be an important site that has both terrestrial and non-terrestrial backhaul for ultra-high reliability.

Figure 10.6 illustrates the hierarchical nature of multi-element NTNs, with both HAPs and UAVs connected using a 5/6G interface for control and backhaul. This architecture would control drones remotely beyond the line of sight of their operator. In some countries, drones must be operated with this line of sight. However, for commercial services, this is likely to change for 6G. Finally, Figure 10.7 shows elements of the radio network onboard NTN elements. This could significantly reduce time-sensitive processing times for advanced 6G services and reduce the traffic on links to and from the gateways.

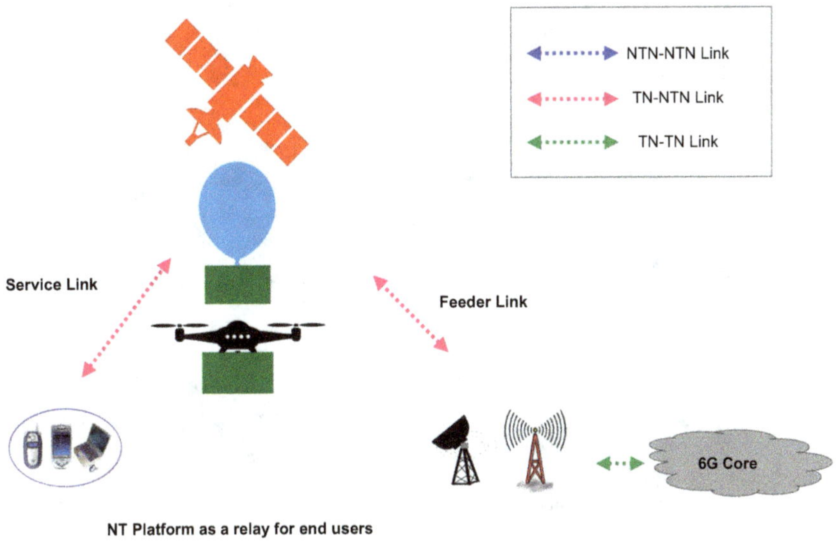

Figure 10.4 NTN architectural options in 6G – NTN as a relay for users

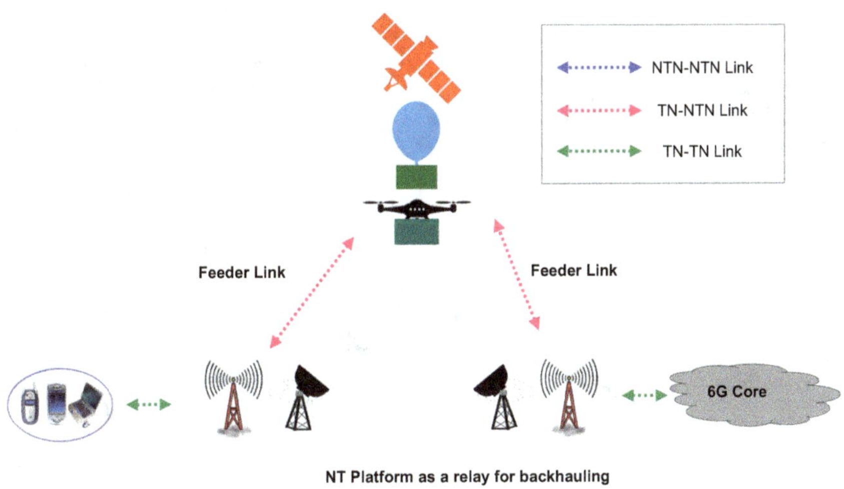

Figure 10.5 NTN architectural options in 6G – NTN as a platform for backhauling

6G NTNs are very suited to temporary situations and areas where traditional cellular networks are not economical. They can provide coverage in areas where macro networks do not extend. But, most valuable of all is the ability to creature a multi-dimensional multilayer architecture (Figure 10.8) combining all of the elements – satellites, HAPs and UAVs to deliver services that TNs will never offer, and we will look at some examples of these in the next section.

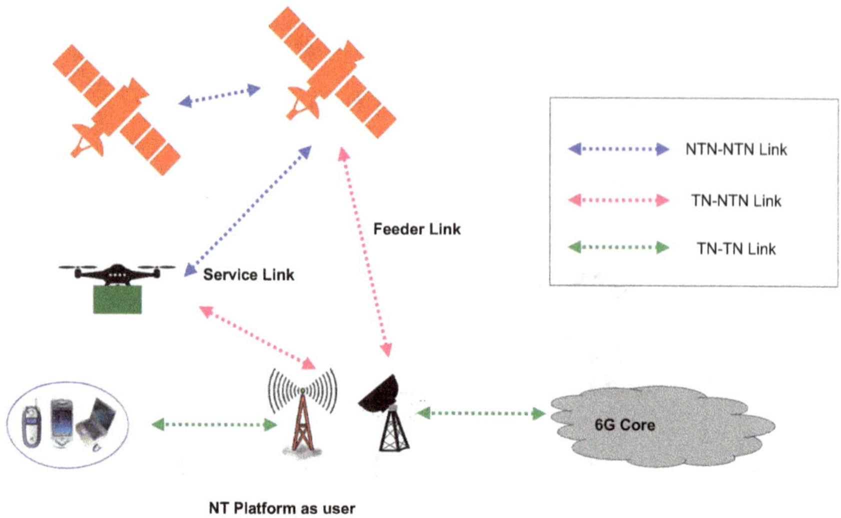

Figure 10.6 NTN Architectural options in 6G – NT as a platform as a user of 6G

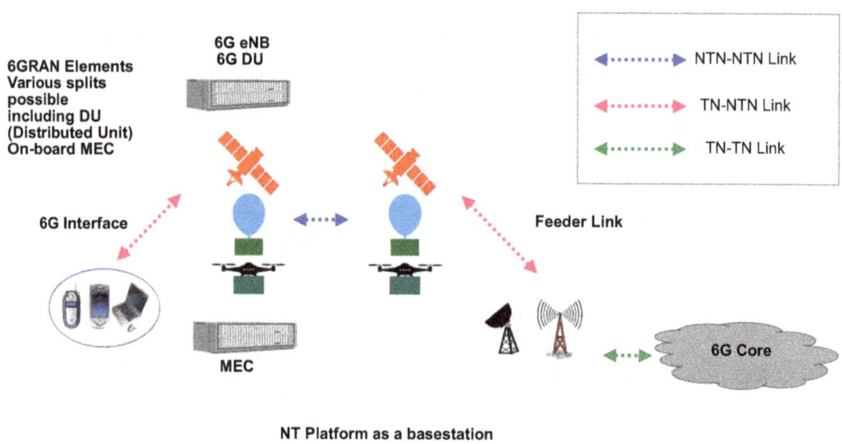

Figure 10.7 NTN Architectural options in 6G – NT incorporating elements of the RAN and MEC

Standards-wise, the current focus for work in Release 18 for NTNs is general and IoT access. New bands over 10 GHz could connect to fixed and moving platforms and building-mounted terminals. 3GPP has also explored protocol and architectural changes to the 5G core needed to support NTN-based gNode Bs. In addition, groups are working on proposals for more spectrum to be allocated to satellites at the upcoming World Radio Conference (2027).

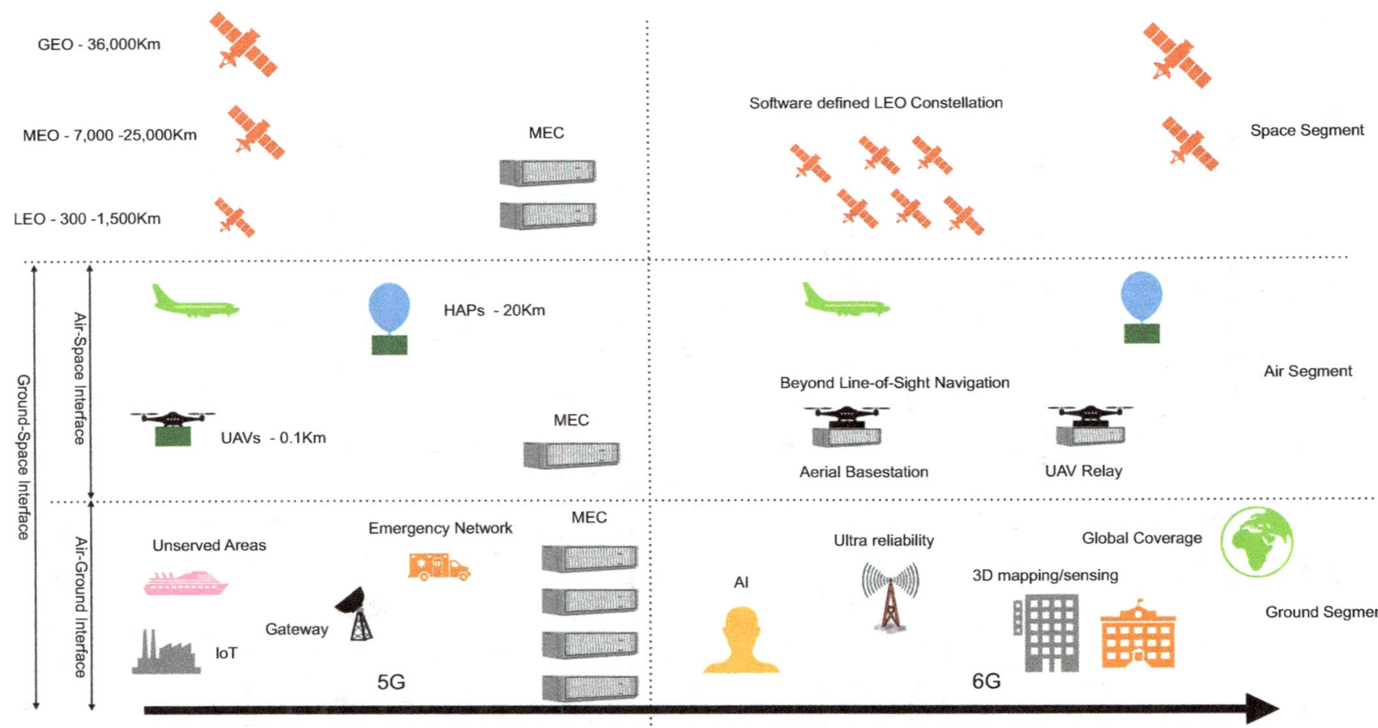

Figure 10.8 All NTN components are unified with TNs in 6G

10.5 NTN applications

Given that NTNs offer limited capacity and have cost constraints when compared to mobile networks. However, they cover vast areas of land and the oceans. This section reviews the best use cases for these devices within these constraints. The section starts with the most obvious services: improved narrowband and broadband coverage. These services will probably start in the 5G time frame and be fully realised in the 6G era. An example might be full national voice coverage initially provided by satellites and later augmented by HAPs to enhance the available data rates and reduce costs.

Several other critical applications for NTNs in 6G have been proposed. One is the ability to direct resources to temporary situations – i.e. improved capacity on demand. An example would be achieving significant capacity for a music festival. Another example would be an emergency response, such as a local private network set up for emergency use in a short time. Internet of Things (IoT) applications often need NTN networks. An example here might be a large farm requiring the network to collect information from sensors in the soil, crops, animals, etc. Typically, these sensors might require ultra-low or zero power consumption – making connection to terrestrial towers impossible (if there is a signal). UAVs could fly over the farm continuously with real-time connection to the 6G network or offline data collection mode. Narrowband IoT (NB-IoT, a low power and low bandwidth 5G interface described in Chapter 2) has been a highly successful LTE technology for IoT connections. It looks like it will also be incorporated into 5G, and there is now a 3GPP study item that can be used to look at NB-IoT via satellite [20].

In most proposed architectures, different elements either offer backhaul/relay or 6G connectivity directly – via at least some RAN elements and functions on the NTNs themselves. Figure 10.8 shows a complicated arrangement whereby some UAVs have a minimal function and are connected to HAPs or satellites for more advanced (and less time-critical) functions. The coordination of resources, beam and frequency management, service coordination and security over this highly complicated set of connection options is one of the significant challenges in 6G [21]. It is also expected that 6G NTNs will provide Mobile edge computing (MEC) and some parts of the core. This would significantly reduce the end-to-end latency for many time-critical services (avoiding the hop back to and from a terrestrial network). Coordinating MEC resources across terrestrial, UAV, HAP and satellite elements might require AI/ML [22]. There are several studies of satellite-based MEC servers: Xie *et al.* [22] outline different architectures and demonstrate that it is possible to offer substantial latency savings compared to cloud computing solutions.

10.5.1 Narrowband coverage

One of the most prominent applications for NTNs is more extensive coverage of narrowband services – such as voice, text and location/tracking. There is sufficient capacity in the new LEO constellations to extend these services to millions of users across the globe. Satellite delivery of these services would cost more than terrestrial delivery. However, it would be attractive for emergency or occasional use, such as hiking in a national park.

There are several possible techno-economic models to deliver a narrowband service. First, an MNO might offer non-terrestrial minutes/texts as a bundle. These could be used, for example, on a cruise liner, when out of terrestrial coverage or in an emergency. Governments might subsidise MNOs to improve coverage by adding NTNs at no extra cost to the user, as happens in the UK with the shared rural network [23]. Users might also be offered an emergency service by a handset manufacturer, MNO or even their government. An example is the Apple/Globalstar emergency service, noted in the earlier section.

Countries may also perceive that 100% coverage (voice, messaging or low-rate data) would constitute a national asset and invest in such a network. As mentioned earlier, that would be a natural extension of the UK shared rural network, aiming for 95% coverage of the UK landmass with 2 Mbit/s by 2025.

10.5.2 Broadband coverage

In the UK, it is reported that 96% of premises can access fixed broadband at 30 Mbit/s or higher [24]. For mobile, the figure is much lower, with the government-subsidised shared rural network aiming to cover 95% of the UK geographic area for at least one MNO with a minimum download speed of 2 Mbit/s. The official definition of 'broadband' has changed in the United States. The minimum speed required to call a connection broadband will rise from 25 to 100 Mbps. However, the genuine capacity limits on even large LEO constellations make it unlikely that they could be used to supply even a fraction of this demand. As we saw in the earlier example of Wales, a single LEO satellite, dedicating all its resources, could serve around a thousand users with a 100 Mbit/s service at a considerably higher cost than fixed broadband or mobile.

It is more probable that HAPs will be used for local capacity increases in sparsely populated regions. The BT trials described above with aircraft also point in this direction. HAPs have better coverage than macro cells but less than satellites. The coverage varies with the height, which can be hundreds of meters for a tethered balloon and up to 25 km for an aerostat (a lighter-than-air aircraft that gains its lift through a buoyant gas – such as helium). It is suggested [25] that 100% coverage of the UK (at 10 Mbit/s+) could be achieved with balloons tethered at 1.5 km height. These are envisaged to stay aloft for many years, just being winched down for occasional maintenance. It has been estimated [25] to require an additional 20,000 towers to give the UK 100 per cent coverage at 10 Mbit/s. The estimate for building this network using terrestrial towers is approximately £10 billion. The same source suggests that the cost of using HAPs might be as low as £1 billion.

Another consideration is NTNs' flexibility and adaptability compared with traditional macro networks. NTNs could provide temporary capacity increases simply by allocating more satellite resources or moving a HAP over the site, for example.

Finally, satellites are very efficient for broadcast services – where the same content needs to reach large numbers of users over an extended area – and as such, they could easily facilitate the transition from digital terrestrial TV broadcast to mobile/fixed networks, thereby freeing up substantial amounts of 'premium' sub 1 GHz spectrum that, in theory, could be used for 6G.

10.5.3 Disaster management

A typical example is maritime search and rescue (Figure 10.9). The initial requirement is to be alerted that persons or assets are distressed and need assistance. The search phase involves determining the location of the sinking boat, and the rescue phase entails coordinating various authorities to deal with the situation.

10.5.4 Urban air mobility

The European Union Aviation Safety Agency has defined urban air mobility (UAM) as 'a new safe, secure and more sustainable air transportation system for passengers and cargo in urban environments'. NASA puts it this way: 'safe and efficient air traffic operations in a metropolitan area for crewed and uncrewed aircraft systems'. The concept embraces passengers and cargo travelling on both crewed and uncrewed craft.

UAM controls the airborne delivery of goods, parts, urgent medical supplies and foodstuffs. It is fully integrated into a multimodal transport system of vehicles, trains, trams, scooters and rickshaws. It also controls crewed and crewless aircraft, balloons and helicopters for passenger transit. Examples of where this would be useful are getting first responders to an emergency in the shortest time or where the use of an air component saves energy.

Realising this scenario requires highly reliable communications, real-time airspace control, remote piloting (whether by AI or humans) and coordination between NTN and TNs.

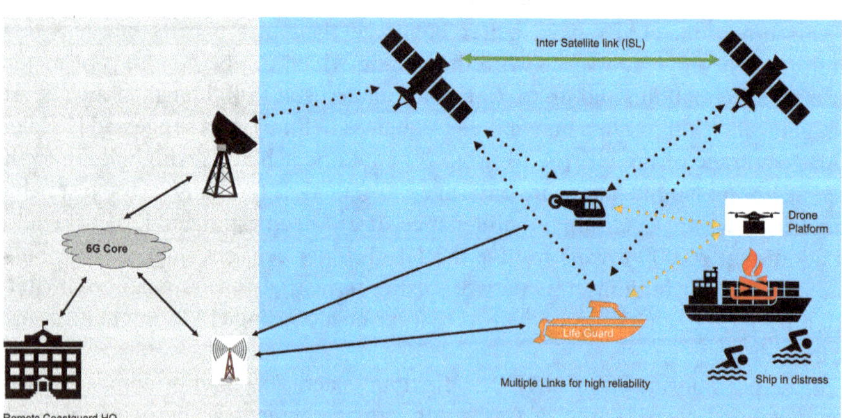

Figure 10.9 NTN disaster scenario example – taken from [26]

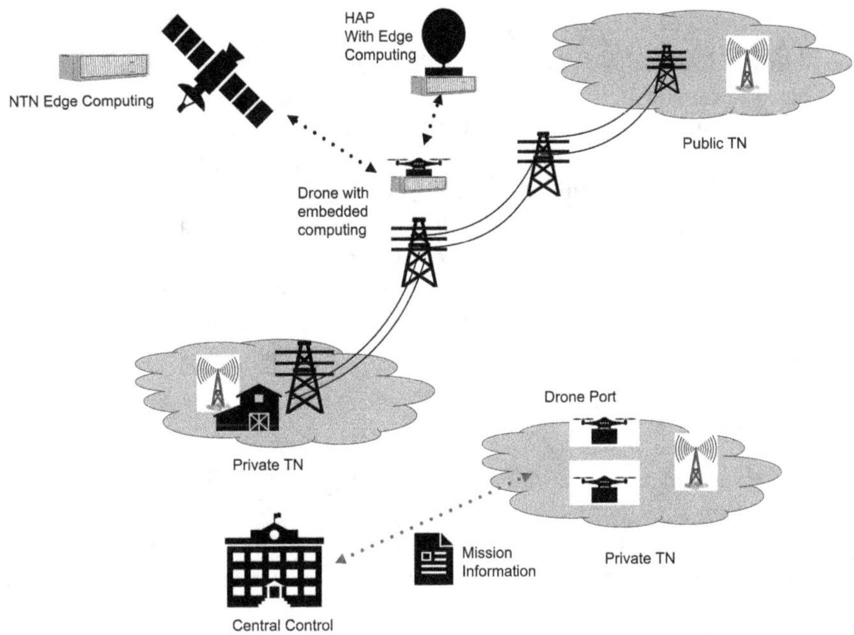

Figure 10.10 Remote power line inspection as an example NTN application (from [26])

10.5.5 Autonomous power line inspection using drones

Beyond line-of-sight control of UAVs (and swarms of them), it is also often considered a significant 6G application. These UAVs might then be used for 3D mapping, which augments that of satellites, to provide more accurate digital twin creation and sensing than is possible from terrestrial measurements alone. There is also a discussion of 'aerial telepresence', allowing eXtended Reality to extend into 3D. One concrete example, described in much more detail in the EU project 6G-NTN, is the autonomous inspection of power lines using drones (Figure 10.10).

The scenario requires the drones to be remotely piloted along the power lines and provide a 3D view of the lines on a live video feed. This feed is then analysed by AI tools running on edge computing to identify any issues that might need maintenance or emergency action (such as object impact or ice build-up). The drones might require connection via either HAPs or satellites or both and would be equipped with sensing (light detection and ranging) and night vision high-resolution cameras. These would necessitate a very high bandwidth connection to the processing computer resources.

10.6 Challenges for NTNs

This section examines some critical challenges for NTNs in the 6G era.

10.6.1 Spectrum challenges

One of the critical issues for satellites is the need to identify the global spectrum. Existing spectrum in the 0–7 GHz bands is heavily used, and obtaining further spectrum for satellites in this range will be a significant challenge. Higher frequency bands are less intensively used and could be allocated to NTN networks with less disruption to existing terrestrial uses.

There is a trade-off between higher frequencies and smaller spot sizes. Smaller footprints achievable with higher frequencies are more useful for constellations looking to provide very high capacity. A typical minimum footprint for the different bands, with a standard antenna size in LEO orbit, are

- C band (4–8 GHz) – 1750 km (1100 miles),
- Ku band (12–18 GHz) – 1000 km (620 miles), and
- Ka band (26–40 GHz) – 350 km (220 miles).

More advanced antennas, such as beam steering with multiple elements, could reduce the beam sizes further.

During the World Radio Conference held in 2023 (WRC-23), work began to determine a regulatory framework for ISLs, and Ka frequencies were allocated for space research, space operation and earth-observing satellite applications. WRC-23 also approved a new primary allocation in the 17.3–17.7 GHz band in Region 2 (Americas). Limits were implemented to protect Region 1 (Europe, the Middle East and Africa) and Region 3 (Asia-Pacific). There are expected to be efforts to identify more spectrum for NTNs in both WRC-27 and WRC-31.

10.6.2 Technical challenges

Numerous technical challenges exist to fully unifying TNs and NTNs in 6G. Xiao *et al.* [21] discuss several outstanding areas that are critical to 6G NTNs:

- Random access – Accessing radio resources to transmit data involves a radio overhead. This overhead can be considerable for IoT applications with small amounts of data to send. A proposal for a much-simplified random access procedure for IoT devices that only need to upload a small amount of data is being considered in Release 18.
- Handover – A challenge for signalling. If the satellite contains an active gNB, handing the signalling state over between satellites is an issue. Signalling 'storms' are created because many UEs must be handed over simultaneously.
- Beam management – Conflicting demands for coverage, signalling, service and feeder link beams exist. In addition, managing handover, interference and QoS for rapidly moving NTNs is a very active research area, and 6G can be expected to offer considerable improvements to current algorithms. AI/ML is seen as significantly improving performance in these areas.
- Doppler resilient transmission – a maximum Doppler shift of 50 kHz creates significant performance degradation for 5G orthogonal frequency division multiplexing (OFDM) systems. New transmission and modulation schemes

have been shown to improve performance significantly but would complicate integration with existing OFDM networks.

- Software-defined networks (SDNs) – Developing an SDN architectural solution across satellites, HAPs, UAs and TNs is a significant challenge. A solution is needed if advanced features – such as network slicing, smart cities, DT and end-to-end QoS are to be implemented. Azari *et al.* [27] describe a solution whereby satellite-based SDN controllers are synchronised with those on the ground.
- AI integration across terrestrial, satellite, HAP and UAV platforms – AI/ML will be a crucial feature of 6G – for both optimising the network and offering AI/ML services via MEC and core functions. The seamless integration of these with NTNs presents considerable challenges.
- THz communications – In the early discussions of 6G, it was often proposed that THz data rates – achieved by utilising huge swathes of mmWave spectrum would be required. Terrestrial data rate expectations have since been scaled back, but NTNs might gain significant performance enhancements by using THz links, possibly optical ones, for inter-NTN and NTN-ground links.
- Sensing and location – Another major component of 6G is sensing/location, and NTNs significantly expand these capabilities. The challenge will be integrating and optimising sensing with terrestrial communications. De Lima *et al.* [28] illustrate a 10 GHz bandwidth radar, centred at 145 GHz, that can resolve range to 30 mm and could be mounted on an NTN platform.
- Intelligent reconfigurable surfaces – Based on metamaterials, these are proposed for use in 6G for beam steering and manipulation of cmWave and mmWave signals. The extension of their use to UAVs and even satellites to save weight, power and cost is being actively researched.

Many projects are also looking at how AI/ML can significantly improve the performance of NTNs – including interference management, beam forming/MIMO, channel measurement/modelling and resource management. The current level of these for NTNs is relatively primitive compared to advanced terrestrial LTE/5G operation and offers excellent scope for increases in efficiency. One concern with HAPs and UAVs is always the flight time before refuelling or recharging. To overcome these limitations, 6G uncrewed platforms could be recharged in flight as part of a fleet/resource management plan.

Readers wanting a comprehensive guide to current NTN R&D, projects and demonstrators should consult Azari *et al.* [2], who have produced an excellent summary of 'Evolution of Non-Terrestrial Networks From 5G to 6G: A Survey' (386 references) and available to download.

10.6.3 Regulatory challenges

A significant issue for NTN systems is the variation of regulations globally. Satellites, with their global footprint, are the most affected by this. Today, some countries do not allow satellite services, and some constellations do not serve certain countries. There are likely to be country-specific restrictions on some services. For example, country B rules that all adult material transmitted over the public Internet must be encrypted and end users age verified. Does that apply to NTNs launched, based, controlled or headquartered there? Or does it apply to any service delivered to a terminal within that country? How is it enforced? Will country B go to the lengths of jamming signals?

Airspace restrictions will apply to HAPs and UAVs, and flight plans must be coordinated. Controlling UAVs and creating no-fly zones, for example, around airports or prisons, is, in fact, a suggested 6G service.

10.6.4 Commercial challenges

NTNs will require new players and a richer ecosystem to deliver fully unified terrestrial and NTNs. MNOs do not own or operate satellites, and it seems unlikely they will operate HAPs themselves directly. If HAPs are used to provide broadband in areas of poor coverage, then it makes far more sense that they are multi-operator in the manner of shared rural networks. One possibility is to reuse the neutral hosting protocols that have been developed to allow operators of shopping malls (and airports, campuses, etc.) to offer mobile services from a range of MNOs. Chapter 9 describes these scenarios in detail, but it is clear how this model might be helpful in 6G. Handset manufacturers might also be part of this richer ecosystem, as witnessed by the deal between Apple and Globalstar for emergency coverage on the newer iPhones. The GVSO concept is introduced in Chapter 16. A GVSO buys capacity from the satellite operator and re-sells it directly to customers. If a handset manufacturer became a GVSO operator, they could equip their handsets with a satellite capability that could be used globally, independent of any MNO subscription. Like Wi-Fi today, there would be no need to provide seamless handover as virtually all applications today do not require this to work effectively.

As more elements of the RAN, including MEC, are located on satellites and HAPs, the integration between the various players gets more profound, and developing the appropriate commercial models and interfaces between them will be a challenge in the future.

10.7 Kessler syndrome and dark skies

The Kessler syndrome, as it is now called, was proposed by NASA scientist Donald Kessler in 1978. It is a scenario where the density of objects in LEO due to space pollution is so numerous that collisions between objects could cause a cascade in which each collision generates further space debris that increases the likelihood of further collisions – like a nuclear chain reaction. In 2009, Kessler wrote that modelling results had concluded that the debris environment was already unstable, ‘such that any attempt to achieve a growth-free small debris environment by eliminating sources of past debris will likely fail because fragments from future collisions will be generated faster than atmospheric drag will remove them’. This is, of course, before the recent increase in LEO Constellations.

One source describes it thus:

> *Consider this scenario: The destruction of a dead spy satellite spawns a swarm of debris in Earth orbit, which wreaks ever-increasing havoc as it zooms around our planet. The cloud destroys several communications satellites, generating more and more debris with every violent collision. It takes out the iconic Hubble Space Telescope and a NASA space shuttle, killing several crewmembers aboard the winged vehicle. It then lines the International Space Station (ISS) up in its crosshairs, destroying the $100 billion orbiting lab with a hail of fast-flying shrapnel [29].*

The threat is to both GEO and LEO satellites, highlighted by the collision of the Iridium 33 and Kosmos-2251 satellites in 2009, which generated a cloud of debris. India also tested an anti-satellite missile in 2019, which resulted in over 400 pieces of orbital debris. Opinions vary greatly as to whether this is or is not a significant threat [30].

A final difficulty facing the new LEO satellite constellations is their possible impact on dark skies. Each satellite creates a trail across the sky, and amateur and professional astronomers have reacted negatively [31].

10.8 Summary and conclusions

It looks as if limited LTE services in existing LTE bands could be offered over LEO satellites shortly. This will be a voice and text service with minimal bandwidth. Apple already offers an emergency messaging service with their iPhones 14 and 15, which is globally available. The Apple service is independent of MNO but with a separate deal between Apple and a satellite provider.

The next level of integration with NTNs is expected to be the transparent relay of the 5G radio interference over satellites, as detailed in 5G release 17. This will allow users access to the full panoply of 5G services such as network slicing, edge computing and quality of service guarantees. These services will be limited to high-value and low-bandwidth applications. They will only operate outdoors (although fixed antennas can be used to create local private networks or Wi-Fi hot spots). These services include voice, emergency scenarios, backup broadband links for extreme reliability, tracking and maritime communications. There is not enough capacity in the planned LEO constellations to offer a general rural broadband service, and, in any case, the cost would be ten times that of a typical terrestrial broadband 5G connection, with the current price much higher than this. There are ways in which satellite costs could be reduced, such as ISLs and rocket launches with greater payloads, but these are not enough to offer 'eMBB from the sky' as a 6G service.

In 6G NTNs, ground, air and space elements will work in unison to deliver a range of advanced use cases. The air element will include HAPs, such as aircraft and balloons, that could provide much greater capacity and smaller cells. These could possibly reuse existing MNO spectrum in rural areas that are coverage but not capacity-limited. These might be used for rural fill-in or temporary capacity. Drones that could be remotely piloted, regulated and coordinated will enhance the 6G sensing and digital twin elements.

In 6G, elements of the radio network and MEC may be located onboard the NTN units. There is a trade-off between weight and capability, but these elements will significantly reduce latency, offer new capabilities and increase efficiency. Example applications in the chapter include complicated maritime rescue and remote power line inspections – both taken from a major study by the EU project 6G-NTN [25].

Formidable obstacles must be overcome before many of the 6G NTN visions are realised. Obtaining global or even regional spectrum for satellite use is difficult. The life span of LEO satellites is estimated to be 7–10 years. The economics of HAPs are still being determined, and the concept has been around since 3G without any real progress. Coordinating airspace with HAPs and UAVs is also a significant challenge. Meanwhile, device manufacturers with no constraints may get there first based on a single supported LEO band requiring no seamless handovers, only roaming. The dawn of the GVSO has begun.

References

[1] Araniti, G., Iera, A., Pizzi, S., and Rinaldi, F., 'Toward 6G Non-Terrestrial Networks,' *IEEE Network*. 2022;36(1):113–20, doi:10.1109/MNET.011.2100191.

[2] Mahdi Azari, M., Solanki, S., and Chatzinotas, S., 'Evolution of Non-Terrestrial Networks From 5G to 6G: A Survey,' *IEEE Communications Surveys & Tutorials*. 2022;24(4):2633–72, doi:10.1109/COMST.2022.3199901.

[3] Dureppagari, H. K., Saha, C., Dhillon, H. S., and Buehrer, R. M., 'NTN-Based 6G Localisation: Vision, Role of LEOs, and Open Problems,' *IEEE Wireless Communications*. 2023;30(6):44–51.

[4] Giordani, M. and Zorzi, M., 'Non-Terrestrial Networks in the 6G Era: Challenges and Opportunities,' *IEEE Network*. 2021;35(2):244–51, doi:10.1109/MNET.011.2000493.

[5] Starlink constellation. [Online]. [Cited Jan 2024] Available from: https://www.starlink.com/technology.

[6] Sputnik. [Online]. [Cited Jan 2024] Available from: https://history.nasa.gov/sputnik.html.

[7] Ashford, E. W.'Non-Geo Systems – Where Have All the Satellites Gone?' *Acta Astronautica*. 2004;55(3–9):649–57.

[8] Science Direct article. [Online]. [Cited Jan 2024] Available from: https://www.sciencedirect.com/topics/engineering/low-earth-orbit#:~:text=Satellites%20in%20LEO%20are%20affected,satellite%20is%207–10%20years.

[9] Apple satellite services. [Online]. [Cited Jan 2024] Available from: https://support.apple.com/en-gb/HT213426.

[10] PC Magazine article. [Online]. [Cited Jan 2024] Available from: https://uk.pcmag.com/networking/150673/starlinks-laser-system-is-beaming-42-million-gb-of-data-per-day.

[11] Ma, S., Chou, Y. C., Zhao, H., Chen, L., Ma, X., and Liu, J., 'Network Characteristics of LEO Satellite Constellations: A Starlink-based Measurement from End Users'. *IEEE INFOCOM 2023 – IEEE Conference on Computer Communications*, New York City, NY, USA, 2023, pp. 1–10, doi:10.1109/INFOCOM53939.2023.10228912.

[12] Pachler, N., del Portillo, I., Crawley, E. F., and Cameron, B. G., 'An Updated Comparison of Four Low Earth Orbit Satellite Constellation Systems to Provide Global Broadband,' *2021 IEEE International Conference on Communications Workshops (ICC Workshops)*, Montreal, QC, Canada, 2021, pp. 1–7, doi:10.1109/ICCWorkshops50388.2021.9473799.

[13] Li, L. K. T., Hofmann, C. A., Völk, F., and Knopp, A., 'Techno-Economic Design Aspects of Satellite Mega-Constellations for 6G Services,' *2021 IEEE 94th Vehicular Technology Conference (VTC2021-Fall)*, Norman, OK, USA, 2021, pp. 1–6, doi:10.1109/VTC2021-Fall52928.2021.9625226.

[14] Starlink costs. [Online]. [Cited Jan 2024] Available from: https://www.satelliteinternet.com/providers/starlink/#:~:text=Starlink costs anywhere from $90 to $2,500 at the highest.

[15] Energy costs for drones. [Online]. [Cited Jan 2024] Available from: https://www.spirent.com/blogs/mno-energy-costs-are-surging-whats-the-plan-to-reach-net-zero.

[16] BT and Stratospheric Platforms trial high-altitude 5G in the UK. [Online]. [Cited Jan 2024] Available from: https://www.datacenterdynamics.com/en/news/bt-and-stratospheric-platforms-trial-high-altitude-5g-in-the-uk/, Accessed March 2024.

[17] How 5G drones will help deliver digital twins. [Online]. [Cited Jan 2024] Available from: https://www.gsma.com/5ghub/drones.

[18] 5G Coverage by drones and challenges. [Online]. [Cited Jan 2024] Available from: https://medium.com/@nmappn/5g-coverage-by-drones-and-challenges-bd21d547262d.

[19] T-Mobile US ready to test satellite-to-cellular service. [Online]. [Cited Jan 2024] Available from: https://www.telecoms.com/satellite/t-mobile-us-ready-to-test-satellite-to-cellular-service#close-modal.

[20] 3GPP TR 36.763, 'Study on narrowband internet of things (NB-IoT)/enhanced machine type communication (eMTC) support for non-terrestrial networks,' 2021. [Online]. [Cited Jan 2024] Available from: https://portal.3gpp.org/desktopmodules/Specifications/SpecificationDetails.aspx?specificationId=3747.

[21] Xiao, Z., Yang, J., Mao, T., *et al.* 'LEO Satellite Access Network (LEO-SAN) towards 6G: Challenges and Approaches', *IEEE Wireless Communications*. 2024;31(2):89–96, doi:10.1109/MWC.011.2200310.

[22] Xie, R., Tang, Q., Wang, Q., Liu, X., Yu, F. R., and Huang, T., 'Satellite-Terrestrial Integrated Edge Computing Networks: Architecture, Challenges, and Open Issues,' *IEEE Network*. 2020;34(3):224–231.

[23] UK shared rural network. [Online]. [Cited Jan 2024] Available from: https://srn.org.uk/.

[24] UK broadband availability. [Online]. [Cited Jan 2024] Available from: https://www.cable.co.uk/broadband/guides/broadband-availability-snapshot/.

[25] Webb, W. and Temple, S., 'Emperor Ofcom's New Clothes: How the Market Approach to Radio Spectrum Failed the Country's Mobile Infrastructure and How to Turn it Around,' 2024, Available from: https://www.amazon.co.uk/Emperor-Ofcoms-New-Clothes-infrastructure/dp/B0CRGSSZSS.

[26] 6G-NTN Deliverable 2.1, Use case definition. [Online]. [Cited Jan 2024] Available from: https://www.6g-ntn.eu/public-deliverables/.

[27] Azari, A., Ghavimi, F., Ozger, M., Jantti, R., and Cavdar, C., 'Machine Learning Assisted Handover and Resource Management for Cellular-Connected Drones,' *2020 IEEE 91st Vehicular Technology Conference (VTC2020-Spring)*, Antwerp, Belgium, 2020, pp. 1–7, doi:10.1109/VTC2020-Spring48590.2020.9129453.

[28] De Lima, C., Belot, D., Berkvens, R., *et al*., 'Convergent Communication, Sensing and Localization in 6G Systems: An Overview of Technologies, Opportunities and Challenges,' *IEEE Access*. 2021;9:26902–26925.

[29] Wall, M. Kessler syndrome and the space debris problem. [Online] 2022. [Cited Jan 2024] Available from: https://www.space.com/kessler-syndrome-space-debris.

[30] Mariappan, A. and Crassidis, J. L., 'Kessler's Syndrome: A Challenge to Humanity', *Frontiers in Space Technologies*. 2023;4:1–12, doi:10.3389/frspt.2023.1309940.

[31] Guardian Newspaper Article, 2023. [Online]. [Cited Jan 2024] Available from: https://www.theguardian.com/science/2023/jan/06/picture-imperfect-light-pollution-from-satellites-is-becoming-an-existential-threat-to-astronomy.

Chapter 11
A radical core network for 6G

The future is already here, it's just not evenly distributed.

– William Gibson, science fiction author

11.1 Introduction

The discussion about the future of 6G is not only limited to new radio hardware but also includes the Mobile Core. This is even more significant if 6G has no new radio interface, as suggested by the Next Generation Mobile Networks Alliance. This chapter explores the two options for the 6G core network, whether it is an evolution of the 5G core network or a completely new system. It provides an analysis of both options and their respective advantages and disadvantages.

The Mobile Core network is the foundation of a mobile network. It is responsible for providing services to customers and managing their accounts. Any issues with the Mobile Core, such as unreliability, lack of responsiveness, security vulnerabilities or poor management, can directly impact the quality of service customers receive.

This chapter describes how the Mobile Core can be improved, starting with the architecture, principles and problems of the 5G Mobile Core network. It also examines the impact of softwarisation on mobile core operations and compares the Mobile Core network to fixed networks.

Several proposed approaches exist for the 6G core network, including a radical proposal for a 'coreless' 6G network. A 'coreless' 6G network mitigates many issues that are present in the 5G core network and allows convergence with Wi-Fi and fixed networks. An alternative revolutionary approach is proposed, combining the 3rd Generation Partnership Project (3GPP) Multi-operator Core Network (MOCN) and trusted gateway functions to create a distributed core 'neutral host' trusted Wi-Fi access point that 'hollows out' the core. If the industry decides to retain a core network for 6G, we propose some protocol enhancements to make it more reliable.

This chapter also explores how Edge Cloud could run 6G services and how this technology could be an opportunity for Edge Cloud to compete with or complement the hyperscalers' clouds. While 6G is an opportunity for mobile network operators (MNOs) to offer Edge Cloud services, several technical challenges must be

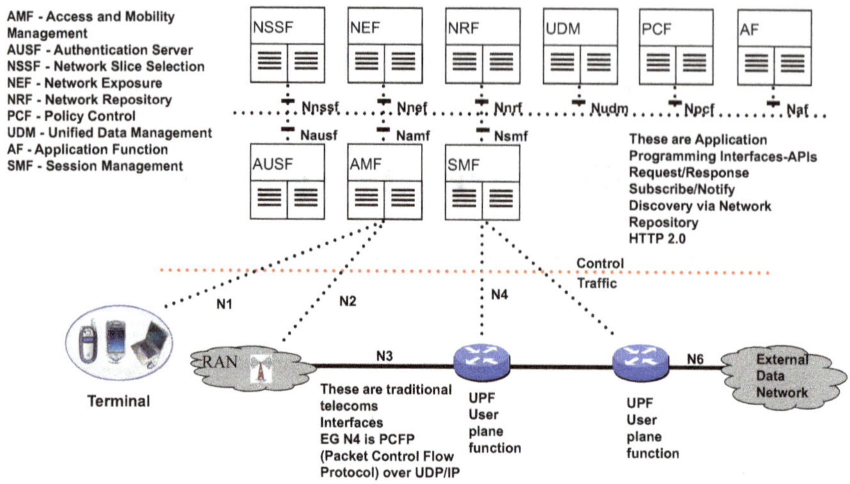

Figure 11.1 5G core network and interfaces

addressed first. Using the network to perform part of the computation could potentially benefit distributed AI techniques.

Furthermore, this chapter examines how fixed networks cope with 6G network volumes and how they keep up with data growth.

The ecosystem security of the 5G radio access network (RAN) has been criticised by the UK and the US governments for being too narrow, with few global vendors offering competitive solutions and little room for innovative start-ups. Chapter 12 discusses this topic in detail. This chapter concludes by exploring the 5G and future 6G ecosystem for the Mobile Core.

11.2 The Mobile Core

The Mobile Core network is responsible for authenticating UEs, authorising the services they are entitled to, managing sessions and managing mobility, i.e. ensuring a seamless service as they move around the globe and communicating with other MNOs' core networks. Figure 11.1 shows the components, called network functions, and interfaces of a 5G core network as specified by the 3GPP, see also Chapter 2. Box 11.1 summarises the 5G core network functions.

Box 11.1: Summary of 5G core network components

5G core network components

AMF – Access and Mobility Management – Receives all connection and session information from the UE and handles the connection and mobility tasks. Manages handovers between gNB.

AUSF – Authentication Server – Authenticates the user to the 5G network and the 5G network to the user. Public networks use a private key stored in a SIM or eSIM, but private networks have a wide choice of authentication methods. Authorises the user to access specific network services.

NSSF – Network Slice Selection – Selects the set of network slices serving the UE and determines the set of AMF to be used by the UE.

NEF – Network Exposure – Exposes network functions to external, possibly third party, applications functions via application programming interfaces (APIs), e.g. for monitoring, provisioning, analytics, setting QoS or charging policy.

NRF – Network Repository – Allows network functions to discover the services offered by other network functions.

PCF – Policy Control – Makes policy decisions, based on subscription information, and provides policy rules for control plane functions.

UDM – Unified Data Management – Manages subscriber information. Generates authentication credentials.

UDR – Unified Data repository – Stores data belonging to UDM, PCF, NEF and others.

AF – Application – Provides application services to the customer, e.g. a video streaming service.

SMF – Session Management – Session establishment and release. Manages tunnels between user plane function (UPF) and RAN. Manages UE IP address allocation, e.g. runs DHCP server.

SEPP – Security Edge Protection Proxy – Enables secure interconnect between different 5G networks' cores.

When is 5G not pure 5G? When it is non-stand-alone (NSA). It is possible to use a legacy 4G LTE core network to expedite the deployment of 5G New Radio (NR), known as NSA. 5G stand-alone (SA) deploys a complete 5G core network, exploiting the full potential of 5G, e.g. network slicing.

The 3GPP adopted softwarisation, network functions virtualisation (NFV) and software-defined network (SDN) concepts for its 5G network architecture. The process that runs traditional network functions as software is known variously as softwarisation, virtualisation, cloudification, cloud-native networks or NFV. The technological leap into network softwarisation has exposed the people who buy network functions, install and operate them inside network operators to a new world of software technologies and management methodologies. These technologies and methodologies are briefly introduced below and summarised in Box 11.2, providing references for those who wish to learn more about each. For the future of 6G, the debates are as much about software methodologies as new radio hardware, perhaps more so if 6G has an optional air interface and uses 5G NR.

The ETSI whitepaper 'Network Functions Virtualisation' [1], which I was the lead author and editor of, introduces NFV, where network components or functions apply the concept of virtual machines and similar virtualisation technologies developed for the Information Technology industry.

SDN [2] is a complementary technology to NFV that is also required to 'softwarise' networks. SDN allows network operators to programme the network, which makes the network more flexible and responsive to changing traffic patterns or requirements.

Further, the 3GPP's 5G core uses a service-based architecture (SBA) [3], see Box 11.2, where the 5G core network functions are implemented as services, using container technologies [4], designed to run cloud-native [5] and communicating via APIs in a microservices approach [6]. A cloud-native approach allows services to be scaled up and down automatically as demands change and self-heal if a physical or software component fails. However, stateless design is an implementation option in the 3GPP 5G core specifications. Stateful network functions make the 5G service more fragile as service availability depends on the underlying computers and network availability. Stateful network functions also inhibit the rapid deployment of software updates as operators cannot remove a stateful network function until all the traffic it holds state for has ceased, which in practice can take weeks. Although the 5G core could be built 'cloud native', the whole 5G network from RAN to 5G core cannot because of the statefulness of some 5G protocols. A stateless design is thus more robust and more 'cloud native'. A significant challenge for 6G is redesigning all 5G protocols to be stateless. Li *et al.* [7] designed and tested stateless core network functions for deployment in satellites to remove the signalling storms caused by low Earth orbit satellites extreme mobility. Li *et al.* redesigned the 5G protocols and functions to move the state to be stored in the UE instead of the core of the network, and this method could be considered for the terrestrial 6G core.

The deployed 5G network functions do not have to be in the same physical location; e.g. the SMF and UPF could be deployed in an Edge Cloud while centralising the AMF. The microservices architecture breaks applications down into multiple smaller autonomous fragments. The microservices approach creates, maintains, tests and runs autonomous fragments independently, increasing the productivity of software writing, integration and testing. We expect this approach to allow the MNOs to offer a diverse and dynamic set of services. The complexity of integrating many components should be more than compensated for by reducing the scope and size of each component.

Microservices work for the IT industry because a range of IT tools automates the specification of software interfaces and the integration and testing of software fragments. DevOps [8], a portmanteau of 'development' and 'operations', is the name given to these tools and methods. DevOps is built on continuous integration/continuous deployment (CI/CD) [9], which means software systems continuously evolve to fix bugs, add new features and address operational issues. DevOps deploys software updates rapidly into operation after automated testing. NetDevOps describes applying DevOps to softwarised networks [10]. NetDevOps addresses the complexity of softwarised networks, which improves the development and operation of 5/6G networks. 5G core networks already apply DevOps techniques. However, most of MNO's network and equipment management systems do not have an SBA nor are microservice-based and do not use DevOps. Legacy management systems mean MNOs cannot leverage DevOps's full speed and flexibility in their services if their network and equipment management systems lag behind.

Box 11.2: Summary of network softwarisation and technologies

Network softwarisation and IT technologies

- **NFV – Network Functions Virtualisation** – Uses IT virtualisation technology to consolidate many network equipment types onto industry standard high-volume servers, switches and storage. The physical network function is replaced by a software function running on a server. NFV can use any form of virtualisation technology, e.g. virtual machines, containers, paravirtualisation and run in the cloud or any general-purpose computer in the network. Physical resources like fibre, copper and spectrum cannot be virtualised. NFV standardisation is led by the ETSI NFV Industry Specification Group [68], which I helped to establish.
- **SDN – Software-Defined Networks** – Refers to a range of technologies that enables networks to be programmable in real time. Initially SDN separated network control and data forwarding using 'standardised' technologies so that third-party software could control the forwarding behaviour of networks and thus optimise it, e.g. OpenFlow [69] and P4 [70]. The Open Networking Foundation [71] led SDN standards. Later SDN also exploited tunnelling techniques [72], e.g. GRE, to enable overlay networks [73] to be built between virtual machines or containers running in the cloud. SD WAN [74] exploits the same technology to connect enterprises' customer-premise equipment to the corporate network or cloud services.
- **Container** – Is a portable unit of software that can be quickly and reliably moved from one computer environment to another. Containers are isolated from each other even though they share the machine's operating system kernel. The sharing of the kernel makes Containers lighter weight, using less storage and memory and being quicker to boot than Virtual Machines.
- **VNF – Virtualised Network Function** – A network function in NFV.
- **CNF – Containerised Network Function** – A Virtualised Network Function using Containers. CNF nomenclature used to differentiate from VNFs using Virtual Machines.
- **Cloud Native** – An architectural software approach that focuses on building and running applications in the cloud such that they can automatically scale up and down and self-heal. Containers are the enabling

technology for Cloud Native approaches. It is also associated with a range of techniques, e.g. DevOps, to permit robust automation of the development, testing, deployment and in-life management of applications. The Cloud Native Computing Foundation [75] leads this.

- **DevOps** – Automates and integrates software development and operations to improve system development. Often seen as complementary to agile software development as it allows for more rapid and safe deployment of code upgrades.
- **NetDevOps** – Modification of DevOps to make it suitable for use in a large network service.
- **SBA – Service-Based Architecture** – Service components are distributed and are accessed and interact remotely through a variety of remote access protocols, e.g. REST, SOAP, JMS.
- **Microservices** – [76] Uses self-contained basic elements of an application that communicates with other elements using standard protocols and APIs that are vendor and product independent. Allows multiple teams to build applications with minimal dependencies on each other.
- **VM – Virtual Machine** – Is an emulation of a computer system that allows multiple isolated VMs to run on one physical computer. Usually considered a heavy-weight approach as each VM needs its own complete operating system.
- **CI/CD – Continuous Integration and Continuous Delivery** – Connects development to integration and deployment using automation. Essential for DevOps.

The IP Multimedia System (IMS) is complementary to the Mobile Core network and essential for Voice over 5G, which uses the legacy VoLTE system for Non-Stand-Alone 5G to access the IP Multimedia Subsystem (IMS). The IMS provides signalling (using SIP) and gateways to connect voice calls to fixed, mobile or VoIP networks. Vo5G, or VoNR, uses a 5G SA core to access the IMS. The 4G IMS and 5G IMS can be the same subsystem.

5G cores are described as 'access agnostic',[1] supporting 3GPP and non-3GPP technologies, so VoWi-Fi using IMS is possible. The IMS is as essential to MNOs as the Mobile Core network. MNOs believe that the IMS, especially Vo5G for Private 5G networks, offers new revenue opportunities for MNOs to sell high-quality voice services to enterprises. It gives MNOs the potential to return to the enterprise voice services market after OTT services displaced them.

Markwide Research [11] and Nokia [12] have proposed that virtual assistants could use Vo5G, e.g. Amazon's Alexa and Apple's Siri, but they do not require an

[1]5G's access agnosticism is only via gateways.

IMS as they do not require high-quality voice connections because they transfer audio samples as files.

Running networks is not an enterprise's core business. If enterprises do not want to buy a 5/6G network slice because they own or share spectrum or MNOs cannot meet their service requirement, they would need their own private 5G[2] core network. The complexity of building and running a private 5/6G core following the SBA approach is overkill. Enterprises want to buy a turn-key solution, and the most straightforward approach is buying a 5/6G core in a box that several vendors offer, e.g. AWS Private 5G [13] or Lenovo [14].

11.3 Open-source activities and standards

3GPP standards alone are insufficient to build a mobile network as they depend on a broad ecosystem of standards and increasingly open-source activities.[3] Although MNOs may not think they are using open-source code, they inevitably and indirectly will because many tools used to create code, e.g. compilers, software source code control, test suites and installers, are open-source. Further open-source may be embedded in their systems, e.g. the LINUX operating system.

5G uses softwarisation techniques which implies using NFV and SDN. Of the many global and national organisations involved in NFV and SDN standards, the two key founding groups are The European Telecommunications Standards Institute Industry Specification Group for Network Functions Virtualisation (ETSI NFV ISG), which initiated the work on global NFV standards in 2012 and the Open Networking Foundation (ONF) that started the SDN movement in 2011.

The softwarisation of networks necessitates network operators developing, integrating, testing and deploying their own- and third-party code into networks. The OSS and IT teams in network operators may be familiar with the software technologies required to do this, but it is striking new ground for network operations teams.

A suite of software technologies is required to implement the softwarisation of networks, see Box 11.2. DevOps is one of the essential practices required to support the softwarisation of networks. DevOps practitioners can choose dozens of open-source tools, Kersten [15] referring to a 'Cambrian Explosion' of DevOps tools. Kersten notes, '*Agile, DevOps, and open source all have something in common: they're driven from the bottom up, with each focusing on empowering the practitioner*'. There are too many DevOps open-source projects to list in a book about 6G. The significant trend is that MNOs will develop and operate 6G Edge Clouds using an extensive array of open-source tools.

There is also a rich portfolio of open-source projects creating programme code that contributes to the softwarisation of 5G networks and future 6G networks. Some projects provide specific 5G or 6G components, such as the Open-RAN project

[2]See Chapter 5 for more details of Industry 4.0 using Private 5G networks.

[3]Open-source code is freely available for anyone to inspect, modify and redistribute. Its supporters claim it encourages collaboration, transparency and innovation. However more than just code can be open-sourced, increasingly hardware designs are being open-sourced.

mentioned in Chapter 2, used in live networks by Vodafone, Rakuten, Dish and other MNOs. The Linux Foundation Networking Projects: LF networking [16] is an example umbrella covering projects that range from high-performance packet handling, orchestration[4] and management, network functions as a service, SDN controller, security and platform as a service that 6G solutions could use. The Linux Foundation also has a '5G Super Blueprint' that '*gives global communications providers and businesses a full architecture of how to deliver a total solution – replete with high-bandwidth, low-latency, scalable and cost-effective digital networking infrastructure – from end-user device to cloud application destination*' [17]. There are many other open-source projects not inside the Linux Foundation, e.g. the ETSI Open-Source MANO (Management and Network Orchestration) project, which is developing a Management and Orchestration (MANO) solution aligned to the ETSI NFV Information Models and the OpenStack cloud computing infrastructure projects which NFV implementations widely use. The Open Networking Foundation (ONF) has open-source projects building a software-defined RAN (SD-RAN) and a software-defined 5G core (SD-Core) with an associated Edge platform (Aether).

'Open-Source' also applies to the specification of hardware as well as software. The term 'White box' is often used to describe a piece of open-source hardware equipment and hardware from an Original Design Manufacturer (ODM) rather than a familiar mainstream OEM brand. AT&T [18] claims that 52% of its production traffic runs on a 'white box' architecture. The Open Compute Project (OCP) '*sets the standards for networking equipment, general purpose and GPU servers, storage devices and appliances and scalable rack designs*' [18], including open hardware and software specifications. The OCP Networking project creates open and disaggregated hardware and software technologies for networks that 6G solutions could use.

11.4 The fixed core

It is essential to understand the architecture of fixed, DOCSIS, DSL and PON networks in a 6G context for two reasons. One to understand the convergence of fixed, mobile and Wi-Fi networks, examined in detail in Chapter 9. Two, to understand the role fixed networks take in supporting 6G ambitions.

In this section, the 'fixed core' refers to the equivalent of the 'Mobile Core' and not the packet and transmission networks that underlie fixed networks, mobile networks and the Internet. Section 11.8 describes the 6G challenges for packet and transmission networks.

Today, the 5G core is the tail wagging the dog because fixed networks carry 80%–90% of data traffic. This section describes today's DSL, Fibre and Cable 'core' network architectures in sufficient detail to compare them to the 5G core network. Fixed networks also have their architecture for offering public Wi-Fi services, as described in Chapter 9.

[4]Orchestration coordinates, connects, monitors and manages platform resources.

Figure 11.2 shows the simplified network architecture providing service for Cable networks using hybrid fibre–coax and 'telephone' networks using copper telephone access lines or optical fibre. The networks between the home router, or residential gateway and the DSLAM/OLT/CMTS is the 'access network', equivalent of the 3GPP RANs, while everything to the right of the DSLAM/OLT/CMTS is the 'fixed core' for comparison purposes with the 'Mobile Core'.

The Broadband Network Gateway (BNG) and Cable Modem Termination System (CMTS) perform similar functions to the 5G AMF and SMF in that they are responsible for connection set-up, handling authentication and authorisation interactions, IP address assignments and managing quality of service (QoS). They terminate the protocol encapsulations over the access network, e.g. PPPoE and IPoE, and place the traffic onto an IP network. The RADIUS or DHCP policy servers store the residential gateways' authentication credentials along with information about the class and QoS for that customer. The BNG and CMTS may additionally generate accounting information, i.e. measure the amount of data each home uses. The RADIUS/DHCP servers perform services similar to the combined 5G AUSF, PCF and UDM.

A Wi-Fi access gateway (WAG) offers public or community Wi-Fi services. The WAG terminates tunnelling protocols, e.g. GRE and IPsec, from the Wi-Fi access point, often incorporated into Home Routers. If the Wi-Fi user has no automatic Wi-Fi service login protocols, e.g. 802.1x, they are directed to a captive web portal to enter their authentication credentials. The policy server contains details of the customer's subscription.

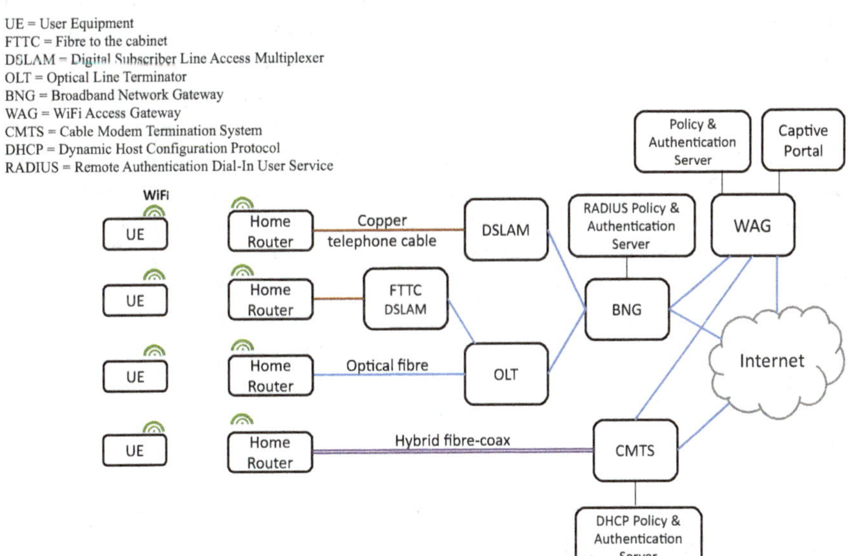

Figure 11.2 Fixed networks architecture

11.5 Alternative 6G core networks

Although the problems are not widely published, organisations building and deploying 5G core networks are experiencing architectural issues. Gold [19], Sofrecom [20] and Dietrich [21] provide some clues to these issues. Meanwhile, innovators, researchers, some standards makers and us are proposing better network architectures for the 6G core.

While MNOs have clarified that 6G must not trigger a hardware refresh of 5G RAN infrastructure and that 5G to 6G upgrades should be by software updates only [19], this does not rule out a new 6G core. Ericsson [22] points out that since 3GPP specifies the 5G core in a future-proof way, the 6G core is an evolution of the 5G core. This section describes how the 6G core network may differ radically from the 5G core network and the opportunities missed if it does not.

There are several ways the 5G core could evolve, through distribution or new protocols to make 5G services more reliable and easier to manage. Alternatively, in the author's opinion, the 6G network could be 'coreless', which, as well as improving reliability performance and reducing costs, enables more ubiquitous services to non-3GPP networks.

11.5.1 RAN-core convergence – a 'hollowed out' core

6G could take the microservices approach, allowing more granular disaggregation and distribution of RAN and core network functions. 'RAN-core convergence' is used [23] to describe an architecture where core network microservices run alongside RAN microservices at the same locations. Reducing latency and core network traffic is the primary motivation for this approach. However, this could enable operation and service innovations. RAN and core functions run similar network functionality for subscriber mobility and session management, especially when running physically alongside each other. Corici [24] proposes their merger.

In the authors' opinion, a further revolutionary step would be implementing a distributed 3GPP Multi-operator Core Network (MOCN), see Chapter 9, on the RAN supporting multiple cores using secure virtualisation methods. MOCN is a 3GPP standard [25] that allows RAN sharing by multiple Core Networks. Then, extend MOCN to support Wi-Fi by adding the Trusted WLAN Integration Function (TWIF) or Trusted Non-3GPP Gateway Function (TNGF) described in Chapter 9. This creates a distributed core multi-MNO, or 'neutral host', trusted Wi-Fi access point. This architecture would 'hollow out' the core, enabling it to scale to millions of femtocells. It would support an 'inside-out' strategy at multi-tenant and residential locations when restricted to a single MNO host at a single site. Figure 11.3 shows this architecture.

11.5.2 Evolving the 5G core for 6G

The 5G PPP Architecture Working Group, as part of its work to identify novel trends and critical technology enablers for 6G [26], has proposed that the architecture of the core network could be 'reshaped' to reduce the number of messages exchanged between network functions, a significant contributor to latency in the control plane,

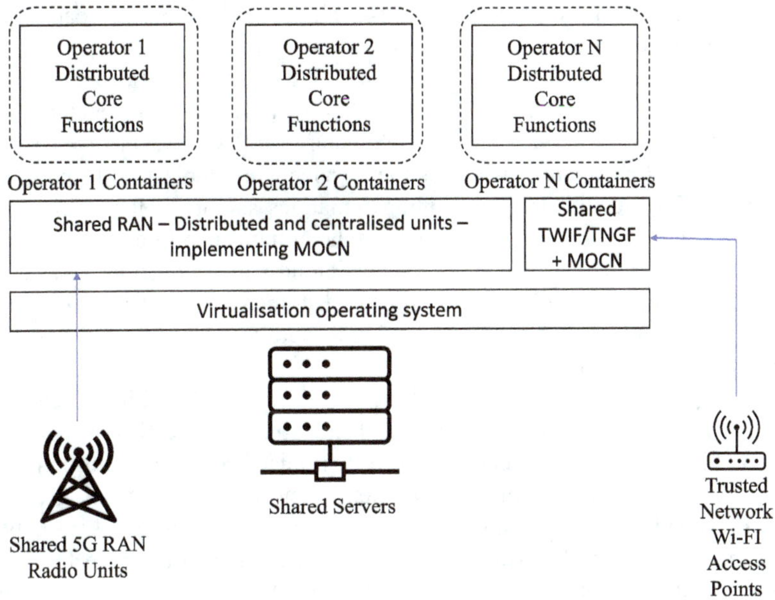

Figure 11.3 MOCN shared 5G RANs and trusted Wi-Fi APs with distributed multi-cores

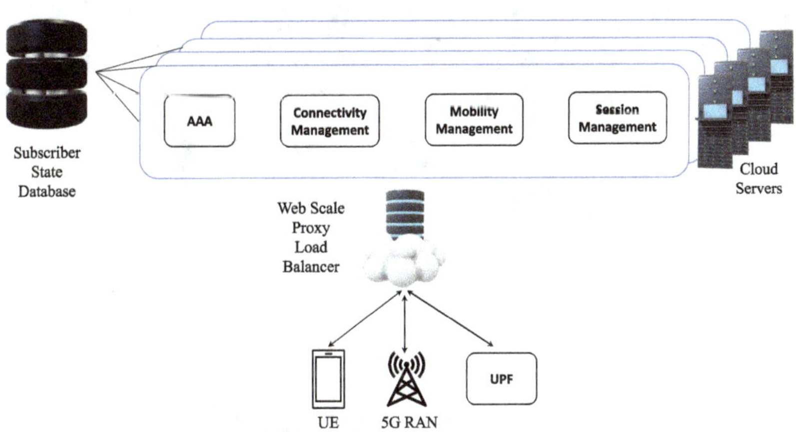

Figure 11.4 Alternative 6G core architecture

and increase flexibility by making network functions stateless. One proposal is to treat the core as one large web service that offers multiple services instead of splitting the functionality into network functions, as shown in Figure 11.4. Here, the 'web' front end has an equivalent role to the AMF, but the services are equivalent to implementing 5G core network functions using web-scale technology.

The scalability of web architectures is well proven, so this approach should scale to support femtocell numbers.

The FUDGE-5G project [27,28] describes the '*scatter landscape of approaches and technologies*' of the 3GPP SBA specifications and implementations and proposes an enhanced Service Based Architecture (eSBA) that '*further cherishes and adopts cloud principles*'. It makes the Service Communication Proxy[5] (SCP) mandatory, aligns it with Internet routing semantics as part of the routing infrastructure for the 5G core and adds resource scheduling capabilities. It introduces a new Who Am I Function (WAIF) to support cloud-native containers that can be deployed many times in different operator or customer domains. The FUDGE-5G project focuses on Non-Public Networks, i.e. Private 5G networks, but 6G should explore its principles for public 6G networks.

Bloxtel [29] has developed a 6G solution called Open6GC with a decentralised mobile network core using asymmetric authentication, Public Key Infrastructure (PKI) and blockchain. Bloxtel claims their decentralised core architecture improves security, prevents SIM cards or their symmetric key from being stolen, gives high scalability and global interoperability with less need to pass keys or tokens between MNOs and improves performance from implicit local breakout. Figure 11.5 shows a photo of Bloxtel's all-in-one 5G access point that hosts both a gNB and a decentralised SIM-compatible 5G stand-alone core.

We make the case for 6G to be 'coreless' in Section 11.5.4.

Figure 11.5 Bloxtel's all-in-one 5G access point (BLXTL AP 200). Courtesy of Bloxtel.

[5]The SCP, introduced as an optional network function in Release 16, routes messages between core network functions to perform load balancing, failover switching, topology abstraction, monitoring and tracing.

11.5.3 Service-based architecture alternatives

The 5G service-based architecture delivers many benefits, especially if a multi-vendor solution is required, but a microservices approach is not always appropriate. The issues with a microservices-based approach are:

1. Increased complexity.
2. More difficult to test.
3. Increased operational overhead.
4. Difficulty to debug.
5. SBA requires teams with specialised skills.
6. Increased latency of the control plane, slowing down network 'login'. Slow logins consume more battery energy for IoT devices that log on every time they send data, as they must keep the radio interface powered up for longer. It also slows down the handling of mobility for all devices.

Microservices are fragile if not implemented fully 'cloud native' and stateless, but using REST[6] for real-time communications is complex. The 5G protocols the AMF uses to connect to UEs and Access Networks are not 'cloud native'; for example, a UE must talk to the same AMF throughout a session and does not readily support stateless microservices, creating a weak link [30]. Fragility in the 5G core implementations makes it challenging to offer Ultra-Reliable services. 6G R&D should examine respecifying external 5G core interfaces using a Representation State Transfer (REST) architecture to enable a more reliable state-less and cloud-native compatible implementation of essential core network functions.

Could the 6G core be recomposed to give a more straightforward core but with the multi-team or multi-vendor benefits of a microservices approach? In the author's opinion, a tool like Bit [31] for web development, which decentralises software development, could be an alternative. This tool would not address the need to scale individual functions independently, which a more dynamic services environment requires where the mix of transactions changes. Or perhaps a new programming language, like Jolie [32], designed for service-oriented programming that abstracts away how many machines run microservices, optimising it to use in-memory communications when possible. we think it is an area for further research.

11.5.4 A radical coreless 6G network

Given the complexity of implementing the 5G core and the bottleneck it creates in KPIs, especially around reliability and latency, we radically question whether 6G could work better without a core network. We uniquely propose a revolutionary

[6]Representation State Transfer (REST) architecture achieves statelessness in the server because clients send relevant session data in every packet such that it can be understood, by the server, in isolation without context information. Typical REST implementations use such standards as HTTP, URI, JSON, XML but these may not be efficient for bandwidth constrained, real time communications and do not have to be used in a 6G solution.

coreless 6G network that would scale faster and easier to support femtocells, be much more reliable, more challenging to target for denial-of-service attacks, reduce latency and support non-3GPP access networks without gateways.

The cost of a 5G core is small compared to the cost of the RAN, but the main commercial motivations for radically redesigning the core network are to make it more reliable, to support URLLC, scalable to support millions of femtocells and more access and device (UE) agnostic. This section examines how the 6G core could be a lightweight ‘hollowed-out’ version of the 5G core to achieve these aims.

A 6G core must perform the following fundamental functions: Authentication, Authorisation, and Accounting (AAA), session management, mobility management, service QoS selection and implementation and management of itself. The following examines the removal or distribution of these fundamental functions to achieve a ‘core-less’ mobile network.

11.5.4.1 Authentication, authorisation and accounting (AAA)

Today, MNOs are responsible for being the ‘identity provider’ via the SIM and authentication details stored in the Mobile Core and the ‘network provider’. If the ‘identity provider’ and the ‘network provider’ could be technically separated, the AAA mechanism could be distributed and robust. It could also create an access network agnostic AAA mechanism. Commercial separation would bring about radical mobile network restructuring, like the impact of fixed access network separation in the United Kingdom, New Zealand and Australia.

Separation of the ‘identity provider’ and the ‘network provider’ is demonstrated by OpenRoaming for Wi-Fi, ‘neutral hosts’ and eSIMs for 3GPP. OpenRoaming uses DNS for network providers to discover identity providers and issues PKI certificates to network and identity providers; therefore, it is scalable to Internet and Femtocell volumes. The user or device authentication procedures use RADsec (secure RADIUS) between the network and identity providers. DNS allows the identity provider to distribute and load balance their RADIUS servers to achieve high scalability. RADsec supports accounting records.

It may not be in the MNO’s commercial interests to separate identity and network providers commercially, but a technical separation provides robust, distributed AAA and a flexible access-agnostic mechanism.

11.5.4.2 Session management, service and QoS selection

Session management manages IP address assignment to UEs and signals the service parameters to the UPF to set up the connection to the Data Network. In wireline broadband services, the Broadband Network Gateway handles session management. IPv6 networks could make Session Management more stateless using a stateless autoconfiguration mechanism for IP address assignment. QoS or session configuration must be held stateful in the UPF unless a best efforts-only service is acceptable, which may be OK for broadband services but not Private 5G enterprise services. If UE or access points could be trusted to mark packets correctly, then a stateless DiffServ approach could be used, removing the need for a state in the UPF. Today, the 5G SMF implementation maintains a session state record using a

24-bit identifier. A research challenge is whether this could be set in the IPv6 flowid or header extension by the UE or AP and trusted in a RESTful style.

Removing the Session Management Function from the 6G core network may be possible because IPv6 stateless autoconfiguration removes the need for IP address management. The session identifier could be in the IPv6 flowid or header extension, while authentication and policy control can use distributed methods like blockchain. The removal of the SMF requires further research.

11.5.4.3 Mobility management

RANs or gNodeBs manage radio handovers between themselves using radio performance measurements made by the UE. At the end of the handover, the RAN signals to the AMF to request that it switch the tunnel connection for that UE to the new RAN.

Is mobility management required with modern Internet protocols? For example, QUIC can manage a client's IP address change as the client roams or moves to a new access node with a new IP address. However, not all common Internet protocols can handle a client's IP address change. Some Internet protocols have 'layer violations' where lower layer information, e.g. an IP address, is embedded in a higher layer message. For example, Voice over IP (VoIP) uses RTP and RTCP, and the latter identifies the RTP connection using an IP address. Therefore, a new VoIP connection must be set up if the IP address changes. Many UEs hide glitches from the user by automatically re-establishing the session to the server when the IP address changes, but in some cases, this may be too slow or fail.

Removing the need for mobility management and tunnels[‡‡] also ensures local break out (LBO) is possible everywhere without additional gateways or complexity, which helps access Edge Cloud. Letting the IP network route packets automatically along the shortest paths rather than via core network sites reduces propagation latency.

The confused semantics of an IP address have driven the need for mobility management. For historical reasons, IP addresses mixed the role of identifying a location in a network with the identification of the user or application, so if a user moved to a different location, their identity changed, confusing many protocols and applications. Identity change when location changed was not a problem when computers attached to the Internet did not move. Over the decades, people have made many proposals for locator/identifier separation to fix this problem, some of which have become IETF 'standards', but networks have adopted none. However, 5G standards support the IETF's locator/ID separation protocol (LISP) [33] over GTP in the user plane.

6G should revisit the need for mobility management in the light of more distributed, stateless solutions.

[‡‡]Tunnels may still be required for non-IP protocols, usually for Private 6G networks, e.g. the Private 6G network may use bridged Ethernet in which case Ethernet packets will have to be tunnelled over the 6G network.

11.5.4.4 Core life cycle management

MNOs must manage the components of a 6G core throughout its entire lifespan. It is not simply a case of buying the software and installing it. Software with unknown bugs and complex distributed systems can fail unexpectedly. The services offered by the components need monitoring. Automatic scaling, failure detection and restarting are preferred. There are failure modes that cannot be detected due to blind spots in the protocols or monitoring procedures. The 6G core will probably break in novel or unique ways that no AI tools will have been trained on, requiring significant human expertise to diagnose and repair. Based on my experience managing a major UK ISP network for several years, I can guarantee that no one wants to look at a Wireshark[7] packet capture in the middle of the night trying to repair a dead 6G network.

The 6G network will be under continuous cyber-attacks requiring significant monitoring, detection and mitigation of security events. AI tools should help automate this.

The core functions need upgrades to fix bugs or deploy new services; however, these upgrades must be performed without interrupting network service. With an excellent stateless cloud-native deployment, deploying newly upgraded functions and turning the old ones off or simultaneously performing A–B testing running different versions is possible. Before network operators deploy new software, they want to see significant testing. NetDevOps tools are required to automate this.

Another incentive to go 'coreless' is that the lifetime cost of managing a core network could be several times more than the original purchase price of the software.

11.5.4.5 A radical coreless 6G network conclusion

A coreless 6G network looks more like an Internet architecture. It might be achievable with sufficient R&D effort, providing significantly more benefits than cost reduction, i.e. ultra-reliable, ultra-scalable and ultra-flexible. An essential requirement for a coreless 6G network is to address the issue of application protocol continuity when IP addresses change for the full suite of protocols required for 6G network services.

11.6 Edge Cloud for 6G

Edge Cloud is used as an infrastructure by 5G network operators to run RANs and Core Networks. This section examines how a private Edge Cloud that runs 6G infrastructure may evolve.

5G Open-RAN, see Chapter 2, requires Edge Cloud to implement its distributed RAN architecture. Network operators, even in small countries, e.g. BT in the United Kingdom, have distributed their 5G core network at several cloud sites

[7]Wireshark is a popular tool used to analyse traffic on networks, often a tool of last resort to figure out why the network or an application is not working.

[34], which are closer to the end-users than the hyperscalers' centralised cloud sites. Rakuten has 330 'Far Edge Sites' to run software RANs and 58 'Edge Sites' with three 'Central' data centres to run their 5G core network [35] for Japan, a country with a population of 127 million across 146,000 square miles. The 'far edge sites' are typically a single rack of servers, while the 'edge sites' are an undisclosed number of servers.

The 'Future of the Internet' Chapter 4 introduces Cloud and Edge Cloud. 5G adoption of NFV, SDN, SBA and cloud-native principles is introduced in Section 11.2. Box 11.3 gives a summary of common cloud terminologies.

Box 11.3: Common Cloud Terminology

Cloud Computing is a shared pool of configurable computers that can be used-on demand, but it comes in many flavours:

- **Cloud or Central Cloud or Public Cloud** – A shared pool of configurable computing resources that can be used on-demand by the public via the Internet. By default, assumed to be a central cloud located in a few (<1000) large (>50,000 servers) data centres. Examples of well-known Public Cloud providers are Apple, Amazon Web Services, Microsoft Azure, Google Cloud Platform and Oracle Cloud.
- **Multi-Cloud** – The use of multiple Public or Private Cloud providers by an organisation.
- **Hybrid Cloud** – Multi-Clouds that are brought together by a technology that allows data and applications to be shared between the different clouds.
- **Edge Cloud** – Cloud infrastructure hosted nearer to the physical location of the user than 'Central Cloud'. Edge Cloud may be public, private, owned by a third party or the using organisation.
- **Extreme or Far Edge Cloud** – Cloud hosted nearer to the physical location of the user than Edge Cloud. Could make use of User Equipment resources e.g. mobile phone, base stations located at mobile masts, a telecommunication's company local office/telephone exchange or other equipment on the customer premise.
- **Private Cloud** – A cloud infrastructure exclusively used by a single organisation. It may be owned by the organisation or a third party and may be located on or off premises.
- **Virtual Private Cloud** – An isolated private cloud hosted within a public cloud infrastructure.
- **On-premises Cloud** – Cloud infrastructure is hosted on the premises e.g. factory, of the organisation running the software or applications on the Cloud. On-premises Cloud is usually a Private Cloud.
- **Fog** – A term invented by Cisco that is like Edge Cloud. The OpenFog consortium defines it as '*A horizontal, system-level architecture that*

distributes computing, storage, control and networking functions closer to the users along a cloud-to-thing continuum', in other words a layer of compute between Edge and Core cloud.

- **Mist** – A similar term to 'Extreme/Far Edge Cloud'. Mist focuses on very lightweight devices whereas Extreme Edge could include more powerful devices like mobile phones.

11.6.1 How might Edge Cloud evolve for 6G?

Does Edge Cloud for 6G differ from Edge Cloud for 5G? Any differences will be due to the evolution of computer virtualisation technologies, Cloud technologies, incorporation of AI into NetDevOps tools and evolution of the vendor ecosystem.

Computer virtualisation and cloud technologies have advanced rapidly, from relative obscurity when VMware launched its virtual machine technology in 1999 to the release of Kubernetes in 2014. Kubernetes is a widely used open-source orchestration system for clouds and 5G Service Based Architecture cores, making it more straightforward to virtualise applications. Virtualisation techniques have become faster and more efficient. Early implementations consumed large amounts of compute resources and were slow. Modern CPUs support virtualisation with little impact on performance, and modern software techniques streamline the compute resources required. The R&D on virtualisation has not stopped; the trends are towards increased automation, increased abstraction of the practicalities of running applications, improved security, reduced overheads and better performance. The design of operating systems is also evolving to support these aims. Computers increasingly use hardware acceleration techniques to off-load packet processing, encryption and data transfer tasks, enabling fewer servers to do more work.

NetDevOps tools and techniques continue to improve to allow faster and more complete integration and testing of systems. AI will contribute more to automating testing, monitoring, fault finding and repairing networks.

The full softwarisation, disaggregation and operation of 5G networks in a NetDevOps approach are not certain because of several challenges that influence how Edge Clouds evolve for 6G. There is a commercial disincentive for incumbent telecom vendors to disaggregate their solutions as this opens them to more competition. Some softwarised 5G solutions may not match the incumbent monolithic solutions' performance. However, Intel and Samsung [36] reported a laboratory test result of a 1 Tbit/s UPF running on a single server with 10,000 active subscribers, demonstrating how the state-of-the-art is advancing. Some softwarised 5G solutions are still evolving to a cloud-native approach. MNOs do not generally have NetDevOps skills, while the hyperscalers have DevOps skills and tools to sell to the MNOs; they lack MNO operation skills. AT&T demonstrated one possible approach to this challenge when Microsoft acquired AT&T's Network Cloud technology in June 2021 to enable AT&T to move its 5G mobile network to the

Microsoft cloud [37]. Hyperscalers have developed toolsets targeted at network operators, e.g. Google Anthos for Telecoms, Microsoft Azure for operators and Amazon AWS for Telecoms. NetDevOps technology and commercial developments continue, and we expect further network operator–hyperscaler partnerships. However, it is strategically vital for operators to retain complete control of the Edge Cloud for RAN to maintain their primary responsibilities.

Hyperscalers running MNO core networks may look to optimise the 3GPP architecture in a proprietary way to make it more efficient to run on their proprietary infrastructure. If 3GPP standards prevent this, hyperscalers may become more active in setting 6G standards.

11.7 6G for Edge Cloud

While 6G solutions can run on an Edge Cloud, there is also the opportunity for 6G to promote or be a catalyst for a more general-purpose public Edge Cloud. It would run new revenue-earning Edge Cloud services for the MNOs and their consumer and enterprise customers. The revolutionary 5G to 6G change is infrastructure moving from a focus on 'private Edge Cloud for 5G' to '6G for public Edge Cloud'.

There is a significant step change of requirements moving from an 'Edge Cloud for 6G' to a '6G for Edge Cloud' paradigm. 'Edge Cloud for 6G' is a private cloud running limited and fully qualified 6G network functions. '6G for Edge Cloud' is a public cloud running unlimited and unqualified applications requiring all the technology of the hyperscalers' public cloud solutions but with much more constrained compute and network resources.

11.7.1 6G for Edge Cloud challenges

A general-purpose public Edge Cloud infrastructure has many more challenges than a private Edge Cloud used only to run 6G network functions. The main challenges are:

1. Establishing common Edge Cloud standards that are widely adopted.
2. Reduced security from running unqualified third-party applications.
3. Mobility support, especially resource management, is necessary when facing a dynamic and mobile user base.
4. Supporting roaming users authorised by third parties.
5. Orchestration across multiple Clouds and networks.

Sections 11.7.1.1–11.7.1.6 explore the details of these challenges.

11.7.1.1 Edge Cloud standards challenge

Establishing common Edge Cloud standards should permit interworking, interoperability and efficient competition at the infrastructure and service levels.

The GSMA Operator Platform Group [38] GSMA recognises the need for commonality in Multi-Service Edge Clouds (MEC) to achieve a global, interoperable scale [39]. The GSMA Operator Platform Group has over 40 network operators and 25 ecosystem partners working on defining API requirements for and running pre-commercial trials of a 'Telco Edge'.

The ETSI Multi-access Edge Computing (MEC) Industry Specification Group [40] ETSI is creating standards for an open Edge Cloud environment that aims to unite the telco and IT-cloud worlds, providing IT and cloud-computing capabilities 'on top' of the mobile network elements. The ISG MEC specifies the elements to host applications in a multi-vendor, multi-access-edge computing environment. ETSI MEC addresses security and life cycle management as well as APIs. The GSMA OPG is focused on those interfaces required between MNOs and uses ETSI MEC APIs where applicable. The MEC has created Federation Enablement APIs [41] to support the GSMA Operator Platform Group requirements, enable inter-MEC system communication and allow 5G operators to collaborate among themselves and with service cloud providers.

ETSI's 3GPP 5G Release 17 has a technical specification TS 23.558 for an architecture enabling Edge Applications [42] focused on an 'Edge Enabler Layer', which connects Edge Application Servers to User Equipment and an Edge Management Layer. TS 23.558 [42] states that alignment with the ETSI MEC specifications is implementation specific. TS 23.558 [42] also identifies overlaps with the GSMA OP specifications. The Linux Foundation (LF) Edge organisation [43] aims to establish an open, interoperable framework for Edge Cloud using open-source projects. ETSI and the Linux Foundation held an Edge Hackathon in Silicon Valley in November 2023 [44] and October 2022 to demonstrate alignment. ETSI MEC and LF Edge have broad industry involvement, but the hyperscalers are notably absent.

TS 23.558, ETSI MEC and GSMA OP specifications are not identical or fully compatible, with unique and overlapping capabilities. Figure 11.6 compares the high-level architectures of these different specifications.

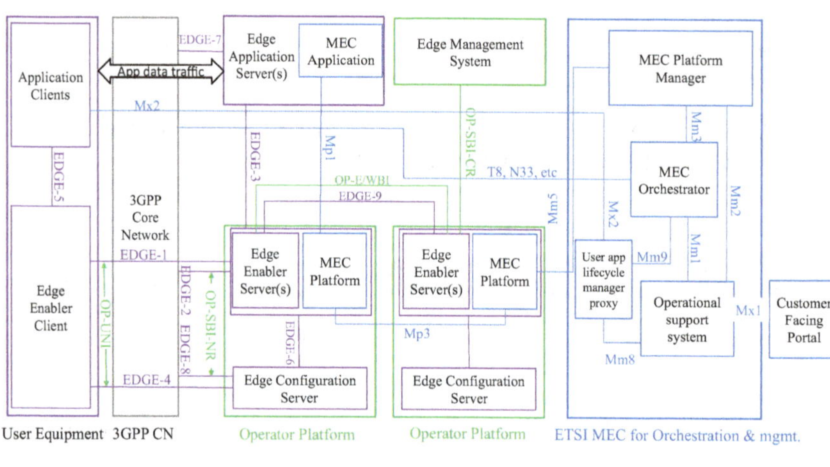

Figure 11.6 3GPP TS 23.558, ETSI MEC, LF Edge and GSMA OP specifications compared

There is an opportunity for 6G specifications to resolve these issues. Edge Cloud standards should be integral to future 6G standards to drive adoption, implementation and critical mass.

Further to the standard interfaces defined by ETSI, GSMA and 3GPP, common programming and communication abstractions are required to function in a highly heterogeneous environment. These abstractions allow a programme to be written once and run anywhere. Many computational abstractions are available, e.g. containers, actors, functions and general-purpose programming languages. Some Edge Cloud in-network functions may require domain-specific languages, e.g. eBPF [45], for packet processing. Achieving standards consensus on programming and communication abstractions is unlikely, but popular ones may emerge much as popular programming languages and operating systems have.

11.7.1.2 Edge Cloud resource management challenge

When the user equipment (UE) or service is required to start an application in the Edge Cloud, the system must discover the preferred Edge Cloud nodes and resources. In many cases, the service runs across several nodes, presenting an NP-hard[9] optimisation problem of embedding the graph of service functions in the infrastructure graph. The placement decision must consider the users' requirements and the Edge Cloud operator's policies. The mechanisms for discovery should be location-independent; that is, the method or protocols should not vary by location and should not require the user equipment to be aware of network topology. The optimum selection of resources is made complex by various factors:

1. Heterogeneous compute resources, e.g. mixes of CPU, GPU and storage types.
2. Heterogeneous applications and service requirements. There are variations in how long the resources need to be held or used, the required resources, responsiveness and importance.
3. Scarcity of resources, especially in smaller Edge Cloud nodes, i.e. lack of 'law of large numbers'.

It should be possible to juggle or optimise resources dynamically without significantly impacting service by designing the applications and services as 'cloud native' or 'Edge Cloud native'. However, third parties design the applications and services, which makes the applications' behaviour unpredictable from the Edge Cloud resource management point of view. It may be necessary for the Edge Cloud provider to qualify the third-party applications. Edge Cloud resource management and optimisation are candidates for AI techniques due to the many variables and combinations. An intriguing possibility is also for AI/ML to learn the behaviour of and qualify third-party applications on the fly; one of many examples of research in this is Khan [46].

[9]NP-hard problems are difficult to solve because they require checking a large number of possible solutions. Usually solved using heuristic algorithms.

11.7.1.3 Mobility support challenge

The resource management challenge becomes even more significant when considering the mobility of user equipment. The optimum Edge Cloud node for an application or service to run may change as the user moves away from one Edge Cloud node towards another; in the worst case, the user may roam onto a different operator's network Edge Cloud. The GSMA Operator Platform refers to this as 'Requirements for Application Session Continuity' [47]. If the application is stateless and contains no permanent data, it can be moved on the fly to a new node without interruption to service. Authentication and authorisation may cause a delay, especially if the user moves between different operators. Preserving the state data of an application moving from one node to another is complicated by the need to move the state. Classical Cloud management systems can move stateful applications, e.g. VMware's VMotion functionality, by copying the running application's live memory. If the data within the application change quicker than state data can be copied, then it is impossible to move the application without some interruption to service. Communication between the application, network and resource management would allow the application to be informed and prepared before the move. However, this requires the standardisation and adoption of a new protocol by application writers.

11.7.1.4 Orchestration across multiple clouds and organisations challenge

Distributed Management or orchestration scales applications by adjusting the amount of resources used by an application (vertical scaling) or adjusting the number of instances of the application (horizontal scaling) using distributed control mechanisms to avoid bottlenecks or single-point-of-failures.

Users, primarily Industry 4.0 users, may require the distribution of applications across multiple networks requiring inter-network orchestration. While Edge Cloud resource management within a single network provider is challenging, meeting these challenges across multiple providers further exasperates trust, security and infrastructure, hardware and software heterogeneity issues.

When public 6G users roam between different network providers, they expect their applications to continue working. The users may not even know their application is running on the Edge Cloud. Roaming extends the mobility support challenge to become an inter-MNO or inter-organisational challenge. Questions arise around whether one MNO can manage the Edge Cloud resources in another MNO, perhaps via a 'slicing' mechanism. Allowing inter-MNO resource management (or slicing) further fragments Edge Cloud resources. Alternatively, MNOs need to trust each other to provide the correct resources to maintain their customers' quality of experience.

11.7.1.5 Security challenges

A public 6G Edge Cloud service must run third-party applications that cannot be trusted. Central Clouds manage this issue well today, and Edge Cloud can apply many central cloud security techniques. Scarcity of resources, physical site exposure and use of lighter weight serverless or functions as a service creates more security challenges for Edge Cloud.

The scarcity of computing resources at individual Edge Cloud nodes makes those individual sites more vulnerable to denial-of-service attacks, although the overall collection of Edge Clouds may be more secure and resilient.

Remote, small facilities or buildings may house Edge Cloud nodes, which are easier to break into than a large Central Cloud data centre, making inserting physically malicious software or hardware easier. The lack of Edge Cloud physical security makes hardware and software attestation essential. Implementations of remote attestation technologies are already in use by Central Cloud providers. Further third parties, e.g. other MNOs or Industry 4.0 organisations, may require proof of attestation of Edge Cloud nodes. Direct anonymous attestation is a method that enables remote attestation of a trusted computer while preserving user privacy, which should satisfy Edge Cloud requirements. Direct anonymous attestation implementations are available within proprietary Trusted Execution Environments, e.g. Intel SGX, Arm TrustZone, AMD SEC and RISC-V [48].

Many 6G Edge Cloud use cases are safety-critical, e.g. autonomous vehicle controls and robots, so Edge Cloud must provide high availability and reliability.

11.7.1.6 Edge Cloud and Industry 4.0 challenges

Industry 4.0 will be a significant user of Edge Cloud, but in this case, it is more likely to be a private on-premises cloud and, therefore, more compatible with a private or virtual private (slice) 6G network than a public 6G network. Edge Clouds powering Industry 4.0 have fewer challenges than public Edge Cloud because of their much-reduced scope. The private on-premises cloud only runs qualified applications, and the system has tight specifications around APIs, operating systems, management, orchestration, DevOps environment and infrastructure. Further, user equipment mobility and roaming are restricted. Cloud service provider selection is a strategic enterprise choice.

11.7.2 Edge Cloud services

6G Edge Clouds allows MNOs to offer more than basic infrastructure as a Service. Box 11.4 describes some typical 'as a service' types. Like the hyperscaler clouds, MNOs could offer more value-added services, such as Platform as a Service (PaaS) and Software as a Service (SaaS).[10] Standardising these services for 6G permits users global roaming while bringing sufficient scale and users to justify the investment in Edge Cloud infrastructure. A distributed Edge Cloud has the potential to offer Edge PaaS that differentiates Edge Cloud from Central Cloud PaaS. Potential Edge PaaS are Consensus as a Network Service, Artificial Intelligence as a Service, microservices and serverless functions.

[10]See Box 11.4: As a Service Terminology.

Box 11.4: As a Service Terminology

Clouds introduced the 'as a Service' terminology to categorise the business model of the services offered in terms of the value adds of the service presented to the customer:

- **MaaS** – Metal as a Service – is designed to automate the deployment of computers or servers (colloquially called 'Metal') in cloud data centres.
- **IaaS** – Infrastructure as a Service – provides fundamental computing resources where the user can run arbitrary software, which can include operating systems and applications.
- **PaaS** – Platform as a Service – supports specific programming languages, libraries, services and tools that can be used by the user created applications. The user controls the applications but not the compute infrastructure.
- **FaaS** – Function as a Service or 'Serverless Computing' a form of PaaS where the cloud provider manages the resources required (hardware and operating systems) on demand. This allows the functions to scale from zero to millions of instances without the customer having to manage the number of servers and instances of operating systems required. Functions should be small, stateless and ephemeral.
- **SaaS** – Software as a Service – provides specific applications running on cloud infrastructure. The applications are accessible from the users' devices through a web browser or a programme interface.
- **AIaaS** – Artificial Intelligence as a Service – provides AI tools that run in the cloud. A specific form of SaaS where the application offered is an AI tool.
- **NaaS** – Network as a Service – Has varying widely used interpretations but generally refers to a service that provides networks or network connectivity on-demand. Often associated with NFV and SDN implementations but sometimes used as a more generic marketing term to describe how a network service is sold and provisioned. In a 5G or 6G context it often refers to how a 'network slice' is provisioned by software.

Dang [49] describes an implementation of a 'Consensus as a Network Service', which delivers a consensus protocol implemented in the network. Consensus protocols implement functions that would otherwise be impossible to achieve in distributed systems, such as consistent agreement on data or decisions preventing conflicts that independent operations can generate. They function even in the event of faults or attacks. They enable joint and fair decision-making. They scale with the number of participants. Implementation in the network would accelerate the speed of reaching consensus while improving robustness.

Example 6G applications of consensus are autonomous vehicles agreeing on positions, velocity and avoidance vectors. Cobots agreeing on the distribution of work and distributed ledgers, e.g. blockchains.

Dang [49] proposes adapting a well-known consensus protocol called Paxos for implementation at ultra-low latency and high speed in the ASICs[11] of network equipment. Section 11.7.3 describes the concept of using networks for compute. Consensus protocols could be implemented in standard computer-driven Edge Cloud to achieve more scalability, lower latency, better privacy and more reliability than in Central Clouds, especially in scenarios where the devices that need consensus are physical neighbours, e.g. cobots and autonomous vehicles.

'Why Cloud Applications Are not Ready for the Edge (yet)' [50] observes issues with the distribution of microservices between Edge Cloud and Central Cloud due to '*The number of transactions multiplied by the network delay between the edge and the remote centralised cloud causes response times to increase dramatically*' which necessitates a redesign of applications and microservices. The microservice architecture is sensitive to network latencies since it involves passing many messages and multiplying network latency many times to slow down the completion of the transactions or applications. Excessive network latency can lead to race or deadlock conditions between the microservices, which can cause the application to crash. A 6G Edge Cloud can implement common microservices close to the users' applications. This reduces the latency of the end-to-end transactions by multiples of the reduced network latency.

Serverless functions are small, stateless, ephemeral pieces of software run in a cloud as a Function as a Service in a style called 'Serverless computing'. The essence of serverless functions is that they are on-demand creations, usually in response to events, perform a specific function in a short time, usually less than a few seconds, and are then deleted. Project Piccolo's [51] vision was that '*Distributed applications should be able to seamlessly use in-network computation without worrying about where it is executed and how – just as endpoints do not worry about how their packets are delivered by the Internet today*' [52]. The Piccolo vision could be adopted in 6G to revolutionise how cloud services are delivered; that is, compute functions would run in the network without the user or application writer concerned about where or which infrastructure provider executed compute functions.

Several researchers have proposed Artificial Intelligence as a 6G service; see Ref. [53], which Chapter 13 describes in more detail.

11.7.3 Using networks for compute

Given the importance of low latency and scalability for 6G use cases, the performance benefits of using network equipment for some compute tasks warrant further research and development. Networks are massive, distributed computers performing the particular task of communication. Imagine if the network was a more general-purpose distributed computer that could participate in edge and centralised Cloud. It could accelerate computations, particularly complex distributed calculations, reduce power consumption and latency, improve robustness and optimise the distribution of computing tasks. It could also reduce data traffic load on the network and data centres.

[11]ASIC = Application Specific Integrated Circuit. ASICs are often used in very high speed network devices.

This section proposes the radical proposition that MNOs could use their network infrastructure to vastly accelerate their own and customers' data processing and compute tasks. Using networks for compute would differentiate the MNOs from and complement the hyperscalers' cloud services.

The idea of network equipment running programmes dynamically carried in packets is not new; Wall [54] proposed 'Messages as active agents' in 1982, while 'Active Networking' was an active research area in the mid-1990s. Active Networking conceived that packets could carry programmes the network switches would implement. It was a method to modify the behaviour of networks. Despite significant research, the industry did not adopt Active Networking as it pre-dated Cloud Networks and SDN, so there was no ready market for it. Now that the use of the Cloud is widespread, where applications run code in places other than the user equipment, there is an opportunity to look again at running programmes in the network. Dang *et al.* [49] describe running a consensus algorithm using a programming language called P4 [55] that can run on some ASICs in network switches. Dang *et al.*'s results show that a Tofino [56] network switch, an Ethernet switch from Intel programmed in the P4 language, can run the Paxos consensus algorithm five times faster with seven times more messages per second than a classical implementation on an Intel server. Another example is a neural network implemented in P4, as demonstrated by Paolucci *et al.* [57].

The IETF had a 'Computing in the Network Research Group (COINRG)' [58], which examined '*deploying processing functionality on networking devices, such as switches and network interface cards*'. This initiative did not achieve any IETF standards.

Compute First Networking [59] is an alternative initiative similar to 'COmputing In the Network' (COIN) with an ambition to provide a general-purpose programming platform distributed over a network. Dagstuhl [60] held a seminar on 'Compute-First Networking' in 2021. The seminar discussed use cases such as privacy-preserving edge video processing, connected and automated driving and distributed health applications leveraging federated machine learning. It discussed research challenges and assessed recent and expected developments in networking and computing platforms. The seminar concluded that Compute First Networking has potential similar to that of cloud computing to transform business but is at least 20 years behind it.

Project Piccolo [51] defined a CFN node to '*deliver an open, low latency, efficient, secure, in-network compute implementation*'.

CFN and COIN are research concepts closely related to Edge Cloud, which may transform the Internet, Cloud technologies and businesses, but they are probably too futuristic to be included in a 6G implementation within the next 10 years.

11.7.4 6G for Edge Cloud conclusion

Some of the technologies that allow MNOs to implement '6G for Edge Cloud' services and reap the value-added rewards exist. Work is progressing on the required standards and remaining technology gaps, as highlighted by the challenges listed earlier. The biggest challenge is whether '6G for Edge Cloud' gains critical mass. Will the hyperscalers and other industry behemoths adopt the developing standards? Will the industry invest in the ubiquitous deployment of Edge Cloud infrastructure? Incorporating the developing Edge Cloud standards in future 6G standards will be essential but insufficient.

A proprietary evolutionary approach driven by cautious investment and partnerships between CSPs and hyperscalers is probably the way forward, although this leads to market fragmentation and inefficiencies. Custom deployment of Edge Cloud for Industry 4.0 applications is likely, but again, this will be led by partnerships between CSPs and hyperscalers, with the Industry 4.0 organisations deciding their strategic partners.

The radical opportunities for using networks as compute are probably beyond 6G timescales.

11.8 Packet transport and transmission

Since the invention of the Internet, there has always been a question of whether transmission and packet switching networks, which most fixed network operators consider their core network, can keep pace with the Internet's growth. We ask the same question for the 6G network: can the transmission and packet switching networks handle 6G multi-gigabit speeds per person?

Measurements show that data volume growth is not outstripping the increase in capacity due to technological advancements. Figure 11.7 shows the broadband data volume shipped annually by Openreach, the UK's fixed fibre and copper network provider. Figure 11.7 shows that the Covid-19 lockdown caused a significant jump in broadband traffic, but this growth has now levelled off. Ofcom, the UK's telecommunications regulator, data show a more linear growth in fixed and mobile broadband traffic growth (see Figures 11.8 and 11.9), while the number of fixed broadband lines is almost static (Figure 11.10).

Figures 11.11 and 11.12 show the increase in equipment interface speeds and packet switching speeds, which outstrips data volume growth.

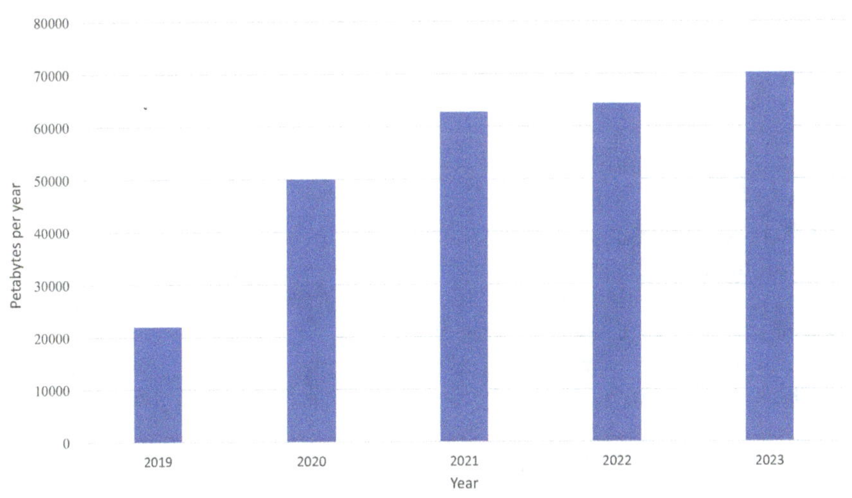

Figure 11.7 Openreach annual broadband traffic volume

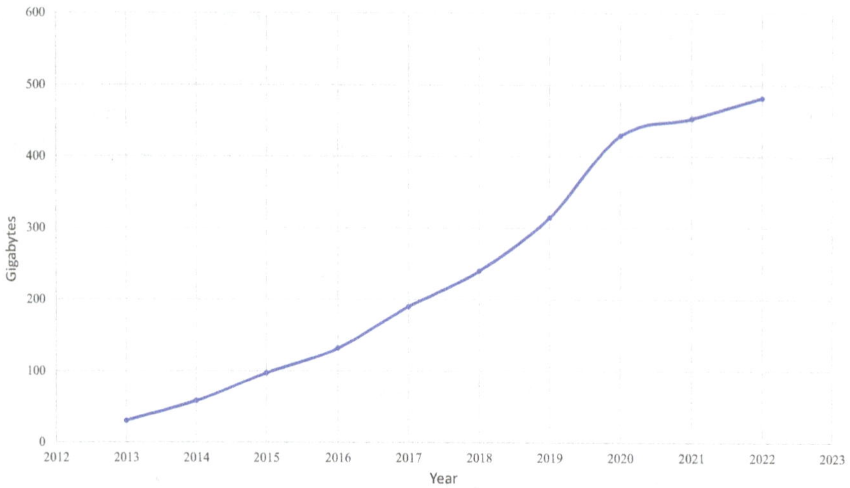

Figure 11.8 Average UK monthly fixed line data usage

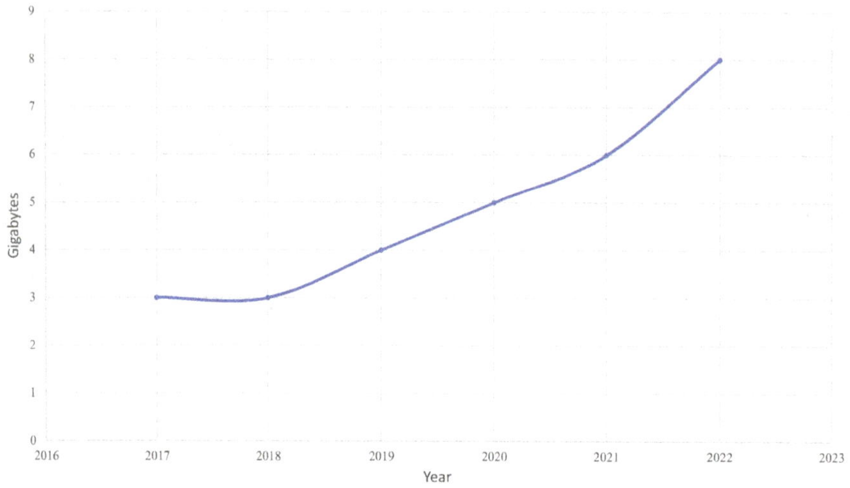

Figure 11.9 Average UK mobile data usage

Laboratory experiments have several times in recent years claimed to have sent more than the global Internet traffic per second down a single fibre, indicating the future potential of optical networks is sufficient. Jørgensen *et al.* [61] measured 1.84 Pbit/s[12] over a 37 core 7.9 km fibre in 2022. Irving [62] claims this is more than the global Internet bandwidth of 1 Pbit/s. Jørgensen, in theory, could achieve 100 Pbit/s. The Japanese National Institute of Information and Communication

[12]One Petabit per second = one thousand Tbit/s = one million Gbit/s.

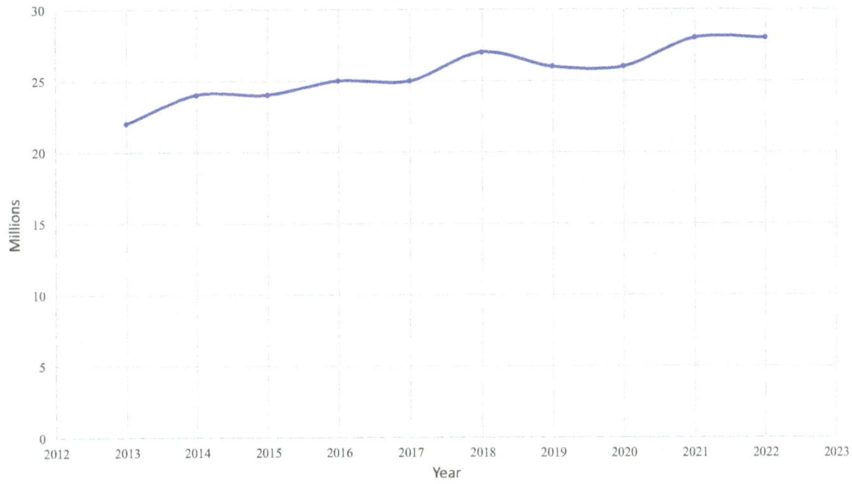

Figure 11.10 Number of fixed broadband lines in the United Kingdom

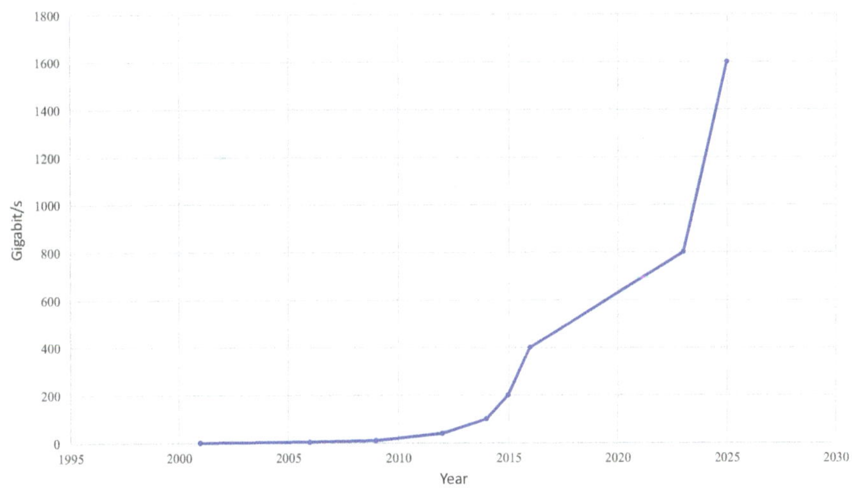

Figure 11.11 Maximum speed of Ethernet interfaces

Technologies (NICT) measured 1.53 Pbit/s in November 2022 using an optical fibre of a standard diameter over 26 km. It is hard to compare how these top capacities have evolved over the years due to the different numbers of fibre cores[13] and the use of different distances. Figure 11.13 plots the claimed rates in Pbit/s over recent years.

[13]Optical fibre cores are the innermost part of an optical fibre that guides the light and hence carries the data signal. All commercial fibres have a single core but experiments are being conducted using fibres with multiple cores to increase a fibre's data capacity.

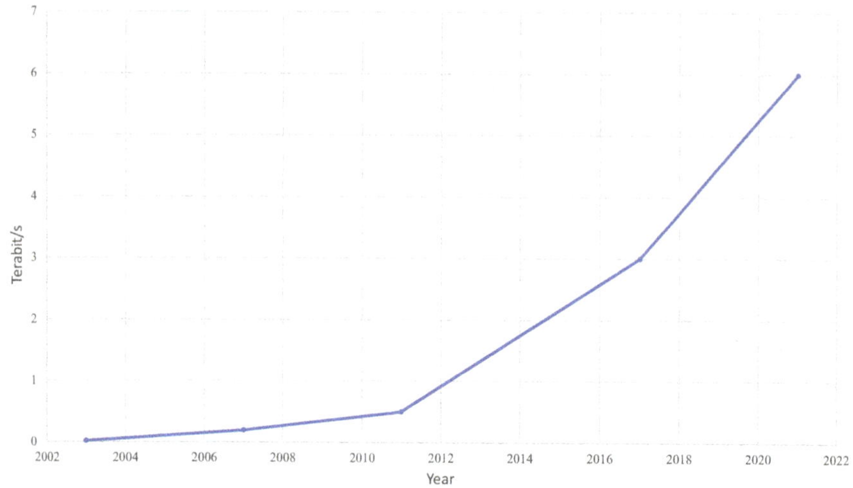

Figure 11.12 Nokia FP chipset packet switching capacity

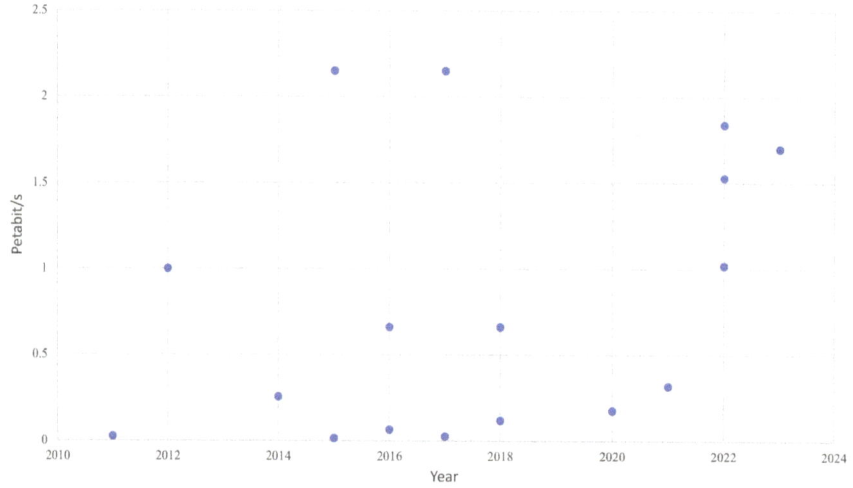

Figure 11.13 Highest capacity optical fibre claims in Petabits per second

Meanwhile, Broadcom [63] has chipsets designed to build an Ethernet switch capable of switching 25.6 Pbit/s; while not designed for Internet traffic, it indicates what electronic packet switching can achieve. If this is too constraining, optical packet switching (OPS) has been a research subject for many years. OPS can switch packets in the optical domain without resorting to the electronic domain. However, no practical and commercial OPS technologies are yet available, and exploiting OPS may require the redesign of existing network topologies and protocols.

11.9 5G core ecosystem and competition

The 5G network core is critical national infrastructure; if the ecosystem providing all its technology is fragile, so is the nation. We therefore ask if the Mobile Core Network ecosystem is healthy. What is the economic impact of Core Networks on MNOs, and what might the ecosystem future look like?

In 2024, there are approximately 1250 MNOs across the globe running approximately 4600 networks [64]. As of June 2023, some 249 of these MNOs have launched commercial 5G services [65], of which 57 have commercial Stand Alone 5G cores. The GSM Association [66] expects operators to invest around $1.5 trillion worldwide in capex between 2023 and 2030. An FCC report estimated that a 5G non-standalone Evolved Packet Core for 50,000 subscribers would cost between $250K and $1.2M [67]. Suppose we assume that a 5G Stand Alone core would cost similarly but with a 75% volume discount. In 2023, the global number of MNO subscribers was approximately five billion; therefore, a full deployment of 5G SA cores across the globe would cost ~$7.8 billion to ~$35 billion or half a per cent to 2.5 per cent of total MNO CAPEX. This is a substantial sized market able to support a diverse Mobile Core ecosystem.

Many vendors are competing to provide 5G SA Cores, including but not limited to (in no particular order) Huawei, Ericsson, Nokia, Samsung, ZTE, NEC, Cisco, HPE, Mavenir, Microsoft, Casa Systems, Radisys, Oracle, Rakuten, Athonet and of course Amazon's private 5G network solution. Open-source projects supplement this with Open5GCore, free5GC, open5gs, Telecom Infra Project's Magma, and 5G Core, to name only a few. For this reason, the UK government notes in its Telecoms Supply Chain Review that '*There is a greater diversity of supply in core network functionality*' compared to RANs.

The challenge for building a 5G core is not the lack of vendors or open sources but in selecting, building, integrating, testing and operating a 5G Core. For these reasons, Microsoft acquired AT&T's Network Cloud infrastructure in 2021. If 6G requires core networks, MNOs may outsource some or all their operations to the hyperscalers. Such a move would also influence the organisations that own and run Edge Cloud infrastructures and perhaps applications, significantly impacting which organisations make the most commercial value from 6G.

11.10 Summary and conclusions

The 6G core could be very similar to the 5G Core, but that would be shortsighted and missing an opportunity to improve its services based architecture and microservices architecture's costs, efficiency and reliability.

Thanks to softwarisation NFV and SDN technologies, upgrading a 5G core to a 6G core may be a software-only upgrade. Continuous Integration and Continuous Deployment, as used by DevOps, make this software upgrade easier than previous generation core upgrades. For those concerned about the performance of softwarised 5G cores, it is rapidly advancing, with Intel and Samsung demonstrating a single server, one Tbit/s UPF, in 2023. However, legacy network and equipment management systems hamper MNOs' evolution journey.

Today's 5G networks are not fully cloud-native; they use stateful protocols between the RAN and the 5G core, creating fragile pinch points. A fully Cloud native 6G network would be much more robust, as required by many Industry 4.0 use cases. R&D must be undertaken to make 6G protocols stateless, perhaps using REST approaches.

Today's services based architecture and microservices architecture have issues with the magnification of control plane signalling they create. These are not just inefficient but limit distribution options due to latency magnification when one transaction requires many messages. The microservice and services based architecture concepts were created to make work easier for programmers and compute resource planners; the correct software tools could perform both tasks.

More radical architectures could deliver more benefits. This chapter explored the following options:

1. The Internet has demonstrated the value of distributed computing. A highly distributed core, where RAN components implement 'core' functions or Bloxtel's decentralised core architecture using asymmetric key, PKI and blockchain, would make the 6G network more robust and scalable.
2. Adding MOCN and TWIF/TNGFs to the distributed RAN and Core components would create a Wi-Fi-neutral hosting infrastructure suitable for the inside-out femtocells.
3. The separation of radio infrastructure providers from authentication and billing providers provides a robust means to scale 6G networks. OpenRoaming for Wi-Fi, 'neutral hosts' and eSIMs for 3GPP demonstrate this separation.
4. Our coreless 6G network might be achievable with sufficient R&D effort, providing significantly more benefits than cost reduction, i.e. ultra-reliable, ultra-scalable and ultra-flexible.

Meanwhile, Private 5G network operators can buy a 5G Core in a box.

Private Edge Cloud provides significant infrastructure for 6G networks and services. Researchers and network operators must develop NetDevOps techniques and tools to enable the operation of Edge Cloud. Given the skill sets of network operators and hyperscalers, there will likely be much collaboration between them for the Edge Cloud.

Using 6G for public Edge Cloud has many significant opportunities and challenges. The central challenge is consolidating multiple Edge Cloud standards and driving broad adoption. Resource management, mobility support and orchestration challenges across organisations are also significant.

More radically, 6G network equipment could become part of the compute fabric offering key compute services, such as distributed neural networks or consensus as a service at fantastic sub-microsecond computation speeds. This would change the art of the possible in distributed computing.

For those worried that 6G networks will swamp the Internet with traffic, the development of optical and packet switching technologies is keeping pace even in a 6G era of gigabits per customer, as today there are petabit solutions.

Politicians and security experts have expressed concern at the lack of a broad competitive ecosystem for RANs, which has motivated projects like Open RAN, but the 5G Core ecosystem, and hence expectations for any 6G core, looks broad and competitive.

Even if radical 6G core options are not adopted, new computer virtualisation and cloud technologies, driven mainly by open-source initiatives, may necessitate a rethink of today's 5G core architecture and modus operandi.

References

[1] Chiosi M, Clarke D, Willis P, *et al.* https://portal.etsi.org/. [Online]. 2012 [cited 2023 Feb 10]. Available from: https://portal.etsi.org/NFV/NFV_White_Paper.pdf.

[2] Wikipedia. [Online]. [cited 2023 July 10]. Available from: https://en.wikipedia.org/wiki/Software-defined_networking.

[3] 3GPP. 3rd Generation Partnership Project; Specification Group Services and System Aspects; System architecture for the 5G System (5GS). 2022. 3GPP TS 23.501 V18.0.0 (2022-12).

[4] Wikipedia. [Online]. [cited 2023 July 10]. Available from: https://en.wikipedia.org/wiki/Containerization_(computing).

[5] Wikipedia. [Online]. [cited 2023 July 10]. Available from: https://en.wikipedia.org/wiki/Cloud-native_computing.

[6] Wikipedia. [Online]. [cited 2023 July 10]. Available from: https://en.wikipedia.org/wiki/Microservices.

[7] Li Y, Li H, Liu W, *et al.* A case for stateless mobile core network functions in space. In *ACM SIGCOMM 2022*; Amsterdam, Netherlands, 2022. pp. 298–313.

[8] Wikipedia. [Online]. [cited 2023 July 10]. Available from: https://en.wikipedia.org/wiki/DevOps.

[9] Wikipedia. [Online]. [cited 2023 July 10]. Available from: https://en.wikipedia.org/wiki/CI/CD.

[10] Shah JA and Dubaria D. NetDevOps: a new era towards networking & DevOps. In 2019 IEEE 10th Annual Ubiquitous Computing, Electronics & Mobile Communication Conference (UEMCON); 2019; New York.

[11] Markwide Research. Vo5G Market Analysis. [Online]. 2023 [cited 2023 Nov 14]. Available from: https://markwideresearch.com/vo5g-market/.

[12] Nokia. Voice over 5G (Vo5G) core. [Online]. [cited 2023 Nov 14]. Available from: https://www.nokia.com/networks/core-networks/voice-over-5g-vo5g-core/.

[13] Barr J. https://aws.amazon.com. [Online]. 2022 [cited 2023 Jul 10]. Available from: https://aws.amazon.com/blogs/aws/new-aws-private-5g-build-your-own-private-mobile-network/.

[14] Lenovo. Lenovo Press. [Online]. [cited 2023 Jul 10]. Available from: https://lenovopress.lenovo.com/lp1483-5g-in-a-box-the-game-changer-for-telco-industry-leaders.

[15] Kersten M. A Cambrian explosion of DevOps tools. *IEEE Software*. 2018; 35 (2): 14–17.
[16] The Linux Foundation. LF Networking. [Online]. 2022 [cited 2023 Feb 10]. Available from: https://lfnetworking.org/.
[17] TL Foundation. https://lfnetworking.org/5g-super-blueprint/. [Online]. [cited 2023 February 3]. Available from: https://lfnetworking.org/5g-super-blueprint/.
[18] Mike Satterlee. How AT&T Sped Up Its 5G. [Online]. 2023 [cited 2023 Nov 14]. Available from: https://about.att.com/blogs/2023/satterlee-open-disaggregated-platforms.html.
[19] Gold K. Unexpected challenges of running Cloud-Native 5G networks. *ICT Today*. 2022: 16–22. Available online from: https://www.e-digitaleditions.com/i/1476549-ict-today-october-november-december-2022/15?
[20] Sofrecom. https://www.sofrecom.com. [Online]. 2023 [cited 2024 Feb 15]. Available from: https://www.sofrecom.com/en/news-insights/technical-challenges-migrating-to-5g-core-network.html.
[21] Dietrich B. https://www.spirent.com/. [Online]. 2023 [cited 2024 Feb 15]. Available from: https://www.spirent.com/blogs/dont-stand-alone-overcome-the-challenges-of-5g-standalone-part-one.
[22] Rune G, Ohlen P, Turanyu Z, and Mildh G. https://www.ericsson.com/. [Online]. 2023 [cited 2024 Feb 15]. Available from: https://www.ericsson.com/en/blog/2023/9/six-talking-points-future-network-architecture.
[23] 5G Americas. https://www.5gamericas.org/. [Online]. 2020 [cited 2023 Feb 10]. Available from: https://www.5gamericas.org/wp-content/uploads/2020/12/Future-Networks-2020-InDesign-PDF.pdf.
[24] Corici M, Eichhorn F, Troudt E, Schreiner F, and Magedanz T. A 6G RAN-core control plane convergence framework. In 2023 26th Conference on Innovation in Clouds, Internet and Networks and Workshops; 2023. pp. 95–99.
[25] 3GPP. https://portal.3gpp.org/. [Online]. 2024 [cited 2024 Feb 19]. Available from: https://5g-ppp.eu/wp-content/uploads/2022/12/6G-Arch-Whitepaper_v1.0-final.pdf.
[26] 5G PPP. [PDF]. 2023 [cited 2023 Jul 10]. Available from: https://5g-ppp.eu/wp-content/uploads/2022/12/6G-Arch-Whitepaper_v1.0-final.pdf.
[27] FUDGE-5G. [Online]. [cited 2023 Jul 24]. Available from: https://fudge-5g.eu/en.
[28] Robitzsch S, Sarathchandra C, Kenan M, *et al.* FUDGE-5G. [Online]. 2022 [cited 2023 July 24]. Available from: https://fudge-5g.eu/download-file/543/iYIVtTS3SKWmXm6QfdQp.
[29] Bloxtel. https://bloxtel.com/. [Online]. 2024 [cited 2024 Feb 16]. Available from: https://bloxtel.com/.
[30] Keliang D, Luhan W, Xiangming W, Yu L, Haiwen N, and Shaoxin H. ML-SLD: a message-level stateless design for cloud-native 5G core network. *Digital Communications and Networks*. 2023; 9: 743–56.
[31] Bit. https://bit.dev/. [Online]. 2023 [cited 2023 July 10]. Available from: https://bit.dev/.

[32] Jolie. https://www.jolie-lang.org/. [Online]. [cited 2023 July 23]. Available from: https://www.jolie-lang.org/.

[33] A. Cabellos. https://datatracker.ietf.org. [Online]. 2022 [cited 2023 Jul 24]. Available from: https://datatracker.ietf.org/doc/rfc9299/.

[34] Juniper. [Online]. 2023 [cited 2023 July 24]. Available from: https://www.juniper.net/gb/en/customers/2023/bt-case-study.html.

[35] Rakuten. https://rakuten.today/. [Online]. 2020 [cited 2024 Feb 19]. Available from: https://rakuten.today/blog/5g-network-reliability-lightreading.html.

[36] Kang I, Hwang K, Ryu S, *et al.* https://samsung.com. [Online]. 2023 [cited 2024 Feb 19]. Available from: https://stg-images.samsung.com/is/content/samsung/assets/global/business/networks/insights/white-papers/samsung-exceeds-1tbps-in-5g-core-user-plane-with-intel/samsung-exceeds-1tbps-in-5g-core-user-plane-with-intel.pdf.

[37] Microsoft News Center. https://news.microsoft.com/. [Online]. 2021 [cited 2023 Feb 10]. Available from: https://news.microsoft.com/2021/06/30/att-to-run-its-mobility-network-on-microsofts-azure-for-operators-cloud-delivering-cost-efficient-5g-services-at-scale/.

[38] GSMA. https://www.gsma.com/. [Online]. 2022 [cited 2023 Feb 10]. Available from: https://www.gsma.com/solutions-and-impact/technologies/networks/operator-platform-hp/.

[39] GSMA. https://www.gsma.com. [Online]. [cited 2023 February 3]. Available from: https://www.gsma.com/solutions-and-impact/technologies/networks/gsma_resources/gsma-operator-platform-group-february-2024-publications/.

[40] ETSI. https://www.etsi.org/. [Online]. 2023 [cited 2023 Feb 10]. Available from: https://www.etsi.org/technologies/multi-access-edge-computing.

[41] ETSI. https://www.etsi.org/. [Online]. 2023 [cited 2023 Feb 16]. Available from: https://www.etsi.org/deliver/etsi_gs/MEC/001_099/040/03.01.01_60/gs_MEC040v030101p.pdf.

[42] 3GPP. Technical Specification Group Services and System Aspects; Architecture for enabling Edge Applications. 2022.

[43] LF EDGE. https://www.lfedge.org/. [Online]. 2020 [cited 2023 Feb 10]. Available from: https://www.lfedge.org/.

[44] Linux Foundation Edge. https://lfedge.org/. [Online]. 2024 [cited 2024 July 11]. Available from https://lfedge.org/the-amazing-2023-edge-native-ai-hackathon/.

[45] eBPF. https://ebpf.io/. [Online]. 2023 [cited 2023 Feb 13]. Available from: https://ebpf.io/.

[46] Khan T, Mehmood A, Riera J, and Song W. Machine learning approach for automatic configuration and management of 5g platforms. In *IEEE 20th Asia-Pacific Network Operations and Management Symposium*; Matsue, Japan, 2019. p. 1–6.

[47] GSM Association. https://www.gsma.com. [Online]. 2022 [cited 2023 Feb 13]. Available from: https://www.gsma.com/solutions-and-impact/technologies/networks/wp-content/uploads/2022/04/GSMA-Operator-Platform-Telco-Edge-Requirements-2022-v2.0.pdf.

[48] Ménétrey J, Göttel C, Pasin M, *et al.* Attestation mechanisms for trusted execution environments demystified. In Eyers D (ed.), *Distributed Applications and Interoperable Systems*. Berlin: Springer; 2022.

[49] Dang HT, Bressana P, and Wang H. P4xos: consensus as a network service. *IEEE/ACM Transactions on Networking*. 2020; 28: 1726–38.

[50] Nguyen C, Mehta A, Klein C, and Elmroth E. Why cloud applications are not ready for the edge (yet). In *ACM/IEEE Symposium on Edge Computing*; 2019; New York. p. 14.

[51] Piccolo. https://piccolo-project.org/. [Online]. 2022 [cited 2023 Feb 13]. Available from: https://piccolo-project.org/.

[52] Grewe D. Use Cases, Application Designs and Technical Requirements. *Deliverable. Robert Bosch GmbH, Bosch Research*; 2020. Report No.: D1.1.

[53] Hexa-X. Deliverable D1.2 Expanded 6G vision, Use Cases and Societal Values. 30/4/21.

[54] Wall DW. Messages as active agents. In *Annual ACM SIGPLAN-SIGACT Symposium on Principles of Programming Languages (POPL)*; 1982; Albuquerque, New Mexico, p. 34–39.

[55] ONF. https://opennetworking.org/. [Online]. 2023 [cited 2023 Feb 13]. Available from: https://opennetworking.org/p4/.

[56] Intel. https://www.intel.com. [Online]. 2024 [cited 2024 Feb 16]. Available from: https://www.intel.com/content/www/us/en/products/details/network-io/intelligent-fabric-processors/tofino.html.

[57] Paolucci F, Marinis LD, Castoldi P, and Cugini F. Demonstration of P4 neural network switch. In *OFC 2021*; 2021; San Francisco.

[58] IETF. IETF Datatracker. [Online]. 2020 [cited 2023 Feb 13]. Available from: https://datatracker.ietf.org/rg/coinrg/about/.

[59] Król M, Mastorakis S, Oran D, and Kutscher D. Compute first networking: distributed computing meets ICN. In ICN '19: Proceedings of the 6th ACM Conference on Information-Centric Networking; 2019. p. 67–77.

[60] Schloss Dagstuhl. https://www.dagstuhl.de/. [Online]. 2024 [cited 2024 Feb 16]. Available from: https://www.dagstuhl.de/.

[61] Jørgensen A, Kong DR, Henriksen M, *et al.* Petabit-per-second data transmission using a chip-scale microcomb ring resonator source. *Nature Photonics*. 2022; 16(11): 798–802.

[62] Irving M. Newatlas. [Online]. 2022 [cited 2023 Aug 2]. Available from: https://newatlas.com/telecommunications/optical-chip-fastest-data-transmission-record-entire-internet-traffic/.

[63] Broadcom. [Online]. 2023 [cited 2023 Aug 2]. Available from: https://www.broadcom.com/products/ethernet-connectivity/switching/stratadnx/bcm88890.

[64] GSMA Intelligence. https://www.gsmaintelligence.com/. [Online]. 2024 [cited 2024 Feb 16]. Available from: https://www.gsmaintelligence.com/data/.

[65] Osio J and Keith E. https://www.spglobal.com/. [Online]. 2023 [cited 2024 Feb 16]. Available from: https://www.spglobal.com/marketintelligence/en/news-insights/research/5g-tracker-97-markets-worldwide-have-commercial-5g-services.

[66] GSMA. GSMA Intelligence. [Online]. 2023 [cited 2023 Aug 3]. Available from: https://data.gsmaintelligence.com/api-web/v2/research-file-download?id=74384059&file=270223-The-Mobile-Economy-2023.pdf.

[67] Dano M. Lightreading. [Online]. 2021 [cited 2023 Aug 3]. Available from: https://www.lightreading.com/open-ran/here-s-how-much-a-5g-wireless-network-really-costs.

[68] ETSI. [Online]. 2023 [cited 2023 Jul 10]. Available from: https://www.etsi.org/technologies/nfv.

[69] Wikipedia. [Online]. [cited 2023 July 10]. Available from: https://en.wikipedia.org/wiki/OpenFlow.

[70] Wikipedia. [Online]. [cited 2023 July 10]. Available from: https://en.wikipedia.org/wiki/P4_(programming_language).

[71] ONF. Open Networking. [Online]. 2023 [cited 2023 July 10]. Available from: https://opennetworking.org/.

[72] Wikipedia. [Online]. [cited 2023 July 10]. Available from: https://en.wikipedia.org/wiki/Tunneling_protocol.

[73] Wikipedia. [Online]. [cited 2023 July 10]. Available from: https://en.wikipedia.org/wiki/Overlay_network.

[74] Wikipedia. [Online]. [cited 2023 July 10]. Available from: https://en.wikipedia.org/wiki/SD-WAN.

[75] CNCF. [Online]. 2023 [cited 2023 July 10]. Available from: https://www.cncf.io/.

[76] Karmel A, Chandramouli R, and Iorga M. NIST. [Online]. 2016 [cited 2023 July 10]. Available from: https://csrc.nist.gov/files/pubs/sp/800/180/ipd/docs/sp800-180_draft.pdf.

Chapter 12
Privacy-enhancing technology and security for 6G

Data is the new oil, and privacy is the new social justice.

– Bruce Schneier

Security is a journey, not a destination.

– Bruce Schneier

12.1 Introduction

Trustworthiness is a theme of many of the 6G research and standards projects described in Chapter 6. Security and privacy are two essential requirements for trustworthiness. Security and privacy are often incorrectly used interchangeably, but they are treated differently in different sections of this chapter. Privacy concerns controlling personal or confidential information, e.g. who knows your bank account number. Security is about protecting IT or network assets from attacks, e.g. your bank's security against hackers.

Telecom fraud,[1] which, according to Talluri *et al.* [1], cost US$32.7 billion in 2019, is a telecom-centric use case that illustrates this. Solving the fraud detection problem requires both the protection of individuals' private data, e.g. who has called you and who you have called, and security to protect the telecom networks.

Section 12.2 examines the use of privacy-enhancing technology (PET) to give privacy in 6G. Section 12.3 examines 6G security challenges and potential solutions.

12.2 Privacy in 6G

12.2.1 Introduction

Privacy is increasingly important as more data about people and systems is collected. For example, the increasing digitalisation of society and industry puts more

[1]Telecom fraud uses telecom products or services to steal money from telecom service providers or their customers. This includes but is not limited to voice phishing (vishing), text phishing (smishing), billing manipulation, SIM swapping and subscription hijacking.

confidential data into the system, e.g. precision healthcare considers an individual's genes, environment and lifestyle in their treatment.

The first reaction is to think this is an issue for greater security to protect the data source and who the data are shared with. However, the value of privacy is not in keeping secrets but in fostering trust. To illustrate this point, consider the UK Post Office Horizon scandal. The scandal involved small sub-post offices run by sub-postmasters who were responsible for preparing their monthly accounts. These sub-postmasters were audited separately by an IT system provided by the Post Office. The IT system, called Horizon, could produce erroneous accounts. The Post Office demanded payment from sub-postmasters, or they faced prosecution by the Post Office using the erroneous accounts as evidence. The Post Office told the sub-postmasters who raised concerns about the faulty system that they were the only ones with a problem. This made them feel isolated and powerless against a large organisation with unlimited funding for prosecution. However, if they had known that there were hundreds of others experiencing the same issue, it would have changed the situation's dynamics. Even if they did not know the identities of the others, it could have allowed for a class action lawsuit against the Post Office. Therefore, privacy is not just about keeping secrets but also about creating an environment of trust where individuals can feel empowered to take action against unjust practices.

PET stems from the solution to the theoretical question Andrew Yao, a computer scientist, asked in 1982, 'Can two millionaires discover who is the richest without either revealing how much they are worth?' Solving the problem of determining the richest millionaire without disclosing their individual net worth to each other involves a solution allowing each millionaire to submit an encrypted statement of their worth to a system that can make a comparison. The calculation process is also encrypted, so those with backend access cannot view the calculation or any related information. Yao proposed a complex cryptographic method, and since then, researchers have proposed many more cryptographic and non-cryptographic solutions. PETs broadly apply to 6G, as discussed here.

PETs are quickly evolving, and Box 12.1 lists some key terminologies.

Box 12.1: Common PET terminology

- **Differential privacy**: When a result of processing the data is released it should not give any more information about an individual than if that individual's data had not been used. Differential privacy allows reasoning about the loss of privacy over multiple queries or transactions.
- **Homomorphic encryption (HE)**: An encryption scheme where it is possible to compute encrypted data without deciphering it.
- **Secure multi-party computation (SMPC)**: Cryptography that enables private distributed computations such that analysis can be performed without the different parties revealing their private input.

- **Trusted execution environment or secure enclaves**: Allows code to be run on a computer processor that cannot be tampered with and its data is protected from the rest of the system.
- **Privacy-preserving computing or privacy-preserving techniques:** Alternative terms for PETs.

PETs are a suite of tools that can help maximise the use of data by reducing risks inherent to data use. Some PETs provide tools for anonymisation, while others enable collaborative analysis on privately held datasets, allowing data to be used without disclosing copies of data. PETs are multi-purpose: they can reinforce data governance choices, serve as tools for data collaboration or enable greater accountability through audits. For these reasons, PETs are also described as 'Partnership Enhancing Technologies' or 'Trust Technologies'.

More generally, many of today's global and societal challenges, e.g. finding cures for diseases, managing ageing populations, climate change, fake news, artificial intelligence (AI) threats, massive fraud and hacking, depend on the aggregation and processing of data. PETs provide methods that aggregate confidential data safely to deliver innovative benefits to society and organisations without losing privacy. Aggregating data has a network effect; the more data aggregated across organisations and nations, the more valuable it is. We think PETs are as crucial to digitising society and industry as 6G; therefore, PETs must be part of the 6G ecosystem.

Increasing concerns about data privacy and sovereignty is another opportunity for PETs and 6G to address, especially given that the critical 6G use cases depend on private or confidential data. Whether personal or specific to an organisation, data have become the currency of the digital world. Data can easily be misused for immoral or criminal purposes if not protected. We can divide data privacy breaches into cyber-security breaches and the intended selling or sharing of confidential information with third parties. An example of the former is T-Mobile's January 2023 data breach, where 37 million customers had personal information exposed. An example of the latter is the scandal where Facebook shared personal data with Cambridge Analytica in the 2010s. Aggregated data are, however, essential to research in many fields, e.g. health, or for optimising the performance of industrial systems, detecting security events and predicting natural disasters. PETs allow value extraction from aggregated personal data without violating anyone's privacy. PETs do not prevent cyber-security breaches, although they reduce the number of places where confidential data needs to be stored, reducing the risks of confidential data loss via cyber-attacks. If 6G operators intend to sell information gleaned from the network, perhaps to drive third-party digital twins and AI models, then 6G operators must not disclose private data about its customers or national critical infrastructures.

The lack of standards for PETs is hindering their adoption. The standards development organisations ISO, IEEE, NIST and others are developing standards for PETs. So why should 3GPP be interested in PETs for 6G? For 6G, PETS would permit 6G operators to sell services safely to third parties using information gleaned from the network, such as usage or sensing data, but not from the customers' data carried confidentially and securely over the network. PETs would also allow 6G operators to pool their knowledge to optimise their networks without revealing information about their infrastructure or customers to each other. These benefits make it critical for 6G to adopt common PET standards.

PETs also complement the following 6G use cases:

1. There are significant benefits to researchers and society from deriving helpful information from personal data for healthcare and assisted living.
2. Biometric information, e.g. face recognition, fingerprints, iris recognition, heart rhythms and voice patterns, could be illicitly obtained from Tactile Internet applications for fraud. Unlike passwords and PINs, it is not easy to change biometric information, so the consequences are more significant if stolen. The Tactile Internet and biometric identification should use PETs to prevent biometric information from being 'stolen'.
3. Digital twins and the Internet of Things (IoT), especially for Industry 4.0, are critical use cases for 6G, and the benefits of sharing these data with researchers and infrastructure managers are significant, but the data could be highly confidential.
4. Smart power grids are a 6G use case where private data, e.g. from smart meters or confidential infrastructure data, would help optimise the power grid.

Many governments have laws to protect data, e.g. the EU's General Data Protection Regulation (GDPR) and the UK's Data Protection Act 2018. These laws place controls on the export of data outside of their jurisdiction. Some organisations may prefer not to export confidential data outside their network or buildings. Edge computing can provide the ability to store and process that data locally, while PETs provide the ability to extract and export valuable knowledge, e.g. insights about the spread of a biological or computer virus.

Secure Multi-Party Computation (SMPC), federated learning (FL) and trusted execution environments (TEEs) are types of PETS that are particularly relevant to Edge Cloud since they are inherently associated with distributed computing. TEEs could give organisations the confidence to run sensitive applications on Edge Clouds operated by third parties. TEEs can be extended to the Extreme Edge, allowing trusted services to be run in the home, office, factory or mobile phone while allowing the users to control the release of their data. The principle is to leave the data where it is and take the processing (software) to the data. Examples include recommender systems, e.g. which movies to watch, what products to buy, insurance quotations and pay-as-you-go charging for equipment, without exporting your data to third parties. For example, see BBC's R&D Databox project [2]. Extreme Edge devices may not be powerful enough to perform data processing or run a trusted execution environment; in this case, a compromise would be to move the data the minimum possible distance for processing, i.e. to the Edge Cloud.

Chapter 13 discusses the AI methods, distributed machine learning and FL used to preserve privacy in more detail.

The rest of this section describes some common PETs and their potential uses in 6G networks.

12.2.2 Trusted execution environments

Would you do your online banking on a stranger's phone? You may trust the hyperscalers, e.g. Google or Apple, to run your program safely in their clouds, but would you trust an unknown third-party computer to run your program, especially if it involved financial transactions or confidential information? Edge Cloud, and especially Extreme Edge Cloud, are potentially physically insecure devices that could have compromised administrators or their hardware or operating systems hacked. They may not be trusted to run programs handling confidential data securely. Fortunately, there is a technology called TEEs that allows them to be trusted. TEEs, also known as secure enclaves, allow secure outsourcing of sensitive data processing to a computer, server, cloud or mobile phone. TEEs work by isolating code and data in an area of the processor and memory protected from the rest of the computer system. The operating system or higher-level administrators cannot access the secure enclave.

TEEs require hardware and operating system support and incorporate a suite of methods to ensure the integrity of the trusted application, the operating system and the hardware on which it runs. Commercial implementations of TEEs that provide efficient computation are widely available, e.g. Intel SGX & TDX, AMD SEV, Arm TrustZone hardware, Apple's iOS Secure Enclave and Android Trusty operating systems. TEEs are even available for low-power IoT devices using energy harvesting with the low-power Arm Cortex-M23 CPU [3].

TEEs could be used in Yao's two millionaires' problem to provide a trusted computing environment in which to run the comparison programme.

TEEs are currently vulnerable to side-channel attacks, e.g. in the past Spectre [4] and Meltdown [5]. A side-channel attack uses indirect information about a computer, such as measuring how long operations take, power consumption and electromagnetic or audio emissions to reveal information that compromises security, e.g. to discover cryptographic keys. Side-channel attacks by attackers with significant resources are a real threat to organisations with highly sensitive data, e.g. government secrets, financial transactions and intellectual property, but they are low risk for everyday computer users. In other words, TEEs may not be suitable for applications handling top-secret data, but they are good enough for personal and confidential data. TEEs may also have a specific role in 6G to allow network operators to trust UEs to run a self-policing function (Chapter 9).

For 6G UE, e.g. smartphones, IoT devices and Edge Cloud should support TEEs. TEEs on UEs would enable a 'move the code to the data' model of processing and transactions, reducing the amount of data shipped across the network and improving privacy.

12.2.3 Homomorphic encryption

Another way to keep your data and computing tasks confidential is to encrypt your data whenever it leaves your device and never have it decrypted until it returns to your device. But how could Cloud computers process encrypted data without decrypting or deciphering it? A technology called Homomorphic Encryption (HE) makes this possible. HE is a property of some encryption schemes, making it possible to compute with encrypted data without deciphering it. HE enables the outsourcing of private data processing, potentially to an untrusted third party, where only the client with the de/encryption key can read the results.

In Yao's two millionaires' problem, each millionaire could encrypt their wealth value with their own homomorphic private key before comparison of their wealth by a third party.

HE is very computationally intensive, so it is slow, expensive and unsuitable for latency-sensitive applications. It also increases the bandwidth required as a homomorphically encrypted data set is larger than the original unencrypted data set. Many HE schemes can only perform limited operations, i.e. do not readily support general-purpose processing, but HE for encrypted machine learning is possible. A compelling use case is outsourcing machine learning using confidential data to the cloud, such as building and using large language models (LLM), e.g. ChatGPT. With today's technology, this is not economically viable, but with LLM compression techniques, improved HE cryptography from ongoing research and dedicated hardware acceleration for HE using FPGAs, it could become feasible [6]. 6G operators may wish to outsource building large AI models using data from their networks to cloud providers without risking the loss of commercial or customer-sensitive information, and HE may be a way to achieve this in the near future. Figure 12.1 shows an example of this.

HE implementations are available and in limited use today. Researchers have demonstrated side-channel attacks, which can extract unencrypted data from a HE system while it is in operation [7].

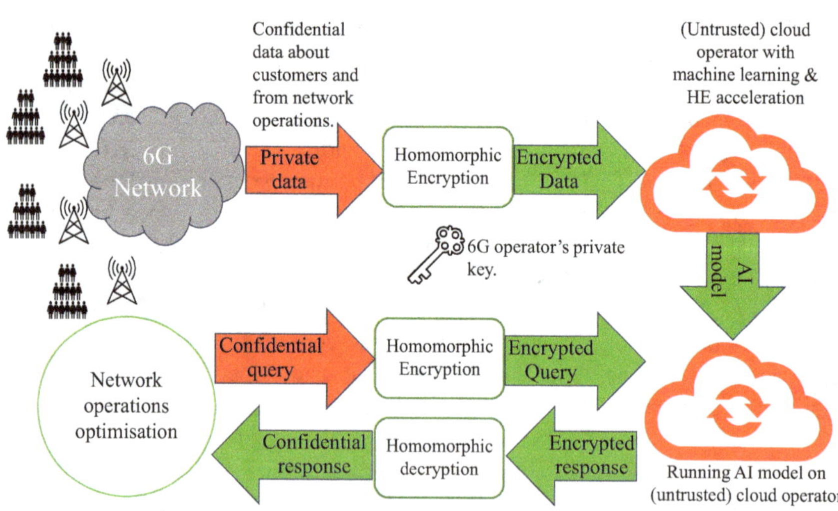

Figure 12.1 Example use of HE by 6G network operator

12.2.4 Secure multi-party computation

There is a set of problems where a group of people or machines require an answer to a problem without revealing their confidential information. It could be trying to income-base split a restaurant bill without revealing how much you earn. A real-life example is the Danish sugar beet auction [8], which uses bids provided by farmers to decide the market price and contracts. The bids are private data as they may reveal a farmer's economic position, and no organisation can be trusted or wants to hold that confidential data. These problems are solved using SMPC.

SMPC enables the joint analysis of sensitive data held by several organisations without revealing that data across the organisations. There is no need for a trusted third party to process the data; the parties involved only obtain the analysis results. SMPC, of course, uses cryptography, so it is highly compute-intensive. It is also communications intensive in that it requires the exchange of many messages between the parties as SMPC fragments data and shares it randomly between the parties.

SMPC is often seen as the solution to Yao's two millionaires' problem, where they only send fragments of their wealth value to each other. SMPC's main difference from TEEs and HE is that it does not rely on a trusted third party; only the millionaires perform the calculations.

Implementations of SMPC are available, but they are at an early proof of concept or pilot stage. Facebook uses SMPC to permit advertisers to optimise their adverts without Facebook revealing individuals' private data; an explanation is available from Meta [9].

Figure 12.2 shows how multiple 6G network operators could use SMPC to optimise their network operations without revealing confidential information about the network or customers to each other.

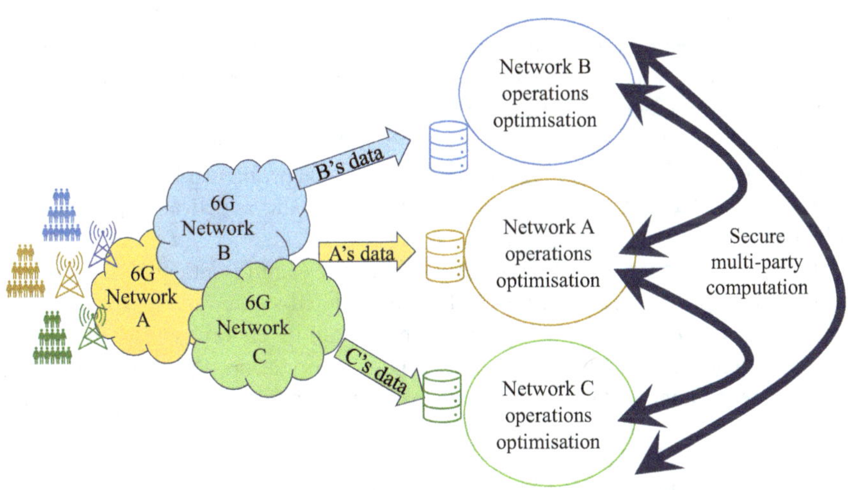

Figure 12.2 Example use of SMPC by 6G network operators

12.2.5 Federated learning

Pooling our confidential personal and industrial data for machine learning and AI optimisation would achieve substantial personal, economic and resource efficiencies, e.g. medical diagnosis algorithms, industry processes, maintenance schedules, traffic routing and fraud detection. Federated machine learning or federated learning (FL) is a technology that achieves this without any loss of privacy or confidential data.

Federated learning trains a model or algorithm by each remote device (e.g. 6G UE) using its data to create its model, which it then forwards to a central server to create a global model. Google's Gboard [10], their predictive text engine for keyboard inputs, is a good example, where each device learns from what you type but does not send what you type to Google but only the model created. Google then combines the models from all devices to create an updated global model, which it can then distribute back to all devices. Chapter 13 also describes federated learning.

According to Talluri *et al.* [1], telecom fraud[2] cost US$32.7 billion in 2019 and with 5G supporting more devices and network functions moving towards the edge, the complexity of fraud detection increases. Talluri [1] proposes using a global federated learning model across multiple MNOs for telecom fraud detection. Wider society and industry could use federated learning for more general fraud detection.

Industry 4.0 could use federated learning to improve the efficiency of industrial processes without individual organisations sharing data with competitors. It could also be used for health to build valuable models, e.g. to predict the spread of disease or improve drugs, without risking sharing patients' private data. Chapter 13 describes how 6G could use over-the-air computation and in-network computing to implement FL for customers at unprecedented speed and privacy.

For 6G, federated learning could build models based on customer application usage. However, end-to-end encryption limits the data available to the mobile network operators (MNOs). Federated learning would allow MNOs to predict traffic demands better and optimise their networks accordingly, without customers having to share their private data with the MNOs. Figure 12.3 illustrates this example of using federated learning by 6G network operators. MNOs can also use federated learning for network fault detection and remediation, where information is learnt from multiple customers experiencing service problems, like an automated and more accurate form of Downdetector [11] with the added ability to suggest fault fixes.

Federated learning is a product in use but may require data spreading across many organisations to be effective, and the distributed system is complex to manage. Federated learning may be vulnerable to 'model inversion', where an attacker can reverse engineer the model to reveal private information. FL is also subject to membership influence attacks where a coordinated campaign can poison the model,

[2]Telecom fraud uses telecom products or services to steal money from telecom service providers or their customers. This includes but is not limited to voice phising (vishing), text phising (smishing), billing manipulation, SIM swapping, subscription hijacking.

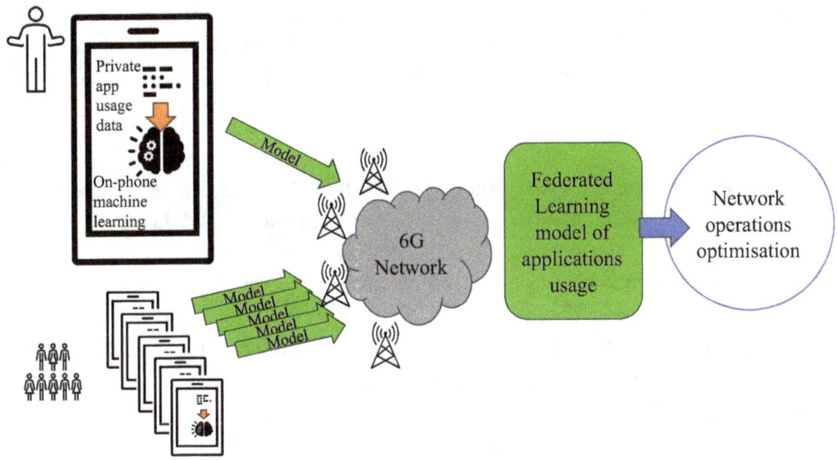

Figure 12.3 Example use of federated learning by 6G network operator

although for a model built from a large number of members, e.g. Google's Gboard, FL is robust to poisoning attacks. How to defend FL from attacks is an area of active research, e.g. Lyu *et al.* [12].

12.2.6 Differential privacy

It is not easy to be sure about the efficacy of any PET without a way to measure and mathematically analyse privacy. Differential privacy enables this, and it is a definition rather than a specific method or technique. Differential privacy is achieved when the result of processing data is released if it does not give more information about an individual than if that individual's data had not been used. Differential privacy allows reasoning about the loss of privacy over multiple queries or transactions.

Yao's two millionaires could use differential privacy to mathematically analyse how effective their algorithm was at keeping the value of their wealth secret.

The usual techniques for achieving Differential Privacy involve adding noise to the dataset or randomised responses. For example, suppose an employer wanted to know how many employees take a day off work sick when they are not sick. Even if a third party conducted the survey, employees might not be honest in answering the survey. If, however, the third-party randomly changed some of the yes answers to no and some of the no answers to yes before recording the data, then the answer to the survey would be correct. If the individual results leaked to the employer, they could not be sure if any individual had taken a day off sick when they were not sick because of the randomised changes to the answers. The employer could, however, find out if specific departments or types of jobs were more prone to this behaviour.

Techniques for achieving differential privacy require large data sets and much expertise to add the right amount and type of noise to get a statistically valid result that protects individuals. Software implementing differential privacy techniques is available, and the US Census Bureau has used it. Differential privacy is subject to ongoing research, and no standards for setting privacy parameters are available.

12.2.7 Privacy-preserving synthetic data

I used to work in the BT Labs, and I lost count of the times researchers from universities or innovators from start-ups wanted confidential data about the network. However, since the network is classified as a critical national infrastructure, these requests could not be satisfied, which probably limited telecommunications research and innovation. The answer to this problem is to create synthetic data about a fictional network with the real network's statistical properties without revealing the actual network's confidential details.

In general, synthetic data creates fictional data representing the original data's statistical properties. Analytics and research can use synthetic data without disclosing data about individuals or systems. For example, a researcher may ask an MNO for data about people's movements to develop more efficient trajectory prediction algorithms for mobile networks. The MNO should not release the raw data as it may allow the tracking of individuals or reveal information about national critical infrastructure. However, it would enable helpful research that could benefit the MNO. Instead, the MNO could create a fictional or synthetic data set with the same statistical properties as the actual data.

The creation of synthetic data is not easy, but AI can help. A method for creating synthetic data using AI is to use a deep learning model, see Chapter 13, which can automatically learn the original data's statistical properties without supervision. Then, AI can generate synthetic data that matches the statistics of the original data. The usefulness of the synthetic data is dependent on the characteristics of the statistics modelled; for example, the synthetic data for trajectory prediction of pedestrians may not be good enough for footfall prediction in a shopping mall.

Software is available to develop synthetic data and has been used to train AI models for languages, autonomous vehicles and fraud detection. Expertise is required to ensure the synthetic data is fit for purpose and does not disclose confidential information.

12.2.8 6G PETs summary and conclusions

PETs enable solutions to many global and individuals' problems by leveraging the value of aggregating confidential information without compromising privacy or confidentiality. PETs foster trust, transparency and best practices that enhance the economy and digital security. The value of PETs to the global economy is enormous, but no one has published an estimate of its value. Yet, Future Market Insights predicts its market size is expected to reach US$ 2.4 billion in 2023 and is projected to reach a valuation of US$ 25.8 billion by 2033 [13].

PETs open new applications and markets, and like the early days of the Internet, we do not know for sure what these are. 6G should at least use PETs, as using common PET standards increases the value of 6G applications and services. The opportunity for PETs in the future digital ecosystem is very significant, and its synergies with 6G ambitions suggest that 3GPP should ensure PETs are part of the 6G ecosystem. 6G network operators could, at the least, use PETs to share 'big data' to optimise network performance and operations while protecting the privacy of their networks and customers, as shown in Figures 12.1–12.3.

We think the evidence presented on PETs in this chapter supports our claim that PETs are as crucial to digitising society and industry as 6G; therefore, PETs must be part of the 6G ecosystem. However, research into using privacy-enhancing technologies for 6G is ongoing [14,15] and not yet on the 3GPP's agenda. In general, the lack of PET standards is hindering its adoption, and the 2023 Royal Society report [16] first recommendation is '*National and supernational organisations, including standards development organisations (SDOs) should establish protocols and standards for PETs, and their technical components, as a priority*'. The report identified the IEEE, NIST, BSI and NPL as example standards development organisations that should identify and convene international expert groups.

12.3 6G Security

In 5G's central use cases, a security failure could be life-threatening or catastrophic, e.g. in health care, remote surgery, autonomous vehicles, Industry 4.0, management of national power grids and other critical infrastructures. Therefore, 5G is the most secure generation yet. 6G needs to be even more secure as 6G's big difference from 5G is 6G's ambition to be the 'network of networks', 'connect the unconnected' and be ubiquitous via the use of non-terrestrial networks, more expansive infrastructure sharing especially Edge Cloud and heterogeneous networks. These ambitions magnify the number of cyber-attack vectors to make 6G not just critical to secure but also challenging to secure. 6G also has a broader user base than previous generations, making it more rewarding to attack.

This section examines some security issues for 6G and potential solutions and starts with a summary of 5G security.

12.3.1 5G security overview

3GGP did much work to make 5G secure at a procedural and protocol level. 5G is secure, and faults in its implementation or operation are the weakest links in the 5G security chain.

The key points of 5G security are a new authentication framework for authenticating UEs to the MNO home networks, privacy protection of the user's identifying information and securing the service-based architecture (SBA) 5G core (Chapters 2 and 11). For the network functions, integrity and security protection of both signalling and user plane data, secure interworking between serving/visited networks and home MNO networks and security for non-public or private networks.

Figure 12.4 shows the functions ensuring a secure connection between users and the 5G networks. 5G authentication and key agreement (5G-AKA) uses

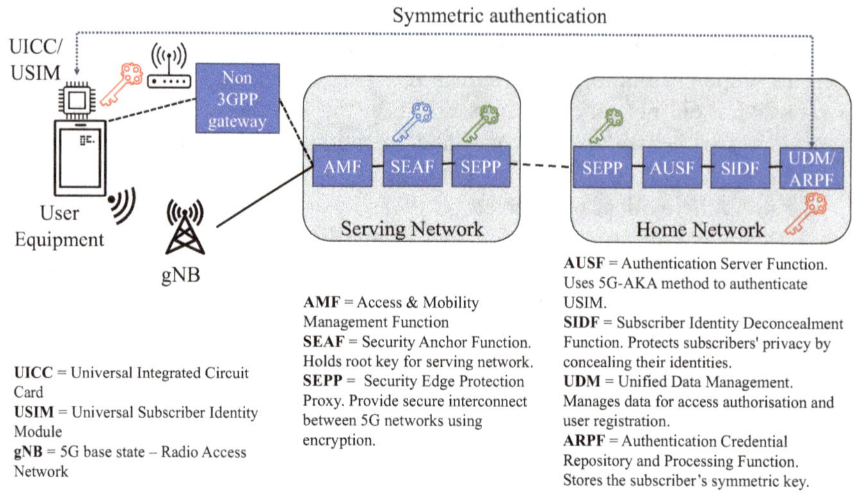

Figure 12.4 5G security functions

symmetric mutual authentication between the UE, using a key stored in the SIM, and the home network's Authentication Credential Repository Function. Mutual authentication ensures that the UE only connects to trusted networks and that only known UE connect to the network. The SIM's (also known as Universal Integrated Circuit Card) secret key is encrypted using the public key of the home network, which prevents any 'man-in-the-middle', including the visited network, from reading the SIM's secret key. This encryption is done inside the SIM to hide the SIM's secret key from the UE. The visited network and the home network both check the authentication of the UE but use a hashing mechanism that prevents the visited network from seeing the SIM's secret key. The visited and home networks form a secure connection using mutual authentication and encryption via Security Edge Protection Proxies (SEPP). The SEPPs also filter and police signalling messages, hiding the MNO's topology from other MNOs, which is essential to prevent anti-competitive behaviours and protect against fraud.

The Subscription Permanent Identifier (SUPI) represents a customer's permanent identity that stays the same throughout the lifetime of the subscription, even if the customer's phone number changes. In 3GPP terms, the Mobile Station International Subscriber Directory Number (MSISDN), the customer's phone number, is not used because it can be changed, reassigned or recycled. The SIM or UE encrypts the SUPI to create a Subscription Concealed Identifier (SUCI); only the home MNO can decrypt this. Further, every time the system uses an SUCI, it changes to prevent an attacker from eavesdropping or pretending to be an MNO network, e.g. a fake base station, to track a specific user. Mutual authentication between the UE and the base station further prevents a fake base station attack. 5G also includes a framework for detecting false base stations based on radio condition information received from UEs.

Network functions (NFs) in the SBA 5G core, see Chapters 2 and 11, need to talk to each other, but these connections must be authenticated and secure to prevent cyber-attacks. NFs perform mutual authentication and secure encryption using IETF's TLS (Transport Layer Security) versions 1.2 and 1.3. Mutual Transport Layer Security (mTLS) based authentication is a method where parties authenticate each other's identities across a secure connection using digital certificates. The OAuth2.0 IT industry-standard protocol passes tokens that authorise an NF to consume the services of another NF. The Web widely uses TLS and OAuth, so these protocols benefit from pervasive real-world experience and security hardening, although at the same time, this lowers the learning barrier for attackers.

Role-based access control (RBAC) manages access to systems and resources based on users' assigned roles, giving granular control over access. In RBAC, users are typically people, but in a microservices architecture, like the 5G core, a user can be a network function. For example, in the 5G core, RBAC would not give the UPF permission to access the SEPP because it does not need to.

5G's softwarisation architecture, see Chapters 2 and 11, introduces security challenges that need addressing. The European Union Agency for Cybersecurity (ENISA) has published 'NFV Security in 5G, Challenges and Best Practices' [17], which lists 60 security challenges and their mitigations.

5G includes standards for authenticating UE to non-public or private networks using UE's that do not have SIMs. However, 3GPP does not specify how credentials, e.g. secret keys, are stored and processed within a UE without a SIM, which could create weak security through poor practices and implementations. UEs connecting to network slices must use a SIM.

12.3.2 Legacy issues

Karakoc *et al.* [18] demonstrated 'bid down attacks' that caused a 5G UE on a 5G network to fall back to 2G, exposing significant security flaws. 2G's encryption was cracked in the early 2000s, so attackers can easily intercept and decrypt phone calls or text messages on 2G[3] networks. If backward compatibility with 4G or 5G is required, attackers can force a UE to fall back to 4G or 5G, and then the 6G UE inherits the potential security issues of 4G and 5G networks. For example, weaknesses in the 4/5G authentication protocols enable energy depletion attacks on low power, e.g. IoT networks [19], impersonation [20] and user tracking attacks [21]. Karakoc proposes protocol modifications that prevent some 'bid down attacks' and suggests that network operators use protocol features that are already available. UE should be extensively security tested before launch. However, Karakoc concludes, '*the sheer variety of attack vectors makes it difficult to fully avoid the threat*'.

Once 6G meets its ubiquitous availability target, perhaps backward compatibility can be removed, but until then, backward compatibility is an essential step in rolling out any new network. Meanwhile, 'bid down protection' defence is

[3]Recent Smart phones have a user configurable option to disable 2G.

required, as well as ensuring up-to-date operating systems and protocol software in the legacy networks and UE to fix known attacks.

12.3.3 Impact of heterogeneous networks

6G's network of networks' ambition to use connectivity via multiple networks, e.g. non-terrestrial networks (NTN), see Chapter 10, non-3GPP networks, see Chapter 9, and infrastructure sharing means 6G must be secure across heterogeneous networks. Multiple types of networks open many more attack vectors, e.g. satellites are more vulnerable to jamming and signalling denial of service (DoS) attacks. There is also the issue of how 6G network operators trust non-MNO and infrastructure providers.

5G (3GPP Release 17) NTN treats the satellite link as a transparent connection to a gNB, so none of the 3GPP protocols are changed other than tweaks to accommodate the increased latency of satellite connections. 3GPP's 5G standards consider various access scenarios: untrusted Wi-Fi, trusted Wi-Fi and fixed and wireline devices, described in more detail in Chapter 9. Trusted Wi-Fi is managed by the MNO and requires strong authentication mechanisms, e.g. EAP-5G. Untrusted Wi-Fi uses standard Wi-Fi authentication, e.g. WPA, and the MNO has limited or no control over the Wi-Fi access point and the fixed network it connects to. Untrusted Wi-Fi requires the UE to use encrypted IPsec, whereas trusted Wi-Fi UE can use null-encrypted IPsec to connect to the 5G Core. In all cases, the devices can have 5G capabilities or not, but all have 5G credentials (Universal Subscriber Identity Module/eUICC[4]) in the form of a pre-shared symmetric key (plus a Subscription Permanent Identifier, public key of the home network and a sequence number). Generally, a different gateway function for each case is implemented in the MNO's core network to connect the non-3GPP access network. 5G does not address untrusted fixed access networks.

There is an opportunity for 6G to increase the network access options, including untrusted networks, while improving security by simplifying the variety of gateways. Too many options and components create confusion and complexity, creating security vulnerabilities. Reducing the number of components, interfaces, procedures and protocols must balance increased functionality. We suggest a generic gateway with a consistent security framework for 6G to minimise security vulnerabilities due to complexity through variety.

6G requires a decentralised approach to its generic architecture as a centralised architecture may not scale and is not compatible with all networks, which is a crucial factor to address in 6G's security architecture. A decentralised or distributed architecture creates security visibility problems; security information must be gathered and correlated from multiple devices.

12.3.4 Softwarisation

Softwarisation, see Chapter 11, introduces the flexibility and speed of response required by MNOs to address cyber-attacks. For example, a compromised NF could

[4]Embedded universal integrated circuit card also known as an embedded SIM (eSIM) which can be integrated in a Trusted Execution Environment.

be quickly removed and replaced. However, softwarisation introduces new security challenges. The ENISA has published 'NFV Security in 5G, Challenges and Best Practices' [17], which lists 60 security challenges and their mitigations. Many of these security challenges would apply whether NFV was used or not, as software always needs maintenance. The main challenges specifically for NFV follow. If attackers breach the virtualisation or cloud system, all NFs running on it could be compromised. If multiple tenants share the same computing resources, there is a risk of data leaks and cyberattacks between tenants. However, most MNOs would not share compute resources with untrusted tenants. Even within a single MNO's 5G network, they use separate security domains to limit the impact of security compromises. For example, the separation of control plane functions from user plane functions from management plane functions and the grouping of network functions with similar security requirements.

Open-source software[5] is increasingly used in 5G networks and is closely associated with softwarisation, from Linux operating systems to complete 5G cores and RANs. Whether open-source software improves or decreases security is much debated, but it does require the MNO to have policies that specifically address the use of open-source software. An ambition of softwarisation is to enable multi-vendor environments, but this requires coordinating security policies and responsibilities across multiple vendors.

Essential security mitigation technologies include:

1. Remote attestation is the ability to check the integrity of the software that the compute infrastructure and the Network Functions use.[6]
2. Trusted computing provides a root of trust to prove that the correct software is running in the right place.
3. Guard rails in the cloud operating systems to ensure no malicious or rogue NF consumes all the resources.
4. Other security mitigation technologies and methods apply, such as anomaly detection or 'security by design'.

The most significant difference with softwarisation is that it opens the black network box. Previously, the black network box vendor was responsible for the security of the software inside the black box. After opening the black network box, security becomes a shared responsibility between the network operator and the software vendors. Even with black network box products, network operators want insight into how the software inside the black box is secured and maintained.

Cloud and virtualisation technologies are evolving. 6G will run on cloud operating systems using different software technologies from today's. It might not be Kubernetes [22], but it could be unikernels [23]. Security software is evolving; for example, remote attestation of running software may become possible.

[5]Open-source code is freely available for anyone to inspect, modify and redistribute. Its supporters claim it encourages collaboration, transparency and innovation.

[6]Current remote attestation technology checks that the correct software is booted but does not check it has not been corrupted after it was booted.

Cryptography is evolving, such as reducing the overhead of homomorphic computing. Because of these evolutions, 6G needs a new security framework, with the possibility that the framework could be less onerous due to improved technologies.

12.3.5 Supply chain

The United States, EU and UK consider the security of telecommunications supply chains essential. The 'UK Telecoms Supply Chain Review Report' [24] describes UK Government policy interventions to create a more competitive, sustainable and diverse supply chain for the UK telecoms sector to reduce the risk of dependency on individual vendors. The USA's 'CHIPS and Science Act of 2022' created a $1.5 billion Public Wireless Supply Chain Innovation Fund to support the deployment of open, interoperable and standards-based RANs to strengthen the USA's telecommunications supply chain. The EU has a similar 'EU Chips Act' with $23.6 billion in funding to support microelectronics and communications technology across the EU supply chain, with projects anticipated in 5G and 6G. Government policies and funding drive 6G towards more open[7] and multi-vendor implementations, reducing the dependency on a small number of 3GPP wireless vendors.

12.3.6 5G-AKA weaknesses

5G-AKA is central to UEs, Serving and Home Networks mutually authenticating each other. While 5G-AKA is more secure than previous generations, its weaknesses have been widely identified and documented. For example, Xiao and Wu [25] list seven, which include:

1. 5G-AKA is vulnerable to replay attacks.
2. Mutual authentication between the UE and the serving network cannot be guaranteed.
3. Attackers can perform denial of service attacks.
4. Under some failure conditions, e.g. lost packets, security also fails.
5. It does not provide perfect forward secrecy. Without perfect forward secrecy, if an attacker obtains the secret key stored in the SIM or the home network, the attacker can decrypt previously recorded messages or future messages.

Xiao proposes a new more secure 5g-ipaka protocol. 6G should consider the various improvements or alternatives to 5G-AKA proposed by researchers. A few of many examples are 5G-AKA-FS [26], 5GSBA [27] and elliptic curve-cryptography [28]. 6G R&D should also consider authentication and key agreements for new services that may run on 6G infrastructure, e.g. digital twins as a service.

5G security depends on Public Key Infrastructure (PKI); the 'public' refers to the key being public, not the infrastructure, and Certification Authorities (CA) that digitally sign key certificates as valid. 3GPP standards address managing PKI and certificates for inter-MNO operations, which are essential to support roaming, but

[7] Although 'open' does not imply common global standards.

the management of PKI and CA within an MNO is outside the scope of 3GPP, and this could lead to weak implementations and operations.

12.3.7 Zero trust

The once-popular perimeter security model creates a boundary between the trusted internal private network and the untrusted external world. Once the user or software function gains entry to the private network, it can try to access any resource it finds. The perimeter security model is weak as any compromised employee or software function, e.g. a worm,[8] can explore the private network to attack vulnerabilities.

The zero trust security architecture (ZTA) takes a different approach and assumes that no user, device, or network function is inherently trustworthy. Authentication should verify all users, devices and network functions before allowing access to any resources, e.g. network services, other network functions and applications and granting only the access necessary to perform their tasks. Employees or software functions cannot explore the private network for vulnerabilities. Networks should be divided into small segments to provide isolation and contain security breaches. Networks and systems should be continuously monitored and analysed for suspicious activity to ensure they are secure, i.e. conforming to security policies, e.g. implanting the correct security configuration and patching software. The zero trust architecture is fundamental to many hyperscalers' architectures, e.g. Google, and a requirement of many organisations, including USA federal agencies.

Olsson *et al.* [29] describe the 3GPP 5G standards as supporting a zero trust approach but identifies further development required in cloud deployments to protect digital identities, attestation of 5G systems, trust in multi-vendor deployments, trust in cloud services provided by third parties, hardware rooted security to address software disaggregation and multi-vendor deployments. 6G requires these developments.

AI has a role in continuously monitoring and analysing 6G networks for suspicious activity and incorrect configuration. AI for 6G is explored more in Chapter 13.

12.3.8 Network slicing

Deloitte [30] states that 5G network slices represent a growth opportunity for MNOs if they can overcome security concerns, with 53% of enterprises surveyed stating that security concerns with network slicing would be an obstacle to adopting the technology.

How secure is a network slice compared to a private 5G network? Network slices, also known as public network integrated non-public networks (PNI-NPN), are implemented on Public Land Mobile Networks, so they must use 5G-AKA and

[8]A software worm is malicious software that explores a network to find hosts that it can replicate itself in. When self-installed in a host it can launch other cyberattacks. Worms do not require human actions e.g. click on links in an e-mail, to spread.

are subject to Legal Intercept requirements. Legal Intercept means that a multinational corporation using network slices has to trust its MNO service providers and the governments of those countries where it uses network slices. A private 5G network would be much more complex for a state actor to tap.

As noted, 5G-AKA does not have forward secrecy, which may be required by many enterprises, especially but not limited to financial transactions. However, 5G Release 16 allows enterprises to run their own authentication and authorisation system, which could have forward secrecy after the UE has completed its primary 5G-AKA.

Network slicing makes the 5G infrastructure multi-tenant; for example, enterprise Alice's network functions share resources with enterprise Bob's network functions. The MNO may have legal contracts with Alice and Bob and, therefore, trust them, but can Alice trust Bob? Thus, implementing network slicing imposes a heavier security burden on the underlying cloud infrastructure, and vulnerabilities to attacks from viruses like Spectre [4] and Meltdown [5] become a concern.

ENEA [31] identifies 5G network slicing protocol weaknesses that would allow a network function in one slice to access information for another slice.

3GPP SA3 has studied the security of network slices with some minor protocol improvements expected in 5G Advanced (Release 18), but the 3GPP has no work planned on network slices for Release 19 or 20.

12.3.9 Quantum safe

A quantum computer uses the principles of quantum mechanics. Quantum computers use qubits, which, unlike regular bits that can only be 1 or 0, can exist in multiple states simultaneously. If you buy into the many worlds interpretation of quantum mechanics [32], you can think of a quantum computer as parallel processing across parallel universes. It is a parallel system of unimaginable scale that could unlock all our digital secrets. From a security point of view, the terrifying fact is that primitive quantum computers exist, and governments, universities and technology companies are investing heavily in their development. 5G security relies on cryptography, so do quantum computers make 5G insecure, and can we make 6G 'quantum safe'?

Cryptography thought to be secure against an attack by a quantum computer is called quantum safe, quantum-resistant or post-quantum cryptography. Today's quantum computers are constrained and suffer high error rates, but the concern is that a future quantum computer could quickly crack today's public key encryption, making the Internet and 5G insecure. Quantum computers of the future may be a risk to some high-value transactions today as a cyber-attacker could record messages today and decrypt them in the future. Other future attack scenarios of concern today include cracking the keys for the digital signatures of important long-lived documents and modifying them, e.g. the owner of an asset could be changed in a digital deed without detection. There has been much research to develop quantum-safe public key cryptography, and the National Institute of Standards and Technology (NIST) will publish standards in 2024.

Symmetric keys of at least 256 bits in length are quantum-safe until after 2050 [33]. Unfortunately, the 3GPP SIM secret key is only 128 bits long. 6G must adopt quantum-safe cryptography.

12.3.10 Physical security

Edge Cloud sites used as 5/6G infrastructure are much more vulnerable to physical break-ins than the central Cloud sites due to their smaller, more distributed nature. If cyber-attackers can physically beak into an Edge Cloud site, then a cyber-attacker can replace or compromise hardware and software in that Edge Cloud site. Attackers may exploit the compromised Edge Cloud site to compromise the whole network. The vulnerability of base stations was illustrated by 5G coronavirus conspirators' attacks [34]. Central Offices or telephone exchanges house Edge Cloud sites today, and although these are more robust buildings than base stations, thieves have broken into them in the past, e.g. BT's Mayfair [35], Stepney [36] and St. Albans [37] exchanges. While MNOs and telcos have robust physical security measures, their Edge Cloud sites cannot be as secure as the hyperscalers' fortress-like sites; for example, see AWS Data Centres [38].

Fortunately, there are software technologies that detect compromised hardware or software. Remote attestation mentioned in Section 12.3.4 can detect compromised hardware or software and stop it from running. Replacing hardware or software without raising anomalous alarms or log messages is also tricky.

12.3.11 Securing AI/ML

6G uses AI, see Chapter 13, which introduces many novel and evolving ways to cyber-attack 6G via the AI vulnerabilities. Some of the most relevant attacks on AI are evasion, information extraction, poisoning and backdoor attacks.

Evasion attacks are where the attacker causes a misclassification during the AI inference phase. The attackers add malicious input to force the AI model to predict the attacker's desired output, e.g. causing an automated vehicle to mistake traffic signs.

Information extraction attacks aim to reconstruct the model or information from its training data, which usually requires knowledge of the training dataset some of which may be publicly available. Attackers use this method to steal models, in which the victim may have invested significant resources in their development. Or infer members or attributes of the training data to find sensitive data such as a user's home address.

Poisoning attacks aim to break the ML model by injecting wrong data into the training set.

Backdoor attacks aim to create a predetermined response from the AI to a trigger input, similar in concept to trigger phrases to activate sleepers in popular spy fiction.

Data engineers can mitigate these attacks by careful control of and statistical manipulation of the training data. They can also use another AI to detect issues in their AI models and training data.

Like any other IT system, an AI must be secured across its design, development, deployment and operational life cycle.

12.3.12 6G security summary and conclusions

5G is very secure for the typical consumer. However, 6G needs to be more secure because its broader use base is more rewarding to attack. Further, 6G's ambition to be the 'network of networks' and use broader infrastructure sharing, especially Edge Cloud, exposes 6G to more threats. This chapter has discussed potential attacks and security weaknesses in 5G that 6G should address, with mitigation suggestions where they are known, which the list below summarises:

1. Bid-down attacks force UEs to fall back to a more insecure generation. Researchers have suggested protocol modifications to prevent some 'bid down attacks', network operators should fully use existing protocol features, and UE should be extensively security tested before launch.
2. Heterogeneous networks are addressed in 5G by various gateways with too many options and components, creating complexity and leading to security vulnerabilities. We suggest that 6G should have a generic gateway with a consistent security framework. 6G requires a decentralised approach to its generic architecture as a centralised architecture may not scale and is not compatible with all networks, which is a crucial factor to address in 6G's security architecture.
3. Softwarisation introduces the flexibility and speed of response required by MNOs to address cyber-attacks but introduces new cyber-attack vectors mainly due to increased complexity. Network operators must implement NFV and SDN security best practices, such as the recommendations in the ENISA 'NFV Security in 5G, Challenges and Best Practices' [17], which lists 60 security challenges and their mitigations.
4. The 5G equipment supply chain is too dependent on individual vendors. Governments have taken actions to support deploying open, interoperable and standards-based equipment, e.g. the USA's 'CHIPS and Science Act of 2022 and the EU 'EU Chips Act', backed with $billions of funding.
5. 5G-AKA is central to UEs, serving and home networks mutually authenticating each other, but it is vulnerable to some attacks. 6G should consider the various improvements or alternatives to 5G-AKA proposed by researchers; a few of many examples are 5G-AKA-FS [26], 5GSBA [27] and elliptic curve-cryptography [28].
6. Implementing a more Zero Trust architecture requires further development in cloud deployments to protect digital identities, attestation of 5G systems and hardware-rooted security.
7. Network slices create a multi-tenant network vulnerable to side-channel attacks. Further network slices can be legally intercepted by state actors. Organisations should encrypt all data using mechanisms under their own control before transmitting it over a network slice.
8. 6G should use the NIST standards for quantum-safe cryptography for asymmetric keys and symmetric keys of at least 256 bits in length for SIMs.

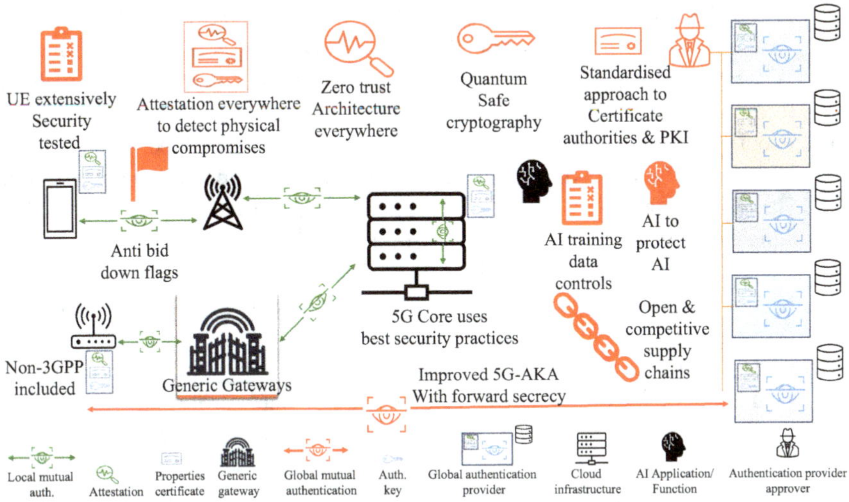

Figure 12.5 6G security recommendations

9. Edge Cloud site physical security weaknesses emphasise the need for remote attestation of software and hardware to guarantee they have not been tampered with.
10. 6G uses AI, which is vulnerable to novel and evolving attacks. Data engineers can mitigate these attacks by careful control of and statistical manipulation of the training data. They can also use another AI to detect issues in their AI models and training data.

6G security should be considered in the context of 6G convergence, described in Chapter 9. Figure 12.5 summarises the above issues and solutions in the context of the 6G convergence architecture presented in Chapter 9. 6G needs methods to secure a highly distributed multi-vendor system, recommended methods for PKI and Certification Authorities, a zero-touch architecture and a secure network slicing architecture that addresses softwarisation vulnerabilities, e.g. side-channel attacks.

References

[1] Talluri S, Zhang Q, and Chen R. A Cloud-Native Federated Learning Architecture for Telecom Fraud Detection. In *InNOMS 2023-2023 IEEE/ IFIP Network Operations and Management Symposium*; 2023. pp. 1–3.
[2] BBC. https://www.bbc.co.uk/. [Online]. 2022 [cited 2023 Feb 10]. Available from: https://www.bbc.co.uk/rd/projects/databox.
[3] Arm Limited. https://www.arm.com. [Online]. 2024 [cited 2024 April 13]. Available from: https://www.arm.com/products/silicon-ip-cpu/cortex-m/ cortex-m23.

[4] Wikipedia. https://en.wikipedia.org/. [Online]. 2024 [cited 2024 Feb 20]. Available from: https://en.wikipedia.org/wiki/Spectre_(security_vulnerability).

[5] Wikipedia. https://en.wikipedia.org. [Online]. [cited 2024 Feb 20]. Available from: https://en.wikipedia.org/wiki/Meltdown_(security_vulnerability).

[6] Hindi R. https://www.zama.ai. [Online]. 2023 [cited 2023 June 23]. Available from: https://www.zama.ai/post/chatgpt-privacy-with-homomorphic-encryption.

[7] Hu Y and Wang F. Cryptology ePrint Archive. [Online]. 2012 [cited 2023 June 23]. Available from: https://eprint.iacr.org/2012/561.pdf.

[8] Toft T. https://csrc.nist.gov/. [Online]. 2012 [cited 2024 Feb 22]. Available from: https://csrc.nist.gov/csrc/media/events/meeting-on-privacy-enhancing-cryptography/documents/toft.pdf.

[9] Gade S, Buddhavarapu P, Yuan S, and Shen M. https://research.facebook.com/. [Online]. 2022 [cited 2024 Feb 22]. Available from: https://research.facebook.com/blog/2022/10/mpc-based-machine-learning-achieving-end-to-end-privacy-preserving-machine-learning/.

[10] Google. https://support.google.com/. [Online]. 2024 [cited 2024 Feb 21]. Available from: https://support.google.com/gboard/answer/12373137?hl=en-GB.

[11] Downdetector. https://downdetector.co.uk/. [Online]. 2024 [cited 2024 Feb 22]. Available from: https://downdetector.co.uk/.

[12] Lyu L, Yu H, Ma X, *et al.* Privacy and Robustness in Federated Learning: Attacks and Defenses. *IEEE Transactions on Neural Networks and Learning Systems*. 2022; 35(7): 8726–46.

[13] Future Market Insights. https://www.futuremarketinsights.com/. [Online]. 2023 [cited 2024 Feb 23]. Available from: https://www.futuremarketinsights.com/reports/privacy-enhancing-technology-market.

[14] Nguyen V, Lin P, Cheng B, Hwang R, and Lin Y. Security and Privacy for 6G: A Survey on Prospective Technologies and Challenges. *IEEE Communications Surveys & Tutorials*. 2021; 23(4): 2384–428.

[15] Ustundag SE and Emrah T. https://www.ericsson.com. [Online]. 2023 [cited 2024 April 13]. Available from: https://www.ericsson.com/en/blog/2023/10/hexa-x-and-data-protection-evolution-in-6g.

[16] The Royal Society. https://royalsociety.org/. [Online]. 2023 [cited 2024 April 13]. Available from: https://royalsociety.org/-/media/policy/projects/privacy-enhancing-technologies/From-Privacy-to-Partnership.pdf.

[17] Nikolouzou E, Milenkovic G, Bafoutsou G, Bryska S, Hajj M, and Loiseaux C. *NFV Security in 5G: Challenges and Best Practices*. European Union Agency for Cybersecurity (ENISA), 2022. Available from: https://www.enisa.europa.eu/publications/nfv-security-in-5g-challenges-and-best-practices/@@download/fullReport.

[18] Karakoc B, Furste N, Rupprecht D, and Kohls K. Never Let Me Down Again: Bidding-Down Attacks and Mitigations in 5G and 4G. In *16th ACM Conference on Security and Privacy in Wireless and Mobile Networks*; 2023; Guildford.

[19] Nguyen VL, Lin PC, and Hwang RH. Energy Depletion Attacks in Low Power Wireless Networks. *IEEE Access*. 2019; 7: 51915–32.

[20] Rupprecht D, Kohls K, Holz T, and Pöpper C. IMP4GT: IMPersonation Attacks in 4G NeTworks. In *Proceedings of the 27th Annual Network and Distributed System Security Symposium (NDSS)*; 2020.

[21] Hussain SR, Echeverria M, Chowdhury O, Li N, and Bertino E. Privacy Attacks to the 4G and 5G Cellular Paging Protocols Using Side Channel Information. In *Proceedings 2019 Network and Distributed System Security Symposium*; 2019.

[22] Kubernetes. https://kubernetes.io. [Online]. 2023 [cited 2023 Nov 1]. Available from: https://kubernetes.io/docs/concepts/overview/.

[23] Wikipedia. https://en.wikipedia.org. [Online]. 2023 [cited 2023 Nov 1]. Available from: https://en.wikipedia.org/wiki/Unikernel.

[24] Department for Digital, Culture, Media and Sport. UK Telecoms Supply Chain Review Report. Department for Digital, Culture, Media and Sport, Telecoms Security and Resilience Team; 2019.

[25] Xiao Y and Wu Y. 5G-IPAKA: An Improved Primary Authentication and Key Agreement Protocol for 5G Networks. *Information*. 2022; 13(3): 125.

[26] You I, Kim G, Shin S, Kwon H, Kim J, and Baek J. G-AKA-FS: A 5G Authentication and Key Agreement Protocol for Forward Secrecy. *Sensors*. 2023; 24(1): 159.

[27] Chow M and Ma M. A Secure Blockchain-Based Authentication and Key Agreement Scheme for 3GPP 5G Networks. *Sensors*. 2022; 22(12): 4525.

[28] Yadav A, Misra M, Pandey P, Kaur K, Garg S, and Chen X. A Provably Secure ECC-based Multi-Factor 5G-AKA Authentication Protocol. In *IEEE Global Communications Conference 2022*; 2022. p. 516–521.

[29] Olsson J, Shorov A, Abdelrazek L, and Whitefield J. Realizing Zero Trust in 5G Networks. *Ericsson Technology Review*. 2021; 5. Available from: https://www.ericsson.com/49a20a/assets/local/reports-papers/ericsson-technology-review/docs/2021/zero-trust-and-5g.pdf.

[30] Hoque EN and Dan L. Deloitte. [Online]. 2020 [cited 2023 Nov 2]. Available from: https://www2.deloitte.com/us/en/insights/industry/technology/5g-network-slicing.html.

[31] ENEA. *A Slice in Time: Slicing Security in 5G Core Networks*. Dublin, Ireland: Adaptivemobile, 2021. Available from: https://info.enea.com/Slicing-Security-in-5G-Core-Networks.

[32] Wikipedia. https://en.wikipedia.org. [Online]. [cited 2024 Feb 20]. Available from: https://en.wikipedia.org/wiki/Many-worlds_interpretation.

[33] ETSI. Quantum-Safe Cryptography (QSC); Limits to Quantum Computing Applied to Symmetric Key Size. *ETSI, Quantum Safe Cryptography (QSC) ETSI Industry Specification Group*; 2017. Report No.: ETSI GR QSC 006 V1.1.1 (2017-02).

[34] Fildes N, Stefano MD, and Murphy H. https://www.ft.com. [Online]. 2020 [cited 2024 Feb 23]. Available from: https://www.ft.com/content/1eeedb71-d9dc-4b13-9b45-fcb7898ae9e1.

[35] Williams C. https://www.theregister.com. [Online]. 2008 [cited 2024 Feb 23]. Available from: https://www.theregister.com/2008/09/11/bt_mayfair_robbery/.

[36] Williams C. https://www.theregister.com. [Online]. 2008 [cited 2024 Feb 23]. Available from: https://www.theregister.com/2008/12/12/bt_reward/.
[37] Williams C. https://www.theregister.com. [Online]. 2005 [cited 2024 Feb 23]. Available from: https://www.theregister.com/2005/03/17/bt_exchange_raid/.
[38] AWS. https://aws.amazon.com. [Online]. 2024 [cited 2024 Feb 23]. Available from: https://aws.amazon.com/compliance/data-center/data-centers/.

Chapter 13
Artificial intelligence for 6G

The future is not about one technology, but the convergence of multiple technologies and disciplines.

– Amy Webb, CEO of the Future Today Institute

AI will be part of every industry, enhancing our abilities in ways we can't even imagine yet.

– Jeff Bezos

13.1 Introduction

The combination of artificial intelligence/machine learning (AI/ML) technology with the upcoming 6G wireless technology brings about a unique synergy that surpasses the individual strengths of each technology. The integration of AI/ML with 6G has the potential to redefine wireless communication by providing real-time data analysis and decision-making abilities, which complement 6G's high-speed and low-latency network. This synergy enables predictive analytics to identify and fix network issues before they impact the service, thereby enhancing the reliability and performance of 6G networks.

Designing, building, operating and optimising cellular networks requires significant skills, time and application of 'rules of thumb'.[1] The multidimensional nature of the problem and the potential amount of data describing the network and its operation makes it difficult for humans to optimise, but it is an ideal candidate for the application of ML and AI. AI can improve network energy efficiency, automate system management, manage radio and compute resources to meet Service Level Agreements (SLAs), detect anomalies caused by faults or cyber-attacks and predict future traffic and network performance. 6G could embed AI in its standards to create better networks. Networks that automatically optimise themselves to meet the requirements of network operators and customers. AI could lower power consumption when demand is low, give higher bit rates only when required and trade-off latency and error rates according to an application's requirements. AI can optimise the design of network protocols on the fly. Semantic

[1]Or non-optical heuristics.

communications are data compression technologies powered by AI that use the context and goals of communications to reduce the amount of data that needs to be transmitted by orders of magnitude.

This chapter summarises some of the critical applications of AI to 6G and its implications. It also examines 6G mobile network operators (MNOs) offering AI services to their customers that exploit the network's topology and hardware to accelerate federated learning, exploit data from joint sensing and communication (Chapter 14) and leverage multi-access edge computing to process data. There are many novel and exciting applications of AI to 6G networks beyond the mundane but essential optimisation and operation of networks also explored in this chapter.

AI is a scientific discipline with a large set of concepts and vocabulary of terminology unfamiliar to telecommunications network people. This chapter focuses on the application of AI to mobile networks and avoids an in-depth explanation of AI. Box 13.1 summarises common AI technologies and terms.

Box 13.1 Common AI terminology

Artificial Intelligence (AI) – The general concept of machines that can reason, learn, solve problems and act autonomously. AI can perform tasks typically requiring human intelligence.

Artificial neural network (ANN) – Computer system inspired by biological neural networks. An ANN is a set of layers of connected artificial neurons. ANNs are usually simply called neural networks.

Autoencoder – An ANN that learns how to encode and decode unlabelled data.

Bayesian networks – A graph or table of conditional probabilities to find out the probability of an event happening. Used in e-mail spam filtering and detecting security events in communications networks.

Convolutional neural network (CNN) – A neural network designed to detect patterns in images, audio or radio signals and similar. It uses multiple convolution layers that slide over the data to look for patterns. Often used in image recognition.

Deep learning – A range of machine learning methods called 'deep' because they use a deep neural network (DNN).

DNN – A neural network with multiple layers between the input and output layers. DNNs can handle complex relationships.

Deep reinforcement learning – Combines reinforcement learning (RL) and deep learning.

Distributed learning – Speeds up the training process by spreading it over many computers.

Feed forward neural network – Data flows from the input layer to the output layer without looping back.

Federated learning – Several learners are trained using their local data and communicate only their model parameters (to protect the privacy of the local data) to an aggregating server, which produces a global model that is sent back to the learners for retraining. Google's Gboard predictive text keyboard uses FL for the next word prediction.

Inference – Deployment of an AI model into the real world where the model infers logical conclusions based on input data.

Loss function – Determines the error between the model's prediction and the real world.

Machine Learning (ML) – Methods used to build models on sample data. These methods assume that algorithms that worked in the past will continue to work in the future.

Model – An AI model is a collection of algorithms trained on data. The purpose of the model is to predict outcomes or make decisions.

Recurrent neural network (RNN) – Data can flow bi-directionally through the layers. Outputs from some nodes can affect subsequent inputs to the same nodes. RNNs in effect have memory of previous inputs.

Reinforcement Learning (RL) – Agents take actions to maximise a reward. RL does not need labelled data and explores the range of actions it can take to maximise the reward. RL could be used, e.g. to maximise signal strength or minimise power consumption.

Q-learning – RL algorithm which iteratively learns to take the correct action and improve over time without requiring a starting model.

Semi-supervised learning – Use a small amount of labelled data with mostly unlabelled data.

Supervised learning – Uses labelled data to build a function that maps features to labels, e.g. learns to recognise animals by each photo being labelled with the animal species.

Transfer learning – Knowledge learned from a task is reused on a related task, used to accelerate learning.

Unsupervised learning – Learns from unlabelled data, e.g. groups together animals that look similar without being told anything about the animal.

13.2 Optimisation of radio coding/decoding

Traditionally, engineers have designed radio coding or modulation schemes to achieve targets such as throughput, coverage and power consumption in the face of impairments in the radio hardware and interference in the radio paths the signals travel over. The engineers optimise the physical layer coding schemes using assumptions or static models of the radio hardware and the radio channels. Standards Defining Organisations (SDO) like the 3GPP or IEEE standardised these coding schemes, which can take years as many institutions and people must reach a consensus. The assumptions about the radio channels will only approximate actual radio channel conditions; hence, the coding scheme is sub-optimal. Developing different coding schemes to meet different performance targets; for example, a scheme to achieve high throughput is not the same as one to minimise power consumption at low bit rates.

Figure 13.1 shows the concept of applying AI to the air interface where the AI component is an ANN, usually a pair of specialised ANNs called an autoencoder, which learns the characteristics of the radio hardware and the radio channel and then optimises the encoding and decoding to achieve the desired properties, e.g. high throughput, low error rate, low power consumption or a weighted compromise of several properties. The decoding ANN provides feedback results to the coding ANN, which may also include feedback from the application, so different applications can have different optimised coding schemes. Research has shown that these systems can learn a new coding scheme in seconds and outperform human-designed coding schemes.

Downey *et al.* [1] demonstrated an autoencoder method that reduced the bit error rate by 42% in a live test on NASA's Tracking Data Relay Satellite System. Cammerer *et al.* [2] showed a system outperforming the 802.11n (Wi-Fi 4) coding scheme by 1.3 dB. Jiang *et al.* [3] demonstrated a turbo autoencoder that outperformed state-of-the-art codes for reliability. Hoydis *et al.* [4] demonstrated an AI air interface using QAM, achieving the same bit error rates as the human-designed system but without using pilot tones, which reduces overheads, increasing capacity for users' traffic.

These methods still have implementation challenges and are the subject of ongoing R&D. Perhaps the biggest challenge is the mindset change required and the impact on standards setting. For example, in 3GPP, the working groups specifying the radio layer one and two specifications would define how the 6G AI system is boot-strapped and the simple starting point protocol. When MNOs introduce new applications or spectrum, the AI radio system automatically optimises layer 1 and 2 protocols without requiring new standards. However, it is reasonable to question if interoperability between different vendor equipment or users roaming between networks could be guaranteed. For example, what would happen if a UE trained in France arrived in Switzerland, where the AI system has evolved a different protocol? The new UE would need to learn a new protocol, which would have to happen quickly enough for seamless roaming. One could imagine scenarios where the French UE caused the Swiss network to adopt a new protocol, necessitating all the Swiss UE to retrain. Would the systems be stable and converge? More

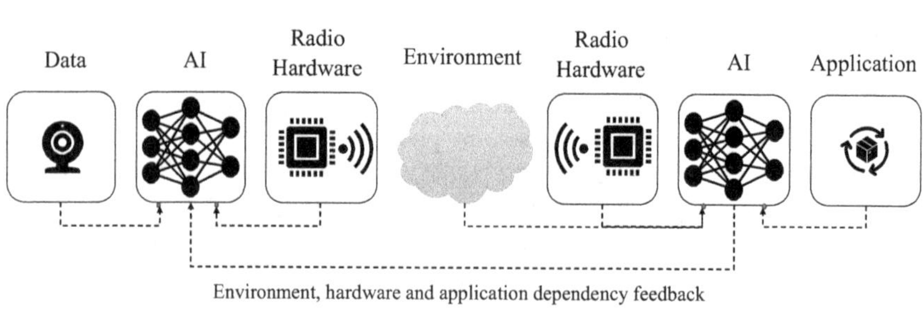

Figure 13.1 AI applied to the air interface

research is required. If 6G is a significant leap forward from 5G, this technology's abilities could drive a new 6G radio.

Spintronics [5] could build CNNs that also work at terahertz frequencies [6], although a spintronic autoencoder working at THz frequencies is a glimpse into the future beyond 6G.

Meanwhile, 3GPP Release 18 includes integrating AI into 5G Advanced to improve the performance of the 5G New Radio (NR) [7]. Use cases considered include improving the accuracy and efficiency of Channel State Information feedback, beam management and positioning precision.

13.3 Protocol-less communications

Protocol-less communication is an approach that uses AI to create optimal radio systems that are highly tailored to their environment and emerge automatically, with no overhead of standards specification, software writing and testing programs.

If we think of protocols as collaborating machines learning a language, then the field of learning to communicate using multi-agent reinforcement learning (MARL) [8] can be applied to training devices to learn communication protocols. This way, devices can learn a human-specified protocol without requiring program writing. The next step would be to give the devices an initial protocol and then allow them to evolve it to become more efficient. Valcarce and Hoydis [9] demonstrate that a Medium Access Control (MAC) signalling protocol can optimally evolve without any built-in knowledge of the target MAC signalling or channel access policy.

13.4 Semantic communications

Semantic communications is a nascent technology that increases the overall mobile capacity without more mobile spectrum or base stations and without changing the laws of physics.[2] As described in Chapters 1 and 7

$$\text{Capacity} = \text{spectrum} \times \text{spectral efficiency} \times \text{density of base stations}$$

Shannon's law describes a wireless channel's theoretical maximum capacity in terms of bandwidth and interference or noise. If MNOs must increase the capacity of their mobile networks and mobile spectrum cannot be increased and increasing the number of base stations is too expensive, then spectral efficiency must be improved. However, since 5G New Radio spectral efficiency is near the Shannon limit, there can be no further increases in overall mobile capacity. Semantic communication comes to the rescue, surpassing Shannon and its bit rate capacity limit. To explain how this is possible, we need to go back to the work of Shannon and Weaver in the late 1940s, who introduced the concept of semantic communications.

[2]Shannon's law and the second law of thermodynamics are related.

Shannon and Weaver [10] stated that there are three levels of communications problems:

1. How accurately can the symbols of communication be transmitted? (The technical problem.)
2. How precisely do the transmitted symbols convey the desired meaning? (The semantic problem.)
3. How effectively does the received meaning affect conduct in the desired way? (The effectiveness problem.)

Shannon's seminal mathematical theory of communications [11] assumed that only Level A, the technical problem of receiving bits transmitted accurately, was relevant to the engineering problem. Level B, testing the conveyance of the desired meaning is a philosophical problem; for example, I could ask, 'Do you understand this section on semantic communications?' and you could answer 'Yes' even if you don't. Level C, testing the effectiveness of communication, is easier to test since the effect of communication, especially for machine-to-machine or human-to-machine communication, can be measured, e.g. a motion controller commands a robot arm to move to a specific position or a person asks a smart speaker or virtual assistant to switch on a light.

The following example illustrates the difference between Shannon's Level A information theory and Semantic Communications Level B. If the transmitter sends the word 'bog' and the receiver sees 'cog', this is only a one-bit (if coded in ASCII) Level A error, but the semantic Level B difference between the words 'bog' and 'cog' is very significant. Any reasonable AI vision recognition system can distinguish between a bog and a cog. If the transmitter sends the word 'bog' in semantic communications and the receiver sees 'swamp', then that is a considerable Level A error but semantically a correct Level B transmission. It is reasonable to assume an AI vision system could label a bog as a swamp. Figure 13.2 compares Shannon's information communication system with the semantic communications system. Figure 13.2 shows that semantic communication depends on the receiver and transmitter sharing some knowledge of the purpose of the communication. The semantic communication system uses coding to overcome semantic noise where the meaning of the transmitted message is corrupted.

Semantic communication means to convey the meaning of a message as accurately as possible. Semantic communication can also encompass the effectiveness of the outcome of the communication. Whereas traditional wireless systems use information theory to create coding schemes that communicate bits as accurately as possible using the least amount of radio spectrum, a semantic communication system creates coding schemes to communicate the meaning as accurately or as effectively as possible using the least amount of radio spectrum.

There are three methods for semantic communication. The first is goal-oriented and only transmits the necessary messages to achieve the goal. Filtering out the redundant messages that are not relevant to the goal achieves semantic compression. The second is to use DNNs to compress source information. The third is to use a shared knowledge base between the source and destination.

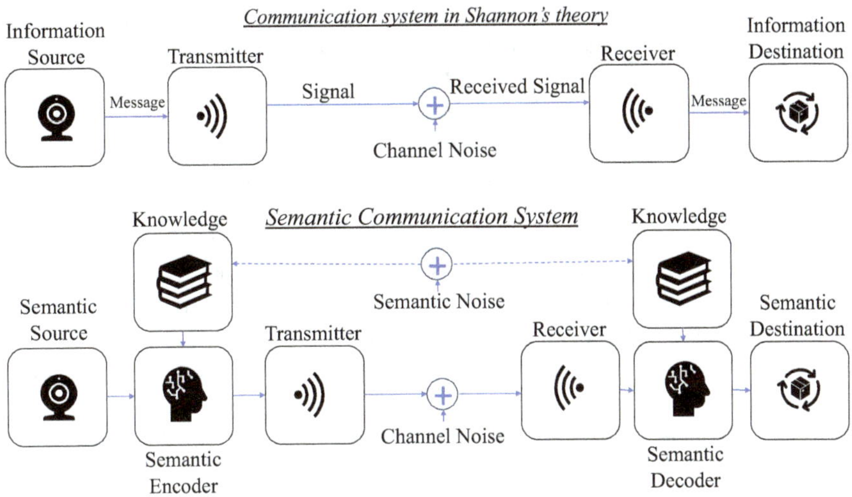

Figure 13.2 Shannon's information communication system and semantic communications system compared

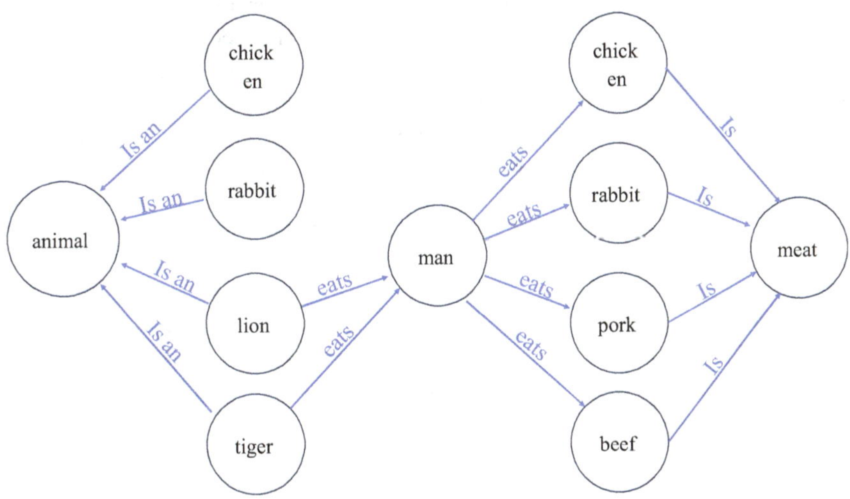

Figure 13.3 Knowledge graph for man-eating rabbit

Semantic communication goes beyond the interpretation and compression of bitstreams, e.g. videos, voice or sensor measurements. One method to improve semantic communication's efficiency and effectiveness is using a knowledge graph. A knowledge graph represents entities in a semantic space and their relationships. Figure 13.3 shows an example of a knowledge graph. If the transmitter and receiver share the same knowledge graph, then meaning can be encoded very efficiently by sending the index of a node in the knowledge graph associated with the message. It can also be very robust as errored messages are matched to the node in the knowledge graph that is most similar. Furthermore, the knowledge graph is used to exclude unreasonable results. For example, if the message 'Man eating rabbit' is sent, there is semantic ambiguity or noise in that message due to poor grammar. It is unlikely that the shared knowledge graph would include knowledge of man-eating rabbits;[3] hence, the semantics of the message can be corrected to 'Man eating rabbit meat'.

For a machine to understand the meaning of a message requires AI; hence, AI plays a crucial role in semantic communications. For example, imagine a crowd-monitoring application. If the application needs to count the number of people, the sending ANN extracts the outlines of the people from a camera feed and sends that; the application needs only the outlines to count people. Semantic communication gives very significant compression in bits transmitted compared to standard video compression. If the application needs to implement face recognition, then the sending ANN extracts just the pictures of faces or a measurement of facial features and sends those. Another example is voice communication; the sending AI could do voice recognition and send text, and the receiving AI could read aloud the text. In such a scheme, the voice heard by the listener would not be recognised as the voice of the talker unless the AI also modelled the speaker's voice characteristics, much like a deep fake. A semantic communication-based video conference system could be primed with a photo of the speaker to enable the receiver to reconstruct the speaker's facial expressions; in this case, the semantic information transferred during the video conference would only be information about the position of the head, eyes and mouth.

The following are some more examples of semantic communication applications. In recommendation systems, which, e.g. recommend videos to watch, rather than transmit all the user's personal information, group the user by historical preferences or content or demographics and then transmit the group the user belongs to. Instead of transmitting an electrocardiogram's complete analogue signal, the electrocardiogram QRS (Q-wave, R-wave, S-wave) frequency [12] waves are sent for biometric monitoring. Distributed learning systems send the data elements that contribute the most to the learning process. Intent-based networking, as seen in section 13.7, is a form of semantic communication that achieves the most effective network configuration.

Lan *et al.* [12] propose that given a specific task, the semantic encoder and the channel encoder, as described in Section 13.2, can be jointly trained to achieve communications more efficiently for that specific task, as shown in Figure 13.4. The disadvantage of this approach is that training the radio network for a specific task causes it to lose its generality.

We can easily misinterpret semantic communication as an advanced compression or coding scheme to transmit bits more efficiently. It is, however, a coding scheme that transmits meaning or achieves an effect efficiently and robustly, exploiting AI techniques. Semantic communication can effectively use the radio

[3]Even if the knowledge graph includes Mony Python's Killer Rabbit of Caerbannog there's no evidence the Killer Rabbit ate its victims.

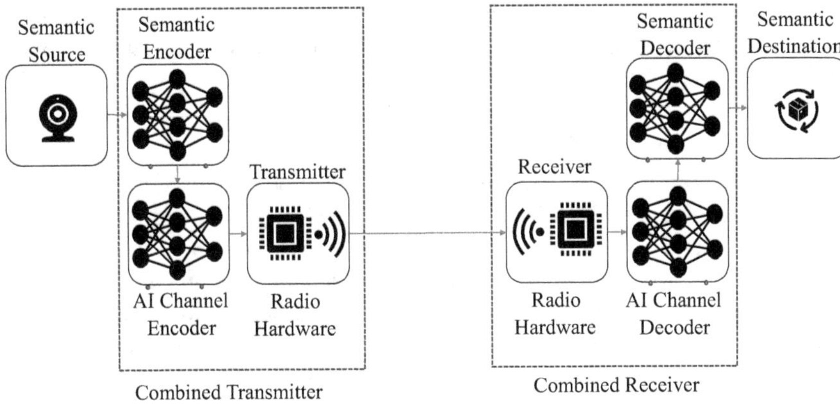

Figure 13.4 Combining the semantic and channel encoders for joint training

spectrum, enabling more effective semantic capacity at the cost of increased complexity for receivers and transmitters.

13.5 Network, resource management and optimisation

Cellular network designers must optimise many variables, e.g. traffic steering, transmit power, network slice resources, channel estimates for MIMO systems, mobile edge computation and edge cache placement. Human designers can create sub-optimal and static designs even with extensive software tools. Handing these tasks to AI may enable a more efficient and dynamic network. AI can also predict short-term traffic demands to allow proactive optimisations. The following examines some examples of the application of AI to network resource management and optimisation.

13.5.1 RAN optimisation

The O-RAN project, as seen in Chapter 2, created the radio intelligence controller (RIC) to allow AI, specifically ML, to be incorporated into 5G RAN management. The gains from applying AI to network optimisation can be significant but not disruptive. For example, Vodafone [13] claimed that in a test lab using 700 MHz spectrum, a RIC running Cohere's [14] multi-user MIMO application that performs accurate channel estimation and prediction doubled the capacity per user compared to a traditional system. Vodafone predicted that, in a 3.5 GHz spectrum network, capacity gains would be four or five times. Given these results, AI should be integral to 6G RAN optimisation. This improvement may seem impossible, knowing that 5G's spectral efficiency is near the Shannon limit, but the performance improvement is due to the improvement in channel estimation and prediction, not the spectral efficiency of the physical layer radio coding.

Li *et al.* [15] propose a federated learning framework for optimising the transmission power of indoor O-RANs. The indoor O-RAN learns its best policy by

RL and periodically uploads its model parameters to a global model. The global model can then be a base model for O-RANs in new environments, improving learning efficiency and O-RAN throughput.

13.5.2 Edge cloud for 6G optimisation

5/6G networks must manage compute resources in the Edge Cloud, supporting multi-access edge computing (MEC) and distributed RANs, and in the core network. Typically, in cloud networks, a container[4] orchestration system, e.g. Kubernetes, determines where to place containers in the available set of computers. Core networks use containers as packages of software that implement network functions. The orchestration system monitors the amount of resources network functions use and determines if a single network function requires more resources to scale it up, e.g. more memory or if more instances of the network function are required to scale it out. This problem compares to playing the game Tetris,[5] trying to place unequal-sized demands within a constrained area without leaving spaces, but with the additional complexity a piece may change its shape or size after placement! In Tetris, when the player reaches the top row, the game ends, but with the core network, the mobile service degrades. Compute resource optimisation is a multi-variable problem that, if poorly done, leads to underutilised compute resources, wasting equipment and power, and poor service. AI could provide optimal solutions here, although such solutions have been under-researched, probably due to the assumption by hyperscalers that compute resources are very large in the centralised cloud data centre, an assumption that does not hold for small Edge Compute nodes.

The other challenge with determining compute resource requirements for a specific network function is that the exact relationship between compute resource and function performance is often unknown. Often, network function vendors specify excessive compute resources to ensure good performance. Further, the traffic patterns can change the behaviour of the network function; for example, a User Plane Function (UPF) requires more computing to forward lots of small packets than a few big packets, although it has the same bandwidth throughput. AI offers solutions to these problems, and it can learn resource requirements and adapt to changing traffic patterns, even using predictions of future traffic to optimise the system for expected demands proactively since network traffic follows cyclic daily, weekly and annual patterns.

When control plane packets arrive at a core network, a Service Communication Proxy (SCP) must determine which network function it should be forwarded to, as there may be many AMFs running. AMF selection is a classic load-balancing problem that could be optimised using AI.

13.5.3 Reference signal received power

The reference signal received power (RSRP) is an essential measurement for managing mobile networks. Network operators gather RSRP data from user

[4]In the Cloud a container is a portable piece of software much like an app on a smartphone.
[5]If you want to see a video of AI playing Tetris, see https://youtu.be/l_KY_EwZEVA.

devices, UE, to assess network performance. However, user presence constrains this technique; it cannot measure RSRP without users, and the data gathered soon becomes outdated. Peizheng Li [16] proposes using neural networks and a digital twin exploiting the RSRP data from UEs to model RSRP. Their technique demonstrates a 20% accuracy improvement using real-world data.

13.5.4 Edge caching and computation offloading

Edge caching, as seen in Chapter 11, can significantly reduce network traffic if the caches are placed in the optimum places in the network and they cache the right content. Zhong [17] uses deep RL to optimise which content is cached and achieves improved cache hit rates by approximately 25% compared to classical caching algorithms in simulations where content popularity changes over time.

When 5G networks implement multi-access edge computing (MEC), it becomes possible to consider offloading compute-intensive tasks from UE to the Edge Cloud. If done optimally, this would reduce UE energy consumption, extend battery life and reduce the time to complete the task. Li *et al.* [18] use a Deep Q Network to optimise the decision to perform the task on the UE or offload to an Edge Cloud. Compute offloading is more likely for non-smartphone-based applications, e.g. facial recognition systems; for smartphone apps, this requires collaboration between the UE vendors app and cloud providers.

13.5.5 Dynamic spectrum assignment

As shown in Chapter 7, much of the radio spectrum is only partially used during the day. Chapter 7 proposes an active weak signal propagation measurement and centralised database system to allow better use of the available radio spectrum. Rutagemwa *et al.* [19] address the same problem using a DNN trained on past radio traffic to predict spectrum use in the next half hour and thus determine which spectrum can be shared. Rutagemwa assumes a standard radio protocol and uses AI to minimise the average wait time for the spectrum to become available. We believe the Chapter 7 method is more explainable, less constrained and more deterministic.

13.5.6 Encrypted traffic classification

Applications encrypt most of the traffic on the Internet and mobile networks for privacy and security. Application encryption prevents network operators from optimising their networks to deliver the right quality of service for applications and makes it more challenging to detect cyber-attacks on their customers. Aceto *et al.* [20] use deep learning to classify traffic and compares the performance of several deep learning methods. Aceto concludes that deep learning is a promising approach to traffic classification, but further research is required to improve its accuracy.

Encrypted traffic classification may be considered an intrusion of privacy by Internet users, but it is an essential technology for 5G Private Networks and Extranets to manage and secure private networks.

13.5.7 Energy efficiency optimisation

According to the GSMA [21], MNOs spent approximately US$17 billion on energy in 2015; therefore, MNOs could make considerable savings if AI could improve energy efficiency. Giannopoulos [22] uses deep RL to show in a simulation that the power consumption of an O-RAN could be reduced by 40% while delivering enhanced throughput compared to the O-RAN without AI optimisation. The AI optimises the O-RAN configuration for experienced throughput and power consumption. It uses UE measurement reports collected from the O-RAN, and the AI maximises energy efficiency by tuning the power allocation scheme for the time-frequency radio resources of all active radio units.

13.5.8 Access traffic steering, switching and splitting (ATSSS)

ATSSS, see also Chapter 9, load balances traffic over multiple access types, e.g. a 5G link and a fixed broadband link. Its purpose is to improve service reliability or increase the total bandwidth available to the customer. Load balancing over paths with significantly different performances, e.g. bandwidth and latency, and potentially dynamically changing performance, is easy to do poorly and challenging to do well. There is an unstable feedback loop in that placing traffic on the highest performance path reduces its performance, which could cause the load balancing algorithms to switch the traffic to the alternative link, which may then experience reduced performance due to increased load, causing traffic to change path again, as shown in Figure 13.5.

Long-term averaging of path performance may stabilise the choice of paths but cannot cope with dynamic paths where performance quickly changes. If the ATSSS function actively measures the performance of the links, then this may cause many

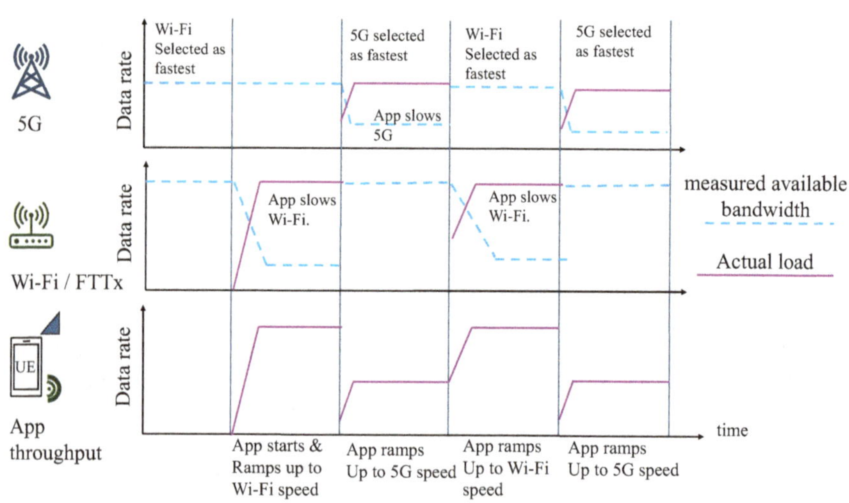

Figure 13.5 Example of ATSSS load balancing badly over 5G and Wi-Fi/FTTx

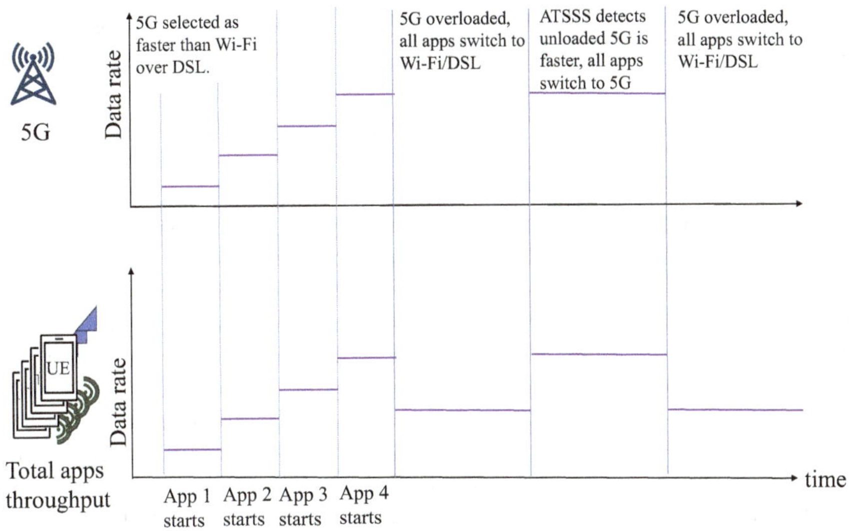

Figure 13.6 Example of ATSSS synchronising multiple UE switching between Wi-Fi and 5G

clients' traffic to switch links simultaneously, magnifying the unstable feedback loop, as shown in Figure 13.6.

Further, different applications respond differently to changing latencies and throughputs. Wu [23] surveyed ATSSS and suggested research into a machine-learning approach to classification and prediction for improved load balancing.

13.6 Network security

The cyber-security of public and private networks and their users is paramount, and many AI techniques have been applied to network cyber-security. Researchers have shown that AI can detect intrusion, denial of service, spectrum anomalies, impersonation, flooding, botnet and malware cyber-attacks [24].

Supervised deep learning can learn patterns from past cyber-security events and generalise them to future event detection. Unsupervised deep learning can recognise patterns different from regular behaviour that may indicate an attack. Thing [25] achieves 99% accuracy in categorising network traffic as legitimate or not using autoencoders. Feng *et al.* [26] use autoencoders to detect abnormal spectrum usage. Saied *et al.* [27] use multi-layer perceptrons to detect known and unknown distributed denial of service attacks. Oulehla *et al.* [28] use neural networks to extract features from botnet attacks.

Deep learning, particularly generative adversarial networks (GAN), can be used to attack networks. GANs can attack federated learning by injecting fake training samples [29]. GANs can also perform password-guessing attacks. Deep learning methods can also break encryption [30].

Chapter 12 explores using federated learning as a privacy-enhancing technology (PET) and the issues of securing AI.

13.7 Autonomic networking (AN) and zero-touch automation

Reducing costs and delivering network services reliably, without human errors, and ensuring that networks are always available requires automation. Modern networks' increasing complexity and size make automation or an autonomous network more imperative. An autonomous network operates according to business goals with no human intervention beyond the specification of its goals. It configures, monitors, diagnoses, repairs, optimises and protects itself. Figure 13.7 shows a high-level view of a closed autonomic control loop. Information processing contains details of network resources, services to be delivered, the processes and patterns to deliver those services, the network policies and contracts and network behaviour models. Context information, metrics and logs from the communications network update this information. The information processing generates knowledge that knowledge analysis processes, which adjusts business policies to meet the business goals of the network operator. A policy processing function turns business policies into configuration. Business policies and configurations are automatically updated to correct deviations from desired network behaviour, ensuring the network meets customer contracts.

Creating autonomous networks involves a broad body of knowledge, skills and standards organisations, and Table 13.1 lists relevant standards activities. The breadth of organisations working on the challenges of network automation illustrates how complex and difficult the challenges are. If you are interested in the summaries of the activities of these organisations, please read Appendix A.

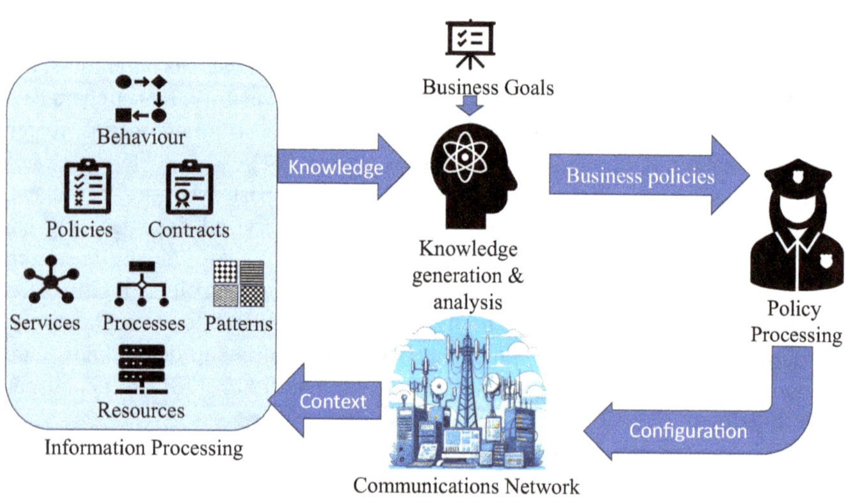

Figure 13.7 Autonomic network and closed autonomic control loop after [31]

Table 13.1 Standards development organisations working on autonomic networks

Organisation	Group acronym	Full name	Reference
ETSI	ENI	Experiential Networked Intelligence	[32]
ETSI	ZSM	Zero-touch network and Service Management	[33]
ETSI	NFV	Network Functions Virtualisation	[34]
ETSI	TC INT WG AFI	Autonomic Management and Control Intelligence for Self-Managed Fixed and mobile Integrated Networks	[35]
ETSI	IPE	IPv6 Enhanced Innovation	[36]
ETSI	MEC	Multi-access Edge Computing	[37]
ETSI	F5G	5th Generation Fixed Network	[38]
ETSI	SAI	Securing Artificial Intelligence	[39]
ITU-T	SG13 FG AN	Autonomous Networks	[40]
IETF	NETCONF	Network Configuration	[41]
IETF	NETMOD	Network Modelling	[42]
IETF	OPSAWG	Operations and Management Area	[43]
IETF	TEAS	Traffic Engineering Architecture and Signalling	[44]
TMForum		Autonomous Networks Project and Intent-Based Management	[45]
NGMN		Network Automation and Autonomy based on AI	[46]
3GPP	SA5	Management, Orchestration and Charging	[47]

Zero-touch automation, also known as zero-touch network or service management, is the concept that a complex network service could be delivered automatically without human intervention. It is driven by necessity as networks and digital services become too complex for people to understand and configure without error.

Autonomic networking (AN) or autonomous networking is challenging because of the complexity and scope of networks. AI is the tool *du jour* for addressing complexity and broad scope, but there are the following challenges in applying AI to AN:

1. Networks must always be available, and there is no room for error in their configuration, which limits the application of some AI techniques, such as RL, that explore the configuration space to obtain the best outcomes.
2. Networks generate lots of data about their state, but it is unlabelled, so inferring whether a specific pattern of data caused a particular event, e.g. a fault condition, is impossible. Reinforcement learning does not need labelled data, but allowing Reinforcement learning to generate fault conditions is unacceptable.
3. Networks face unexpected conditions. My experience in managing networks is that the bugs and faults never seen before cause the most extensive outages and

are the hardest to fix. The unexpected conditions would not have been included in the AI training, so that it could have responded inappropriately. Robust safety mechanisms must be in place to prevent this.

4. A digital twin of the networks could be used to train AI, but that only works if the digital twin is highly accurate, in some cases accurate enough to replicate the bugs in the software of the modelled system; for example, the digital twin may correctly determine the route of traffic from A to B according to a prescribed algorithm, but the network software may not implement that algorithm accurately and behave differently in 'real life'.
5. The AI training time may take too long to keep up with the dynamics of the networks.
6. AI creates a model to predict the consequences of certain input conditions. The model is trained on data to minimise the error between the model's predictions and the actual results. For example, the AI model could take the traffic demand as input and output the required network resources. The function used to calculate the error is called the loss function. Most AI scenarios use a generic loss function such as mean square errors. However, for networks, the loss function must be tailored by experts since generic loss functions do not capture the network engineering rules; in other words, the past design of networks is not necessarily a good indicator of the future design of networks.

Research on autonomic networks continues developing techniques to address some of the above challenges. There are few examples of real-world deployments of AI in mobile networks. Nokia and China Mobile had a trial of AI-powered RANs in 2021 to forecast bandwidth demand in Shanghai and detect anomalies with 90% accuracy across ten thousand cells on China Mobile's network [48]. Viettel High Tech urban 5G trial network in Hanoi used AI for physical layer signal processing on 5G ORAN equipment in 2024 [49]

13.8 Generative AI

A McKinsey report [50] claims that generative AI could capture $100 billion in incremental value and $140 billion in productivity gains for telcos. Seventy per cent of the value is in using generative AI for customer service and marketing. Eighty-five per cent of telcos focus on using generative AI for customer service.

Generative AI is a type of AI that can create new data, such as text, images or code, based on the information it was trained on. It also enables effective communication between humans and machines using natural language. Examples of Generative AI are ChatGPT, Google's Gemini and Microsoft's Copilot. Generative AI has a wide range of use cases in MNOs; some examples are:

1. In customer service, chatbots and recommendations for call centre agents.
2. For marketing and sales, content generation and customer sentiment analysis.
3. Analysis of social media feeds to detect network problems through customer sentiment analysis.

4. For IT, software development and synthetic data generation.
5. For support functions, procurement optimisation, writing and summarising procurement documents, knowledge management, HR Q&A and general workplace productivity.

These examples are independent of the cellular network generation.

13.9 6G for digital twins as a service

MNOs need to find new revenue services, and since the focus of 6G is the digitisation of industry and society, MNOs offering or playing a revenue-earning role in enabling AI services is an obvious opportunity. AI as a service has been proposed by several researchers as a 6G service, also known as '6G for AI'. The concept combines the Edge and Core Cloud infrastructure used to run 6G network functions and RANs with the rich data from the network, including sensing data, see Chapter 14, to offer AI services to enterprises and consumers. Can MNOs compete with or complement the AI services the hyperscalers offer? How can MNOs differentiate or complement the AI services from the hyperscalers?

One technical differentiator is the highly distributed nature of the MNO's networks. The MNOs could exploit the similarity between an ANN and their networks to distribute ANNs over the network. This distribution achieves greater privacy as data is abstracted from its source earlier. It is faster than sending all the data to a central cloud and would be much faster if network hardware could implement the ANNs, see Paolucci *et al.* [51]. Another radical example of in-network computing is 'Over-the-air computation' [52], where an analogue algorithm exploits the superposition of wireless signals[6] for federated learning. The analogue approach performs better than traditional digital, especially in low-power and low-bandwidth applications.

Distribution also enables compression and caching of the data in the network, essential in an intensive IoT scenario to reduce data volumes on the transport network and the Internet.

The MNO infrastructure may suit collaborative, federated, distributed and split learning more than the hyperscalers' infrastructure. These technical differentiators, however, may not hold if the hyperscalers run the MNOs' edge clouds and are permitted to offer their services from the same sites. The incompatibility between hyperscalers' AI services, e.g. APIs and MNO services, is an obstacle without global standards. MNOs are partnering with a range of hyperscalers to address this.

The MNOs and 6G could focus on 'Digital Twins as a Service' (DTaaS) that would exploit the rich data available to 6G networks as a targeted service for specific vertical industries. Mobile networks can accurately track phones carried by people and in vehicles, and with JCAS (Chapter 14), even those without phones, but selling this data would be an intrusion of privacy. Instead, the MNO could use

[6]Superposition is a fundamental feature of wireless and allows multiple signals to be transmitted and received simultaneously. The received signal is the sum of all the transmitted signals plus noise.

the data to create a digital twin modelling the volumes of movement of abstracted people and vehicles, and this would be helpful planning data for local government and retail organisations. Although Google or Apple can also do this with the GPS data they collect from phones. MNOs could build digital twins of their customers' applications, which would be helpful for technology, sociology research and managing networks. mmWave technology could perform electrocardiograms (ECGs) remotely. Imagine MNOs being able to build a digital twin of a nation's health. Google's Fitbit Care product can do similar, but only for those wearing Fitbit devices.

If an enterprise deploys a Private 6G network slice inside a factory or warehouse or across a port or freight yard, then the MNO can use the tracking data to input it into the enterprise's digital twin. However, would this data already be available to the enterprise via other means?

In summary, competitive differentiation for the MNOs from the technical advantages of JCAS and distributing AI across a network has yet to be proven to have commercial value. Further, for a digital twins service, the MNOs must find data they possess in a 6G, public or private, network that the hyperscalers or enterprises do not already have.

13.10 Challenges of using AI for 6G

AI could significantly improve 6G and enable new services, but several challenges must be addressed first.

Foremost, the AI system must be trusted. Networks must always be available; any configuration errors lead to network outages. This means AI, particularly RL, has reduced scope to learn from its network configuration mistakes. Operations are unlikely to trust an AI that is not resilient and cannot explain why it took specific actions. The R&D and maintenance of a robust AI follows similar practices to robust software, including risk management, technical documentation, quality management system, testing and certification. The AI for 6G systems described in this chapter is complex, and it is difficult for a human operator to interpret and understand what they are doing. An active field of study is Explainable Artificial Intelligence (XAI) [53], which develops methods to make AI transparent and allow human users to understand and trust that the AI is making good decisions. Network operators are already familiar with limited dynamic automation in networks in the form of routing protocols. From my experience of managing networks, the behaviour of routing protocols must be well understood, i.e. they must be predictable and explainable. Network operators lose sleep when routing protocols become unpredictable and unexplainable. Their experience with routing protocols may pave the way for broader AI-driven automation.

Supervised learning systems require large amounts of labelled data. For network operations, this would require network conditions or state data to be associated with outcomes, e.g. failure to meet an SLA and the actions performed by humans to fix the failure. Such data are typically unavailable from legacy networks and is not yet available for 6G. Bayesian networks, a graph model of probabilities, also known as 'probabilistic networks' or 'belief networks', are often applied to network events, e.g. alarms, to make predictions, e.g. associate network state with specific fault conditions. Labelled or unlabelled data can train Bayesian networks.

Researchers have access to limited data sets from MNOs, which limits the applicability of some AI research to general mobile networks, i.e. they have been tested on a minimal set of network data. If MNOs could share network data with each other and researchers, this would improve the training of AI models and their applicability to more MNOs. However, sharing this data is fraught with difficulties surrounding commercially confidential information, individuals' privacy and national critical infrastructure security concerns.

Reinforcement learning does not need labelled data, but it works by exploring the network configuration space, which could lead to service outages. An analogy is allowing a network apprentice to figure out how to configure a network by trial and error without training or the help of instruction manuals.

The time it takes to train an AI model may mean they are not practical for real-time use. The network behaviour may change faster than AI models can be trained, i.e. they always act on out-of-date learning. Even when the network environment and design change slowly, the AI models become less effective over time, making less accurate inferences requiring regular model refreshes.

Robust safety mechanisms are required to handle unexpected network conditions, which the AI models would not have been trained on. Humans in the loop (HITL) is one form of safety mechanism where the AI makes recommendations to a human, and then the human performs the recommended actions. However, HITL depends on the AI being able to explain its recommendations and for the human to have enough skills and knowledge to understand the consequences of their actions. Many network issues are caused today by human error, and HITL does not prevent dual AI and human errors.

AI regulation is evolving, creating challenges for a single global AI for 6G solutions. The European Commission is creating a 'European Artificial Intelligence Act'. The US has a blueprint for an AI Bill of Rights [54]. The UK has a non-statutory AI framework for managing AI risks and opportunities. It allows regulators, e.g. OFCOM for telecommunications, to exercise their expert judgement when applying the framework.

13.11 Summary and conclusions

AI for 6G has both revolutionary and evolutionary elements.

The revolutionary category includes ANNs for radio coding and decoding, multi-agent reinforcement learning (MARL) for protocol-less communications, semantic communications, dynamic spectrum assignment and 6G for Digital Twins services. Using ANNs for radio coding and decoding can only marginally improve the capacity of 5G New Radio because it is so near to the Shannon limit. However, it offers radical flexibility, the ability for radio coding schemes to optimise on the fly for different applications without requiring years of work in standards-defining organisations and software development. Using MARL to allow networks to evolve their higher-layer protocols has a similar radical impact on flexibility and the need for standards and software development. Semantic communications can radically change the effectiveness of communications, and this technology optimises communications around achieving the tasks that are the purpose of communication. It may achieve orders of magnitude improvement in robustness and efficiency of spectrum use, not by improving the spectral efficiency of radio coding schemes but by improving the efficiency of the communications semantics. There are exciting and radical services 6G could offer for Industry 4.0 digital twin building. They leverage joint communications and sensing (JCAS) data from the networks. Over-the-air computation and in-network computing could implement AI services, e.g. federated learning, for customers at unprecedented speed and privacy. AI has been proposed for dynamic spectrum assignment by predicting the requirements for the next half hour; however, we believe that using a weak signal propagation analysis method in a CBRS 2.0 architecture, as explained in Chapter 7, is more robust. These revolutionary technologies require MNOs and mobile vendors to have radical mindset changes. However, all these technologies are nascent and require further R&D and testing.

In the evolutionary category, there are technologies that make significant to marginal improvements in the performance and operation of mobile networks. For RAN optimisation, Vodafone has claimed that using AI to perform accurate channel estimation has doubled traffic capacity in the lab. Optimising the placement of functions and the number of resources assigned to them is complex, and AI optimisation methods make this tractable. For example, edge caching hit rates have been improved by 25%. AI has been used to reduce O-RAN power consumption by 40% while delivering enhanced throughput. BT has used machine learning to predict traffic levels enabling it to reduce capacity in EE's mobile network, when not required, to save 4.5 million kWh per year [55]. AI can also manage traffic flow, classify encrypted traffic for optimum treatment and load-balancing traffic over multiple access types (ATSSS). AI is useful in detecting cyber-attack patterns or other abnormalities currently reported to HITL but could lead to automatic mitigations in the future. Network configuration and maintenance can be automated using AI. Network automation is more of a necessity than a cost-saving exercise for MNOs' public networks as they employ few people to design and operate public mobile networks. Using AI for designing and operating private networks or network slices is necessary due to the complexity and amount of customisation required to meet enterprises' dynamic requirements, especially multinational corporations. The extensive number of standards-defining organisations and collaborative forums working on network automation illustrates how important and challenging network automation is. Generative AI applied to customer service, chatbots and call centres and marketing and sales content generation are the short-term leading and most lucrative use cases for MNOs.

There are significant technical challenges in applying AI to mobile networks and their services that require further research to address. Foremost network operations must trust AI, and AI must be explainable to achieve trustworthiness. AI requires large amounts of network data to learn from, and although networks generate large amounts of operational data, e.g. traffic patterns, equipment operating parameters and environmental conditions, these data are generally unavailable to external AI researchers. Networks evolve, the topology and the traffic patterns change over time, and the AI models built on past data become outdated, leading to inefficiencies or service outages. The AI system does not work if the data the AI is trained on changes quicker than the time it takes to train the AI model. AI mechanisms that do not require labelled data, such as RL, must be limited to the configuration space they can explore to prevent network service outages. Robust safety mechanisms are required, especially to handle unexpected network conditions. Placing humans in the loop may not be robust enough due to human error, noting that human error has been a significant cause of network outages.

There are clear rewards for MNOs from applying AI, whether for revolutionary or evolutionary purposes, but significant technical and mindset challenges must be overcome first. The potential AI rewards explain AI being the centre of attention at Mobile World Congress 2024 [56].

Appendix A Standards development organisations working on autonomic networks

The ETSI Zero touch network & Service Management (ZSM) Industry Specification Group (ISG) [57] is studying the challenge of zero-touch network management and has published several reports covering requirements, frameworks, solutions and demonstrated proof of concepts. The ETSI Experiential Networked Intelligence (ENI) ISG [58] focuses on improving network operations using closed-loop AI mechanisms called 'cognitive network management'. The ETSI Network Functions Virtualisation (NFV) ISG [34] focuses on the softwarisation of network functions, a prerequisite for AN. The ETSI Autonomic Management and Control Intelligence for Self-Managed Fixed and Mobile Integrated Networks (TC INT WG AFI) [35] working group develops use cases, requirements, test specifications and frameworks for autonomic networks. The ETSI IPv6 Enhanced Innovation (IPE) ISG [36] focus includes automated end-to-end IPv6 networking. The ETSI MEC ISG [37] aims to create standards for integrating applications across multi-vendor Edge Computing platforms. Autonomic networks must manage MEC resources and use MEC resources, especially when Edge Compute becomes an integral part of 5/6G networks. ETSI 5th Generation Fixed Network (F5G) [38] studies the evolution of fixed networks needed to match and enhance the benefits of 5G. It includes efficient network operations using intelligent fault management and defines an autonomous F5G network using intent-based management as defined by the TM Forum. The ETSI Securing Artificial Intelligence [39] ISG focuses on using AI to enhance security, mitigate against AI attacks and secure AI from attack. The ITU-T's Focus Group on Autonomous Networks (FG-AN) [40] drafts technical reports and specifications for autonomous networks. It also identifies relevant gaps in standards. The IETF's Network Configuration (Netconf) working group [41] is responsible for developing and maintaining widely adopted protocols and data models for managing networks, including zero-touch provisioning. The IETF Network Modelling working group (NETMOD) [42] defines a network modelling language called YANG and core data models used as basic building blocks by other IETF working groups. The IETF Operations and Management Area (OPSAWG) working group [43] picks up *ad hoc* proposals for developing operations and management standards, driven by network operators' interests, that are not within the remit of other IETF working groups. The IETF Traffic Engineering Architecture and Signalling (TEAS) working group [44] defines protocols that enable operators to control how specific traffic flows are treated in a network, used to guarantee quality of service and service level assurance. The TMForum's autonomous networks project [45] '*aims to define fully automated zero-wait, zero-touch, zero-trouble innovative network/ICT services for vertical industries' users and consumers, supporting self-configuration, self-healing, self-optimising and self-evolving telecom network infrastructures for telecom internal users: planning, service/marketing, operations and management.*' The NGMN 'Network Automation and Autonomy based on AI' project [46] researches requirements and implementation architecture to use AI technology for network automation. The 3GPP's SA5 working group has studies on 'Management, Orchestration and Charging', which addresses the management and orchestration of 5G networks. SA5 [47] introduced the communication service management function, the network slice management function and a network slice subset management function to provide the necessary abstraction for automation.

References

[1] Downey J, Hilburn B, O'Shea T, and West N. Machine learning remakes radio. *IEEE Spectrum*. 2020; 5: 35–9.

[2] Cammerer S, Aoudia F, Dörner S, Stark M, Hoydis J, and Brink ST. Trainable communication systems: Concepts and prototype. *IEEE Transactions on Communications*. 2020; 68(9): 5489–503.

[3] Jiang Y, Kim H, Asnani H, Kannan S, Oh S, and Viswanath P. Turbo autoencoder: Deep learning based channel codes for point-to-point communication channels. *Advances in Neural Information Processing Systems*. 2019; 32: 2758–68.

[4] Hoydis J, Aoudia FA, Valcarce A, and Viswanathan H. Toward a 6G AI-native air interface. *IEEE Communications Magazine*. 2021; 59(5): 76–81.

[5] Leroux N, Riz AD, Sanz-Hernández D, Marković, Mizrahi A, and Grollier J. Convolutional neural networks with radio-frequency spintronic nano-devices. *Neuromorphic Computing and Engineering*. 2022; 2(3). Available from: https://iopscience.iop.org/article/10.1088/2634-4386/ac77b2.

[6] Walowski J and Münzenberg M. Perspective: Ultrafast magnetism and THz spintronics. *Journal of Applied Physics*. 2016.

[7] Lin X. https://arxiv.org. [Online]. 2023 [cited 2024 April 15]. Available from: https://arxiv.org/ftp/arxiv/papers/2308/2308.05315.pdf.

[8] Pieter A, Igor M, Ryan L, Jon G, and Jack C. OpenAI. [Online]. 2017 [cited 2023 Jul 27]. Available from: https://openai.com/research/learning-to-communicate.

[9] Valcarce A and Hoydis J. Towards joint learning of optimal signaling and wireless channel access. *IEEE Transactions on Cognitive Communications and Networking*. 2021; 7(4): 1233–43.

[10] Weaver W and Shannon C. Recent contributions to the mathematical theory of communication. *ETC: A Review of General Semantics*. 1953; 10(4): 261–81.

[11] Shannon C. A mathematical theory of communication. *Bell System Technical Journal*. 1948; 27(3): 379–423.

[12] Lan Q, Wen D, Zhang Z, *et al.* What is semantic communication? A view on conveying meaning in the era of machine intelligence. *Journal of Communications and Information Networks*. 2021; 6(4): 336–71.

[13] Maistre RL. TelecomTV. [Online]. 2021 [cited 2023 July 27]. Available from: https://www.telecomtv.com/content/open-ran/vodafone-boasts-5g-open-ran-milestone-with-ric-evaluation-41605/.

[14] Dolan R. Cohere-tech. [Online]. 2021 [cited 2023 July 27]. Available from: https://www.cohere-tech.com/blog/spectrum-multiplier-vmware-cohere.

[15] Li P, Erdol H, Briggs K, *et al.* Transmit power control for indoor small cells: A method based on federated reinforcement learning. In *IEEE 96th Vehicular Technology Conference*; London, UK, 2022. pp. 1–7.

[16] Li P, Wang X, Piechocki R, Kapoor S, Doufexi A, and Parekh A. Variational autoencoder assisted neural network likelihood RSRP prediction model. In *IEEE 33rd Annual International Symposium on Personal, Indoor and Mobile Radio Communications*; Kyoto, Japan, 2022. pp. 554–9.

[17] Zhong C, Gursoy M, and Velipasalar S. A deep reinforcement learning-based framework for content caching. In *52nd Annual Conference on Information Sciences and Systems*; Princeton, NJ, 2018. pp. 1–6.

[18] Li J, Gao H, Lv T, and Lu Y. Deep reinforcement learning based computation offloading and resource allocation for MEC. In *2018 IEEE Wireless Communications and Networking Conference*; 2018; Barcelona. pp. 1–6.

[19] Rutagemwa H, Ghasem A, and Liu S. Dynamic spectrum assignment for land mobile radio with deep recurrent neural networks. In *2018 IEEE International Conference on Communications Workshops*; 2018; Kansas City. pp. 1–6.

[20] Aceto G, Ciuonzo D, Montieri A, and Pescapé A. Mobile encrypted traffic classification using deep learning: experimental evaluation, lessons learned, and challenges. *IEEE Transactions on Network and Service Management*. 2019; 16(2): 445–58.

[21] GSMA. https://www.gsma.com/. [Online]. 2016 [cited 2024 Mar 7]. Available from: https://www.gsma.com/solutions-and-impact/connectivity-for-good/public-policy/wp-content/uploads/2015/02/GSMA2016_Brochure_MobileEnergyEfficiencyOverview.pdf.

[22] Giannopoulos A, Spantideas S, Kapsalis N, *et al.* Supporting intelligence in disaggregated open radio access networks: Architectural principles, AI/ML workflow, and use cases. *IEEE Access*. 2022; 10: 39580–95.

[23] Wu H, Ferlin S, Caso G, Alay Ö, and Brunstrom A. Survey on multipath transport protocols towards 5G access traffic steering, switching and splitting. *IEEE Access*. 2021; 9: 164417–39.

[24] Zhang C, Patras P, and Haddadi H. Deep learning in mobile and wireless networking: a survey. *IEEE Communications Surveys & Tutorials*. 2019; 21 (3): 2224–87.

[25] Thing VL. IEEE 802.11 Network anomaly detection and attack classification: a deep learning approach. In *IEEE Wireless Communications and Networking Conference*; 2017; San Francisco. pp. 1–6.

[26] Feng Q, Zhang Y, Li C, Dou Z, and Wang J. Anomaly detection of spectrum in wireless communication via deep auto-encoders. *Journal of Supercomputing*. 2017; 73: 3161–76.

[27] Saied A, Overill R, and Radzik T. Detection of known and unknown DDoS attacks using artificial neural networks. *Neurocomputing*. 2016; 172: 385–93.

[28] Oulehla M, Oplatková Z, and Malanik D. Detection of mobile botnets using neural network. In *Future Technologies Conference*; San Francisco, CA, 2016. pp. 1324–26.

[29] Hitaj B, Ateniese G, and Perez-Cruz F. *Deep models under the GAN: information leakage from collaborative deep learning*. In *2017 ACM SIGSAC Conference on Computer and Communications Security*; Dallas, TX, 2017. pp. 603–18.

[30] Maghrebi H, Portigliatti T, and Prouff E. Breaking cryptographic implementations using deep learning techniques. In *Security, Privacy, and Applied Cryptography Engineering: 6th International Conference*; 2016; Hyderabad. pp. 3–26.

[31] van der Meer S, Fleck J, Huddleston M, Raymer D, Strassner J, and Donnelly W. Neutral principles and concepts for autonomic networking. In: Calisti M, van der Meer S, and Strassner J. (eds.), *Advanced Autonomic Networking and Communication*. Birkhäuser Basel, 2007, 1–25.

[32] ETSI. ETSI. [Online]. 2023 [cited 2023 July 28]. Available from: https://www.etsi.org/technologies/experiential-networked-intelligence.

[33] ETSI. ETSI. [Online]. 2023 [cited 2023 July 28]. Available from: https://www.etsi.org/technologies/zero-touch-network-service-management.

[34] ETSI. ETSI. [Online]. 2023 [cited 2023 July 28]. Available from: https://www.etsi.org/technologies/nfv.

[35] ETSI. ETSI. [Online]. 2023 [cited 2023 July 28]. Available from: https://portal.etsi.org/TB-SiteMap/INT/INT-WG-AFI-ToR.

[36] Liu W. ETSI. [Online]. 2021 [cited 2023 July 28]. Available from: https://www.etsi.org/newsroom/blogs/technologies/entry/etsi-isg-ipe-off-to-a-good-start.

[37] ETSI. ETSI. [Online]. 2023 [cited 2023 July 28]. Available from: https://www.etsi.org/technologies/multi-access-edge-computing.

[38] ETSI. ETSI. [Online]. 2023 [cited 2023 July 28]. Available from: https://www.etsi.org/committee/f5g.

[39] ETSI. ETSI. [Online]. 2023 [cited 2023 July 28]. Available from: https://www.etsi.org/technologies/securing-artificial-intelligence.

[40] ITU-T. ITU. [Online]. 2023 [cited 2023 July 28]. Available from: https://www.itu.int/en/ITU-T/focusgroups/an/Pages/default.aspx.

[41] IETF. IETF datatracker. [Online]. 2018 [cited 2023 July 28]. Available from: https://datatracker.ietf.org/wg/netconf/about/.

[42] IETF. IETF datatracker. [Online]. 2018 [cited 2023 July 28]. Available from: https://datatracker.ietf.org/wg/netmod/about/.

[43] IETF. IETF datatracker. [Online]. 2013 [cited 2023 July 28]. Available from: https://datatracker.ietf.org/wg/opsawg/about/.

[44] IETF. IETF datatracker. [Online]. 2020 [cited 2023 July 28]. Available from: https://datatracker.ietf.org/wg/teas/about/.

[45] TMForum. TMforum. [Online]. [cited 2023 July 28]. Available from: https://www.tmforum.org/collaboration/autonomous-networks-project/.

[46] Zhang Y, Deng L, Verspecht T, *et al.* NGMN. [Online]. 2022 [cited 2023 July 28]. Available from: https://www.ngmn.org/wp-content/uploads/NGMN_AAAF_v1.0.pdf.

[47] 3GPP. 3GPP. [Online]. 2023 [cited 2023 July 28]. Available from: https://www.3gpp.org/3gpp-groups/service-system-aspects-sa/sa-wg5.

[48] Nokia. https://www.nokia.com/. [Online]. 2021 [cited 2024 Mar 13]. Available from: https://www.nokia.com/about-us/news/releases/2021/01/21/nokia-and-cmcc-trial-ai-powered-radio-access-on-live-5g-network/.

[49] Business Wire. https://aithority.com/. [Online]. 2024 [cited 2024 Mar 13]. Available from: https://aithority.com/machine-learning/viettel-high-tech-and-deepsig-announce-ai-ml-based-neural-receiver-operating-in-live-5g-network.

[50] Stephen C, Ignacio F, Tomás L, Victor T, and Benjamim V. https://www.mckinsey.com/. [Online]. 2024 [cited 2024 Mar 8]. Available from: https://www.mckinsey.com/industries/technology-media-and-telecommunications/our-insights/how-generative-ai-could-revitalize-profitability-for-telcos.

[51] Paolucci F, Lorenzo DM, Castoldi P, and Cugini F. Demonstration of P4 neural network switch. In *OFC 2021*; 2021; San Francisco.

[52] Amiri M and Gündüz D. Over-the-air machine learning at the wireless edge. In *IEEE 20th International Workshop on Signal Processing Advances in Wireless Communications*; 2019. pp. 1–5.

[53] Saleem R, Yuan B, Kurugollu F, Anjum A, and Liu L. Explaining deep neural networks: A survey on the global interpretation methods. *Neurocomputing*. 2022; 513: 165–80.

[54] White House Office of Science and Technology Policy. https://www.whitehouse.gov. [Online]. 2022 [cited 2024 Mar 10]. Available from: https://www.whitehouse.gov/wp-content/uploads/2022/10/Blueprint-for-an-AI-Bill-of-Rights.pdf.

[55] BT. BT.com. [Online]. 2024 [cited 2024 July 20]. Available from: https://newsroom.bt.com/bt-group-rolls-out-energy-saving-cell-sleep-technology-to-ee-mobile-sites-nationwide/. [Online]. 2024 [cited 2024 Mar 13].

[56] GSMA. https://www.mwcbarcelona.com. [Online]. 2024 [cited 2024 Mar 13]. Available from: https://www.mwcbarcelona.com/themes/ai?date=2024-02-26.

[57] ETSI. ETSI.org. [Online]. 2023 [cited 2023 July 28]. Available from: https://www.etsi.org/technologies/zero-touch-network-service-management.

[58] ETSI. ETSI.org. [Online]. 2022 [cited 2023 July 28]. Available from: https://www.etsi.org/technologies/experiential-networked-intelligence.

Part III

Visions of 6G – evolution and revolution

Chapter 14
Joint communication and sensing (JCAS) in 6G networks

Synergy – the bonus that is achieved when things work together harmoniously.

– Mark Twain

14.1 Introduction

As 5G technology advances, the new 6G network is expected to improve upon the three principal features of enhanced mobile broadband (eMBB) for immersive communications, critical communications requiring ultra-low latency and reliability (URLLC) and massive machine-type communications. However, some vital differences will set 6G apart from its predecessor. One of the most revolutionary features of 6G is radio sensing and positioning capabilities. Although defined in 5G standards, they were never enacted at scale. Nevertheless, 6G is expected to offer even better positional accuracy on its launch than the 5G standards. The importance of sensing is seen as a bridge between the three worlds of Human, Physical and Digital – the vision of the 6G as defined by one of the many significant research programmes, the EU – Hexa-X.

Hexa-X 6G vision is to connect the worlds and revolves around their interactions: A human world of our senses, bodies and intelligence, and values; A digital world of information, communications and computing; a physical world of objects and organisms

EU flagship 6G research programme Hexa-X [1]

Nevertheless, as we saw in Chapter 6, this vision is also shared in various forms across the global 6G research communities.

Sensing can form the bridge between human, digital and physical worlds and profoundly change how people and machines 'see' the world. Joint Communications And Sensing (JCAS) combines communication and sensing within a single technology with minimal impact on each function. It builds on key themes identified in 5G for Industry 4.0, such as immersive extended reality, control of robots and drones and use of digital twins (DT). Sensing and localisation would allow the rapid creation of three-dimensional (3D) spatial maps of environments. The Doppler shift in the received echo could measure any moving object's velocity. It would inevitably create exciting and revolutionary new applications for the iPhone 6G in 2030, such as 3D photography, possible spectroscopic scanning for food allergens and even a contactless electrocardiogram [2].

The ability to 'see' the environment using radio waves rather than visible light means that these images are unhindered by darkness, smoke or fog but would look crude alongside a photographic image. At the lower frequencies of mobile networks, the accuracy/blur would be measured in metres, while towards 100 GHz, this could increase to the millimetre range.

Mobile networks are a great starting point for enabling a massive sensor network, as they already have extensive coverage on day one operation, with transmit/receive nodes providing full area coverage with interconnection between these nodes for enhanced accuracy. In many cases, sensing is complementary to using DT to create an accurate model of the physical world. This technology combination could be a differentiator for the mobile industry's use, particularly in macro and microcell environments, as the sensing can be delivered almost 'for free' on day one.

Furthermore, integrated sensing and positioning in a mobile communication network are essential motivations for mobile operators. Gathering information about the environment can ultimately enhance communication performance. For instance, a DT of the environment created through sensing can be beneficial in aiding communication functions such as radio resource management, beamforming, mobility management, minimisation of driving tests, etc. [3].

The EU Hexa-X project, described in Chapter 6, is one of many worldwide research projects researching sensing and its uses in the 6G. As part of its vision, it aims to merge the human, physical and digital environments into a singular entity as part of the 6G. They see the ability to locate, track and sense physical objects as crucial to bridging these worlds. From the beginning, it sees localisation, radar and sensing as essential components of 6G. This approach will not only provide the high performance needed to support location accuracy and low-latency requirements for identified use case families that are far beyond the current capabilities in 5G but also allow for the seamless integration of radar, communication, computation, localisation and sensing at both the hardware and software levels.

Sensing is also a key technology in the US ATIS Next G Alliance report on 6G technologies [4] under Radio technologies for communications – sensing fusion, where they see sensing for situational awareness as becoming essential in many new commercial and industrial applications and services such as smart-home/factories/cities, interactive gaming, extended reality/AR, health monitoring and automotive safety. This aligns well with the EU and China 6G research projects in JCAS, where spectrum and hardware are shared for both activities.

Nevertheless, sensing and location are already nascent in other wireless technologies, such as Wi-Fi, Bluetooth low energy (BLE) and ultra-wideband (UWB), used in consumer devices such as smartphones or VR goggles. For example, the recent release of the Apple Vision Pro supports a virtual keyboard where people can type in the air in front of their goggles. Regardless of whether people enjoy such an interface, it can recognise gestures based on sensing technologies. In this chapter, we delve into the potential use cases and performance of JCAS in various wireless technologies, including 6G and non-mobile technologies such as Wi-Fi, Bluetooth, light direction and ranging (LIDAR) and UWB. The aim is to provide a better understanding of the opportunities that arise when these technologies are combined with communications, especially for mobile operators and device manufacturers. We also examine how Over-The-Top competitors can leverage these technologies to compete with mobile operators.

14.2 How does it work?

Positioning or location determination refers to estimating a device's location by analysing radio signal measurements, such as received signal power, time-of-flight between transmitter and receiver, direction of the signal or a combination of these factors. There are two fundamental modes for locating and sensing. In the first mode, called monostatic, the transmitter and receiver are located together (as shown on the right-hand side of Figure 14.1), based on Ali Behravan *et al.* [3]. This is mainly used when the object is not part of the wireless communications system. The transmitter and receivers are separated in the second mode, called bistatic (two) or multi-static (more than two). Using two separate receivers can improve the accuracy of the positioning determination process.

Figure 14.1 shows several base stations (BS) positioned around a User Equipment terminal (UE) on the left-hand side. In contrast, a car is being sensed on the right-hand side, which also includes an onboard monostatic-based radar system for car/pedestrian safety. The UE's position can be determined using the time difference of arrival (TDoA) method from the three BS, represented by red lines. This method constrains the UE to lie at the intersection of two hyperbolas in two dimensions. Alternatively, directional link angle of arrival (DL-AoD) or up link angle of arrival (UL-AoA) lines can be formed, intersecting at the UE's location. It is also possible to combine both TDoA and AoD methods. On the right-hand

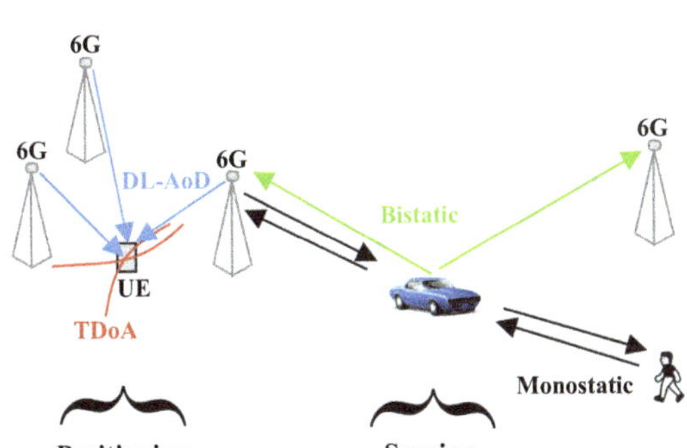

Figure 14.1 Positioning of UE with several BS based on [3]

side, you can see that the BS (in DL) and the scattered signal (in UL) perform monostatic sensing on the car, represented by black lines. This is used for objects which do not include a radio transmitter, as mentioned previously. On the other hand, bistatic sensing is represented by green lines at two BS.

While sensing and location seem to be important differentiators for 6G, there is already work within 3GPP Rel. 16–18 on 5G standards. The positioning work in 3GPP over New Radio (NR) Rel. 16 is aimed at indoor and outdoor use cases. It exploits 5G features such as high bandwidth and massive antenna systems within its network architecture. The 5G specification provides a detailed positioning reference signal (PRS). Bistatic positioning can be carried out in both the downlink and uplink using a combination of time-difference of arrival (TDoA), angle of arrival (AoA) and angle of departure (AoD). It can take advantage of the scattered signal from the object for multi-static operation. Hence, timing synchronisation between the BS is crucial for TDoA. However, this can be replaced by round trip time (RTT) at the expense of a higher signalling load. The base station locations are known precisely with backhaul connections to either a mobile core or Mobile Edge Compute, where calculations and spatial maps are created with DT.

14.3 Frequencies of operation and beamwidth

The operation frequency will also influence the performance and accuracy of positioning and sensing, as will the channel bandwidth, which is inevitably higher with higher frequencies of operation. Each generation of mobile has exploited higher frequencies of operation in the search for new spectrum and higher capacities – the 'spectrum staircase', as we describe it in Chapter 8. 5G was the first to use the 3.5 GHz bands for wide-area coverage and to explore the cm and mmWave bands for extreme communications towards 100 GHz in hot spots. 6G will likely move further into the sub-terahertz region, exploiting 100–300 GHz during its lifetime. Nevertheless, sensing as an integrated capability of the mobile network is of interest throughout the frequency range used by mobile communication networks, starting as low as 700 MHz, where the lowest time-division duplex (TDD) bands are located, moving upwards to the sub-terahertz region where the reduced wavelength fundamentally allows better resolution [5].

Beamwidth is crucial in positioning and sensing applications [6]. Phased arrays, especially in high cm and mmWave, allow the formation of very narrow (pencil) beams of coherent signals that are concentrated in a specific direction controlled by the phased array. This narrow beam can be steered in a desired direction or allow pencil beam-like scanning of the environment for high-definition 3D images. The enhanced beam formation in 3D space is advantageous for overcoming the higher path loss at mmWave and mitigating interference from different directions with these very narrow beams. As seen in Chapter 8, increasing the frequency of operation allows for increases in the size of the phased array within a constant area with the resulting reduction in the wavelength of operation or reductions in the physical size of the array. The increased phased array size directly reduces the beamwidth. Therefore, phased array technology goes hand in hand with increases in the frequency of operation and reductions in beamwidth for more accurate positioning and sensing applications.

The wide spread of frequencies in mobile networks will inevitably lead to differences in the accuracy between urban macro (UMa), urban microcells (UMi) and small cells for Indoor Open Office (IOO). Nevertheless, the cellular network cell hierarchy may correlate well with individual use cases. For example, there is no need for a building located in a macro cell to be found within 1 cm accuracy. At the same time, mmWave may allow this level of accuracy when controlling a robot in a factory. GPS is the outdoor elephant in the room for positioning, particularly for outside applications, but the use cases in 6G are much more comprehensive.

14.4 Possible levels of performance

Without a 6G network design, we can use 5G as an example to consider the performance levels for positioning and sensing using the 5G PRS, which supports many techniques to achieve UE positioning [7,8]. In the work of Dwivedi *et al.* [8], simulated conditions using UMa and UMi scenarios consisting of seven hexagonal cells, each comprising three sectors. The UMa and UMi had 1600 m × 1600 m and 500 m × 500 m areas, respectively, where the inter-site distances are 500, 200 and 20 m in UMa, UMi and IOO scenarios, respectively. The evaluations are done for UMa and UMi in FR1 (2 GHz), and the IOO scenario is evaluated in FR1 and FR2 (28 GHz).

As shown in Table 14.1, the evaluations show an technology accuracy potential of a few meters outdoors and a few decimetres indoors for 5G positioning, which is highly applicable to many different use cases we discuss below, using 2 GHz outdoors and 28 GHz indoors. In comparison, only GPS receivers for outdoor applications can provide horizontal accuracy better than 3–4 m. 6G is expected to offer higher performance levels with outdoor and indoor deployments using the frequency bands in the 7–16 GHz and 100–300 GHz regions for indoor use cases. This gives it significant advantages over GPS, which only works well outdoors.

The results obtained from these simulations can be compared to the recent location work within 3GPP described by Taylor [9] for targeted use cases in Rel. 16 and Rel. 17 for scenarios of 80% of UEs, increasing to 90% of UEs in Rel. 17,

Table 14.1 Accuracy of simulated 3GPP scenarios

UMa 2 GHz	UMi 2 GHz	IOO
UMa interference (7.9 m)	UMi interference (1.9 m)	IOO with interference (1.4 m)
UMa no interference (3.4 m)	UMi no interference (1.5 m)	IOO with no interference (1.4 m)
		IOO multi-RTT (0.7 m)
		IOO FR2 (0.2 m) 28 GHz

Table 14.2 3GPP positioning support as per reference [9]

Release	Horizontal accuracy indoors	Horizontal accuracy outdoors	Vertical accuracy indoors	Vertical accuracy indoors	Latency level
16 (80% of UEs)	<3 m	<10 m	<3 m	<3 m	<1 s
17 (90% of UEs)	<1 m	<1 m	<3 m	<3 m	<100 ms

see Table 14.2. This technology aims to support indoor and outdoor use cases by utilising the advanced features of 5G, such as high bandwidth, massive antenna systems and evolved network architecture. They see these features as crucial in supporting the arrival of many IoT devices in the market.

Ericsson has partnered with NXP semiconductors to investigate new use cases for a network that integrates communication and sensing functionalities into the same transmission/reception nodes [10]. They are exploring the implementation aspects of future systems and evaluating technical challenges and opportunities for JCAS use cases. In particular, the compromise between the choice of parameters for sensing versus those for communications. It is part of the Hexa-X EU research project and has become integral to the recent and ongoing standards, such as in 3GPP and IEEE [11]. As mentioned above, there is also interest in the United States and China in this area.

14.5 Use cases and requirements

The Hexa-X project proposes representative 6G use cases [3,7,12]: sustainable development, Immersive telepresence, massive twining, cooperating Robots (cobots) and local trust zones, while JCAS can be used within the network itself to improve performance. Figure 14.2 summarises the requirements for accuracy, maximum latency and availability for different use cases based on [3] with the 5G Rel. 16 and Rel. 17 target figures. The color-coding of use cases represents their availability requirements. The red use cases of telesurgery and placement of medical devices in bodies cannot be met by 6G but are likely to be met with specialised medical surgical equipment anyway. Patient tracking could also be met by GPS/cellular in most cases. Similar applications are sold today to track personal pets. Table 14.3 shows the requirements for location accuracy, range resolution and velocity resolution for sensing use cases [3].

14.5.1 Sustainable development

This category of 6G use cases focuses on sustainable development while minimising the environmental impact of various industries. It explores how positions and sensing can be used as an adjunct to other technologies in delivering new services for sustainable development. For example, remote healthcare is made

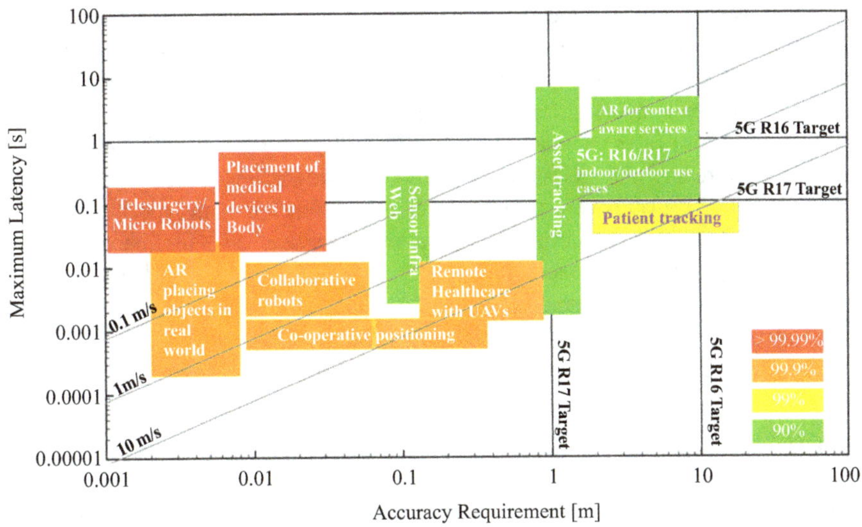

Figure 14.2 Accuracy, latency and availability requirements (based on Reference [3] with 5G R16/17 targets included)

Table 14.3 Requirements for sensing use cases: from Reference [3]

Use case	Location accuracy (m)	Range resolution (m)	Velocity resolution (m/s)
Gesture recognition for human–machine interface	0.01	0.01	0.3
DT for manufacture	0.01	0.01	0.5
Traffic monitoring	0.5	0.5	0.5
Robots to cobots environmental mapping	0.01	0.01	0.5
Robots to cobots object sensing	<0.01	<0.01	0.1

possible by technologies such as drone deployment for medical sample collection, which requires accurate location data. In this case, a service guarantee can be achieved with moderate availability and relaxed latency requirements. Other examples include monitoring weather conditions, tracking biodiversity and asset tracking, all of which have relatively relaxed accuracy and latency requirements.

14.5.2 Immersive telepresence

In this group of use cases, the merging of cyber-physical worlds, mixed reality co-design and merged reality game/work use cases can greatly benefit from using radio as a sensor. Precise and efficient localisation and Simultaneous Localisation and Mapping (SLAM) can help bring together the digital and physical worlds for real-time interaction between users far away. These particular use cases demand innovative methods of interaction between humans and machines, which necessitate accurate gesture detection. This category of use cases includes gesture recognition for human–machine interactions and augmented reality (AR). Gesture recognition requires acceptable range and angular accuracy, while AR may have varying requirements depending on the specific use case. For example, AR intended to provide context-aware services, such as a shopping mall experience, requires moderate location and angular accuracy. AR intended for placing an object in the real-world demands cm-level location and tight orientation accuracies.

14.5.3 Local trust zones for humans and machines

Today, we usually think of mobile communications as being exclusively cellular. However, many new use cases require local or private communications for highly sensitive information. These use cases may need to be tightly integrated within the wide area cellular network. This involves network topologies beyond cellular and security beyond classical architectures, as discussed in [12]. To ensure the protection of individual or machine-specific information and independent sub-nets, local trust zones are used. For example, body area networks or automatic guided vehicles (AGVs) must be dynamically integrated within a wide-area cellular network or stay within a private on-premises network.

One part of this family includes use cases with very tight location and orientation, availability and latency requirements, such as futuristic telesurgery, localising micro-robots within the human body and placement of medical equipment on the body. Here, while the requirements may be beyond that positioning accuracy within 6G and even if delivered within a specialised medical machine, the implementation for the local trust zone for this information remains the same.

Another aspect is a sensor infrastructure web. For example, a simple autonomous vehicle (with no onboard sensors) moves around an environment. While relying on third-party sensors as if they were on board. Alternatively, this could be used for patient tracking and monitoring. A sensor infrastructure web supports people or devices without sensing capabilities in both cases.

14.5.4 Massive twinning

DT for manufacturing are an essential component of the 6G platform. A digital representation of a physical manufacturing environment can be created by locating and mapping all the scattering objects. These can be any object within a factory floor, whether stationary or in motion. This can be done using the extensive network of established radio links to support communication in an industrial setup ‘for free’.

This particular use case family also extends to immersive smart cities. Creating a digital replica of a city makes it possible to obtain real-time data on traffic, pollution and utilities, among other things, which can be used to improve the city’s overall liveability. Achieving this use case requires careful localisation and monitoring of the city and all its entities.

Monitoring traffic is a practical application of future sensing technology. In such scenarios, the ability to measure the speed of objects is just as important as precise positioning. For instance, two or three Tx/Rx nodes could monitor an intersection in a city. Identifying a cyclist approaching a blind intersection could be used to warn the car's collision avoidance system and moderate its speed to avoid an accident. While the vehicle itself would have on-board radars, the reaction time is improved with pre-warning.

14.5.5 From robots to cobots

6G builds on the use cases identified in 5G for Industry 4.0 with enhanced performance, providing a communications framework with localised control and computing for all aspects of a future factory, supporting the robots. In this group of scenarios for robots to cobots, consumer robots, mobile robots, cooperative robots and flexible manufacturing processes can significantly benefit from localisation and SLAM. To provide effective and reliable sensing and localisation for these scenarios, the future radio access network must ensure the high availability and reliability of localisation and SLAM services. Additionally, the network must be a core platform to support these services and guarantee high integrity of sensing and localisation estimates [7].

14.5.6 Use within the mobile network itself

In addition to the above use cases, location and sensing can improve the mobile network's performance. For example, the sensing technology can identify objects that temporarily obstruct the direct propagation path between a transmitter and a receiver. With this information, the network can quickly adjust the direction of the transmitter beam to utilise either a reflected beam or switch to a different transmitter for communication with the device [10].

In addition, creating a DT of the radio network environment could yield valuable information to simulate or predict network performance. For example, it can evaluate novel network algorithms or plan the deployment of a new transmitter node [5]. It could help choose the optimum QAM scheme for that environment with up-to-date knowledge of the actual physical situation.

So, what are the opportunities and challenges?

14.6 Technical challenges and enablers

As discussed in this book, wide area coverage deployments will not happen in mmWave or sub-THz but below 5 or 6 GHz. Therefore, it is essential not to limit sensing to mmWave but also to extend its use to other frequency bands used in UMa and UMi to address all the use cases of 6G. As mobile network operates on multiple frequency bands, from low band to high band, sensing results from multiple carriers could be combined to improve performance further.

Furthermore, sensing, particularly at mmWave frequencies, should not be solely limited to determining object locations and spatial mapping but should also consider other use cases, such as spectroscopic sensing of objects. For example, this could be used to scan your meat in a restaurant for potential allergens against your device's health profile or in the context of a smart city to detect pollution levels.

The same hardware used for communication should be reused for location and sensing with no or minor modifications so that they can rely on a whole network of sensors and transmitters from the 6G launch using the same spectrum and hardware – building a separate sensing system using a spectrum dedicated to that function would reduce the overall spectrum efficiency and incurring additional hardware costs in UEs and BS. Therefore, JCAS will use a single hardware system that transmits data to communication users and detects target objects using backscattered signals with the same signal or using dedicated signals for each service. JCAS faces multiple challenges in hardware implementation, signal processing, information theory and interference mitigation.

14.6.1 JCAS air interface waveforms

One major challenge for JCAS is designing a new air interface waveform that allows outstanding spectral efficiency for communications with accuracy in sensing with limited time–frequency resources. Multi-carrier schemes, such as orthogonality frequency division multiplexing (OFDM), are attractive due to their flexibility and have been used since 3G. However, one of the challenges in high mobility situations, such as for vehicles and drones, is that for OFDM radars, the inter-carrier interference destroys the orthogonality of sub-carriers and degrades the performance of fast Fourier transform algorithms [7,13,14]. Single carrier waveforms, such as those used in Wi-Fi 802.11ad, are more attractive from a hardware perspective but must be designed to avoid high side lobe levels [15].

Traditionally, point-to-point communication and radar system performance have been extensively studied with Shannon's information theory, where the data speed, error rates and channel capacity are well understood. However, the 6G JCAS network needs to operate closed-loop processes, including sensing information, information flow processing and action instructions, where there is more significant interaction between parameters and their consequences. Therefore, the design of JCAS is much more complex as it needs to explore close-loop information theory and key performance metrics for the JCAS [16].

While the above outlines traditional static solutions with pros and cons, a promising alternative approach is to design an adaptive waveform that can jointly adjust communications and sensing performance based on artificial intelligence (AI) and machine learning (ML). This can build on the environmental mapping, including static and moving objects. Nokia, NTT and Docomo are working on an AI-native air interface that allows radios to learn. Currently, this is being used only to improve the spectral efficiency, but authors believe this could be used to address the sensing problem.

14.6.2 Multiple-input multiple-output (MIMO) for JCAS

Massive phased arrays (massive MIMO) will be essential in communications and sensing. In Chapter 8 on mmWave, we explained how the communications link budget is improved with MIMO technology. In sensing applications, resolution in angle is obtained using massive phased arrays. These are particularly attractive above 60 GHz, where the arrays can be integrated with the electronics on a chip in system-on-a-chip (SOC) implementations; for lower carrier frequencies, massive phased arrays can provide extended range at the cost of physically larger arrays. Therefore, massive phased arrays are particularly attractive for mmWave and THz frequencies for both communications and sensing with progressive miniaturisation of components. Especially in the higher sub-terahertz region in 6G (100–300 GHz), these massive antenna arrays will allow fine spatial (angle and delay) resolution, enabling highly directional sensing and imaging applications such as gesture detection and 3D mapping while being less susceptible to ambient light and weather conditions than light and infrared technologies. These also support highly directional beams of less than a degree beam width to allow high-definition 3D mapping of the environment, which, when combined with DT, will find many applications in industrial settings.

14.6.3 Reconfigurable intelligent surfaces (RIS)

RIS is an emerging technology that can modify how an incoming electromagnetic signal is reflected [17]. At its most basic level, the electronic control of surface properties can change how signals are reflected. Think of it as a mirror whose surface can be bent to change the direction of the reflection.

In more complex configurations, hundreds of elements on a single RIS can be individually controlled, allowing for unprecedented reflection steering, albeit at a relatively low rate of up to 60 Hz [7]. RIS can be implemented through reconfigurable metamaterials and conventional discrete antennas. Metamaterials can dynamically and artificially adjust the physical properties of transmitted, received or impinging EM waveforms to obtain desired electrical or magnetic characteristics that are not available in nature. Discrete antenna solutions adopt antenna elements and act as an independent unit that modifies the behaviour of the wave in the desired manner. When many tiny antennas are adopted on large, intelligent surfaces, they represent a natural evolution of MIMO technology. An RIS can allow a signal to reach its destination when no line-of-sight can be achieved.

Although the potential of RIS in wireless networks is significant, it does face new challenges that need to be efficiently addressed [18]. These challenges include reflection optimisation, channel estimation and deployment from communication design perspectives.

14.7 Other types of positioning and sensing technologies

While the current focus is on JCAS in 6G, other radio systems such as Wi-Fi, BLE and UWB have had positioning and sensing available for several years, combining both communications and sensing modes. In addition, LIDAR, which uses a laser, is also available for ranging-type applications such as photography, where it is used to add camera focus in low-light conditions. UWB and LIDAR are used in the latest iPhones, such as the iPhone 15, for potential content creation for the Apple Vision Pro VR Goggles.

14.7.1 Wi-Fi positioning system

The Wi-Fi ecosystem is a widely used wireless system that can also be used for positioning systems in indoor locations for 'almost free', as in the case of 6G. Wi-Fi indoor positioning solutions [19] utilise pre-existing Wi-Fi access points or sensors to detect and locate transmitting Wi-Fi devices such as smartphones or tracking tags. This allows organisations to leverage their existing Wi-Fi infrastructure to enable location-aware applications with no additional hardware. The location data gathered from either sensors or access points or transmitted from access points to client devices is then processed by various locating applications to generate information for different indoor location-aware use cases.

Wi-Fi-based positioning systems can adopt different techniques to determine the location of devices, with most relying on received signal strength indicator (RSSI)-based methods. However, some applications can utilise more advanced Wi-Fi positioning methods, such as time-of-flight and AoA [20]. Wi-Fi fingerprinting is also an RSSI-based method. Here, Wi-Fi positioning via fingerprinting involves creating a database that records the location and signal strengths of surrounding access points and the coordinates of a Wi-Fi device; however, making the fingerprinting database requires a time-consuming calibration process requiring updates for installing new access points [21].

Wi-Fi is still a practical option for indoor positioning, although it is typically less accurate than competing positioning technologies like BLE and UWB [19]. These have found broader uses in smartphone applications, such as the iPhone 15, where UWB is used to find other Apple products. Yet most of these indoor applications are nascent.

14.7.2 Bluetooth low energy

While Bluetooth and BLE share a similar name, they are regarded as different technologies. BLE explicitly targets markets with ultra-low power demand rather than high throughput. In BLE, data communication with a low-energy radio happens in short bursts that do not need to be very frequent [22]. BLE was released in 2010 as part of version 4.0 of the Bluetooth Specification. BLE is a separate protocol from Bluetooth (Bluetooth Classic), which is incompatible.

Although BLE and Wi-Fi primarily use RSSI to detect the location of people, devices and assets, BLE achieves higher location accuracy (Table 14.4) [19]. BLE requires significantly less power than Wi-Fi, allowing for more flexible hardware options and applications.

14.7.3 Ultra-wideband

UWB is a short-range radio technology that accurately determines and measures the distance between it and other UWB-equipped devices. Unlike BLE, which

Table 14.4 Summary of positioning and sensing indoor radio technologies [19] with optimal conditions and deployment

	Wi-Fi	**BLE**	**UWB**
Location accuracy	<10 m	<5 m	±40 cm
Range	Optimal 0–50 m Up to 500 m	Optimal 0–25 m Up to 100 m	Optimal 0–50 m Up to 200 m
Latency	Typically, 3–5 s to get location	Typically, 3–5 s to get location	<1 ms to get location
Power consumption	Moderate	Very low: option for embedded cell battery in select hardware options	Low: option for embedded cell battery in select hardware options
Frequencies	2.4 GHz, 5 GHz	2.4 GHz	3.1–10.6 GHz
Data rates	Up to 1 Gbit/s	Up to 2 Mbit/s	Up to 27 Mbit/s
Cost	$$$ (however low with existing APs)	$$	$$

measures signal strength between devices, UWB measures signal time between devices. UWB-enabled devices exchange pulse shapes of information to enhance the ranging accuracy, which can be used for a ranging exchange. UWB requires low power to operate, making it an energy-efficient technology. UWB tracking is so accurate that it can pinpoint an object's location below 10 cm. Major manufacturers in the automotive industry believe UWB-equipped devices may serve as digital keys [23]. The UWB in the Apple iPhones transmits on two frequencies – 6.24 and 8.2368 GHz while the new UWB chip (U2) in the iPhone 15 can locate your friends up to 60 meters away with Precision Finding in Find My Friend.

14.7.4 LIDAR – 3D imaging and photography

LIDAR is a sensing method that uses light as a pulsed laser to measure ranges. It has been built into iPhones since 2020, and its primary use has been to improve focusing in low-light conditions for photography. It also measures distances.

One of the exciting developments recently has been in the iPhone 15 Pro Max, where LIDAR is the fourth lens and is also used for 3D scanning to build a 3D model of an environment for its use with Apple's Vision Pro Goggles in 2024. This could allow 3D photographs, the first significant change to the format for over a hundred years, and it could also allow its use in user-generated VR/AR and games. So, while it may be argued that following the original iPhone in 2008, all the significant upgrades have been incremental: with better processors, better displays, better camera technology, additional networking, payment, etc., this could be a real defining moment for the future of the iPhone and other smartphones.

14.7.5 Summary of performance of Wi-Fi, BLE and UWB

Table 14.2 summarises the performance of Wi-Fi, BLE and UWB technologies for indoor positioning applications [19]. Considering the simulated performance of 5G in Table 14.1, only UWB offers similar performance to a likely 6G system.

14.8 Summary and conclusions

The ability to locate, track and sense physical objects bridges the human, physical and digital worlds. Mobile networks provide an excellent starting point for establishing a massive sensor network, as they have extensive coverage from the beginning of their operation. Transmit and receive nodes have interconnection between the nodes, enhancing accuracy. Hence, the sensing can be delivered almost 'for free'. All major global research programmes for 6G technology aim to support JCAS as a feature that sets it apart from its predecessor. This feature is expected to revolutionise many applications and services. However, this capability has also appeared within 5G standards, albeit with lower performance levels. Nevertheless, the higher speeds of 6G will also support faster information transfer from sensors to DT, enabling the creation of 3D spatial maps of environments within its ecosystem. It could find uses in the iPhone 6G described in Chapter 15.

Sensing in 6G is not solely confined to mobile systems; other wireless systems, such as Wi-Fi, BLE, LIDAR and UWB, have offered location and sensing technologies for several years. BLE, LIDAR and UWB have found their way into smartphones for developing applications, which could grow consumer interest in sensing and location technology. It will be interesting to see how Apple's use of LIDAR, which combines the iPhone and Vision Pro goggles for 3D photography and imagining, plays out over the remaining decade before 6G. Today, only UWB will likely offer a similar position accuracy to a future 6G JCAS.

The 6G JCAS provides an alternative approach and enabling platform for more widespread use across industry and society, such as 3D spatial mapping supported by DT coupled with local trust zones within its architectural approach. Thus, the monetisation of JCAS is likely to be around the application/service using the whole platform.

Although GPS is extensively used today in cellular apps very successfully, it is often problematic in many city centres, where several satellites are not visible, particularly inside buildings, as there is no satellite coverage. 6G is set to offer better coverage and accuracy for JCAS, providing a metre or so accuracy in macrocells, increasing to the centimetre range for indoor 6G deployments and is likely to be implemented across all the cellular bands in 6G. It is a key feature not only for Industry 4.0 but also for many other use cases in consumer, health, smart cities and vehicle safety. Sensing could also evolve to support spectroscopic scanning for new smartphone Apps and pollution measurement in cities for improved liveability.

JCAS is supported by worldwide research not only in the EU Hexa-X project but in the Unites States in the ATIS Next G alliance, as well as in China, Korea and Japan as their possible inputs to 3GPP. In the Hexa-X project, vision, localisation, radar and sensing are intrinsic parts of 6G from the outset. Extreme performance is needed to support location accuracies and latencies foreseen in the identified use case families. This will lead to a tight integration (at hardware and software levels) of radar, communication, computation, localisation and sensing. Massive twining will be an integral part of the new world of services.

The level of accuracy will differ in different cell types. In many respects, MIMO in the sub-terahertz region is an up-and-coming solution as it can combine extreme miniaturisation for an SOC with low-cost, high degrees of angular resolution with a pencil-like beam for 3D mapping and spectroscopic scanning. However, different approaches are required for macro and microcells using frequencies below 6 GHz.

JCAS faces many challenges, mainly because it aims to use the same spectrum and hardware for communications and sensing. It will face multiple challenges in hardware implementation, signal processing, information theory and interference mitigation. While the above outlines traditional static solutions with pros and cons, a promising alternative approach is to design an adaptive waveform that can jointly adjust communications and sensing performance based on AI and ML. This can build on the environmental mapping, including static and moving objects.

References

[1] Hexa-X EU. Expanded 6G vision, use cases and societal values – including aspects of sustainability, security and spectrum. [Online]. 2021. Available from: https://hexa-x.eu/wp-content/uploads/2021/05/Hexa-X_D1.2.pdf.

[2] Chen J, Zhang D, Wu Z, Zhou F, Sun Q, and Chen Y. Contactless electrocardiogram monitoring with millimeter wave radar. [Online]. 2021. Available from: https://arxiv.org/pdf/2112.06639v2.pdf.

[3] Behravan A, Yajnanarayana V, Keskin MF, *et al.* Positioning and sensing in 6G: Gaps, challenges, and opportunities. *IEEE Vehicular Technology Magazine*. 2023; 18(1): 40–48.

[4] ATIS Next G Alliance. Next G alliance uploads. [Online]. 2022. Available from: https://nextgalliance.org/wp-content/uploads/dlm_uploads/2022/07/TWG-report-6G-technologies.pdf.

[5] Andersson H. Joint communication and sensing in 6G networks. [Online]. 2021 [cited 25 October 2021]. Available from: https://www.ericsson.com/en/blog/2021/10/joint-sensing-and-communication-6g.

[6] de Lima C, Belot D, Berkvens R, *et al.* Convergent communications, sensing and localization in 6G systems: An overview of technologies, opportunities and challenges. *IEEE Access*. 2021; 9: 26902–26925.

[7] Wymeersch H, Shresta D, de Lima C, *et al.* Integration of communications and sensing in 6G: A joint industrial and academic perspective. In *2021 IEEE 32nd Annual International Symposium on Personal, Indoor and Mobile Radio Communications (PIMRC)*, Helsinki, Finland, 2021, pp. 1–7, doi:10.1109/PIMRC50174.2021.9569364.

[8] Dwivedi S, Shreevastav R, Munier F, *et al*. Positioning in 5G networks. *IEEE Communications Magazine*. 2021; 59(11): 38–44.
[9] Taylor C. 3GPP new radio (NR) positioning support. [Online]. 2023 [cited 30 August 2023]. Available from: https://www.3gpp.org/technologies/nr-positioning-2.
[10] NXP. Joint communication and sensing in 6G networks. [Online]. 2021 [cited 21 November 2021]. Available from: https://www.nxp.com/company/blog/joint-communication-and-sensing-in-6g-networks:BL-SENSING-IN-6G-NETWORKS.
[11] 3GPP. Study on NR positioning enhancements – TR 38.857. [Online]. Technical Specification Group Radio Access Network, Release 17, March 2021. Available from: https://portal.3gpp.org/desktopmodules/Specifications/SpecificationDetails.aspx.
[12] Hexa-X. Hexa-X deliverable 1.2. [Online]. 2021 [cited 30 April 2021]. Available from: https://hexa-x.eu/wp-content/uploads/2021/06/Hexa-X_D2.1.pdf.
[13] Hakobyan G and Yang B. A novel intercarrier–interference free signal processing scheme for OFDM radar. *IEEE Transactions on Vehicular Technology*. 2017; 67(6): 5158–67. doi:10.1109/TVT.2017.2723868.
[14] Keskin MF, Wymeersch H, and Koivunen V. MIMO-OFDM joint radar-communications: Is ICI friend or foe? *IEEE Journal of Selected Topics in Signal Processing*. 2021; 15(6): 1393–1408.
[15] Grossi E, Lops M, and Venturino L. Adaptive detection and localization exploiting the IEEE 802. 11ad Standard. *IEEE Transactions on Wireless Communications*. 2020; 19(7): 4394–4407.
[16] Feng Z, Wei Z, Chen X, Yang H, Zhang Q, and Zhang P. Joint communications, sensing and computation enable 6G intelligent machine system. *IEEE Network*. 2021; 35(6): 34–42.
[17] Wu Q, Zhang S, Zheng B, You C, and Zhang R. Intelligent reflecting surface aided wireless communications: A tutorial. *IEEE Transactions on Communications*. 2021; (69): 3313–3351.
[18] Kisseleff S, Chatzinotas S, and Ottersten B. Reconfigurable intelligent surfaces in challenging environments: Underwater, underground, industrial and disaster. *IEEE Access*. 2021; (9): 150214–150233.
[19] Inpixon. Wi-Fi RTLS, location tracking & positioning. [Online]. Available from: https://www.inpixon.com/technology/standards/wifi.
[20] Tomažič S. Indoor positioning and navigation. *Sensors*. 2021; 21(14): 4793.
[21] Labs L. Can Wi-Fi access points be used for RTLS asset tracking? [Online]. 2022. Available from: https://www.link-labs.com/blog/can-wi-fi-access-points-be-used-for-rtls-asset-tracking.
[22] NovelBits. Bluetooth low energy (BLE): a complete guide. [Online]. 2022 [cited 6 September 2022]. Available from: https://novelbits.io/bluetooth-low-energy-ble-complete-guide/.
[23] Medium: Digital Car Keys Step into Era of UWB. [Online]. 2024. Available from: https://medium.com/@marketing_1114/digital-car-keys-step-into-era-of-uwb-3408895bd725.

Chapter 15
iPhone 2030 (6G) and other devices

Every once in a while, a revolutionary product comes along that changes everything.

– *Steve Jobs*

Every decade, the arrival of new mobile generations brings fresh capabilities, but the most significant moment in wireless communications for consumers occurred on 29 June 2007. It was the day when Steve Jobs, in his legendary keynote at the MacWorld Expo, unveiled the original iPhone as 'an iPod, a phone and an internet communicator'. This game-changing device altered the course of mobile usage and prompted UK mobile network operators to explore Wi-Fi offload at public hotspots due to the exponential surge in mobile data. The iPhone emerged independently from the traditional cellular industry and demonstrated that a revolutionary end device could shape the direction of wireless communications for well over a decade.

This chapter delves into the technological trends that may define an iconic smartphone in 2030, envisioning a future iteration like the hypothetical Apple 'iPhone 2030 6G'. This exploration provides insights into the potential user experience and sheds light on the network requirements that may arise for both 6G and Wi-Fi 8/9.

15.1 A review of the Apple iPhone history and its impact

The original iPhone featured an aluminium and plastic body, a 3.5-in. LCD, a microphone, headset controls and a 2-megapixel camera. Before the iPhone, smartphones primarily relied on physical keyboards and had significantly smaller screens. However, the iPhone revolutionised the industry by introducing a touch keyboard, which allowed users to navigate the Internet via its Safari browser and would become the standard input method even for larger devices like iPads. Moreover, the iPhone played a pivotal role in changing the prioritisation of wireless networks, shifting the focus from cellular to Wi-Fi. Doing that paved the way for 4G to replicate the Wi-Fi experience.

Upon its launch on the 2G Edge network, which offered speeds of up to 384 kbit/s, the original iPhone emphasised Wi-Fi as the preferred network experience over cellular connectivity. Wi-Fi networks then and today deliver faster connection speeds, usually by a factor of 10. In contrast, while other smartphones also incorporated Wi-Fi capabilities, they typically prioritised cellular use due to their association with mobile operators and their sales channels. Wi-Fi was commonly supported at a second or third layer within the device operating system, which required managing network selection and passwords.

Numerous estimates indicate that approximately 70%–80% of mobile data usage today occurs through Wi-Fi networks, highlighting Wi-Fi's significant role in mobile data consumption.

In the fourth quarter of 2006, a year before the iPhone's announcement, the global smartphone market was relatively small, with just over 20 million devices sold. Nokia was the market leader with half of these sales, followed by Canadian BlackBerry maker RIM, Motorola, Palm and Sony-Ericsson. These players may not be recognisable to most readers today. For example, RIM BlackBerry, which had a third of the market, focused on professional email with email systems and devices. Their smartphones had a physical keyboard, and they initially underestimated the importance of consumer-focus devices with touch keyboards, cameras and elegant operating systems. This partly led to its eventual demise.

While the original iPhone, introduced in 2007, was a revolutionary concept, it did have some minor technical issues associated with the mobile phone part. The subsequent release of the iPhone 3G the following year addressed these issues and introduced new connectivity options, such as 3G data and Global Positioning System (GPS), which formed the basis for Google Maps. Before this, mobile operators had experimented with location services, which could have been based simply on the cell ID. Still, they were focused on a more sophisticated technical solution, which lost their first-mover advantage.

Additionally, the iPhone introduced the App Store with an initial offering of around 500 applications. This move marked a significant turning point, and Apple's app strategy, later adopted by Android, played a crucial role in their success. Over time, the App Store grew to around 2.2 million apps in 2017, but it gradually declined to 1.6 million by 2022 due to Apple's initiative to remove outdated or 32-bit apps.

Within a year, Apple transformed the dynamics between device vendors and cellular operators. Previously, mobile devices were developed in close collaboration with mobile operators, relying on their support for sales. However, with the iPhone, operators started competing to stock and offer the iPhone first.

Since 2007, the smartphone industry has undergone significant transformations. New players from consumer electronics backgrounds have replaced the old guard of 'mobile device companies'. While user capabilities have evolved, they have not undergone revolutionary changes until perhaps 2023/24 with the introduction of three-dimensional (3D) imaging capabilities for the Apple Vision Pro Goggles. For example:

- Smartphones have physically grown a little larger with superior display technology, but we still use touchscreen keyboards, and images appear similar unless compared side-by-side. Recent trends to 120 frames per primarily benefit gamers.

- The App Store's importance has grown, and it now hosts 1.6 million apps as of 2022, solidifying the smartphone's position as a mobile computer in your pocket for almost everything.
- Smartphone technology advancements have rendered many entry-level cameras obsolete, with only professional Digital SLR cameras remaining. Still, the latest smartphone cameras are increasingly challenging these, even for professional use, particularly video with an action movie Midnight, filmed using the iPhone 15 Pro.
- Location-based services have evolved since the original 3G iPhone. They feature '3D-looking' maps and integrate car navigation with live traffic updates. They also include access to public transport, with arrival and departure times shown for journey planning.
- The smartphone ePayment system is beginning to replace cash following COVID-19, where many retailers specified cashless payments.
- Siri, initially released as a standalone app in February 2010, was acquired by Apple and integrated into the iPhone 4S upon its release in October 2011.

While the iPhone's user characteristics have evolved, there have been revolutionary changes to its internal architecture. Apple transitioned from a 32-bit operating system to a 64-bit for iOS 7 in 2013, improving processor handling and significantly expanding memory-addressing capabilities for future advancements. In 2016, Apple introduced the Apple File System, optimised for solid-state drive storage, replacing the previous Hierarchical File System in 64-bit iOS devices. Perhaps the most revolutionary change was the introduction of Apple's ARM-based system-on-a-chip (SoC) in 2022 with the M1 chip, followed by the announcement of the M2 chip and, more recently, the M3 in the latest iPhone 15 models. These chips integrate the central processing unit (CPU), graphics processing unit, unified memory architecture, AI engine and more into a single chip, boasting the world's fastest CPU core in low-power silicon. This approach eliminates communication bottlenecks between various elements of a modern computer, resulting in lower latency. It provides Apple with a cross-platform SoC computer engine for iPhones, iPads, desktops and laptops, allowing for new form factors and device convergence.

So, while changes have been evolutionary to date, the most significant recent change to the iPhone since its launch is the inclusion of stereoscopic and 3D imaging capabilities in the iPhone 15 Pro for use with its Vision Pro Goggles in 2024. This is the first change to the photographic format in over a hundred years. A combination of the ultra-wide angle and standard lenses offers stereoscopic images. At the same time, light detection and ranging (LIDAR) will allow the creation of 3D spatial maps using its range-finding capabilities. This technology was initially incorporated into their models in 2020 to aid in focusing range measurements in low-light photography. *This could be a precursor to the Joint Communications And Sensing (JCAS), Chapter 14, that 6G could bring with its combination of Digital Twins.*

In summary, the original iPhone introduced a range of innovative features based on the touchscreen and app store while shifting the focus to Wi-Fi networks as the preferred choice for connectivity. This shift in prioritisation, along with Wi-Fi's significant speed advantage over cellular networks, has dramatically influenced the landscape of mobile data usage since.

15.2 Future technological advances for the 'iPhone 2030 6G'

The smartphone has become an essential item for millions worldwide, just as Steve Jobs predicted when he said it would be 'An iPod, a phone and an internet communicator'. However, the essential functions of smartphones were established over 10 years ago, and although technology has improved, the core capabilities remain the same. However, the next revolution in wireless communications could again emerge with a revolutionary wireless terminal or a novel and game-changing feature integrated into smartphones that will unlock various innovative applications. For instance, this could include the advent of virtual reality (VR) goggles directly connected to mobile and Wi-Fi networks, transforming how we perceive and interact with digital content for augmented reality (AR). This digital content could be partially generated with smartphone camera technology such as LIDAR or with JCAS in 6G. This evolution holds the potential to reshape the landscape of wireless communication and unlock unprecedented opportunities for enhanced user experiences. Looking ahead to the future of the iPhone in the 6G era, what are the potential changes that could bring about a revolutionary shift? While it is challenging to predict with certainty, several possibilities could significantly transform the smartphone experience:

15.2.1 mmWave-enabled innovation.

The potential of an iPhone 2030 (6G) holds several captivating features, especially its ability to capture 3D images of the surrounding environment using either the smartphone's mmWave communications system based on JCAS or other technologies such as LIDAR or ultra-wideband, described in Chapter 14. Furthermore, the associated multi-100 Gbit/s communications system allows transmitting these significantly larger image files to a digital twin within the 6G network, which embodies the spatial map. This might be terabytes in its storage requirements. This breakthrough opens a plethora of applications.

In the iPhone 2030 (6G), the mmWave communications system will likely employ frequencies surpassing 100 GHz, utilising phased arrays comprising thousands of antennas. It is important to note that the resolution of a lens in air is determined by dividing its wireless wavelength by 2. Thus, with frequencies at 100 GHz and beyond, the system can achieve a minimum resolution of 1.5 mm, which falls below 0.4 mm at 400 GHz. Leveraging beam scanning techniques, it becomes possible to systematically monitor received mmWave signal signatures from various angles with millimetre-level accuracy, creating detailed 3D images of physical spaces [1]. Real-time positioning enabled by this technology finds applications in manufacturing and logistics but will inevitably find surprising consumer applications.

In addition, the camera could enable 'seeing' in low-light conditions or even fog or smoke, replacing the need for a torch on smartphones introduced over a decade ago. In fire rescue situations, it could enable vision in smoke-filled rooms, even detecting people hidden behind walls using reflections of its mmWave beams. Moreover, it allows capturing 3D images that find utility in VR and AR, autonomous driving and assisted living technologies.

Beyond smartphone usage, this technology has broader implications in healthcare. It can facilitate behavioural monitoring to assess individuals' physical and mental health, provide fall detection capabilities to alert caretakers regarding older citizens, enable geofencing for individuals with dementia and even serve as a navigation aid for the visually impaired. For the latter two applications, goggles worn by the user could be implemented, utilising the newfound 3D imaging capability. The advent of 6G technology and its associated mmWave communications system has the potential to spawn numerous innovative applications with the development of future Apps.

The integration of these advanced imaging capabilities, supported by the 6G mmWave communications system, holds tremendous potential for transforming various industries and unlocking new possibilities for enhanced user experiences and practical applications.

15.2.2 Light field recording

Despite the remarkable advancements in smartphone camera technology, which have greatly improved image processing capabilities, traditional digital photography remains limited to capturing images in two dimensions. However, light field photography, exemplified by Lytro cameras, offers a transformative approach by enabling the capture of images in three dimensions with a simple button click. The focus point of the photo can be determined later in the software. This revolutionary technique moves the emphasis from optics to computation.

Light field cameras [2] have thousands of micro-lenses positioned between the primary lenses and the digital photographic sensor. These micro-lenses scatter incoming light before it reaches the sensor, capturing data that retains the direction of each light ray. This fundamental shift in image capture allows users to adjust focus and perspective after taking the photo. Naturally, this results in larger image files and reduces photographers' need for extensive skills. It is highly likely that future AI-powered photography programs, such as Luminar™, will be able to identify the optimal focus position within the image file and handle tasks such as cropping and overall image processing automatically. The large data files will be able to utilise the mmWave communications of the 6G device.

Although light field technology has existed since Lytro cameras were introduced in 2012, its adoption has yet to be widespread. However, notable developments have occurred, including Google's acquisition of Lytro in 2018, which led to the integration of light field technology into projects like Project Starline [3]. Project Starline explores an experimental video communication method called a 'magic window', offering a simulated face-to-face experience with someone not physically present. Furthermore, Apple has been granted a light field patent for gesture capture.

While it remains uncertain whether an iPhone 6G or other 6G smartphones will incorporate light field cameras, the underlying technology could still find applications within the 6G device ecosystem. It can enhance emotional video communications or enable gesture-based interactions, bringing users new dimensions of expression and interaction.

15.2.3 Ultra-low light camera

Integrating a digital camera into smartphones presents inherent physical constraints on the size of the digital photographic sensor. As pixel density increases to achieve higher image resolution, the sensor's sensitivity for low-light photography diminishes. While electronic flash can compensate for this limitation in general photography scenarios, its practicality is limited in smaller smartphones. However, including specialised cameras that excel in low- or no-light conditions could overcome this significant obstacle and greatly enhance indoor photography capabilities.

The ISO speed of a camera refers to its light sensitivity. Digital SLR cameras usually have a maximum ISO setting of 51,200, determining their low-light performance. Older readers will remember buying 35 mm film for their cameras with an ISO of only 400. However, Canon's ME20F-SH camera has a top ISO of 4 million, allowing it to capture images even in complete darkness. Furthermore, smartphones nowadays often come with a night-sight feature that uses machine learning and advanced algorithms to produce exceptional low-light photos.

By incorporating such innovative camera technologies into smartphones, the limitations of indoor photography due to low light can be significantly overcome. This allows users to capture high-quality images even in challenging lighting conditions, expanding the creative possibilities and ensuring memorable moments are not limited by environmental constraints.

15.2.4 A new human sense – smell camera

Since their invention, cameras have been a visual medium, but future cameras may add a sense of smell to the images. This could create even greater emotion from a photo. Great chefs like Heston Blumenthal have recreated smells to engage the senses for their meals. Manisha Mohan from the MIT Media Lab outlined a 'Smell Camera,' which consists of a pump connected to a smartphone [4]. The user would control the pump to capture smells in a gelatin capsule and then 'play' the memory later, complete with smells.

15.2.5 Holographic displays – Project Echo holographic displays

Undoubtedly, one of the most ground-breaking advancements in the iPhone 2030 6G would be the inclusion of a holographic display, both for the iPhone itself and other displays such as TVs. As referenced in [5,6] and exemplified by Project Echo from Realfiction, this technology would offer an unparalleled visual experience that resembles looking through a window into a scene. The holographic display would provide a sense of depth, allowing for perspective shifts and the ability to see past foreground objects that obstruct those in the background. It aims to deliver an immersive visual experience that closely mimics reality.

Unlike VR, where users wear goggles and need to look around physically, a holographic display on the iPhone or TV would enable viewers to observe different perspectives by moving their heads or adjusting the device's position. Each viewer would see a slightly different image optimised for their specific line of sight, like looking through a window. Multiple images are computed and transmitted in real time to cater to everyone's unique viewing positions.

We often encounter pseudo-3D images and data on two-dimensional displays or pages. However, the transformative potential lies in experiencing accurate volumetric pictures on the future smartphone, as depicted in the science fiction series 'The Expanse.' However, achieving this level of holographic display poses significant technical and economic challenges that have been pursued for many years. Computational, transmission and rendering requirements are immense, with calculations demanding 6.6×10^{15} flops, 3×10^{15} bits/s data rates and 1.6×10^{12} phase pixels [5]. Including a holographic display in the iPhone 2030 6G would be a significant advancement, changing how we see and engage with visual content. However, the technical and economic obstacles are genuinely daunting.

15.2.6 The 'iPhone 6G' is a real-time health hub

Integrating various body sensors, described in the 6G concept as the Internet of Bio Sensors, for health monitoring could be a key feature of the iPhone 2030 (6G), making it a vital device for preventive medicine and healthcare. As a future for the existing health app in iOS that stores personal information, the iPhone could gather comprehensive health data from these body sensors, which are downloaded each night or transmitted in real time to your local doctor or hospital. AI algorithms within the app could determine medical emergencies with real-time transfer of body sensor information to your doctor's systems for urgent treatment.

In addition, in an emergency room, the NFC++ technology could facilitate the transfer of this information to the paramedic through a simple touch using high-speed near-field communications. This streamlined process would enhance the seamless sharing of health data between the iPhone and healthcare providers, ensuring a more efficient and effective healthcare experience.

15.2.7 'iPhone 2030 6G' networking capabilities

As a premium smartphone device, the iPhone has consistently embraced the latest network technologies, including advancements in cellular connectivity, Wi-Fi, Bluetooth, near-field communications, ultra-wideband and satellites. This trend is expected to continue with the iPhone 2030 (6G), which will strive to support cutting-edge network capabilities with 6G and Wi-Fi 8, possibly 'Wi-Fi 9'.

6G will likely embrace centimetre waves within the 6.5–24 GHz range, which would be supported in the iPhone for primarily indoor applications and small cell scenarios such as sporting stadiums, allowing speeds of over 10 Gbit/s. These may also support low Earth orbit (LEO) satellites in outdoor use with similar speeds outside normal cellular coverage, see Section 15.2.8 below.

The iPhone 6G is anticipated to also incorporate mmWave communications above 100 GHz, enabling multi-100 Gbit/s data transfer rates. This will likely involve utilising GaN technology, initially exploring the lower part of the 100–200 GHz range, as described in Chapter 8. At the same time, other semiconductor materials may enable operation up to 400 GHz during the lifespan of 6G.

The iPhone 6G is expected to offer Wi-Fi 8 (launched around 2028), which might offer theoretical speeds of 100–200 Gbit/s, close to the Wi-Fi AP with low latency for advanced gaming (Chapter 3).

Regarding 'Wi-Fi 9' expected around 2030, the iPhone 6G could pioneer using the 127 GHz ISM band, opening a 7-GHz channel for Wi-Fi. For example, leveraging a 16 cm^2 GaN phased array at 127 GHz would enable an array with over 4096 antenna elements and an impressive equivalent isotropic radiated power of 65 dBm (over 3 kW) capability. (*This could be useful for non-communications applications such as spectrometry.*) With a channel bandwidth of 7 GHz at 127 GHz, 'Wi-Fi 9' could support communication speeds of up to 35 Gbps or higher. Meanwhile, with 20 GHz channels, foreseen for 6G, it would support over 100 Gbit/s. These could be available in 'information petrol stations', as described in chapter 16, found in public places such as Coffee shops for ultra-high-speed downloads such as 4K movies in a few seconds.

For normal cellular on 6G wide area networks, the sub-5 GHz bands could provide over 300 Mbit/s as a typical user experience.

15.2.8 Outside with Global Service Coverage

The iPhone 14 was the first to incorporate satellite networking for emergency calls, and the iPhone 6G will likely offer satellite broadband access worldwide using LEO satellites (chapter 10). It has the advantage that it only has to work with one LEO operator and its frequency to cover the earth as a new form of service provider – Global Virtual Satellite Operator. So, you will always have broadband access outside in oceans and vast land masses not covered by cellular.

Apple may sell this service directly, as it does with its current emergency satellite offering. So, while the 6G theme of 'Global Service Coverage', as described in Chapter 6, implies that it is a service integrated with cellular, this could also be provided in a similar way that Wi-Fi is today, outside the mobile ecosystem. The advantage for the smartphone vendor is that they could agree with the LEO provider to cover their spectrum band within the device for exclusive access to their network.

While the need for Gbit/s communications in 5G and 6G, particularly with Wi-Fi 7/8 advancements, may be questioned for smartphones, it is essential to consider future possibilities. Even though current smartphone activities like streaming 4K UHD movies require less than 5 Mbit/s, advancements in 3D imaging could justify higher speeds above 10 Gbit/s.

15.2.9 iPhone 2030 6G personal food allergens detector

One notable potential feature of the iPhone 2030 6G is the integration of mmWave phased arrays for sensing applications. By exploiting wider channel bandwidths above 100 GHz, it becomes possible to detect the presence of specific chemical elements based on frequency scanning spectroscopy. For instance, users could use smartphones to check their meats in restaurants for allergens. Additionally, mmWave phased arrays in smartphones can enable radar-based gesture detection, which could benefit Light Field technology and a touchless iPhone experience.

In summary, the iPhone 6G will likely continue its tradition of embracing the latest network technologies, offering enhanced connectivity options, improved data speeds and exploring innovative applications to deliver an exceptional user experience.

15.3 New device form factors – smart glasses

The focus so far has been on the iPhone's form factor, which, while it was a reinvention of previous smartphones before 2007, was a revolutionary device. Undoubtedly, it will continue to evolve in the coming years before the arrival of 6G, with many exciting features – some have already been described, while some have not. But what if a new device form factor emerges? In 2024, Apple launched their Vision Pro Goggles, where users see two 3660 × 3200 pixel 1.41-inch (3.6 cm) micro-OLED displays with a total of 23 megapixels, usually running at 90–100 Frames per second through the lens [7]. These use Wi-Fi 6 on today's broadband networks. While the first iteration of the Vision Pro does not support cellular connectivity, this is not a precedent for the future exclusion of the technology from this burgeoning device category. So, could this become the new mass market form factor for devices?

Figure 15.1 Man with smart glasses (Getty Images)

Like the iPhone in 2007, these goggles do many things. They combine a display for viewing 4K movies and potentially replace the computer display. With a virtual keyboard, the device supports gesture recognition to replace the mouse.

Figure 15.1 shows futuristic smart glasses that are much more fashionable to wear. They combine sensors for 3D imaging, which could be used to 'see in the dark' or through smoke. If, between now and 2030, we can develop display technology marrying all the capabilities of the first-generation technology of the Apple Vision Pro with much better fashionable glasses, this could replace the desktop computer and the smartphone.

15.4 Finally, the 1987 Knowledge Navigator revisited

In 1987, Apple developed a concept called the Knowledge Navigator, described by its then CEO John Sculley in his 1987 book, 'Odyssey: Pepsi to Apple' [8]. It represented a futuristic device that could access an extensive network database and used a software agent to assist the individual in the search for information. Apple produced some videos showing the technologies. The most famous was shown on a BBC Horizon programme [9]. In this video vignette, a university professor arrives home and switches on his computer, which is in the form of a tablet the size of a large-format book that includes a foldable display. On the screen appears a bow-tie-wearing butler who informs him that he has several calls waiting. Although he ignores most of these calls, including one from his mother, he uses the system to compile data for a talk on deforestation in the Amazon Rainforest. While he is doing this, his computer informs him that a colleague is calling, and they exchange data through their machines while holding a video-based conversation.

Another video shows a young student using a smaller handheld system version to prompt him while he delivers a class presentation on volcanoes. The system even sends a movie of an exploding volcano to the video 'blackboard'. In the final video, a user scans a newspaper by placing it on the screen of the full-sized version and then uses it to help him learn to read by listening to his checked results and prompting him when he pauses.

In many respects, these influential concepts foreshadow Apple's future technology. People have dated the speech-to-text assistant prediction in the video to September 2011, from the video concept of 1987. Apple launched Siri in October 2011, and its voice-activated personal assistant software is similar.

Today, looking at this video from 1987, we would recognise many aspects as possible within the next 5–6 years. Using AI/ML future systems such as ChatGPT would enable understanding the spoken text, searching for information on deforestation in the Amazon Rainforest and writing a first draft of the article for review by the professor and their colleague. It could produce graphs based on spoken requests and analyse the data for their production. It could include videos within the article and show the results on any display device. So, in many respects, the video foretells the possibilities of an iPhone 2030 (6G), which can use AI as a network service within a 6G network. One day, it could have a holographic projection of the 3D graphs and data, well in the year 2100!

Apple plans to introduce its first AI system, iOS18, in 2024. This could profoundly impact how we use the iPhone over the coming years to 6G. It may reduce the need for the number of apps in the App Store, as a single AI App can replace many of them. Apple's close integration with the iPhone 6G hardware will significantly impact its usage and future implications.

15.5 Summary and conclusions

The hypothetical iPhone 2030 (6G), an exemplary device, could play a significant role in any wireless communications revolution. However, the need for Gbit/s communications in the iPhone 2030 (6G) will primarily depend on the requirements of the display technology and 3D camera imaging. While there is no necessity for 8K displays, as the human eye may not discern the difference from 4K displays, emerging VR/AR goggles are becoming standalone networked terminals that do not rely on smartphones. These devices require up to 1 Gbit/s communication for high resolution and no motion sickness, while smartphones become the 3D camera for their content, requiring perhaps 100 Gbit/s speeds to transfer to Digital Twins.

While we have focused on a futuristic iPhone to highlight its features and technology, one of the most significant impacts for the future of 6G would be the emergence of a new mass-market device form factor. Perhaps VR/AR goggles could be the revolutionary end device that could shape the direction of wireless communications for over a decade as the original iPhone did. Or smart glasses, as shown in Figure 15.1. Could such devices replace the smartphone? Apple launched their Vision Pro Goggles in January 2024; these use Wi-Fi 6 on today's broadband networks. However, the weak spot is not the broadband network but how that is distributed within the home. Consumers will demand improved Wi-Fi connectivity to support these devices, and operators must ensure they offer full-home, reliable, fast, low-latency connectivity to help them [10]. While the first iteration of the Vision Pro does not support cellular connectivity, this is not a precedent for the future exclusion of the technology from this burgeoning device category. So, could this or smart glasses become the new mass market form factor for devices?

While the precise changes that will shape the iPhone 2030 (6G) are yet to be revealed, these potential advancements highlight the ongoing quest to push the boundaries of technology and elevate the smartphone experience to new heights. mmWave technology will not only support communications speeds of 100 Gbit/s but also form the basis of a new millimetre wave sensing for 3D images and spectrometry for scanning for food allergens. This technology sees in complete darkness and fog. The iPhone 6G may introduce Apple as a Global Virtual Satellite Operator with coverage of the earth for any app when out of range of cellular. What is certain is that its camera technology will evolve beyond what we have today, perhaps even replacing professional cameras. The iPhone 6G will support not only 6G but future generations of Wi-Fi, such as Wi-Fi 8 or 9. It could become a health hub with body sensors monitoring your medical condition in real time.

Fundamental to the future iPhone will be AI, and in particular, its AI interface. An AI app within iOS 18 will play an important role here. Its evolution and support of 6G technology in the iPhone 6G will likely profoundly impact the user experience, finally realising all the aspects and more of the original Apple knowledge Knowledge Navigator from 1987!

References

[1] Aladsani M, Alkhateeb A, and Trichopoulos GC. "Leveraging mmWave imaging and communications for simultaneous localization and mapping," *IEEE International Conference on Acoustics, Speech and Signal Processing (ICASSP)*; Brighton, UK, 2019. pp. 4539–4543.

[2] picturecorrect.com. "The future of light field cameras and how they work" [Online]. Available: https://www.picturecorrect.com/the-future-of-light-field-cameras-and-how-they-work/.

[3] Google. "Project Starline: Feel like you're there, together" [Online]. Available from: https://blog.google/technology/research/project-starline/ [accessed 18 May 2021].

[4] MIT Media Labs. "Smell camera: record, play, rewind" [Online]. Available from: https://www.media.mit.edu/projects/smell-camera2/overview/.

[5] Blanche PA. "Holography, and the future of 3D display" [Online]. 2021. Available from: https://www.light-am.com/article/doi/10.37188/lam.2021.028.

[6] Ingle R. "The future of 3D displays and holographic televisions" [Online]. 2022. Available from: https://www.azooptics.com/Article.aspx?ArticleID=2124.

[7] Wikipedia. "Apple Vision Pro" [Online]. 2024. Available from: https://en.wikipedia.org/wiki/Apple_Vision_Pro.

[8] Wikipedia. "Apple Knowledge Navigator" [Online] 1987. Edited May 2024. Available from: https://en.wikipedia.org/wiki/Knowledge_Navigator.

[9] Apple Knowledge Navigator [Online]. 1987. Available from: https://youtu.be/umJsITGzXd0.

[10] Mason A. "The Apple Vision Pro's mixed reality headset makes good Wi-Fi connectivity more important than ever" [Online]. 2023. Available from: https://www.analysysmason.com/research/content/articles/apple-vision-pro-launch-rdvs0-rdmb0/.

Chapter 16
What could 6G be?

The R&I [Research and Innovation] projects will develop smart communication components, systems, and networks for 6G following both an evolutionary path through further enhancements of 5G Advanced technology, as well as a more revolutionary path by investigating the benefits of promising technological enablers.

– Shaping Europe's Digital Future, European Commission [1]

16.1 Introduction

The book aims to answer one of the most significant questions in mobile: what will the upcoming sixth mobile generation (6G) be like? Could it be an extension of the already existing 5G Advanced, or will it be a completely new technological revolution that supports an array of new applications and services? The European Union has proposed this question as the foundation of its research project portfolio, which six other regional global partners are addressing. The goal is to contribute to the technical definition of 6G in 2027, which will be released as part of Release 21 of 3GPP. And, of course, we must remember the economic challenges facing mobile network operators (MNOs), particularly today in 5G.

The *raison d'etre* for 5G was for a broader digitalisation of society and industry while continuing to advance the traditional consumer proposition of faster mobile broadband access with lower latency. It allows the mobile sector to tap into a new revenue stream typified by Industry 4.0 but with much broader applicability. Success here could pay for the investments in 6G, whereby it only claims a few per cent of the value created from Industry 4.0. It provides a licensed spectrum for its control infrastructure, free from interference, and all under a single standards framework with guaranteed low latency and resilience. However, 5G has taken the usual mobile siloed approach, embracing only cellular solutions, and more is needed to achieve a broader digitalisation. Nevertheless, the commercial success of 5G is crucial for the confidence and timing of further investment in 6G, even if it only occurs in the latter half of this decade.

While it might take into the middle of the decade for 5G to reach critical mass for deployment globally, the public, academics, governments and private sectors are already shifting their focus to 6G, which is expected to be launched in 2030. 6G is building on the foundations of 5G Advanced while potentially introducing new and exciting services based on new technologies. It would also provide sensing as a bridge between the physical, human and digital worlds, supporting completely new ways for design, control and manufacture, which we only see today in Sci-Fi movies. This would find broader applicability in consumer applications. This book has explored the groundwork needed for their development by analysing these technologies' strengths, weaknesses and timing.

Collaboration between various standards organisations is vital to ensure the complete interworking of technologies for the broadest digitalisation. It is essential to consider 6G technology and other access technologies for a 'win–win' scenario. It is necessary to recognise that 6G alone cannot achieve broader digitalisation. Interworking with the 'fixed' world and other wireless technologies is crucial to enhancing the size and timing of this opportunity. Here, 6G could benefit commercially from using its 'network intelligence' rather than access technologies.

The state of the fixed broadband network is also crucial as it moves towards Gbit/s speeds. Many applications are fixed network-based and do not necessarily need a wireless connection. Further, the movement of 6G indoors with massive speeds will require a fixed network to support it.

Future options for 6G must consider the role of Wi-Fi and its evolution over this decade and the broadest interworking with IoT, as today's cellular technology only has about 25% of this market. While Industry 4.0 will drive broader digitisation there, the future of Internet applications on the fixed network will drive many consumer trends for new applications and services in home environments. Wi-Fi is the dominant wireless access technology there, with a roadmap to Wi-Fi 8 that supports many aspects of 6G. Therefore, 6G should embrace it for its future home services. The EU flagship Hexa-X has a key theme, 'Network of Networks', which would be revolutionary if it achieves these aims. It would be the most revolutionary feature of 6G!

We start this chapter with a brief review of the history of mobile generations in Section 16.2. While the past does not predict the future's direction, it provides at least context, as the old Russian Proverb says:

> *If you dwell on the past, you will lose the sight of one eye. If you forget the past, you lose the sight of both eyes.*

As 5G marks the beginning of broader digitalisation, we must review what we learned from this in Section 16.3 with our 11 key lessons noting that even-numbered generations are usually considered a 'fixer' for the previous odd-numbered generations. This will be crucial when determining the evolutionary path that 6G will take.

Looking ahead, there are high expectations for the new technologies and services that 6G could bring. In November 2022, the ITU published its report on future technology trends and services towards 2030 and beyond. This report will be a crucial input for the organisation of the 3GPP standards. In Section 16.4, we critically review the predictions for future services. In contrast, we have already reviewed the key 6G white papers in Section 6.7, which brings a consensus view of what 6G brings. While the previous chapters of this book have dwelt on the technologies, this chapter will focus on the service scenarios of the future of the Internet, Industry 4.0, and devices to help determine the most cost-effective implementation of 6G and its

Key Performance Indicator (KPI)/Key PerformanceValue Indicators. Once we have done this, we can take an evolutionary view of 6G in Section 16.5 and explore the revolutionary options it could bring. This will also involve challenging some long-held assumptions and proposing radical reforms to the approach to the spectrum, which may be difficult for the mobile community. This chapter aims to present key messages from these elements and offer scenarios for a holistic future of 6G. This will enable many operators, over-the-top (OTT) companies and device manufacturers to provide valuable services to their customers.

And, on a cautionary note, there is a naive assumption that because we have had five mobile generations, there would naturally be a sixth, ten years after 5G in 2030: In the 1980s, the Japanese government established an extensive research programme, over $300m, on Fifth Generation Computing Systems (FGCS) [2], which failed. The four preceding generations were: the first – vacuum tubes, which Tommy Flowers used to build Colossus to break the German enigma codes in the Second World War; the second – transistors and diodes; The third – integrated circuits; and the fourth, microprocessors. Their programme was based on the widely believed idea that the basis of massive performance gains would only be achieved via vast numbers of parallel CPUs rather than increasing the number of logic elements in a single processor. Similar projects in the United States and Europe were established to research parallel processing. The programme ran from 1984 to 1992 and failed mainly because the speed of their solutions was surpassed by less specialised hardware based on Intel x86 and Sun workstations.

5G might be the last mobile generation to use a new air interface that supports spectrum ranges reaching mmWave, 1 ms latency, massive improvements in reliability and resilience, approaching 200 Mbit/s broadband speeds for wide-areas cellular applications beyond current needs, a wireless local loop that challenges Cable and FTTH with Gbit/s speeds, massive IoT deployments, and softwarisation based on low-cost hardware. These advancements should be enough for the next few decades, during which we can expect a series of releases from 3GPP, each with minor incremental improvements converging on generation '5.99'. This is the view taken by Santiago Tenorio, network Architecture director at Vodafone, which we quoted in Chapter 1 [3]. Essentially, this is the approach taken by the Internet. Nevertheless, we believe new revolutionary applications and key architectural improvements will push '5.99G' to 6G. However, the justification for a new air interface for 6G remains unclear against its economical cost of upgrading the radio access networks.

Finally, we summarise this book with its key messages in Section 16.7 after the short conclusions in Section 16.6.

16.2 A brief history of mobile generations – evolution and revolution

Each mobile generation, from the first to the fourth, has brought a revolutionary change, and they have been successful because they have excited their consumers, making the smartphone one of the most prized personal devices. The first generation (1G) introduced analogue mobile phones to the world, albeit as a costly device and initially confined to cars because of their size and weight. Portable versions were called bricks for apparent reasons. However, this established mobile voice calling as a desirable service. The second generation (2G) introduced us to the pocket-size mobile device and democratised voice calling based on digital technology. This technology allowed improved spectral efficiency to enable a mass-market consumer product. It added limited message and data services featuring tens of kbps data rates. The third generation (3G) was the first to lead with mobile Internet. It introduced the promise of video calls and mobile TV, which excited the public's imagination with their potential but largely failed as the chosen air interface did not support the capacity requirement for these services. Before 3G, there were competing cellular standards, code division multiple access (CDMA) in the United States and 2G Global System for Mobile Communications (ETSI) in Europe, released as global standards. 3G was the first global standard based on CDMA.

However, the critical event that shaped today's industry was 2008, when Apple introduced its 3G iPhone, forever changing the smartphone experience. It led to using Wi-Fi as a premium communications experience, offering much higher data speeds than cellular. It helped establish that even today, 70%–80% of mobile data is consumed on Wi-Fi. It also introduced the App Store with 500 Applications. This affected the mobile industry in two ways. First, it demonstrated the superior experience of Wi-Fi for broadband access, paving the way for 4G. Second, it positioned smartphone manufacturers working hand in hand with app developers as a source for innovation, killing the idea of the mobile industry developing a killer app.

In its handset business around that time, Nokia had approximately one-third of the world's handset market and even had its smartphone operating system, Symbian. However, by the second quarter of the 2012 financial year, Nokia had sold only 600,000 Symbian and Windows phones combined in North America. In the same period, 26 million iPhones and 105 million Android phones were sold [4].

The fourth generation (4G) made mobile broadband Internet feasible with average speeds of 10s of Mbit/s, even in fully occupied cells. This opened the opportunities for video content in a way that was impossible with 3G. Key to its technical success was the improved spectral efficiency and wider channel bandwidths using orthogonal division multiple access (OFDM), which is still the basis of the air interface for 5G. Now, there were no limits to the use of video on smartphones and other mobile devices.

However, the people who gained the most since the launch of 3G or 4G were not the MNOs but companies like Apple, Samsung, Google, Netflix, Skype, etc., working on new types of smartphones and Apps. Here, the 'generation' arrived every year with new hardware developments followed almost immediately by hundreds of new apps. Their smartphones were – mobile computers agnostic to the networks, with App stores selling thousands of new applications and unlimited innovations for search, content, video streaming, gaming, etc. Apple and Samsung took over the smartphone sectors, with Huawei joining them recently in time scales traditional industries could not imagine. Fundamentally, 4G established a broadband dumb pipe to a very intelligent pocket device, which allowed others than mobile operators to innovate, which they did brilliantly.

Some have argued that the even generation has fixed the previous odd generation. 2G fixed 1G analogue voice to make its mass market, while 4G fixed 3G data speeds to enable what we have today. Hence, the argument that 6G will fix 5G. However, another way to think about it is that the odd generation has established the business and user cases for the next generation based on the actual market and provided its technical roadmap to enable its provision at scale. However, all four generations have targeted only various consumer/professional markets.

The fifth generation (5G) is different from the other four that preceded it, as (i) The targeted use cases are mainly in domains of machine control while the others were in the domains of human interaction; (ii) as a consequence, it did not just follow a philosophy of only capacity increases but introduced new parameters such as latency, reliability and resilience; (iii) it introduced softwarisation to the network architecture to reduce CAPEX and OPEX costs; (iv) it did continue with OFDM as there was no better air interface and (v) it was the first to pioneer the use initially of cm wave and later mmWave spectrum bands to provide ultra-high-speeds. It changed the focus of the use case from humans to a broader digitalisation of society and industry, which may be a precedent. From our estimate in Chapter 5 on Industry 4.0, if the adoption of wireless technology as the communications framework for management and control could claim only 2% of the value of the Industry 4.0 global productivity improvements, this would be worth US$36B annually, which alone could justify the investment in 6G. It has a fantastic business case in a global sense. Of course, the devil in the detail and execution is critical.

5G has also made an essential strategic call by moving solely from a philosophy of only consumer capacity increases. For example, towards the end of 4G, in 2019, countries had average connection speeds ranging from 22 Mbit/s to over 50 Mbit/s, while watching a 4K UHD video would only require a data connection of less than 5 Mbit/s on either smartphones or tablets. We have reached a point where high data speeds are no longer essential in smartphone use, and 4G networks could support up to 100 Mbit/s with 4G Advanced. These are typical user speeds we refer to here. Nevertheless, 5G continued consumers' trend of offering even higher speeds and greater capacity via the extensive use of the 3.5 GHz band with high modulation schemes. Therefore, perhaps not surprisingly, consumers have yet to be impressed with the higher speeds of 5G. The average revenue per user (ARPU) has mainly stayed the same as 4G, with little interest in paying more for higher rates. This has dramatically affected the confidence of MNOs in 5G and the ambition for 6G.

5G was the first to major as a gateway to a broader industry digitalisation, providing the communications fabric for Industry 4.0 to support controlling machines with low latency and higher reliability networks. Strategically, you can see this as moving away from a 'dumb pipe' to providing all the critical infrastructure for industry and society. It also included Mobile Edge Computing (MEC) and Cloud to facilitate these use cases. 5G is more than just an agnostic bit pipe as

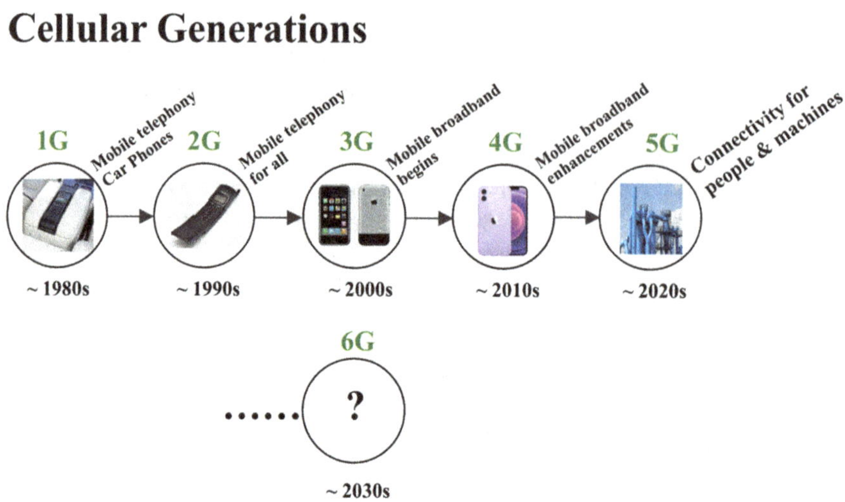

Figure 16.1 Cellular generations based on Hexa-X-Erik Dahlman (Ericsson) Keynote: EuCNC & 6G Summit 2023

the generations before were. Beyond Industry, it supports applications in similar ways for agriculture, the construction industry, the energy sector with smart grids, and the Internet of Things (IoT). In these, it introduced network splicing for wide area cellular support of new services to provide a higher quality of service (QoS), which is necessary. The new approach for widened digitalisation is achieved through the concept of *softwarisation.* While the previous generations were implemented with dedicated hardware, 5G is based on software running mainly on general-purpose hardware for low cost, interchangeability, and readily available. The new softwarisation concepts such as software-defined radios, software-defined networking (SDN), and network functions virtualisation (NFV) are going to allow network operators to keep operational Expenditure (OPEX) and Capital Expenditure (CAPEX) low. These platforms will enable future network upgrades, such as 5G Advanced and even the sixth generation (6G), to be implemented in software.

Using the cm and mmWave bands in 5G allowed mobile operators to offer fixed wireless access (FWA) and extremely high speeds in metro microcells. It incorporated satellite integration for global coverage to help address extended coverage beyond cellular networks for new uses, such as drone control.

5G has pioneered the broader digitalisation of society and industry for 6G in many ways; see Figure 16.1 for the roadmap from 4G to 6G.

16.3 Eleven lessons influence the 6G direction

The 5G vision, published in 2015 by the ITU, set the scene for commercialising 5G in 2020 with the aim of a broader digitalisation of society and industry. There was an expectation that the key features of 5G would be available for its visions from day one of its launch. However, 5G comes in releases during its lifetime, and most of the critical features for its vision arrive in 5G Advanced from 2024 in phase 3 in Release 18 [5] with a standalone core. The initial phase of 5G (Phase 1) was launched with a non-standalone core and offered little differentiation to the 4G, so, approaching half of the 5G lifetime, it effectively provides the same services as a 4G advanced network! MNOs have become frustrated with 5G with little return after their investments in CAPEX to deploy it, and there is now a reluctance to further invest in the standalone core network. However, outside the MNO community, others, such as Apple, Huawei and Samsung, have continued to sell smartphone updates with more advanced cameras and microprocessors, and App stores continue to thrive. The OTT service providers, such as Netflix, Apple, Microsoft, Google, etc., are doing equally well.

Enterprises have yet to adopt many 5G use cases, such as smart factories, connected cars, new real-time IoT, etc. This is partly due to the late arrival of the critical features needed but also a failure to recognise that more than 5G alone is required to enable change of this scale for industry and society. The lifecycles of many of these new use cases are not synchronised with the dawn of a new mobile generation, and the supply chain required to deliver the infrastructure is more complex. Furthermore, many new services, such as autonomous driving, urban air mobility (UAM) – future aerial mobility for passengers and cargo, extended reality (XR), holograms, and digital twins, were in the early stage of development in 2015 and have yet to live up to their expectation after a decade. Some believe that real holographic displays will arrive around the year 2100! So, there was a massive expectation gap for 5G between the public and some senior managers in mobile operators, and the reality.

However, there has been some early success in 5G new use cases involving private licensed spectrum and dedicated local infrastructure rather than network slicing. Analysys Mason predicts that the number of private LTE/5G networks worldwide will soar from over 4,000 in 2022 to more than 60,000 by 2028. As per their report, enterprise spending on private networks will increase from $1 billion in 2022 to $9.2 billion in 2028. These numbers suggest a remarkable 1,400% growth in total network deployments during this period and an 800% increase in annual network spending [6]. However, Analysys Mason predicts that the spend on private networks in 2028 would be less than 5% of the equivalent spend on public networks, per their market predictions for the public cellular market. Part of this delay could also be associated with a much more complex supply chain and a need for more trust in using the public network coupled with 5G coverage for critical applications.

So, how does this affect 6G? Will it slow development or steer its R&D with much better expectations management? The following sections explore the critical lessons from 5G and how to address these for 6G.

16.3.1 The case for radical spectrum reform

Spectrum fuels the mobile generation; the more you have, the more you can travel. The ‘spectrum staircase’ has driven us to seek even higher use of the radio spectrum for each successive generation to support increased channel bandwidth needed for high speeds, with the perception that no spectrum is available at lower frequencies below 3 GHz. 5G is the first mobile generation to explore spectrum use in a wide area of cellular technology at 3.5 GHz using massive MIMO arrays. Previously, it was thought that spectrum above 3 GHz was not economically viable for wide-area use.

In Chapter 7, 'A radical approach to spectrum management', we showed that despite the perception that there is a shortage of prized spectrum below 3 GHz suitable for wide area cellular networks, the utilisation of the first 3 GHz is only between 10% and 20% in most countries as indicated in Figure 16.2, Achieving an efficient usage of this prime spectrum could yield circa 2 GHz additional mobile spectrum for 6G without resorting to the cm and mmWave bands for most of their use cases. This would, in particular, enable wide area cellular use and good inbuilding coverage from its present cellular infrastructure. Here, we proposed a radical approach to spectrum management using weak signal propagation reporting within a US Citizens Broadband Radio Service (CBRS) like system but without the need for propagation models with their inherent inefficiencies. The system's ability to handle interference in real time is crucial. However, outside this, moving broadcast TV to the Internet would yield close to almost 400 MHz of prime spectrum for mobile (470–854 MHz), and there are plans from WRC 2023 to do this in the long term.

Spectrum sharing is going to be increasingly crucial for 6G. The 4G/5G CBRS is a ground-breaking initiative addressing issues related to mobile spectrum allocation. Unlike the normal allocation of mobile spectrum across a country or a state with a single MNO, CBRS localises mobile spectrum at a county level and shares it amongst different user groups. It uses a Spectrum Access Sharing (SAS) system to allow shared spectrum use in three priority tiers. The first can be used to protect existing incumbents. The second will enable auctions for priority access, not spectrum bands, while the third can support free use, all with no interference.

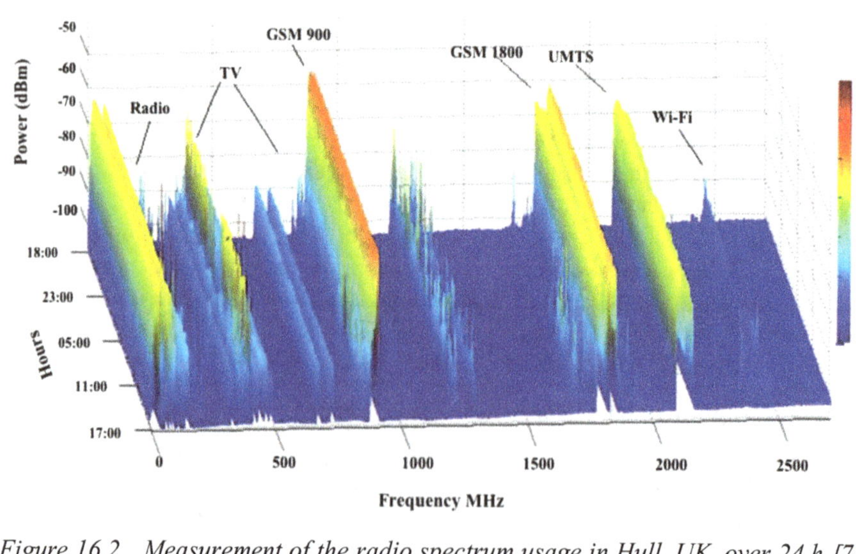

Figure 16.2 Measurement of the radio spectrum usage in Hull, UK, over 24 h [7]

It democratises the mobile spectrum, opening use to non-mobile operators, which will be increasingly crucial for Industry 4.0.

However, because of the perception of the spectrum shortage, 5G followed the spectrum staircase to even higher frequencies, and 6G has already proposed using 6.5–24 GHz and above 100 GHz.

Spectrum is not a scarce resource but rather a resource unwisely used. We argue that enough spectrum is available for 5G and 6G below 3 GHz apart from extreme use cases!

16.3.2 Mobile data traffic is slowing and may plateau in the 5G decade

One of the key reasons for moving to higher frequencies in the 'spectrum staircase' is to support the general mobile data growth, which, towards the end of the last decade, has a compound annual growth rate (CAGR) circa 50% in most regions. The mobile data traffic characteristics will highly influence the spectrum requirements for 6G in the next decade. 5G mainly uses the 3.5 GHz band for smartphones, while FWA uses cm and mmWave band frequencies and will have a different data usage pattern.

Focusing on smartphone data, a recent report by Ericsson [8] shows the mobile traffic associated with smartphones in Table 16.1.

This shows a significant decline in the CAGR of data growth from around 50% in most of the last decade, with Ericsson predicting that the CAGR for 2022–2028 will stabilise at around 20% in North America and Western Europe, with a slightly higher rate of 25% in Latin America as it catches up with data usage. Other regions have similar figures.

The growth rates in video consumption on smartphones and tablets are expected to plateau as there is a limit to how many videos people watch daily. However, the growth of high-bandwidth applications from Industry 4.0 could change this trend. William Webb predicted in his book published in 2018 that video viewing would plateau around 2027. He recently updated his prediction that this could happen at a lower plateau about 1–2 years earlier in the UK.

In that case, with data growth slowing further or even plateauing, 5G using 3.5 GHz and expanding to 4.2 GHz would not only fulfil the needs of this decade but also establish a strong foundation for 6G technology in the Low and Mid band

Table 16.1 Mobile data traffic per smartphone: based on data from Ericsson Report [8]

Region	**2018**	**2019**		**2020**		**2021**		**2022**		**Ericsson prediction 2022–28**
	GB/m	**GB/m**	**CAGR (%)**	**GB/m**	**CAGR (%)**	**GB/m**	**CAGR (%)**	**GB/m**	**CAGR (%)**	**CAGR (%)**
North America	8.4	10.1	20.2	12.3	21.0	13.2	16.1	19.6	23.6	**20**
Latin America	2.6	2.6	38.5	5.6	46.8	7.8	43.7	10.9	43.1	**25**
Western Europe	5.1	7.4	45.1	11.3	48.9	15.7	44.9	19.7	23.6	**19**

Frequencies for wide-area cellular. At the same time, there is already more than enough cm and mmWave spectrum for any growth in FWA.

16.3.3 The spectrum dichotomy

5G has opened a 'spectrum dichotomy' associated with the frequency bands and the likely application use, which could profoundly affect 6G use cases. It could change how we partition the use of the spectrum bands and the concept of seamless mobility for applications.

Figure 16.3, as presented by Erik Dahlman from Ericsson, summarises the spectrum use of 5G and the proposed spectrum for 6G. Before 5G, there were gradual increases in the spectrum staircase below 3 GHz and in 5G up to 3.5 GHz. However, adding cm and mmWave technology with its associated spectrum was like jumping straight to the top of the stairs from the first few steps. This has significant consequences for various use cases, as shown in Figure 16.4, which explains the spectrum utilisation and their outcomes for each band. Frequencies below approximately 4–5 GHz enable economic cell densification of traditional cellular coverage from external cell sites using MIMO, 'outside-in', which could also be complemented with internal small cells – femtocells. Like WLANs, there could be campus-like scenarios for contiguous coverage outside, such as in factory yards, shopping centres and sports arenas. Outside these high-capacity use cases, the wider cellular network would support the handover with lower speeds on the lower frequency bands, changing the user experience.

Of course, cm and mmWave are great for FWA-type use, which deliberately reduces the field of view to increase the antenna gain, enhancing the link budget. Therefore, mobile operators today primarily use cm and mmWave to compete with

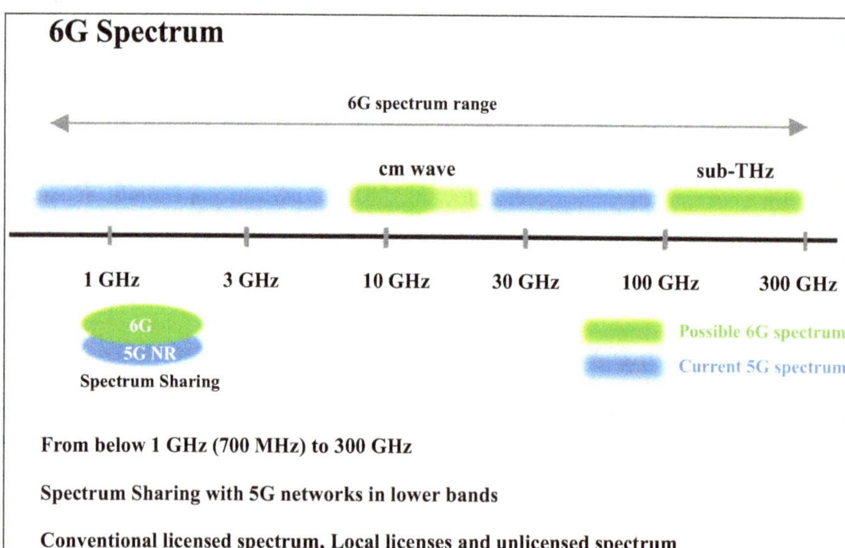

Figure 16.3 5G and proposed 6G spectrum based on Hexa-X-Erik Dahlman (Ericsson) Keynote at EuCNC & 6G Summit 2023. Redrawn.

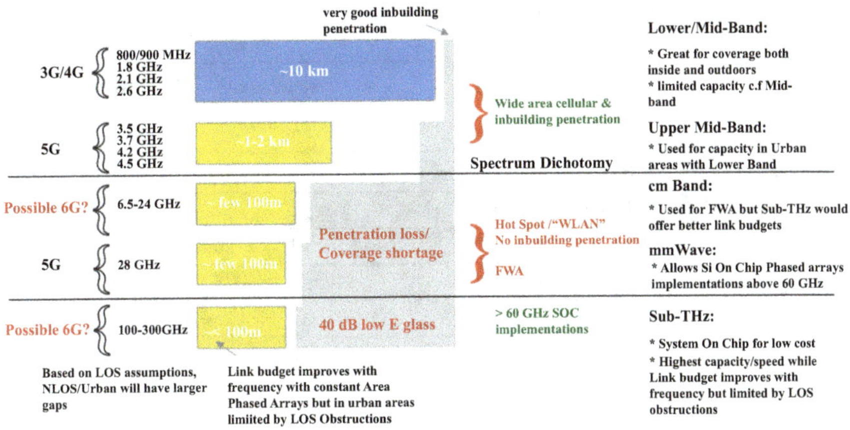

Figure 16.4 The use of spectrum for cellular systems – The spectrum dichotomy (see Figure 6.14 for expanded detail)

fixed broadband, particularly FTTH and Cable operators with FWA. It is also used in sports arenas, where video replays are highly demanded during games.

Now, 5G/6G wide-area cellular may use low and mid-band spectra up from 700 MHz to 4.5 GHz due to the high costs of cell densification above this range. Notably, as more spectra become available below 3 GHz, whose spectrum occupancy is typically only 15%–20%, these lower parts of the low and mid-bands will offer good in-building coverage for service continuity. Typically, these wide areas cellular networks provide hundreds of Mbit/s, while cm and mmWave come into their own when new predominately indoor applications demand speeds greater than 1–10 Gbit/s. However, as discussed in Chapter 2, as ultra-reliability low-latency communications (URLLC) places a heavy load on the spectrum efficiency, cm or mmWave may be used in factories, e.g. to support 1 ms latency. There will be no widespread outdoor cm or mmWave cellular deployment apart from places such as factories, with a campus-like deployment, machines such as robots are required to move between the yard and the factory floor. There is no possible handover between these cm and mmWave bands and the low and mid-bands, as the application speed disparity is too great.

16.3.4 What do applications need?

The insatiable demand for mobile data has been primarily based on the use of video, with circa 70–80% of mobile data used on Wi-Fi networks, predominately indoors. Previously, we discussed how mobile data growth is slowing and potentially plateauing without a new generation of bandwidth-demanding applications that would exploit the cm and mmWave spectrum for massive speed and capacity.

Let us understand how future application requirements influence the types of mobile network deployments for 5G today and what that means for 6G. To do this, we reviewed applications' telecommunications characteristics across a vast field in Chapters 4 and 5 on the future of the Internet and Industry 4.0, as shown in Tables 16.2 and 16.3, identifying their data speed and latency requirements.

Table 16.2 Applications requirements focused on home and smart cities

Category	**Use case**	**Latency (ms)**		**Bandwidth (Mbit/s)**	
		From	**To**	**From**	**To**
AR/VR/gaming	AR/VR motion-to-photon	1	17	5	1000
	Collaborative gaming	1	50		1000
Healthcare	General	1	20	2	100
	Accurate tactile feedback	1	5	0.01	2
	Tele-surgery	10	72	20	10,000
Education and culture		5	20		1000
Tactile Internet	Audio feedback	20	70	0.1	1
	Video feedback	30	85	2	18
	Uncalibrated tactile feedback	5	50		
	Simulcast	1	20	2	18
Linear TV			1000	2	18
Automotive	Wireless roadside backhaul	10		1000	10,000
	Sensor sharing	<20		0.02	25
	Platooning	<10	20	0.01	
	Coop collision avoidance	<10	100	0.01	
	Coop lane change	<10	100	0.01	5
	Emergency trajectory alignment	<3			
	Remote driving info sharing	<5			
	Video sharing	<10		10	700
	Info sharing for automated driving	<100			
	Telemetry offload		1000	0.01	60,000

Table 16.3 Applications requirements focused on Industry 4.0

Category	**Use case**	**Latency (ms)**		**Bandwidth (Mbit/s)**	
		From	**To**	**From**	**To**
Mobile robots	Coop motion control	1	50	1	200
	Cooperative driving	10	50	5	100
	Real-time video streaming	10	100	2	18
	Video operated remote control	10	100	15	1500
Factory	Motion control	0.25	<10	1	16
	Assembly robots or milling machines	4	8	1	2
	Mobile cranes	12		0.33	1
	Process automation	>50			<5
	Process automation monitoring	50			<5

(Continues)

Table 16.3 (*Continued*)

Category	Use case	Latency (ms)		Bandwidth (Mbit/s)	
		From	To	From	To
	Control to control comms	<10			<5
	Discrete automation	10			<5
	Video-assisted app	10		10	25
	Sensor	10	30	1	
	Video		100	1	10
	Voice		100	0.02	
Transport and logistics	Field sensor/instrumentation		10,000	0.01	
	Emergency safety notification		10,000	0.0001	0.02
	Asset tracking		10,000	0.0001	0.02
	Time-critical sensing	<30	100	0.0001	0.02
	Remote drone operation	10	30	15	50
	Real-time control for discrete automation	<1	<10	1	16

First, most use cases require less than 100 Mbit/s with 10 ms or above latency, whereas outside use could be supported using a wide-area cellular with a spectrum up to 4.2 GHz. This influences our argument on 6G KPIs, 16.5.1.10, when discussing the need for 6G improvements. Those inside could be supported on Wi-Fi 6 for pure data requirements. Of course, future 6G applications may exploit digital twins, MEC, AI cloud, etc.

16.3.4.1 Implications for industrial settings

There is a natural emphasis on the ultimate delivery or access methods to end-users and their devices. Today, we prioritise connectivity through Wi-Fi, private 4G/5G, public cellular neutral-host and DAS (distributed antenna systems), public safety, IoT and audio-visual systems. These access technologies' specific roles and use cases differ significantly across various sectors. For instance, an industrial setting like a factory will undoubtedly have different requirements than an office or a residential property. Each industry will have its combination of technologies and delivery partners, but they all share the goal of achieving gigabit speeds while minimising latency.

With emphasis on ultra-low-latency applications that require a response time of 1–10 ms and Gbit/s speeds, these applications will most likely use cm and mmWave frequencies for speeds and sufficient spectrum needed for low latencies. For this purpose, small cells with a private 5G spectrum are expected to be used. In addition, other essential requirements in these settings include the use of digital twin, MEC, and a local core network provided by the 5G architecture. While industrial applications within the building require low latency, applications outside, such as driverless cars with a response time of over five milliseconds and video speeds of less than 100 Mbit/s, could still be supported on a 5G cellular network.

This indicates that most of the improvements for 6G would accrue from the architectural changes supporting a 'Network of Networks', sensing and AI.

16.3.4.2 Implications for home settings

Typically, Wi-Fi is used in the home with either FTTH or Cable for gigabit speed capabilities. Nevertheless, Apple's Vision Pro, an exemplar of XR, uses Wi-Fi 6 with refresh rates of 100 Hz and 12 ms latency, supports 4K video and is supported on sub-100 Mbit/s broadband. Future versions of this technology will likely support mobile access, while Ericsson has demonstrated that XR applications can be supported on external 5G cellular networks with 30–100 Mbit/s rates, see Section 16.4.5. Therefore, it is likely that Wi-Fi 7/8 will support the majority of 6G use cases in homes. Femtocells could be used, but when using the lowest cm band of 6.5 GHz, several would be required to provide coverage of a home at a significant additional cost to Wi-Fi already deployed. This is an opportunity for mobile operators to embrace Wi-Fi technology for future 5G and 6G services.

While some people have speculated that future 6G applications require 100+ gigabit speeds, this would require massive improvements to the broadband networks at enormous cost, not to mention the many years it would take to roll this out.

16.3.4.3 General conclusions to application requirements

The analysis described in this book shows that 5G's current URLLC 1 ms performance ambitions meet all future predicted Internet and industrial requirements, excluding some outliner Industry 4.0 cases. In those, there are questions about whether reducing latency further goes beyond the point of diminishing returns where there are already fixed network alternatives, as these machines are invariably static. As a reflection of these requirements, in an interview of 150 enterprises in North America, only 3% wanted less than 10ms latency, see Chapters 4 and 5. However, today's real-world 5G, Wi-Fi and fixed networks fall short of meeting these requirements in some niche situations, e.g. remote control of robots in dangerous environments, transmitting raw uncompressed UHD 360 degrees video for VR requires more than 10 Gbit/s. However, accurate or calibrated remote tactile feel requires latencies less than 5 ms where 1 ms will suffice.

Exceptionally, cm and mmWave-based services could be used in venues such as open space sports arenas, as mentioned in the previous section, to offer multi-Gbit/s speeds for the high demand of video clips of replays, extremely high peak demand during games, etc. Overall, in this secondary use case, they do not support or probably do not need handover to the external cellular network as their use is self-contained, such as in a manufacturing plant. These factors relentlessly drive against the economic case for widespread cellular deployment of cm and mmWave in 6G.

All of this supports the argument that future applications for 6G are primarily met using circa 100 Mbit/s and 10 ms latency with 1 ms as a suitable ambition during its lifetime. These can be supported on a 5G Advanced cellular network or using Wi-Fi inside buildings for raw speed and latency.

Of course, 6G delivers more than just speed and latency, as we will cover in the next section.

16.3.5 Industry 4.0 architectural requirements

In many respects, 5G's *raison d'etre* supports use cases in Industry 4.0, smart cities, farming and construction, where the architectural tenets are interoperability between multiple organisations and technologies, modularity, digital twins and flexibility – a Network of Networks. 5G has the same tenets for its internal construction but viewed from the Industry 4.0 architect's point of view, it does not readily integrate with multiple network technologies. 5G is too mobile-centric (siloed) to deliver on all its promises.

For example, extranets are outside its scope, the modularity options are difficult to understand and not plug and play, and it appears more ossified than flexible, especially concerning spectrum. To be Industry 4.0 ready, 5G/6G and Wi-Fi 7 need more proven reliability, cross-technology inter-operability and cross-organisations integration capability. There is scope for 6G to simplify the 5G telecommunications options to deliver the implementation and maintenance modularity and flexibility required for Industry 4.0. If 5G Advanced and Wi-Fi 8 address these issues, then 6G could repackage 3GPP standards to make them more consumable by industry, especially by Operational Technology architects.

5G Advanced must deliver more interoperability with non-3GPP wireless systems, especially Wi-Fi, Ethernet field buses and the plethora of Industrial Internet of Things (IIoT) protocols. 5G Advanced must also deliver on non-terrestrial network IoT (NTN-IoT) to track assets on circa 85% of the Earth's surface with no cellular coverage. Wi-Fi 8's promise of ultra-high reliability will be a leading choice for Industry 4.0 solutions.

While 5G does not have low enough latency for the wireless control of servo loops for high-precision machine tools, which requires latencies of 5 μs–0.5 ms, which could be addressed in 6G, it does lead to a question of whether there are diminishing returns for this particular use case as these machines are invariably static with existing fixed network solutions in place already as pointed out previously. Therefore, any gains would only apply to new installations if needed.

The ongoing research and standardisation of carrier-scale network automation can be exploited for Industry 4.0, as network configuration needs to be automated for the vision of a reconfigurable factory.

5G NB-IoT is designed for a density of 1 million devices per square kilometre but allows only one thousand mobile networks per country. If 5G Private networks and IIoT are successful, we should consider millions of private mobile networks per country [9].

MNOs face challenges and opportunities. One challenge is exploiting synergies with fixed VPN operations for private 5G networks. Opportunities include being part of integrated ecosystems for digital twins and AI solutions, especially for SMEs. MNOs also need to work with other MNOs and vendors to build Industry 4.0 extranets.

Industry 4.0 is about seamlessly working with others, which 3GPP and MNOs need to do to make their services attractive to the industry and unlock the value of Private 5G networks.

Let us consider the task of the operational technology (OT) architects when asked to architect an Industry 4.0 implementation; they must select the technologies

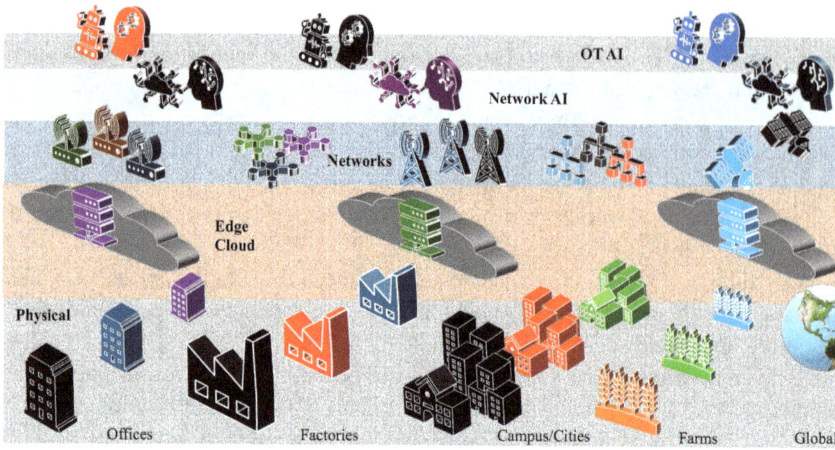

Figure 16.5 Industry 4.0 6G ecosystem

to build solutions that connect the global multi-organisational supply chain, use wireless to create a flexible, modular factory and integrate with legacy technologies. 5G provides more tools for the OT architect's kit bag, but more guidance on using 5G is required. The OT architects need vertically integrated blueprints and an open ecosystem that does not lock them into product vendors or network operators. SMEs do not have OT architects, so they need off-the-shelf Industry 4.0 solutions, including networking.

Figure 16.5 shows the 6G ecosystem for Industry 4.0, which must cover use cases ranging from small offices to global coverage, where supply chains cross multiple organisations and types of organisations, multiple network technologies, especially wireless, will be used. The network service provider artificial intelligences (AI) will have to manage the complexity of these networks and cooperate with the operational technology of artificial intelligences, which will be instrumental in delivering the efficiencies promised by Industry 4.0. An infrastructure of private and public Edge Cloud is required to run all of this.

16.3.6 5G launched too early

The purpose behind developing 5G technology is to enable a more extensive digitalisation of society and industry. As a result, it would provide the mobile sector with a new revenue stream characteristic of Industry 4.0 but with much broader applicability. However, the implementation of 5G technology comes in phases, and at the time of launch, it only supported a non-standalone core, which offered little differentiation from a 4G network.

According to the Dell'Oro Group's report released in 2024 [10], deploying 5G Standalone networks (SA) is slower than expected, hindering market growth. As of January 2024, only fifty 5G SA enhanced mobile broadband (eMBB) networks have been commercially deployed by MNOs worldwide. Of these, 18 were launched in

2022, and only 12 were launched in 2023. They estimate it will take until 2028 for 75% of 5G subscribers to be on 5G SA networks. The report highlighted that the MEC market segment will lead the charge for growth rates in the Mobile Core Network (MCN) market with an expected 26% CAGR in the 2023–28 period, followed by the 5G MCN market segment at a 7% CAGR.

As mentioned earlier, Section 16.3, Analysys Mason's report highlights that private LTE/5G networks are showing early success in new use cases, particularly those involving dedicated local infrastructure and private licensed spectrum, instead of network slicing. However, the report also predicts that the spend on private networks in 2028 would still be less than 5% of the equivalent spend on public networks, as per their market predictions for the public cellular market.

This indicates that 5G was launched before it was ready, which has created a negative view among MNOs of the opportunities in this area. 6G must learn these lessons.

16.3.7 Wi-Fi versus cellular in commercial and industrial settings

In a large commercial or industrial setting where many access points or cells are required using mid-band spectrum, the 4G/5G radios will offer superior coverage to Wi-Fi with the support of MEC, digital twins and a local core network. An example here was BMW, where the company used a private 5G network to enhance its automotive manufacturing operations at their factory campus in Regensburg, Germany. It would also have the advantage of supporting handover between indoor and outside environments for campus-type settings. Therefore, 5G/6G cellular could offer low costs and excellent technical capabilities in these large industrial settings, particularly around reliability, control, security and scale [11].

The balance may tip towards Wi-Fi in smaller facilities requiring only a few access points and having different concerns or needs for reliability and performance. Wi-Fi 8 plans to support similar use cases to Industry 4.0 with lower latency and higher reliability.

Moving into the next decade, we may see Wi-Fi 9 supporting mmWave bands in the sub-terahertz region, first in the USA 124 GHz ISM band and perhaps with 20 GHz channels for 100 Gbit/s speeds. At this point, it is reasonable to wonder what the differences between Wi-Fi and cellular using mmWave would be other than the support of specific bands. While Wi-Fi uses unlicensed ISM bands and cellular uses licensed ones, Wi-Fi has remarkably created global standards in its ISM bands. Of course, cellular will be able to use all the mobile spectrum below 4.2 GHz for mobility with seamless handover in wide areas complementing its mmWave use for internal applications for Industry 4.0. Today, it will offer a solution embracing MEC, Digital Twins, and a Core Network, where 'Wi-Fi' needs to develop or adapt to the 3GPP elements.

16.3.8 Security and privacy

For the typical consumer, 5G is highly secure. However, researchers have identified the following potential attacks and security weaknesses in 5G, which should be addressed in 6G; 'Bid down attacks' that could force 6G UE to use earlier, possibly more vulnerable Generations, 5G-AKA vulnerabilities,[1] especially the lack of forward secrecy, quantum computer cracking which requires 6G to upgrade to quantum-safe cryptography and increase symmetric keys length. 6G needs holistic methods to secure a highly distributed multi-vendor system, a holistic approach to PKI and Certification Authorities, a zero-touch architecture, and a holistic secure network slicing architecture. 6G will be operating with evolved virtualisation and security technologies, which should improve security but may require a rethink of the 5G security framework.

Privacy-enhancing technology (PET), as discussed in Chapter 12, solutions to many global and individual' problems by leveraging the value of aggregating confidential information without compromising privacy or confidentiality. Privacy-enhancing technologies (PETs) foster trust, transparency, and best practices that enhance the economy and digital security. The value of PETs to the global economy is enormous, but no one has published an estimate of its value. However, Future Market Insights predicts the PETs' market size is expected to reach US$ 2.4 billion in 2023 and is projected to reach a valuation of US$ 25.8 billion by 2033 [12].

PETs open new applications and markets, and like the early days of the Internet, we do not know for sure what these are. 6G should at least use PETs, as using common PET standards increases the value of 6G applications and services. The opportunity for PETs in the future digital ecosystem is very significant, and its synergies with 6G ambitions suggest that 3GPP should ensure PETs are part of the 6G ecosystem. 6G network operators could, at the least, use PETs to share 'big data' to optimise network performance and operations while protecting the privacy of their networks and customers.

16.3.9 Reaching actual sustainability goals

Since 3G, each generation has addressed sustainability with reductions in energy per bit, and data centres with 5G have offered lower energy per bit than 4G. However, this is only a part of the total energy use of 5G. Consider the softwarisation aspects in 5G using NFV. NFV hardware in the field has not achieved the efficiency demonstrated in the laboratory environment. Generally, an NFV 5G solution currently uses ten times the electricity of a black box solution. At the same time, BT Research Laboratories reached parity in the laboratory and achieved much lower power consumption with NFV for some specific black box solutions. It all hinges on the complexity of writing efficient software, primarily open-source components. Therefore, 5G and future 6G need to monitor the whole network's energy usage.

[1]5G-AKA is a technique used in 5G to ensure mutual authentication between the subscriber and the network. It also facilitates key agreements that protect NAS, RRC, and user plane traffic. 5G AKA is similar to EAP AKA, but with enhancements to improve roaming security: https://www.mpirical.com/glossary/5g-aka-5g-authentication-and-key-agreement#.

16.3.10 Are non-terrestrial networks part of the MNO experience?

3GPP plans to integrate satellite networks into the 5G and 6G cellular networks. Yet mobile operators do not own these satellite networks, which include geostationary and low earth orbit variations. Including these networks is not necessarily solely in the domain of the mobile operators. For example, Apple, in the iPhone 14 and 15, uses satellites to provide coverage in emergencies independent of the device's mobile contract. This is done through satellite messaging (SOS) services, automatically connecting to the satellite network when Wi-Fi or cellular is unavailable. However, to connect to a satellite, you must be outside with a clear view of the sky and horizon. This is an interesting example of using non-terrestrial networks provided by the handset vendor, which could play out further in 6G for other services, including voice and Internet access. This is not dissimilar to the device support of both Wi-Fi and cellular as independent networks with no real need for seamless handover for applications that share both.

Whereas there is considerable coverage of low Earth orbit (LEO) satellites, there are roles for geostationary satellites. It also plays to the strength of satellites here – covering areas not covered by other traditional networks and, in this case, for an important use case for rescue services. They do not require a seamless handover, and roaming between networks is fine; hence, this questions the level or need for integration within 3GPP standards. Nevertheless, some mobile operators may wish to include satellite coverage to extend cellular coverage to remote areas with low usage density, as this offers better economics than further cellular buildout. Here, they would use LTE bands.

One of the objectives of 5G and 6G is to help address the digital divide for less developed countries. Here, LEO satellites could provide greater bandwidth for broadband, but as discussed in Chapter 10, but they are generally more expensive than cellular or fixed wireless access. LEO satellites must identify the global spectrum for economical smartphone use or with the use of external antennas on buildings.

Exemplary satellite services in developed countries included controlling drones across large areas and extending economic coverage of cellular services to areas with low densities of users, such as national parks. We have had these classical use cases for satellites for decades! They do not provide 100% coverage, nor does any access technology, and they do not need to be integrated within 3GPP standards for their use.

16.3.11 The battle for the home and convergence: a network of networks

In the early days of 2G and 3G technology, the focus was on voice and low-speed data through low- and mid-band spectra. Outside cellular base stations were the primary means of building the network, and the usage model was mainly for outdoor and indoor voice calls, with limited low-speed data usage. These frequency bands covered '90%' of homes and offices, but the data speeds were slower than

Wi-Fi. People, therefore, used Wi-Fi in those places, offering at least ten times faster rates.

In 4G, mobile data became much more critical, and people started using OTT apps for voice, texting and video. Since mobile operators typically followed a mobile-only strategy, they invested heavily in their macro-cellular network using higher powers and higher orders of MIMO but at higher frequencies to support data use outside with lower speeds inside homes and offices. An additional 17–20 dB link budget is typically required in mid-band to provide a reasonable level of indoor coverage, which is often difficult to achieve. There were no prospects for small cells inside buildings as Wi-Fi was invariably supplied with the home broadband connection.

However, several factors have changed. Many mobile operators have become fixed network providers, while fixed network operators have also become mobile operators through either merger or acquisition or via MVNO arrangements. This is the beginning of telecom convergence, first at the corporate level and second possibly affecting the design of the future access network, as we explored in Chapter 9. Nevertheless, using a single-core network for all fixed/mobile traffic per se is not economical. What change can bring about convergence, then? The evolution of 6G services in the home and broader industry, supporting the new applications requiring higher speeds, lower latencies, and integral support of other 6G network assets, such as Digital Twins and MEC, would establish a greater need for a convergence network. Yet, here, the converged operators would choose between supporting small cells in the home using licensed 6.5–24 GHz spectrum or fully embracing Wi-Fi as a full member of the 6G family. Clearly, WiFi makes more sense! At the same time, cellular options may win out in larger industrial settings.

However, an alternative to "outside-in" for 6G coverage is inside-out, where small cells in homes, using Low and Mid band spectrum would not only lead to a much better consistent coverage indoors, where 80% of data is used, but it also serendipitously cover adjacent city buildings, extending its usefulness. It is green as the energy per bit is at least 17–20 dB lower.

The other great merit for operators is that no costs are associated with site acquisition, rental, or electrical power. However, it does require a change in mental attitude – moving away from direct ownership of where your assets are deployed with network planning to an arrangement where they are placed in a home or office, and perhaps a software client directs the best placement. This helps enable convergence and extends the opportunities for inside-out coverage scenarios.

16.4 A critical analysis of the ITU's 'Future Technology Trends International Mobile Telecommunications (IMT) Future Technology Trends towards 2030 and Beyond'

Strategy is figuring out what not to do.

– Steve Jobs [13]

In Section 16.3, we discussed the lessons learned from 5G, which will impact the development of 6G. This section will explore the proposed new use cases for 6G to evaluate what can be achieved through evolution and what will require a revolutionary approach. We will take a critical stance to ensure that the development of 6G is not overly complex and represents the best value for money.

Nevertheless, several factors will influence the development of 6G services and technology. The development of 6G will face financial and technical limitations, but it will also present new opportunities for emerging services. In November 2022, the ITU released a report titled 'Future Technology Trends International Mobile Telecommunications (IMT) Towards 2030 and Beyond,' which outlines its vision for 6G technologies based on its views of the new 6G services. This report, in collaboration with other organisations, such as the Next Generation Mobile Networks Alliance (NGMN), which includes contributions from operators, vendors, and research parties, plays a significant role in defining the requirements for the future mobile network based on its views of services. Here, we critically review the extended use cases to the original three use case types in 5G: eMBB, massive machine-type communications (mMTC), and URLLC. This allows us to specify the envisioned 6G KPIs in a more realistic light to avoid over-engineering the next generation. It will present an insight for mobile operators to see what is technically and commercially practical and to develop their evolution strategy from 5G.

16.4.1 3D holographic display applications

One key underlying driver for many of the improvements in the KPIs for 6G from 5G is the belief that 3D holographic applications will emerge between 2030 and 2040, which further requires a massive increase in communication speeds and a reduction in latency from those specified in 5G. This, in turn, requires extending the spectrum requirements of 6G towards 3 THz to find the channel bandwidths needed for this application. This one application profoundly changes many of the 6G KPIs, yet is it real?

The idea of holographic displays builds on one of the verticals identified in 5G, augmented reality and virtual reality (VR). Unlike VR, where the user wears goggles and needs to move to see a different perspective, in holographic displays, users are not required to wear any goggles. They see other views as though they were looking through a window. However, the computational, transmission, and rendering requirements are truly immense, with Pierra-Alexande Blanche, in his excellent paper on the challenges of the Holographic displays [14], estimating that they require 6.6×10^{15} flops, raw uncompressed data rates of 3×10^{15} bit/s, and 1.6×10^{12} phase pixels in one example.

To consider how daunting these figures are, first, consider the computing requirements, which is the easiest! Apple's M2 GPU, on a 5 nm process, has a theoretical performance of 3.6 Teraflops (3.6×10^{12} flops), which is almost 200 times slower than the holographic computing requirement. If Moore's law continues to apply, it will not be at these transistor geometries; it will take until circa 2045 to get this performance level, which is likely beyond the time frame for 6G. The other two requirements are even more challenging than the computing target of the holographic display.

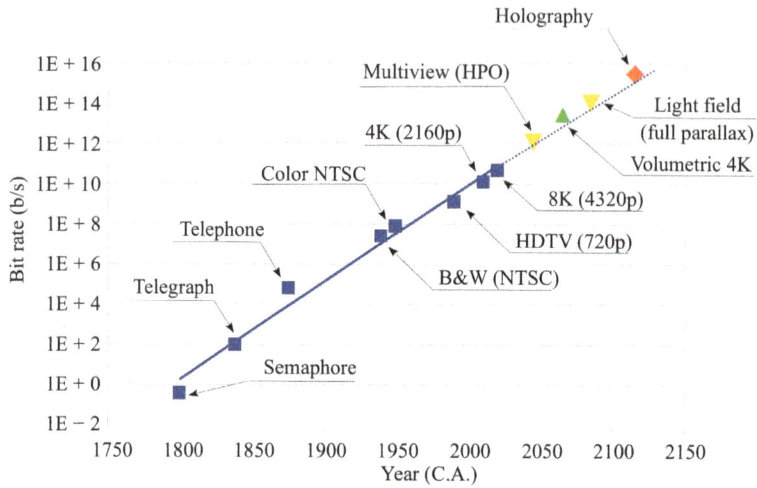

Figure 16.6 Stairway to Holography: Pierre-Alexandre Blanche, Light: Advanced Manufacturing published 2021 2 [14] (Open Access Rights)

In Figure 16.6, Blanche extrapolates the emergence of commercial displays by the year 2100! These numbers illustrate the daunting task of Holograms, and the authors believe they will not be realised in 6G or even 7G! Nevertheless, it does bring out a vital omission identified above. If holographic displays did emerge within the lifetime of 6G, they could require, even with compressed video images, data rates in the Terabit/s range with main indoor uses connected to a fixed network. While in the past, the evolution of the access networks, ADSL to DOCSIS and FTTH, has more than matched the mobile data speed, the increase with the emergence of Tbit/s Holographic displays would tax the most advanced access network evolution. This would be a significant barrier to their adoption regardless of the immense technical challenges.

There is also a naïve view that people need holographic displays, which are much better than VR. There have been many false dawns in display technologies with issues of associated content creation. How many times have 3D displays/movies been tried in theatres and TVs with no success? Even the world's first 8K television was unveiled by Sharp at the Consumer Electronics Show (CES) in 2012, and more than a decade later, there are no 8K movies or TV programmes to watch.

Without holograms, many of the 6G key KPIs, such as communication speeds, spectrum requirements, latency, and jitter, are not required during the system's lifetime, which has a profound impact.

16.4.2 Tactile and haptic Internet applications

5G was the first to propose Tactile and Haptic Internet applications where human operators can monitor remote machines in VR and are aided by tactile sensors with haptic feedback for enhanced control. This led to the latency requirement of 1 ms in 5G URRLC. In Chapters 4 and 5, we explored the latency requirements for human control of machines across many fields, including remote surgery, as shown in Table 16.2. We found that 1 ms represented the lowest requirement, with most applications requiring over 10 ms latencies. This is unsurprising as human reaction times are more than 100 ms, so 1 ms generally represents an essential threshold for future developments.

While 6G may come to specify a lower figure of, say, 0.1 ms, following the tenfold improvements norm from previous generations, this would not be necessary for remote surgery or any other human control application and is only required where a machine's control loop was disaggregated with computer control moved to a MEC away from the machine itself. This is cited as the main reason for reducing the latency target within the IMT report. As we pointed out in the previous section, EtherNet has already met this requirement, and there are no actual mobile requirements as these machines are static.

It is reminiscent of the advocacy for V2X vehicles under 5G control. Intelligence is remoted to a central computer with additional latencies required for the transcoding and over-the-air transmission in both directions to achieve a targeted low latency! Now, there is a single point of failure, which needs its reliability to be two orders of magnitude greater than the objects it manages with the associated expense of this. At the same time, there are already simpler and cheaper ways to achieve the same commodity. The authors do not believe there is any significant technical or commercial advantage to this, so we suggest retaining the 1 ms low latency figure. This is important as reducing latency impacts spectrum efficiency.

16.4.3 Information 'petrol stations'

The ITU report identifies a new type of hot spot or 'Information petrol station' access points (APs) placed typically in coffee shops or transport hubs, such as airports, train and bus stations, shopping malls, and other public places with extremely high data rates, as shown in Figure 16.7. In these places, people would

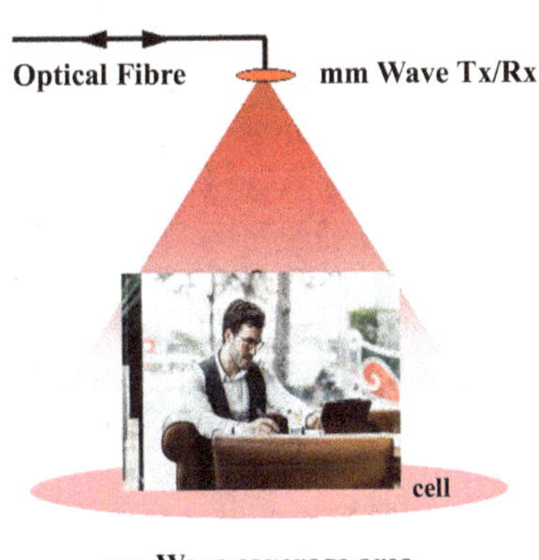

Figure 16.7 Information 'Petrol Station' in a Coffee Shop covering an area with 100–200 Gbit/s access

use these to quickly ‘fill up’ their devices with high-capacity content, such as movies. Or upload extremely large files. We believe this ambition could be fulfilled using the mmWave spectrum with 100–200 Gbit/s speeds in the 100–200 GHz range spectrum. This could mean, for example, that a tablet-like device could download a 4K movie in less than 10 s. These could also support mmWave back-haul. The ITU report flags co-existence with cellular services and security as a significant issue requiring further attention.

16.4.4 Connectivity for everything

This is a new and exciting scenario for 6G – Connectivity for Everything and builds on the IoT theme in 5G. These scenarios already include real-time monitoring of buildings, cities, cars, transportation systems, roads, the environment, and critical infrastructure such as water and power. It is extended to the Internet of bio-things through intelligent wearable devices and intra-body communications achieved via implanted sensors, which take the scenarios of mMTC and put them in a human context. This could be very important for people, and the UK National Health Service, for example, already offers pacemakers with nightly downloads of their data at home. The IMT report cites this as an important future cellular use case, but the issue here is that ‘bio-things’ have small batteries and require low-power radios such as BlueTooth Low Energy. As everyone using these will have a smartphone, the smartphone could form its ‘smart hub’ and integrate all the low-power communications into its health app, such as Apple’s Health App built into its iOS. It could use Bluetooth to secure radio access to these sensors and report the data via the app to your health provider or third-party monitoring service. The cellular provider provides the data pipe as part of the standard data connection. While we do not doubt the value of the Internet of Bio is very high for people, the value in this proposition is from the provision of the sensors, monitoring, and App ownership but not in the cellular connections other than as a bit pipe.

Cellular has always seen itself as providing the infrastructure for IoT, which has been the primary driver for a 5G KPI connection density of $10^6/\text{km}^2$ in 5G. However, cellular has always had a small part of the IoT market, shared with many different IoT systems, each optimised for its particular use case and market. While the total IoT connections are predicted to increase by 162% in the period to 2028 [15], the cellular market share is expected to fall with the 2G/3G switch-off based on Ericsson data: The cellular market share will drop from 26% of the IoT base in 2022 (2.7 Billion out of 13.2 billion connections) to close to 19% in 2028 (5.4 billion out of 28.7 billion). To put the 5G KPI of connection density in context, the UK ranks number 21 in the list of countries by population. The population density in the United Kingdom is 280 per km^2, implying that for every person, there is a capacity within 5G of over 3500 IoT devices. However, the 2028 prediction of the total IoT market of 28.7 billion represents only about two devices per person, with a cellular share of 0.4 per person. Therefore, the 5G KPI of connection density will not need to be increased for the Connection for Everything, even in cities, unless there is an additional four orders of magnitude increase.

For the other uses of Connectivity for Everything in buildings, cities, etc., listed above, the asset owner and their system integrators will largely determine the types of solutions. There are many to choose from, each highly optimised for its use case. Many use the existing Wi-Fi infrastructure and have access to electrical power; some require low frequencies for range extension to reach the places deep inside buildings, and some require very low power with batteries that are difficult to replace with very short ranges. Cellular strength is the mobility and tracking of objects that move or are deployed in remote locations with no fixed network but within cellular coverage.

16.4.5 Extended reality (XR) – interactive immersive experience

This builds on the 5G vertical of augmented reality (AR)/VR, where XR provides a more interactive immersive experience that can blend virtual and real-world environments seamlessly while offering a new multi-sensory experience to users [16]. In the Ericsson technology review referenced, they see 5G virtual reality being supported with 30–100 Mbit/s in the downlink and less than 2 Mbit/s in the uplink with latencies of between 5 and 20 ms, which allows XR to be supported in both wide-areas cellular and indoor cellular deployments with seamless handover or on home Wi-Fi. For example, Apple launched their Vision Pro Goggles in January 2024, where users see two 3660 × 3200 pixel 1.41-inch (3.6 cm) micro-OLED displays with a total of 23 megapixels, usually running at 90–100 Frames per second through the lens [17]. These use Wi-Fi 6 on today's broadband networks. However, as mentioned earlier, the weak spot is not the broadband network but how that is distributed within the home. Consumers will demand improved Wi-Fi connectivity to support these devices, and operators must ensure they offer full-home, reliable, fast, low-latency connectivity to help them [18]. While the first iteration of the Vision Pro does not support cellular connectivity for 4G or 5G, this is not a precedent for the future exclusion of the technology from this burgeoning device category.

6G XR is expected to provide higher resolution, a larger field of view, and higher frames per second. As graphics rendering moves to the cloud, this will translate into higher demand on the transmission data rate and end-to-end latency. This combination could drive 6G XR towards 1 Gbit/s in the downlink from below 100 Mbit/s today. However, if device manufacturers see slight improvements in today's technology, they will be reluctant to build these improvements because it increases their device costs, and many people will have sub-Gbit/s broadband.

The category of XR devices is still in its early stages, which makes it difficult to predict its evolution and future telecom needs. There are also several barriers to the mass-market adoption of XR devices and services and metaverse-type visions. The cost barrier is a significant part of establishing a self-sustaining business case for mass-market adoption.

It's uncertain what proportion of use cases will be focused on indoor usage and if low latency (1–10 ms) network attributes will be needed. Significant differences will likely exist between the network demands of indoor and outdoor use cases, with indoor use cases tending towards VR with better Wi-Fi home distribution and outdoor use cases tending towards AR. Today's FTTH and cable networks will meet broadband speed requirements (0.1–1 Gbit/s) but need better Wi-Fi solutions.

16.4.6 Multi-dimensional sensing

This is a fascinating area identified within the IMT report for 6G. Sensing based on measuring and analysing wireless signals will allow various opportunities for mobile devices. These include high-precision positioning, mapping and environment reconstruction, and ultra-high-resolution imaging. They could support gesture and motion recognition on the user interface side of the future mobile smartphone.

The ability to localise, track, and sense physical objects is the bridge connecting the human, physical, and digital worlds. This is likely a revolutionary 6G theme, differentiating it from 5G, although its specifications were included in 5G. It is a key feature not only for Industry 4.0 but also for many other use cases in consumer, health, smart cities, and vehicle safety. It could eventually support spectroscopic scanning for new smartphone Apps and pollution measurement in towns for improved liveability.

Mobile networks are a great starting point for enabling a massive sensor network, as they already have extensive coverage on day one, with transmit/receive nodes providing full area coverage and a good interconnection between nodes. Joint Communications and Sensing (JCAS) is likely to be implemented across all the cellular bands in 6G. Hence, the sensing can be delivered almost 'for free'. However, sensing technologies have already been developed for Wi-Fi JCAS, and other approaches, including LIDAR and Ultra-Wideband in the iPhone 15, have been developed for 3D imaging for Apple Goggles, see Chapter 14. Therefore, we expect different solutions supporting this range of sensing technologies, from cellular JCAS to device sensing.

16.4.7 Machine-type communication

IMT 2030 sees the 5G trend of enabling machine-type communication (MTC) as continuing in importance for 6G. As 5G did, it sees MTC as a significant driver behind its IoT plays and towards the future digitalisation of economies. MTC encompasses both the critical MTC (cMTC) and massive MTC (mMTC) scenarios where cMTC targets mission-critical connectivity with stringent requirements on key performance indicators (KPI) such as reliability, latency, dependability, and synchronisation accuracy. Meanwhile, mMTC addresses connectivity needs for many potentially low-rate and low-energy simple devices where connection density and energy efficiency are the most critical KPIs. However, as we have explained, the current 5G KPI connection density is sufficient for the needs of 6G, and an unconstrained increase in IoT devices would challenge 6G sustainability aims, see Sections 6.72 and 6.73.

16.4.8 Global seamless coverage

This builds on a theme in 4G and 5G previously: connecting unconnected people and providing them with a high-quality mobile broadband service with arguably little direct commercial or technical success. In 5G, this includes the interconnection of terrestrial and non-terrestrial networks to bring advanced mobile services to some of the poorest countries and enable them to access advanced broadband services. While this is a very laudable objective, there has already been significant progress in developing countries towards mobile smartphone use with associated services, such as mobile banking, promoting trading between remote communities. Mobile ownership follows similar trends to those of more developed countries, albeit a few years behind. Data analysis shows this results from access to cheaper old-generation mobiles and smartphones. According to the GSM Association (GSMA), by 2030, smartphone adoption is forecasted to soar to 87% in Sub-Saharan Africa, up from 51% in 2022. The report attributes this rapid growth to the declining prices of smartphones and the increasing number of young individuals familiar with the latest digital technologies. With most new users being digital natives, smartphones are used for many activities beyond traditional voice calls and SMS.

Developing a global satellite network based on LEO technologies with seamless integration with cellular where satellite coverage is provided to high-end smartphones will do nothing to extend mobile ownership in poorer countries. Furthermore, LEO's access costs will further increase the mobile subscription, making it unaffordable for the poorest people.

Developing countries will continue to use 4G and 5G networks with low-cost devices. Back in the day, they introduced, e.g. innovative services around the use of SMS, such as money transfers for payment of goods for rural farmers. As the GSMA report establishes, the younger generation with access to 4G or 5G networks will find that their access speeds will meet all their innovative activities. Hence, 6G will likely be deployed much later than in developed countries. They need older networks with previous-generation devices with lower costs, not new technologies.

More generally, satellite opportunities are essential in 6G, which we covered in Chapter 10, in extending coverage to areas where it is uneconomic to build cellular networks and across 2/3 of the world's surface for sea and aviation. However, the device vendor can provide many use cases with the satellite operator, independent of the cellular operator.

16.4.9 Vehicle to vehicle (V2V) and vehicle to infrastructure (V2I)

In the future of connected cars, it was argued that 5G would profoundly affect the automotive industry. 5G would be used for vehicle-to-everything (V2X). Not only could the car be driven by the network, but it would be able to communicate with other vehicles, sharing information such as speed and braking, that the network could orchestrate cars to come together in convoys with very short distances between each vehicle, reducing traffic congestion and even eventually removing the need for traffic lights. There could be integration with smart cities CCTV cameras, V2I, identifying pedestrians and bicycles approaching a blind road crossing to warn approaching vehicles. Every major 5G exhibition showed some form of this to demonstrate what 5G could do.

A contrarian view was that the automotive industry had yet to plan to do anything as expansive as this with 5G. It could take up to 10 years to influence the design cycle of their new car models, and they needed much earlier engagement. In an early 5G vision for V2X, the network drives the car and has mobile connections to everything. This would significantly increase uplink speeds for video, capacity requirements and coverage beyond network economics. However, while mobile networks generally cover 99% of the population, they cover much less of the country, so the cars would need to be able to be driven by an onboard car computer when they suddenly lose coverage. Does that sound different from the function of a Tesla or Google driverless car, which needs little help from the network to drive it at all? However, the automotive industry had alternative plans in the broader context of safety. Many premium cars today have very advanced safety systems based on mmWave radars and control of both the steering and braking systems for emergencies, determined on board without network latency. Here, drivers pay directly for these advanced features, but these migrate to standard models in a few years.

For broader V2X communications, in the United States, for example, the automotive industry had access to 5.9 GHz allocated for intelligent transportation systems (ITS) – specifically, for Dedicated Short-Range Communications (DSRC) vehicle safety technologies, so it did not need 5G radios with intrinsically higher latency. While this spectrum was not used, this indicates the lack of appetite for V2X and V2C (cars).

6G focuses more on the 'Connected car' while implying high data speed requirements. While the modern car may generate large volumes of data, as does a computer within its hardware, virtually none needs to be shared. Today's premium cars have 4G radios for software upgrades, vehicle tracking, real-time road traffic conditions and access to databases of information relevant to driving, such as electric charging points and nearby petrol stations, which generate very little network traffic. Most of these information databases can be side-loaded. Over time, these features will migrate to the standard car models.

Therefore, the authors see little in either 5G or 6G than that already achieved in today's 4G networks for the connected car with increased features built directly into the vehicle.

16.4.10 Data speed

The IMT 2030 report states that for 6G, we will need to find more spectrum in the low and mid bands and the mmWave bands towards THz for wider channels to support higher data speeds.

For the highest data speeds, the report states that this trend will continue leading up to peak rates approaching Tbit/s indoors by the end of the decade as the basis for even higher speeds in 6G. This is hard to justify for many reasons previously mentioned above. First, the only foreseeable application requiring Tbit/s is holographic displays outside the 6G or even 7G time frame. If this were to happen, these would likely be connected directly to the fixed network, where there are no foreseeable plans to reach even widespread use of more than 100 Gbit/s by the decade's end. In the authors' view, achieving a ubiquitous 10 Gbit/s broadband connection would be a higher priority and challenging. This demonstrates a naive approach, where it is assumed that fixed networks can support this if there is a mobile requirement. Second, for mmWave devices described in Chapter 8, reaching Tbit/s by the decade's end is unlikely as this would require channel bandwidths of circa 200 GHz based on the low QAM schemes needed because of transistor non-linearities. Such large channels could only be found by moving to frequencies beyond 300 GHz, towards 2 THz, and using new semiconductor technologies which have yet to be developed in commercial foundries. However, it may be possible to reach circa 100 Gbit/s by the end of the decade in the sub-300 GHz spectrum if channels of circa 20 GHz become available. Their likely use would be for fixed wireless access. Therefore, 100 Gbit/s represents a realistic extreme experience for 6G in the sub-300 GHz spectrum.

16.4.11 THz and mmWave for non-communication activities

Of course, mmWave will have many other applications besides communication, such as sensing and imaging systems with potential sub-centimetre resolutions. The availability of this technology in handsets and tablets could facilitate radical new 'spinoff' services described in the second scenario in a radical evolution, Section 16.5.2, where digital twins will play an important role here. For example, the digital twin will represent that 3D physical world. Digital twins were initially proposed in 5G for the Industrial community for IoT in the context of Industry 4.0, where the design of products can be enhanced and managed via integration with the product's digital twin. Beyond Industry 4.0, the concept of the Digital twins has been extended from non-living objects to humans. The closer association of the Digital twin in these and other mobile contexts will help the MNOs market this service, where the Digital twin is in MEC resources.

Similarly, the exploitation of mmWave technology towards 1 THz will further support FWA applications by providing a wireless interface to an optical fibre that could be within reach within 20 years.

16.4.12 6G spectrum requirements for extreme performance

It is expected that 5G technology will use the sub-tera spectrum up to 100 GHz, while 6G will explore a spectrum beyond 100 GHz with a realistic limit of 300–400 GHz. *(It will also likely use high parts of the 6 GHz band and parts of 7–24 GHz.)*

As stated earlier, to achieve a speed of 100 Gbit/s, it is necessary to have channels of around 20 GHz, which is feasible using the higher frequency range. However, if we want to achieve a speed of 1 Tbit/s, we will need a channel of at least 200 GHz, while the maximum spectrum use for 6G is likely to be limited to 300 or 400 GHz later. Therefore, to achieve speeds of up to 1 Tbit/s, we would need to use all the spectrum available between 100 and 300 GHz!

However, these high frequencies require an indoor wireless network limited to a single room, with no handover to an external network. Coverage of a home would require many access points, while Wi-Fi in the 5 GHz band has good coverage with options of cheap extenders, and Wi-Fi 7/8 will support peak multi-Gbit/s data rates near the access point. Therefore, 6G home consumer services are likely to be supported by Wi-Fi rather than 6G femtocells, even using spectrum as low as 6.5 GHz. However, in industrial settings, small cells are expected to be deployed supporting 100 Gbit/s type speeds using 100–300 GHz.

External wide-area networks will likely operate up to 4.2 GHz for economic coverage. Hot spots, such as information petrol stations offering 100 Gbit/s, using 100–300 GHz, could exist in popular places like Wi-Fi hot spots today. Nonetheless, these 4.2 GHz networks could support application speeds per user of 200–300 Mbit/s in congested cells with spectrum reform and wider channels. Thus, introducing 6G mmWave technology may resemble a WLAN scenario limited to a single room, while the 6G external wide-area cellular networks will take us beyond 5G today.

The indoor environment is the most challenging for 6G and its spectrum.

16.4.13 Spectrum sharing technologies

In Chapter 7, we described a novel approach to spectrum sharing which used weak signal beacons as propagation probes within a spectrum sharing system. Explicitly, this is intended to avoid the interference and inefficiencies of propagation planning by replacing this with real-time measurements. It has the potential to improve spectrum utilisation massively. The IMT report identifies spectrum-sharing technologies as necessary for 6G. We will discuss this further in the following central section on a 6G revolution. If this approach were fully utilised, it would be transformative to the spectrum use below 3 GHz and go a long way to provide enough spectrum for 6G there.

16.4.14 Network and computing convergence

The ITU report further supports the 5G use of MEC in future IMT networks for low latency services, where the network will use the nearest edge computing site. They give examples of applications to exploit this, which include AR/VR rendering, autonomous driving, and holographic-type communications. Still, we remain sceptical of the last two. We previously discussed that Ericsson believes XR could be supported on 5G networks today. Nevertheless, it does pave the way for various forms of convergence, not the old fixed/mobile form, but rather around 6G networks with MEC, digital twins, sensing and cloud and computing resources from hyperscalers as described in Chapter 9.

16.4.15 Mobile edge computing

MEC, first cited in 5G, will likely be further enhanced in 6G for IMT networks. It is an integral requirement for low-latency services and supports using digital twins, particularly for Industry 4.0 applications. This could emerge as a critical differentiator for mobile operators, as there are natural synergistic advantages with combinations of low latency, MEC and digital twins.

We believe that 6G MEC will default to an evolution of 5G MEC, but the revolutionary opportunity is to make MEC more general-purpose and competitive against the hyperscalers. Using 6G for Edge Cloud has significant research and standardisation challenges. More radically, 6G networks could become part of the computing fabric offering key computer services, such as distributed neural networks or consensus as a service. However, most of the technology needed to implement '6G for Edge Cloud' services exists, and work is progressing on the required standards and technology gaps. Incorporating the developing Edge Cloud standards in future 6G standards will be essential but insufficient to be revolutionary. To be revolutionary, the hyperscalers and other industry behemoths must adopt these developing standards and invest in the ubiquitous deployment of Edge Cloud infrastructure. Then, there could be services involving network resources and opportunities for MNOs to help fund their deployments in partnerships, see Chapter 9.

The revolutionary universal 3GPP Edge Cloud would compete/partner with hyperscalers for low latency, security, privacy, reliability, and global coverage with high data localisation via explicit user control. Edge Cloud would be supplemented with in-network computing, using network elements as part of the computer, e.g. network nodes could be neural nodes. It could become the default for on-premises Industry 4.0 solutions, enabling more joined-up Industry 4.0 supply chains across organisations with minimal integration costs. It would be the infrastructure that enables MNOs to offer value-added services such as digital twins as a service.

16.4.16 Digital twin

Digital Twin was identified as a critical application within 5G. The digital twin is a digital representation of the physical world that allows for interactions in the real world through interactions with the digital twin. It has many use cases in both 5G and 6G. These range from manufacturing, where the Twin models a manufactured object and variations in tolerances can be understood as variations in performance, to 3D representations of the physical environment for radio propagations or smart cities. This theme is continued in the IMT report. While they have not identified any new requirements yet, these will emerge from any successful deployment in 5G. We have covered these in Chapter 11.

16.4.17 Proliferation of intelligence

With advances in AI and Machine Learning (ML), machines using these technologies will transform the data collected by wireless networks into reasoning and actionable decisions. This will help people understand and make better decisions or help networks optimise their performance, helping to achieve better performance and sustainability. AI and ML are expected to be distributed across all the network layers to achieve this.

This proliferation of intelligence will be fundamental in helping to achieve coupling between the human, physical, and digital worlds of 6G. 6 G provides an opportunity to provide a framework to enable real-time control and trustworthy intelligence while maintaining the human in control [19].

Similarly, as in Digital Twins, there are opportunities to combine real-time distributed learnings and collaborations between intelligent robots. This will also emerge during the latter stages of 5G but represents a much wider area than just robots and provides a more important theme in 6G. It could, for example, use the channel sensing data and 3D positions of the environment to optimise the performance of robots.

16.4.18 Our conclusions to a critical analysis of IMT 2030

If the three use cases of (i) Holographic displays requiring Tbit/s, (ii) disaggregation of the control loop requiring sub ms latencies, and (iii) control/supporting of driverless cars requiring 10's Mbit/s per car fail to materialise, this has a profound effect on the conventional key KPIs for 6G leaving them broadly similar to 5G hence the quote at the beginning of this book that all we need is 5G. We also remain sceptical about the need to expand the number of IoT devices supported per cell site. However, that takes the view that each generation is purely about faster/higher physical layer characteristics and a new air interface. It misses the fundamental changes in the proliferation of intelligence – the use of AI and ML across network layers, MEC, Cloud, virtualisation and core networks, which could, if successfully implemented in a 'Network of Networks', profoundly change the role of 6G and make it the most consequential cellular generation since 1G. It also misses the point that the investment in 6G could come from mobile operators and hyperscalers, bringing a new form of network convergence.

In Section 16.4, we have presented our analysis of many of the proposed key service features of 6G, which we will discuss in the next section on the evolution and revolution pathways for what 6G could be.

16.5 Scenarios for 6G – evolution and revolution

After reviewing the IMT report covering proposed use cases, the key 6G white papers in Section 6.7 and analysing the technical performance of 6G's foundations in the previous chapters and summarising them in Section 16.4, we are ready to discuss our vision for 6G's development in both an evolutionary and revolutionary framework. Figure 16.8 presents a schematic representation of a 6G network, which

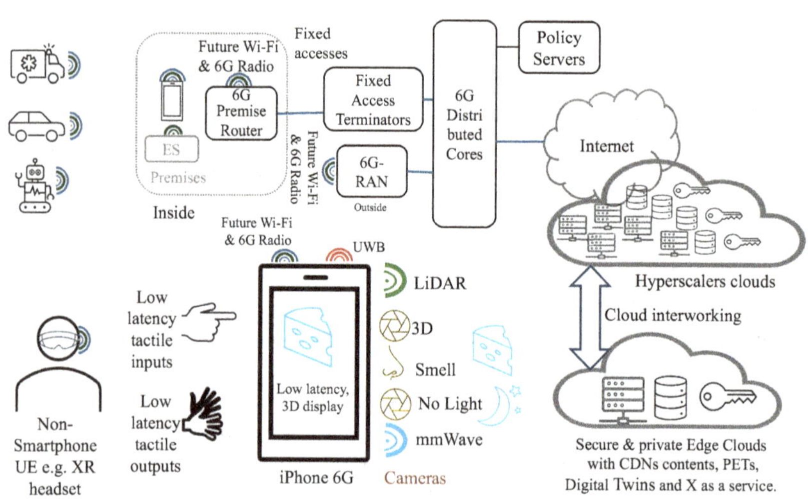

Figure 16.8 Future 6G devices and networks and relationship to hyperscalers

is device-centric, embracing 6G cellular and Wi-Fi, which we will explore in detail here in these scenarios.

The approach taken by 6G towards the digitalisation of industry and society builds upon the approach taken by 5G with significant improvements. It uses AI, ML, and sensor technology, built into six new themes: Network of Networks, Connecting Intelligence, Trustworthiness, Global Coverage, Sustainability and Extreme Experience. The core of 6G will be based on 5G, using softwarisation and virtualisation. These themes address many of the issues faced by 5G, as previous even number generations of cellular technology have done in the past. However, the implementation of 6G in 2030 poses financial challenges when calibrated based on the progress of 5G in its mid-decade. The NGMN groups have suggested that a new air interface should be optional to reduce its deployment costs. So, while a revolutionary scenario will be based on the key aspects and performance of 6G evolution, it is intended to explore revolutionary paths to broader digitalisation. Figure 16.8 is a 'taster' of the two approaches we explore below with a futuristic iPhone 6G.

16.5.1 The first scenario is '6G evolution'

The 6G evolution scenario aims to continue the strategy of 5G, which involves the digitisation of society and industry, connecting the Human, Physical, and Digital worlds through software-based delivery platforms. This supports the enormous range of applications and services mentioned in Tables 16.2 and 16.3 from Chapters 4 and 5. The primary objective of this approach is to enable more extensive use of AI and ML across the mobile infrastructure, with Digital Twins serving as the digital representation of a physical product, system, or process. Therefore, 6G is intended to build upon and improve the shortcomings of 5G. While 5G has already set ambitious targets for its infrastructure, evolving from Release 16 to 18.

As discussed in Chapter 6 on 6G standards activities, the current 6G 'directional vision' from various regional parties of 3GPP is emerging. The approach confirms that 5G's eMBB, URLLC, and mMTC technologies will continue to evolve, while AI and sensor convergence technologies will likely be applied to communications. The 6G network is expected to grow significantly, with the ability to learn and act autonomously. Its core implementation could be based on the 5G core used in 5G advanced, with SDN and NFV technologies following a softwarisation strategy.

The challenges 6G faces are being addressed by projects such as Hexa-x, amongst others, in the six areas of Connecting Intelligence, Network of Networks, Sustainability, Global Service coverage, Extreme experience and Trustworthiness. There will be a need for 6G networks beyond just providing connectivity and, for example, using digital twins for spatial mapping based on sensor measurements. With the advent of technologies like the Internet of Senses and Cyber-Physical Systems, 6G systems will be designed to offer a much more integrated network computer fabric. This will transform the network into a pervasive, globally interconnected platform optimised for efficiently handling application components while giving the impression of locality. Security and privacy are critical to this pervasive use of personal information across many networks. Our discussion in Chapter 12 around Privacy Enhancing Technologies provides a possible solution.

In the following sections, we define some of the service aspects 6G will address in its evolutionary path from 5G Advanced:

16.5.1.1 Artificial intelligence – an evolution

Integrating AI/ML technology with the upcoming 6G wireless technology can create a powerful synergy that surpasses the individual strengths of each technology. This combination can redefine wireless communication by providing real-time data analysis and decision-making abilities, complementing 6G's high-speed and low-latency network. This synergy can improve the reliability and performance of 6G networks by enabling predictive analytics to identify and resolve network issues before they affect the service. AI and ML are expected to be distributed across all network layers to achieve this. We see this as the primary path for AI in 6G evolution, where the 6G MEC will evolve from that of 5G. Additionally, in Digital Twins, there are opportunities to support collaboration between entities such as intelligent robots using the channel sensing data and 3D positions of the environment to optimise the performance of robots, see Sections 16.4.14–16.4.17.

16.5.1.2 Network of networks – an evolution

The decision that 6G makes in this field is critical to its future. It can follow the traditional evolutionary approach, where it deals with its technologies as separate entities, hoping that its solutions will be adopted by all those who require broader digitalisation of industry and society and that MNOs will thrive. However, it overlooks the fact that many other access technologies, both wireless and wired, are better optimised for specific use cases. Therefore, replacing them with a generic cellular device and associated access network is incredibly challenging, even with its other 6G assets in MEC, DT and Core network. We discuss the benefits of embracing a broader definition of a network of networks in Section 16.5.2.1 using a revolutionary approach, which would lead to faster digitalisation, more widespread and arguably a better future for MNOs.

16.5.1.3 Future IoT

While it is claimed that 6G will support Trillions of embedded devices for the IoT, this faces immense challenges in energy harvesting and multiaccess schemes, as discussed in Chapter 6. Nevertheless, sustainability, and in particular, the reduction of energy in 6G networks, is of paramount importance. However, the SK Korea white paper warning could be remarkably prescient about the expectation gap a 6G era could create to reality if it does not address the other critical factors in developing and delivering new services other than just the communications elements. We outlined our views in this area in Section 16.4.4, Connectivity for Everything.

16.5.1.4 Latency requirements

According to the IMT reports, a significant portion of traffic in the industry vertical will be based on small data related to measurement or actuation. They believe this will require ultra-low latencies in control loops, significantly below 1 ms. However, our findings indicate that the lowest latency requirement is 1 ms for 6G, balancing application requirements vs cost, with most applications needing over 10 ms. It is worth noting that machines requiring sub-1 ms latency are likely to be static and already served with Ethernet solutions, reducing the demand even further for any wireless solution. Therefore, we do not deny that there is a sub-ms requirement; it is only that this does not need to be supported directly in 6G in its air interface but rather through a sub-network in a possible Network of Networks architecture in the revolutionary future.

16.5.1.5 6G spectrum

The upcoming 6G spectrum is expected to be more versatile (see Section 16.4.12). It will use cm wave, particularly in the 6.5–24 GHz range, to cater to high-capacity mobile use cases, especially for indoor applications or campus-type coverage similar to WLANs. However, it is more suitable for industrial settings than for home use, where Wi-Fi is expected to continue dominating, particularly with Wi-Fi 7 and 8 with improved latency and reliability. Wi-Fi 9 is expected to arrive in the early 2030s, but it has yet to be defined.

16.5.1.6 Data speed requirements

While the IMT report asserts that higher data rates will continue leading up to 1 Tbit/s indoors by 2030, giving rise to sub-terahertz communications, we believe this is not viable as it would require channel bandwidths of 200 GHz, see Section 16.4.10. We believe that there will be no holographic displays, so such high speeds are unnecessary. Therefore, mmWave in the sub-THz region, 100–300 GHz, will address 100+ Gbit/s speeds for extreme communication needs in both evolution and revolution. Still, mmWave will find uses other than for communications, such as in sensing, helping achieve 1 cm resolutions. Over time, this technology could find revolutionary applications in spectrometry, such as scanning your meat for allergens.

16.5.1.7 Extended reality

6G XR builds on the 5G vertical of AR/VR, where XR provides a more interactive immersive experience that seamlessly blends virtual and real-world environments while offering users a new multi-sensory experience, see Section 16.4.5. Therefore:

- 6G is set to improve using cm or mmWave technologies for higher resolutions and frame rates, potentially up to 1 Gbit/s and lower latencies approaching 1 ms, particularly in industrial settings.
- 6G can offer extremely high performance in indoor settings with handover to the wide-area network when required by downgrading to 5G performance levels, 100–200 Mbit/s.
- Many Network infrastructure providers will have mobile and fixed assets as 'converged' operators, so there is a much greater incentive to use Wi-Fi, with lower costs, in the home to support Gbit/s XR.
- The broadband access network will support multi-Gbit/s rates within the present plans for the role out of 10 Gbit/s FTTH and Cable with DOCSIS 4.0.

16.5.1.8 Vehicle to infrastructure and vehicles (V2X)

While the support for V2X would drive dramatically higher up-link speed requirements in 6G, we find no evidence to support this emergence or even need apart from some outliner safety cases. Today, cars that use onboard sensors and computing have made significant progress since V2X was first proposed in 2015. These offer significantly lower latencies and avoid cellular networks' reliability and coverage requirements. Generally, connected car scenarios do not need high-speed data requirements, see Section 16.4.9.

16.5.1.9 Connectivity for everything

This new and exciting scenario for 6G – Connectivity for Everything, as discussed in Section 16.4.4, builds on the IoT theme in 5G. These scenarios include real-time monitoring of buildings, cities, cars, transportation systems, roads, the environment and critical infrastructure such as water and power. This is also extended to the Internet of bio-things through intelligent wearable devices and intra-body communications achieved via implanted sensors, which take the scenarios of mMTC and put them in a human context. This could be very important for people, and the UK National Health Service already offers pacemakers nightly downloads of their data at home.

We covered this theme in detail in Section 16.4.4. While we do not doubt the value of the Internet of Bio is very high for people, the real value in this proposition is the provision of the sensors, monitoring, and App ownership, where the main role for MNOs would be as a data pipe. Still, it could support applications with sensing and digital twins.

16.5.1.10 Global service coverage

A network with '100%' coverage appears inspirational, but it is most useful for communication over oceans or vast land masses without cellular networks, as discussed earlier in Section 16.4.8. Since satellite-based networks require an unobstructed sky view, this is not generally a significant limitation. This technology will, for example, benefit emergency services and control robots and drones in these areas. There will be many logistical type applications, such as tracking objects worldwide. Geostationary or LEO satellites will be used, depending on the application. Some mobile operators will likely use LEO satellites to extend coverage outside their cellular networks in both the evolutionary and revolutionary scenarios of 6G. The traffic density would be very low in these areas, and expanding the cellular coverage would be uneconomic. The level of integration between cellular and satellite is questionable as these applications do not require seamless handover but fast roaming. Still, it may be necessary for drones/robots to control using both networks.

In both the evolutionary and revolutionary scenarios of 6G, new commercial models for satellite coverage could emerge in smartphones independent of the mobile operator (see Section 16.4.8).

Using satellites to address the global digital divide will depend on affordability in monthly access and device implementation costs; however, it does not look promising. As we discussed earlier in Section 16.4.8, even in less developed countries, mobile ownership is predicted to reach similar levels to developed countries towards the end of the decade, albeit using older smartphones and 4G networks.

16.5.1.11 6G AI–native air interface

Each generation of mobile technology introduces a new air interface, so the evolutionary path would expect to include this. In the next session, we debate this assumption as OFDM used in both 4G and 5G has taken us close to the Shannon limit. What could a new air interface be?

ML has been used in everything from image recognition to robotic systems, and more recently, significant advances have been made with ChapGPT. AI could be used to engineer radios that could adapt their signalling schemes to achieve the best performance for any situation. Nokia believes [20], a dynamic AI/ML air interface, will be a crucial component of 6G. The radio will use AI to learn and set up the best bespoke waveforms and their constellations to improve performance. The primary use cases will be in private networks, which will benefit the most. Nokia believes that its AI-native air interface can improve the throughput by between 20% and 30%. This level of improvement achieved by AI coincidentally demonstrates how close we are to attaining Shannon's limited performance. Mobile operators will need to consider the economic benefits of a new air interface that requires massive changes to their access network for any relativity small 6G improvements in spectral efficiency. The NGMN has stated that a new air interface should be an operator option to implement or not. Furthermore, JCAS is already supported within the 5G standards, so it is yet to be proved if the air interface needs changes to support JCAS further.

The following section presents our analysis of the envisioned performance and KPIs for 6G, which would meet all the likely requirements of 6G at the lowest costs.

16.5.1.12 Envisioned performance and KPIs for 6G evolution and revolution

5G has laid a solid foundation for eMBB, mMTC, and URLLC, which are ahead of their time. After analysing the new use cases and extensions in the IMT 2030 report in Section 16.4, we have identified the following KPIs as necessary to meet future 6G requirements:

- Spectrum: extension up to 1THz?
  The upcoming 6G technology will likely focus on extending the spectrum from 100 GHz to 300 GHz (possibly 400 GHz) in the mmWave region using GaN, SiGe, and Si CMOS technologies. The first phase will concentrate on the 100–200 GHz range and will be the main activity area for communications during the 6G lifetime, while 300 GHz (400 GHz) will find uses in sensing and spectrometry.
- Throughput/data rates up to 10–100 Gbit/s (Ten times the original 5G KPI target)
  This is driven primarily using mmWave in the 100–200 GHz spectrum bands. This region can support 100 Gbit/s data speeds for campus, hot spots, FWA and industrial settings, with channel bandwidths of around 20 GHz.

  Wide-area cellular networks will likely use frequency bands up to 5GHz (4.2GHz), with spectrum reforms of the sub-3 GHz region and the use of 700 MHz. This will provide a typical user experience of 200–300 Mbit/s.[2]

- End-to-end latency 1–10 ms (the same 5G KPI target)
Perhaps controversially, we are not convinced that significant use cases need less than 1 ms as this is beyond human perception, and disaggregation of control loop functions is either a niche use case with an existing solution or not required. Reductions in latency come at the cost of spectral efficiency.
- The density of connected devices is 1×10^6/km (the same 5G KPI target)
The original 5G target would meet even the most realistic increase in mobile IoT or the connection of everything. Based on the belief that everything is in the mobile domain, this might increase, representing more than ten mobile devices per metre squared in dense areas!
- Global Service Coverage (using satellite networks)
The prime focus of using satellites is likely to be extending coverage outside regular cellular networks across vast land masses or oceans where cellular networks cannot exist for initially niche use cases or low traffic densities. Nevertheless, it will support important Industry 4.0, construction and agricultural use cases. However, we need to be more sceptical about addressing the digital divide between developed and developing countries due to high access costs, which are intrinsically challenging to scale capacity for mass demand, and high terminal costs.
- Frame error rate 10^{-6} (the same 5G KPI target)
While some will push for an increase to 10^{-9}, while Japan proposes 10^{-7}, this comes at the cost of spectral efficiency and is not required for Industry 4.0 applications where, in general, the main issue is not the frame error rate for control of machines but the repeated loss of packets.
- Localisation precision 1 cm in 3D (ten times the original 5G KPI target)
MmWave, particularly towards 300 GHz, will find exciting applications in positioning and 3D imaging systems reaching 1 cm resolutions and scanning applications in many use cases. Localisation will likely exist across all the cellular bands with lower resolutions at lower frequencies.
- Very high energy efficiency (the same 5G KPI target)
The reasons behind the increasing energy demand for mobile systems are multifaceted. While decreasing the energy consumption per bit is crucial, this is countered by the significant surge in data usage, resulting in an overall rise in this energy consumption. However, some mobile data applications replace activities that require physical movements, which consume far more energy and associated carbon emissions. For instance, a Zoom call can replace the need for travel. Nevertheless, it is crucial to reduce energy consumption in mobile systems globally. However, we must monitor the whole network's energy use to achieve this.
- Possible Modest improvement in Spectrum efficiency from 5G based on an AI air interface.

We summarise these in Figure 16.9.

[2]Based US Q4 2023 Ookla (all operators combined 3.5GHz (n48) – 200Mbit/s 3.7GHz (n77) – 250Mbit/s.

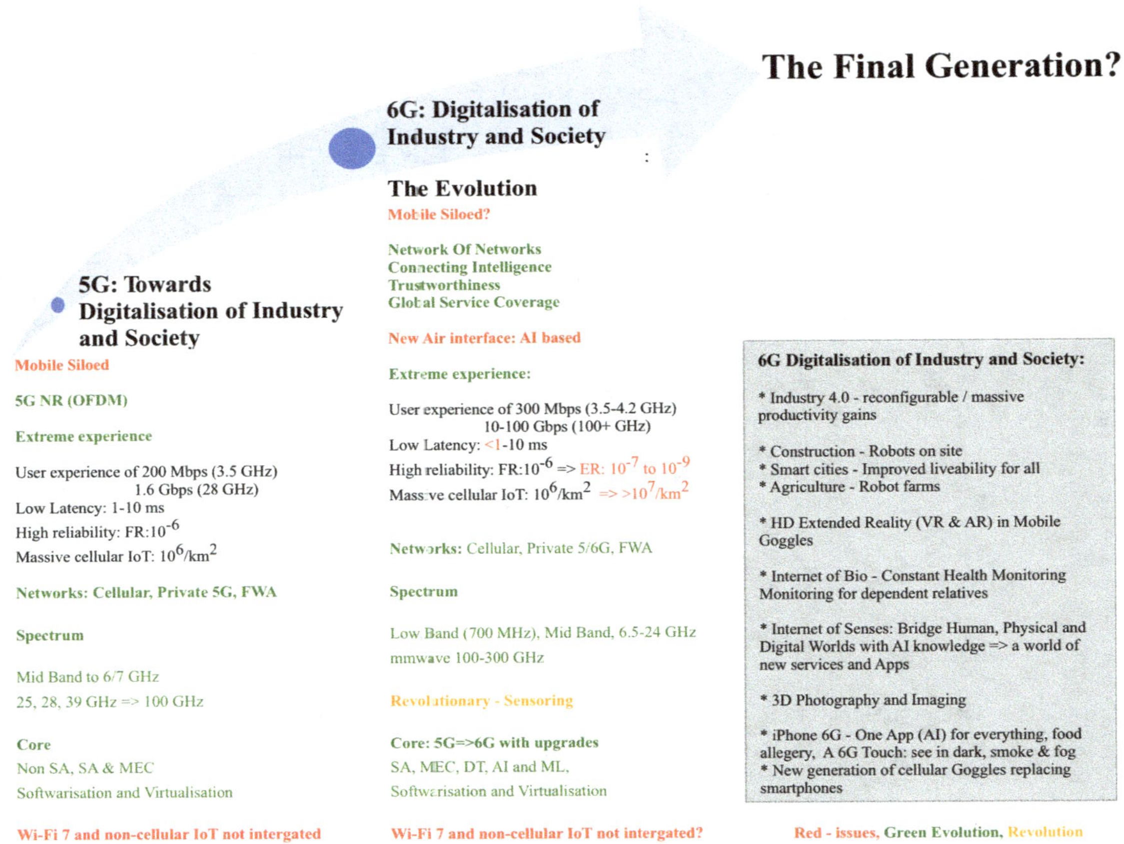

Figure 16.9 Summary of 6G evolution with likely KPIs and applications

16.5.2 The second scenario – '6G revolution with ten revolutionary ideas'

> *A technological revolution is a period in which one or more technologies are replaced by another novel technology in a short time. It is an era of accelerated technological progress characterised by innovations whose rapid application and diffusion typically cause an abrupt societal change.*
>
> – *Technological revolution, Wikipedia*

The intention of the *6G revolution* scenario is not to replace the evolutionary approach of moving from 5G Advanced to 6G, as we covered in the previous section, but rather to complement it with possible revolutionary paths based on either the use of new technologies, extensions to architectures or solving challenges to the mobile industry's approach. Therefore, the KPIs for revolutionary scenarios would be the same as those for evolution outlined in Section 16.5.1.12. Here, we explore the revolution's key features; see Figure 16.10 with its key differences to 6G Evolution explained in Table 16.4.

16.5.2.1 Network of networks – a revolution with convergence

The development of 6G should focus on creating a Network of Networks, which includes Wi-Fi, non-cellular IoT, and fixed networks as sub-networks within the overall architecture. This approach will accelerate digitalisation and open new opportunities for MNOs. Currently, the cellular industry's ability to address the broader digitalisation of society and industries is limited, as indicated by its percentage of IoT. IoT is an essential feature related to all the various execution sectors of the broader digitalisation of society and industry. In 2024, the cellular industry is projected to have less than 25% of all IoT, expected to fall to less than 20% by 2028 when 6G is defined in standards. Therefore, cellular networks must embrace over three-quarters of other systems and support them with management and control using MEC, DT, Core, AI and ML systems. This would place cellular networks in a position of strength. This is why large network operators provide both retail services and wholesale use of their networks, as the sum of these revenue sources is more significant than a pure retail play. By extension, this approach also supports other key themes in 6G, such as Global seamless coverage, Connecting Intelligence, and Trustworthiness, making its offering more comprehensive.

In particular, Wi-Fi 8 is expected to support similar use cases for Industry 4.0 with lower latency and higher reliability, and it will be a crucial technology in 6G homes and smaller industrial settings. However, it must incorporate other elements of the 6G architecture, including DT, MEC, AI, ML and its Core for future services. Similarly, multiple non-cellular IoT networks support highly optimised use cases and would benefit from inclusion in 6G with access to its assets. Moreover, fixed networks provide mobile backhaul and support ultra-low latencies with Ethernet. Even during the lifetime of 6G, factories using these already will be reluctant to replace them.

Such an approach will enable 6G to work with the hyperscalers with interconnection to their AI and cloud infrastructures. This approach provides the infrastructure for the broader digitalisation of industry and society much earlier, and it is a win-win scenario. It also provides opportunities for joint funding of 6G, as Caroline Gabriel, Analysys Mason, have pointed out.

> *The telecoms, media and technology (TMT) industry needs to make progress with 6G because the true digital transformation will not be enabled by a siloed mobile network, but rather by one that is built from the outset around the convergence of various access technologies, and of networks and cloud.*
>
> – *Caroline Gabriel, Analysys Mason*

One of the advantages of the convergence model facilitated mainly around the Intelligence layers for 6G services is that it provides MNOs a new approach to building mobile coverage since the first-generation era. 1G pioneered a 'Macro-cellular first – outside in' method, which led to significant increases in macrocell transmission powers and energy use in subsequent generations as both frequencies and cellular broadband speeds increased. This was done to overcome the challenges of reaching mobile users where 70%–80% of mobile data is consumed – indoors. This trend will likely worsen with 6G's even higher transmission frequencies, as explained in our Spectrum dichotomy, and greater emphasis on indoor applications – homes and factories. However, converged operators can satisfy these and other wireless use cases indoors by using indoor Wi-Fi access points (APs), Wi-Fi 8, or small cells to support most of the new use cases for 6G. Moreover, this approach can also provide serendipitous coverage of adjacent buildings and outdoors, as described in Chapter 9.

6G is uniquely positioned to drive the broader and earlier digitalisation of society and Industry, not as a mobile silo but as a general platform playing to its strengths in mobile.

16.5.2.2 AI – a revolution

The revolution path for AI in 6G builds on the evolutionary path (see Section 16.5.1.1). AI is internalised mainly within the 6G network to optimise its use and improve performance (see Chapter 13). The revolutionary path has two main branches associated with the Network of Networks (see Sections 16.5.2.1 and 16.3.5) and its relationship with hyperscalers, who will have their AI and cloud platforms. In the first branch, 6G AI will be used to manage the complexity of the Network of Networks and offer broader AI services. The first elements could be repackaged in the second branch with hyperscaler AI services. It is one area that strengthens the *raison d'etre* for a broader collaboration with hyperscalers in 6G. Applications builders will use both to create more significant synergies. Of course, the fruits of these collaborations could be unimaginable new applications of AI beyond that foreseeable in its use within networks.

One of the new areas for the 6G revolution use of AI within its network will be to help manage a 'Converged' network of networks while providing more significant insights across these domains. It would also extend to the security and privacy aspects of 6G as outlined in Chapter 12 on Trust and Privacy.

In many ways, the six themes of 6G would be attractive to hyperscalers in collaboration with the MNOs based on a movement away from a siloed approach, as explained in Chapter 9. At the same time, the revolutionary aspects of the Core and Edge for Cloud, covered below in Sections 16.5.2.6 and 16.5.2.7, are integral to this revolutionary future.

16.5.2.3 A new air interface is optional for revolution!

In the evolution path, we discussed that a new 6G air interface would likely be based on AI with relatively modest improvements in spectrum efficiency compared with previous generations. At the same time, the NGMN has already stated that a new air interface should be an operator option to implement or not. So, does 6G need a new air interface? That's a good question, as circa 75%–85% of the network upgrade costs are associated with the radio access network. Keeping the present 5G air interface would allow the wide area cellular networks to be unchanged for 6G with a typical user experience of 200 Mbit/s sufficient for Extended Reality. At the same time, sensing is already supported by 5G standards. The cm and mmWave spectrum can support 10 Gbit/s speeds, while new modems for higher mmWave spectrum do not need a new air interface, and these could allow 100 Gbit/s with 20 GHz channels. So, against this backdrop and a difficult financial situation in the Telecoms industry, it is hard to make the case for a new air interface. The real revolution is what 6G enables, not how it does it. We believe that virtually all the exciting applications for 6G, such as Extended Reality, including VR and AR, sensing applications, smart cities, and industrial use cases in factories, construction and agriculture, can be supported by the present 5G air interface. The other use cases, such as Holograms, are illusionary or extreme outliners. So, while it may seem perverse, not having a new air interface would be revolutionary!

16.5.2.4 JCAS in revolution

JCAS is likely a critical revolutionary feature of 6G technology, differentiating it from 5G, although its specifications were initially included. It would also be available in an evolutionary approach. The ability to localise, track and sense physical objects is the bridge connecting human, physical, and digital worlds and hence is a crucial feature of new 6G services. Mobile networks are a great starting point for enabling a massive sensor network, as they already have extensive coverage on day one operation. Hence, the sensing can be delivered almost 'for free.' We described the main opportunities and features in Section 16.4.6.

It is a key feature not only for Industry 4.0 but also for many other use cases in consumer, health, smart cities, and vehicle safety. It also supports spectroscopic scanning for new smartphone Apps and pollution measurement in towns and cities for improved liveability. Its uses are not limited to user applications. It will fill a role in cellular networks. For example, in beam steering in MIMO, the cellular sensor network can detect blockages in real-time and move the beam past these faster.

The level of accuracy will differ in different cell types. In many respects, MIMO in the sub-Tera hertz region is an up-and-coming solution as it can combine extreme miniaturisation for a system-on-a-chip (SoC) with low-cost, high degrees of angular resolution with a pencil-like beam for 3D mapping and spectroscopic scanning. It can provide cm-level accuracy while wide area networks using sub 5 GHz spectrum likely allow circa 1m resolutions.

So, there are opportunities for various solutions where JCAS majors on wide-area support on day one and in machines already incorporating 6G cellular for applications such as Industry 4.0, Smart Cities, Robots, etc. The openness of its sensing capacity would differentiate it from the evolution scenario. It could use other sensing technologies, such as UWB, in parallel with cellular.

16.5.2.5 Spectrum reform – 6G below 3 GHz

Spectrum is the fuel of the wireless generation. The more you have, the further you can travel. Each generation has required higher speeds and capacity for data, which has inevitably led to moving up the spectrum staircase from low band (<1 GHz) to mid band (1–6 GHz), while 5G was the first to exploit high band (>6 GHz) with cm and mmWave using frequencies above 24 GHz. Yet the propagation characteristics radically change in usage scenarios with associated network deployment costs.

Earlier, we discussed a spectrum dichotomy where frequencies in the low to mid band went hand in hand with the types of service, voice to data, with seamless handover using an outside-in cellular network. In contrast, using the high band (cm and mmWave) would require both an external and indoor network for seamless handover using these frequencies that support the same channels and speeds. However, the small cell sizes for high bands make cellular densification uneconomic. Nevertheless, the canon behind this evolution is that spectrum is a scarce resource, and only moving to higher frequencies will afford frequencies with larger channels to support higher speeds and capacity.

The frequency range of up to 3 GHz is the most valuable for cellular, offering good coverage, speed, and capacity possibilities. Not long ago, mobile operators believed that circa 3 GHz was the limit for economic cellular deployment. Yet, in many countries, spot studies of spectrum occupancy in the 30–3000 MHz have found typical utilisations of 10%–20%. There is good reason to believe that a figure in this range is representative at a national level. Spectrum regulators acknowledge that spectrum utilisation is unknown, while large areas have low spectrum occupancy. Assuming an occupancy spectrum utilisation of 80% of the 3 GHz Band and a lower limit of 600 MHz for cellular, there are close to 2 GHz of additional spectrum that could be sufficient for 6G. This would probably meet '99%' of the new services of 6G '99%' of the time. Also, remember that data usage may plateau in the late 2020s. Of course, this does not preclude using cm and mmWave for FWA or extreme data usage confined to predominantly indoor applications.

The IMT 2030 report has recognised the importance of spectrum sharing. The US CBRS is, in many ways, a revolutionary attempt to redress some of these issues by making mobile spectrum available at a county level and using an SAS to share the spectrum. It is a first step towards democratising the mobile spectrum, opening use to non-mobile operators, which will be increasingly crucial for Industry 4.0. However, it still bakes a conservative propagation model within its design and cannot identify and control interference. The use of a propagation model has shown that it can 'lose' between 40% and 75% of the gained spectrum, so there is a high price to pay for its use. In the case of sharing, this encourages discord as sharers of the spectrum endlessly debate the model's merits, and incumbent operators argue for even greater protection while newcomers less. We have shown that a new system based on weak signal propagation measurements (reporting) within a CBRS-like architecture, CBRS 2.0, could address all spectrum management issues. It will provide a means to share the spectrum widely with many different types of users for the first time, removing the need for conservative propagation models and addressing interference in real-time, one of its most essential features. The detailed considerations of sharing the C-band between mobile and satellite providers exemplify its more comprehensive exploitation. Using such an approach for Low-band and Mid-band frequencies could dramatically increase the use of this spectrum, potentially providing at least 1.9 GHz, below 3 GHz, for 6G with great network deployment economics. At the same time, the 3.5 GHz band will likely be extended to 4.2 GHz with the satellite C-band on this base station.

A vital part of CBRS's success has been using mobile bands, which overcomes the issues of previous TV white space type systems' specialised hardware. So, using CBRS 2.0 would require including its bands at a future WRC conference. This overall approach allows using the other elements of the 6G architecture. Now, that would be revolutionary.

16.5.2.6 A radical core network

The 6G core is anticipated to be similar to the 5G core in its basic structure but with enhanced efficiency, scalability, and reliability. Although the 5G core cost is relatively small compared to the RAN, redesigning the core network can significantly improve its reliability and scalability. This is critical for supporting millions of Wi-Fi AP/small cells and enabling URLLC. The main objective of such a redesign is to make the core network more agnostic to the access network and device (UE). This is a vital part of the Network of Networks discussed above.

Certain essential functions could be carried out in the RAN components instead of the core. For instance, in 5G, both the core and the RAN handle mobility, and this could be handled just once in the RAN. However, other core functions need to be assessed as well. For example, the need for mobility management and tunnelling to prevent UE IP address changes should be questioned. Modern IP protocols and applications should be able to handle IP address changes without disrupting the service. Nevertheless, the challenge is to ensure continuity in application protocol when IP addresses change for the complete range of protocols required for 6G network services.

Some protocols used by 5G UE and RAN to the core are stateful, making them less reliable and unsuitable for distribution in a true cloud-native fashion.

Therefore, it is necessary to review and change these protocols for 6G. We can use DiffServ for QoS and RESTful protocol styles to achieve a more distributed and stateless approach. We can also use an Open Roaming approach to AAA, enabling a 6G 'core-less' architecture.

A coreless 6G architecture, as seen in Chapter 11, may look more like an Internet architecture, but it might be achievable with sufficient R&D effort. It would provide significantly more benefits than cost reduction, i.e. ultra-reliable, ultra-scalable, and ultra-flexible.

16.5.2.7 6G for Edge Cloud

6G Mobile Edge Cloud (MEC) will default to an evolution of 5G MEC, but the revolutionary opportunity is to make MEC more general-purpose and competitive against hyperscalers. Using 6G to offer general-purpose Edge Cloud services has significant research and standardisation challenges. Central will be consolidating multiple Edge Cloud standards, addressing the challenges of resource management, mobility support and orchestration across organisations. More radically, 6G networks could become part of the computer fabric offering critical computing services, such as distributed neural networks or consensus as a service.

Most of the technology needed to implement '6G for Edge Cloud' services exists, and work is progressing on the required standards and technology gaps. Incorporating the developing Edge Cloud standards into future 6G standards will be essential but insufficient to be revolutionary. To be extreme, the hyperscalers and other industry behemoths will have to adopt these developing standards and invest in ubiquitous deployment of Edge Cloud infrastructure.

The revolutionary universal 3GPP Edge Cloud would compete with hyperscalers for low latency, security, privacy, reliability, and global coverage with high data localisation via explicit user control. Edge Cloud would be supplemented with in-network computing, using network elements as part of the add, e.g. network nodes could be neural nodes. It could become the default for on-premises Industry 4.0 solutions, enabling more joined Industry 4.0 supply chains across organisations with minimal integration costs. It would be the infrastructure that enables MNOs to offer value-added services such as digital twins as a service.

16.5.2.8 PETs – as a service in the 6G revolution

Trustworthiness, a recurring theme in 6G research and standards projects, is crucial for its evolution. Security and privacy, the twin pillars of trustworthiness, are becoming increasingly important as more data about people and systems is collected. For instance, the digitalisation of society and industry is leading to the accumulation of more confidential data. Precision healthcare, which factors in an individual's genes, environment, and lifestyle in their treatment, is another example. We are optimistic that PETs will play a pivotal role in the 6G evolution, as outlined in Chapter 12. The growing concerns about data privacy and sovereignty present another opportunity for PETs and 6G to address, particularly considering that the critical 6G use cases rely on private or confidential data.

Homomorphic encryption enables computation on encrypted data; the user can encrypt their confidential data and send it to the MEC or hyperscaler's cloud for processing, and only the data owner can decrypt the results. Secure Multi-Party Computation (SMPC) works by randomly splitting data into little pieces and scattering it across multiple parties so that no one party can reconstruct the data of any other party. SMPC allows organisations to share data analysis without sharing confidential information. Trusted Execution Environments allow organisations to trust third parties, e.g. MEC or hyperscaler clouds, to run programs securely where their confidential data cannot be accessed, even by the hyperscalers.

In a 6G Revolution, PETs-as-a-service could support the need for greater privacy in the 6G use cases and those from the more comprehensive Network of Networks. PETs-as-a-service requires technical capabilities from UE to hyperscalers. For example, homomorphic encryption requires the UE to perform the encryption and the MEC or hyperscaler cloud to run on computers that can work on homomorphically encrypted data, which is more computationally intensive than regular computing. Federated Learning requires the UE to build local models and the hyperscaler to build a global model from the local models. Secure Multi-Party Computation could involve only UE or a combination of UE, MEC and hyperscalers. Trusted Execution Environments (TEEs) could be run on UE to enable service providers to trust the UE to run programs uncompromised, while MEC and hyperscalers can run TEEs so organisations can trust their data and programs run in the cloud are not compromised.

PETs-as-a-Service would bring sophisticated capabilities to allow organisations and individuals to liberate value from confidential data without losing privacy, e.g. optimising disease treatments. PETs-as-a-service would be enabled by 6G convergence as it requires capability across the whole information technology communications (ICT) ecosystem.

16.5.2.9 iPhone 6G and new device formats

When we compare the 3G iPhone, released in 2008, with the iPhone 15 Pro, launched in 2023, we realise there has been little change on the surface. With the advent of 6G technology, can we expect significant improvements? Or will 6G devices with the same impact as the original iPhone shift the direction of 6G? The most promising areas for change in cellular usage are smartphones and new devices associated with XR. For instance, the Apple iPhone 15 Pro and Vision Pro goggles already support stereoscopic video and 3D imaging.

Earlier in Chapter 15, we described our take on what could happen in an 'iPhone 6G' as an exemplary take on smartphones in 2030. The iPhone 6G will have massive, phased arrays in the mmWave bands, potentially reaching 300 GHz. Setting aside the apparent communication speeds, these antenna arrays could find uses in a much more comprehensive range of applications, which, in the term, could exploit the cellular deployment of MEC and Digital Twins. These applications could become available from an app store where developers outside the cellular industry could create them. Remember that access to the cellular array using its band would most probably be in the domain of the cellular operator. Still, it is in their interest to allow this. Otherwise, this could move to a potential 'Wi-Fi 9' mmWave band.

Earlier, we gave an example of using the 127 GHz band with 4096 antenna elements, which could realise an EIRP of over 3 kW in a smartphone. This could be used for several applications, such as scanning spectroscopy to identify allergens in food based on your health App profile and MEC, building 3D images of the environment as a new form of photography again using network-based computing or even seeing in-the-dark touches. The use of these frequencies would enable exact positioning applications. On the user interface, there could be gesture recognition within Accessibility for Sign language recognition with a new language model stored with the iPhone 6G. These are just some examples of what is possible to bring about a revolutionary change in the use of smartphones for 6G. Many apps could use the MEC and Digital Twins within the cellular infrastructure for consumer-type applications.

There are also likely to be new device formats for the first time, such as the Apple Vision Pro goggles, which will combine both Wi-Fi and cellular access. The emergence of 'Smart glasses' would be revolutionary. They will be available in both evolutionary and revolutionary scenarios of 6G. The primary use cases may differ between those access networks. Since these devices will combine other technologies, as in the case of the iPhone 6G, these could also create revolutionary apps optimised, particularly for the goggles format.

Within the concept of Connectivity for Everything, extending the smartphone as a hub for body sensors could have critical applications for health in much wider use than today, where this is nascent.

For 6G or Wi-Fi communications, this new smartphone could support data speeds of 35 Gbit/s or above in Information Petrol Stations, allowing massive download speeds of movies or other content. The mid band and Wi-Fi spectrum will allow typical download speeds of Gbit/s.

16.5.2.10 6G virtual satellite operators – global coverage

In many respects, satellite NTN access is like Wi-Fi in that it is an independent network to cellular access. It is integrated into mobile devices that allow the same application to be used without seamless handover. In the previous 6G evolution theme for Global coverage (refer to Section 16.5.1.9), we discussed the importance of achieving global coverage in 6G technology. This is to address the issue of the regional digital divide and provide coverage to areas not reached by cellular networks across vast areas of land or oceans. 3GPP is currently working on integrating this feature into 6G technology, which is an improvement from what is presently available in 5G. This could help extend cellular network coverage, as a T-Mobile pilot in the United States demonstrated.

However, it is important to note that this view is based on the assumption that 6G LEO technology is a natural extension of current cellular technology. However, a significant difference in 6G is that many use cases can be supported without direct integration into mobile networks and through non-MNOs. As a result, a new type of virtual operator could emerge – the Global Virtual Satellite Operators (GVSOs).

The key players in this field could be the makers of high-end smartphones and laptops/tablets, such as Apple and Samsung, who can become GVSOs, utilising the new generation of LEO satellites. Their advantage is that they can directly work with one of the LEOs and support only their frequency band in their high-end devices for global sales, bringing significant traffic to the chosen LEO operator. This does not have to be an LTE frequency band but the current bands used by the LEO. Apple has already done this for a satellite-based SOS text service; see Chapter 15, where people who wish to retain the service will probably be charged monthly after the first two years. This would bring in additional monthly revenues after the device sale for manufacturers, creating a more direct relationship with their customers. As the applications for this use are likely to be niche or infrequent, it is in many respects an ideal market for them to enable – they do not bear insignificant start-up costs in each country as an MVNO, and they only need to support one satellite radio in their devices to reach a global audience for their network. It still leaves the LEO operators to sell directly into markets, such as remote broadband services not reached by standard terrestrial network operators.

Technically, this usage model addresses the complexity of the satellite's frequency bands (see Chapter 10), which 3GPP must deal with both for direct cellular LTE use and other use cases with the LEOs using new or present LEO frequency bands. Not to mention all the handover issues as the satellites move across the earth in their orbits, covering different countries with different spectrum regimes. Equally, the LEO operators can deploy similar satellite antennas.

Other uses, such as controlling drones across vast areas of land or even seas, similarly do not require both networks; if they do, their application will support standard automatic roaming between networks.

This is an example of the revolutionary use of 6G NTN and a new form of 6G operator.

16.5.2.11 Contrasting evolution with revolution

Figures 16.9 and 16.10 summarise the 6G evolution and revolution paths, explaining the key elements in Table 16.4, while Figure 16.11 describes the benefits.

6G requires a revolution that is not based on new themes but rather logical extensions to the existing themes and architectures. These extensions likely happen over the decade. The most significant is to initially make 6G more open to collaborations with hyperscalers and explore joint funding for its implementation. Also, to finally embrace Wi-Fi as part of a 6G convergence based on the Intelligence layer. As many of the leading players in 6G will have both fixed and mobile networks, they will be able to see the merits of both mobile and Wi-Fi technologies as a portfolio. They are also committed to the rollout of FTTH or Cable enabling 10 Gbit/s access networks supporting the appropriate use of Wi-Fi in the home and small cells in industrial settings. Having both access and Intelligence within its network, such as MEC for low latency, should make it attractive to hyperscalers to use as wholesale-type services, as described in Table 16.4.

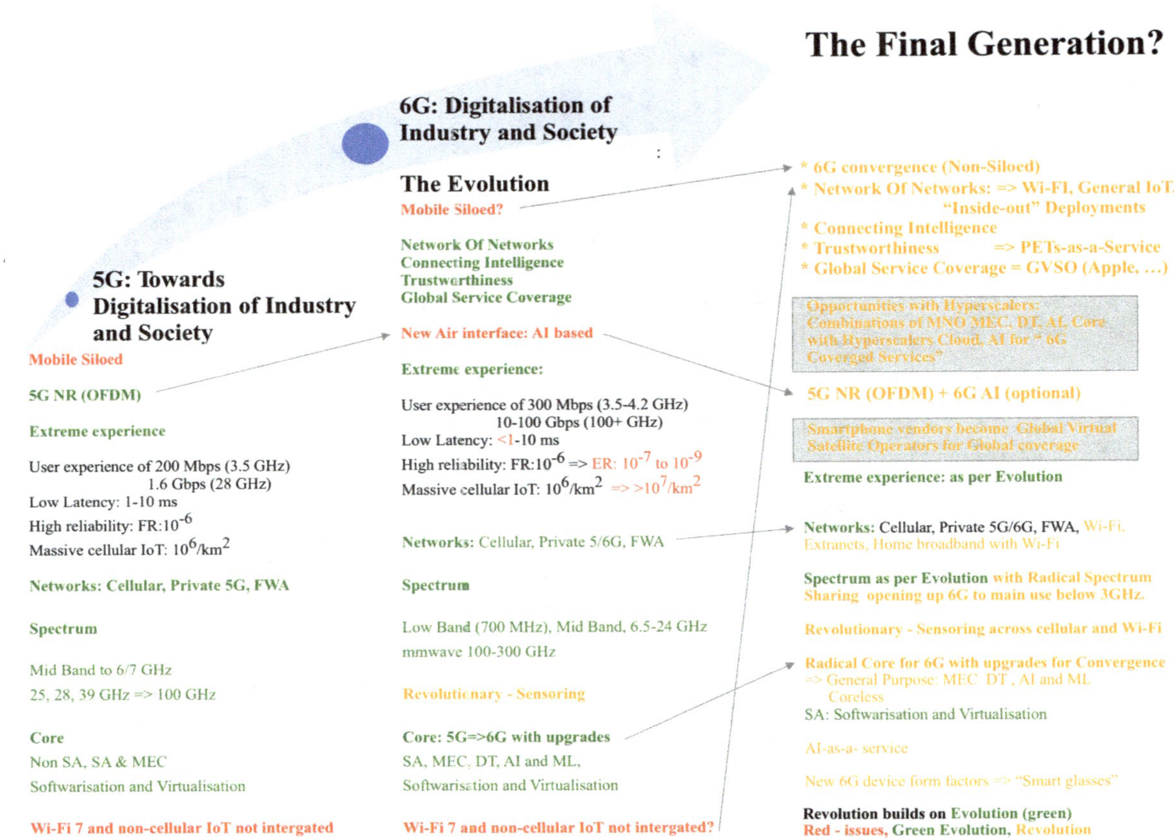

Figure 16.10 Summary of 6G revolution scenarios built on evolution

Table 16.4 6G evolution and revolution options, see Figures 16.9 and 16.10

5G	6G Evolution	6G Revolution
eMBB, URLLC, and mMTC	These Technologies will continue to evolve	These Technologies will continue to evolve
Mobile siloed	**Network of Networks:** Possibly following the 5G siloed approach	**6G Convergence:** Supporting Wi-Fi, non-cellular IoT, Extranets with 6G technologies: MEC, DT, AI, Core services Plus potential collaboration with hyperscalers.
	Connecting Intelligence: AI and ML inside to enhance performance	**6G Convergence:** external use of AI and ML for Apps
	Global Services Coverage with Satellites, extension of cellular to areas considered non-economic coverage, emergency services	**6G Convergence:** Plus, Global Services Coverage from third parties as part of smartphone partnerships
	Trustworthiness	**6G Convergence:** Wide-scale use of MNOs PETs for third-party Apps
No sensing	Sensing introduced across cellular bands ranges from 1 cm to approx. 1–2 m	Plus, extension to device Apps for others including spectrometry such as Food Allergies detection, see in fog/smoke 'touches' etc.
mmWave cm: up towards 100 GHz for communication only	Communications up to 100 Gbit/s with 100–300 GHz spectrum, higher speed FWA, use with MNO sensing for 1 cm resolution	New uses in non-communications applications such as radar, e.g. to see in dark, smoke or fog, food allergy detection in smartphones
Air interface	**New AI air-interface**	**Optional: with 5G option**
Core	Evolution of 5G Core	Radical coreness – similar to the Internet.
Mobile Edge Computing	Evolution of MEC	General purpose to support Network of Networks
Digital Twins	Digital twin for MNO services	External DT for Apps
Small cells	Small cells/femtocells	Wi-Fi 8 for 6G Home services and small cells for industrial/ office/
Spectrum: public cellular and private 5G spectrum focusing on 3.5 GHz and cm for canvas and Fixed Wireless Access.	Plus: More diverse: 700 MHz, 3.5–4.2 GHz, 8–24 GHz, 100–300 GHz. The eventual use of TV bands in the 2030s. Limited spectrum sharing	Radical spectrum sharing using Weak Signal Propagation reporting to avoid use of propagation models
Devices	New device formats for XR	New revolutionary apps based on device format for XR
Sustainability		Whole network monitoring

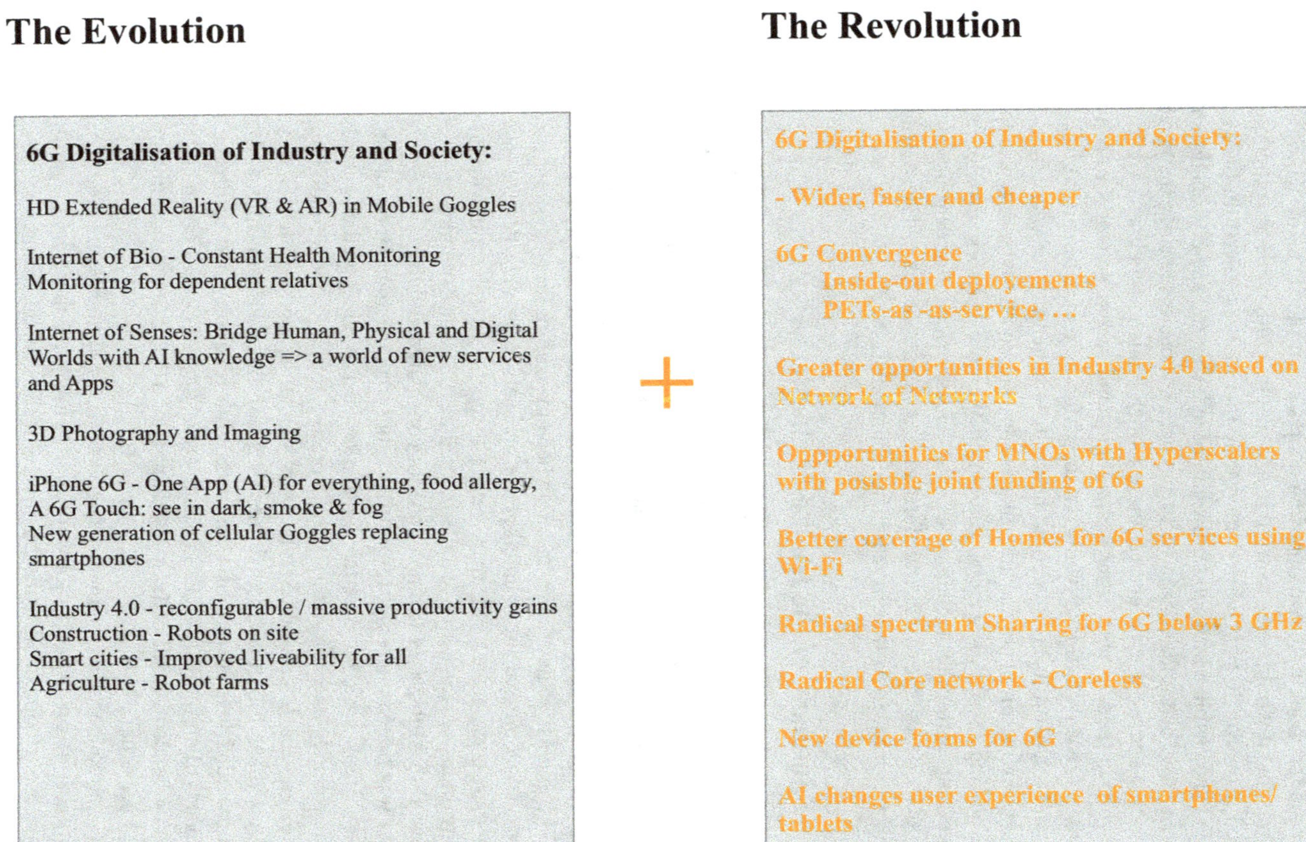

Figure 16.11 Contrasting evolution vs. revolution in 6G

16.6 And, finally

We began this book with two opposing quotes about the status of 6G in 2030. One view was that it would come to fruition as usual in 2030, while the other questioned its need. However, we aimed to demonstrate that there is a need for 6G to improve upon 5G and enable a broader digitisation of industry and society. 6G does not need to be ten times better than 5G in KPIs, but rather, it needs to bridge the gap between what is real and what is perceived and establish realistic timelines. Additionally, it must work in tandem with other technologies and access networks to fully realise its potential as a Network of Networks supported by its intelligence, global service reach and trustworthiness. 6G could be the first generation of wireless technology that does not require a new air interface on day one in 2030. Instead, the focus will be on making revolutionary changes within the core network, promoting greater interworking with other technologies. The widespread use of sensing and AI will make 6G one of the most transformative generations of wireless technology. However, it may arrive a few years later than usual.

16.7 Summary of the book and chapter

The digitalisation of society and industry is an inevitable trend that will help connect the physical, digital, and human worlds, bringing numerous benefits to society and the next stage of industrialisation, Industry 4.0. A crucial aspect of Industry 4.0 is the communications framework that allows the independent exchange of information, triggers actions, and controls the systems while ensuring that humans can interact within the systems through extended reality and haptic feedback for control. The intensive development of cyber-physical process systems is heading towards developing artificial intelligence, i.e. self-improving devices and objects, on an increasing scale. Yet, there needs to be a general framework for implementing Industry 4.0 with a detailed schedule. This also requires a 'Network of Networks' that embraces technologies not considered cellular.

5G was the first attempt to position cellular as the foundation of the communications framework for broader digitisation. It brought together various aspects such as robotics and automation, collaborative robots (cobots), cloud computing, the IoT, and the Internet of Services. Its standards, from 5G-to-5G Advanced in 2025, provided the fundamental elements for the communications framework. These included low latency of 1 ms, ultra-high reliability and resilience, average data speeds of around 100 Mbit/s in wide areas, and 10 Gbits/s in cm and mmWave frequency bands in small cells. It also supports IoT devices, connecting millions of devices per square km.

5G introduced new elements, such as MEC, digital twins, and a new Core network based on network softwarisation. However, there is a significant gap between expectations and reality, as many of the critical features required for the strategy are expected to arrive midway through the decade. Additionally, consumer appetite for even faster mobile broadband has diminished, with no applications demanding it. This has challenged the mobile operators' confidence not only in 5G but also in their commitment to 6G.

6G builds on the foundations of 5G and is vital for the broader digitalisation. It also has the potential to introduce revolutionary new technologies and bring new network-based applications and services, such as sensing, which, combined with the Digital twins, would allow digital models of the physical world for numerous applications. Yet, it is not the only access technology needed for this new world, as Wi-Fi, non-cellular IoT and fixed networks are equally required.

3GPP, through its coordination with other standards organisations, is attempting to address its role within the Internet, fixed networks and Wi-Fi. Yet, while you can point to visions such as the IMT 2030 view of cellular, there is no overall view of the communication framework for society and industry, which is also struggling to do this.

This is not meant to sound negative as, in many respects, there has been significant progress with tailored solutions in different sectors; hence, 6G associated with other technologies need to maintain flexibility.

> *There are industries where technological progress is realised faster and easier, as well as industries that are more difficult to implement new solutions (studies PwC, Deloitte, London, UK).*
>
> – *Prof. Bożena Gajdzik et al. [21]*

This book has taken an unusual approach to '6G – Evolution or revolution?' in that it started with analysing the needs of society and industry to identify the critical technology elements required. It is agnostic to whether these are fulfilled by cellular, Wi-Fi, other IoT standards or a combination of these. It also reviewed how and why the mobile industry has developed to consider radical approaches to the future, which may take time to be adopted but would have a profound effect. The wireless spectrum and its management and control are fundamental to this. The present spectrum staircase takes us to the same place of smaller cells, using progressively much higher frequencies where the use cases for cellular and Wi-Fi become indistinguishable. Yet the elements of the 6G network, such as MEC, digital twins, AI and Core, all implemented with softwarisation, are universal.

So, what have we learned in this book that describes a holistic approach to 6G?

(I) With 5G approaching the mid-point of its decade, the initial focus has been extending the consumer offering rather than on Industry 4.0. This is because of the late arrival of critical elements such as standalone cores in 2024 and other features in 5G Advanced. Not surprisingly, this has led to no significant increase in ARPU, and time is running out to demonstrate new business opportunities for a massive investment in 6G as early as 2030.

    5G has much work to do to demonstrate the required investment in 6G. Otherwise, it could have a slow start to its launch in 2030. However, its KPIs will take it on a long journey into new areas, as these are well-defined for 6G, too.

(II) Wi-Fi will also be involved in the digitalisation of industry and society, which is inevitable. It plans to embrace this within Wi-Fi 8 circa 2028, while Wi-Fi 9 will likely include mmWave above 100 GHz for extreme communications speeds.

(III) 6G is the evolutionary path from 5G where the EU Hexa-X project summaries it well:

> *The Hexa-X vision is developed to connect the physical, digital, and human worlds through future wireless technology and architectural research. The Hexa-X vision aims to create an x-enabler fabric of connected intelligence, networks of networks, sustainability, global service coverage, extreme experience, and trustworthiness [22]*

The present 6G 'directional vision' is merging where 5G's eMBB, URLLC, and mMTC technologies will further evolve. In contrast, AI and sensor convergence technologies will likely be applied to communications. The 6G network is set to grow significantly to learn and act autonomously. Its Core implementation could be based on the 5G Core used in 5G Advanced with SDN and NFV technologies.

(IV) Nevertheless, new revolutionary devices could emerge that dramatically change the dynamics for new services based on mmWave, Extended Reality, sensing, positioning, etc., enabling new opportunities for MEC, AI and its core network with Digital Twins.

Sensing is emerging as a new revolutionary 6G service which could be implemented across all the cellular bands, allowing different levels of accuracy. The ability to localise, track, and sense physical objects is the bridge connecting human, physical, and digital worlds. It also supports spectroscopic scanning for new smartphone Apps and pollution measurement in cities for improved liveability. Other positioning and sensing technologies, such as Wi-Fi, LIDAR, BLE, and UWB, have existed for several years. BLE, UWB and LIDAR have found their way into smartphones for developing applications, which could grow consumer interest in sensing and positioning technology.

(V) 6G needs to avoid the expectation gap created by 5G.

We have reviewed the IMT 2030 report on technology trends and services for 2030 and key white papers on 6G, which are seminal contributions. Many of the proposed KPIs for 6G assume Hologram Displays, disaggregation of control loops, and network control of cars. We remain sceptical about all three for the reasons explained. We presented our view of 6G KPIs without these, allowing a lower implementation cost.

In particular, our review of application requirements supports the original 5G use cases, and we find little to support a latency requirement of less than 1 ms, nor application speeds significantly more significant than 5G. It is always essential to avoid the law of diminishing returns by building new specifications for outliner cases. For example, for control loops, latencies of 0.25 ms or less are already addressed by Ethernet, and the need for wireless is only for new builds for a cableless factory.

(VI) The legal definition of wireless has been increased from 275 GHz to 3 THz in recognition that we can explore this new region with oscillators, power amplifiers, antennas, detection, and signal processing. Yet, our focus on using this valuable resource has been below 10 GHz for over a hundred years.

Meanwhile, 5G opened up the cm and mmWave bands for use with circa 28 GHz and mmWave towards 100 GHz. 6G is set to explore 7–24 GHz cm bands and the 100–300 GHz region during its lifetime to enable 100+ Gbit/s systems.

mmWave, based on new semiconductor material systems with phased arrays, challenges the commonly held belief that it is the worst to operate at higher frequencies. The feature sizes of antennas in the sub-Tera Hertz region will allow SOC of all base station electronics.

mmWave has the potential to support massive bandwidths for FWA, but its ability to compete economically as an overbuilder to existing broadband networks such as cable or FTTH is limited. Nevertheless, it could enable revolutionary new services for cellular-based devices.

(VII) One of the biggest challenges for 6G remains spectrum, especially in the bands below 5 GHz. However, with reform and spectrum sharing, there is sufficient spectrum below 4.8 GHz to enable virtually all 6G usage in the wide area with these bands.

Our proposal for CBRS 2.0 with weak signal propagation reporting has the potential to allow enough wide-area spectrum, driving up the utilisation of spectrum below 3 GHz from circa 15% to 80%. It could provide 1.9 GHz of additional 6G spectrum. It could be the basis of a spectrum-sharing system extended across all bands from Low to High.

(VIII) The indoor wireless environment is one of the biggest challenges for 6G. While industrial settings/offices will more easily allow small cells, particularly in large areas, using parts of 8–24 GHz spectrum with appropriate backhaul connections, this is different for the home. Wi-Fi in 7/8 is likely to be the critical technology here.

(IX) Extended reality using higher frame rates and higher resolution inside homes and offices could give cellular operators a foothold for the first time inside and start generating a new inside-out deployment strategy for consumers based on dual-band small cells.

Yet, with the emergence of converged operators offering both fixed and mobile services to their customers, they may find that Wi-Fi provides them with new opportunities in the home for their services. When combined with new network assets such as AI, MEC and DT, this supports a single-core network for both fixed and mobile. Here, there is potential for collaboration with hyperscalers to bring a new form of convergence for the first time.

(X) The objective of '100% coverage' using LEO Satellites only directly works outside with a clear view of the horizon.

The use of this technology to help address the digital divide is questionable. The LEO Satellites will likely have high monthly access costs and require expensive smartphones.

Nevertheless, this provides new opportunities for controlling drones, emergency access, and cellular extension to areas of low-density usage.

Smartphone manufacturers like Apple and Samsung could enter satellite space as GVSO. They can support a single LEO band with one of the LEO satellite operators to offer various services, including Voice, text, and broadband data.

(XI) Sustainability is of paramount importance, of course. The proposal within 6G to monitor and control use cases to avoid unsustainable development is exciting but appears to need an enforcement regime.

For example, the proposal within 6G to extend IoT to one trillion devices for Industry 4.0 would put its sustainability target into question; therefore, energy harvesting is proposed. While this has a clear role in all future wireless systems, it is questionable if this can provide sufficient power for all IoT use cases and if it would threaten the sustainability role.

Building all wireless networks where data is consumed can help address the energy consumption problems of a macro-cellular-based 'outside-in' approach to coverage. The support of dual-band radios that cover the external wide-area frequencies and the high-band frequencies provides the basis for the evolution of inside-out coverage, explained in Chapter 9.

(XII) We have proposed radical core, edge and PETs for security and privacy. 5G is a significant step forward in the softwarisation of networks by adopting a cloud-native paradigm for the 5G core network. 6G should extend this to make all network functions cloud-native to improve 6G service reliability and scalability. This requires modifying all 5G protocols to be web-scale and stateless. Edge Cloud will be fundamental to lowering transaction latencies and creating opportunities for network operators to sell new services; however, this requires global Edge Cloud standards to build scale and enable interworking between network operators and hyperscalers. Industry 4.0 and 6G networks will generate Zettabytes of confidential and private information that, if shared, could create industrial and network efficiencies. If 6G systems implemented Privacy Enhancing Technologies, this value could be unlocked, but the privacy and confidentiality of the data owners would be safe. It could be offered as a 6G service.

(XIII) One of the critical questions for 6G remains whether it needs a new air interface and for what. NGMN has suggested it should be optional. While each new generation of mobile technology brings a new air interface, OFDM has taken us very close to the Shannon limit over the past five generations. An 'AI air interface' currently promises only 20%–30% improvements in spectral efficiency. Keeping the present 5G air interface would allow the wide area cellular networks to be essentially unchanged for 6G as it supports frequencies bands including cm and mmWave while the spectrum dichotomy implies that the 7–24 GHz and above 100 GHz are small cell-based with campus deployments or industrial deployments.

The focus would then be on the radical core and Edge and interworking with all relevant network technologies to provide the flexibility to meet the current ill-defined timescale and scope of society and industry 4.0.

(XIV) The 6G themes of Network of Networks, Connecting Intelligence, Trustworthiness, and Global service coverage could form the basis of a new form of convergence and enable collaboration with hyperscalers to help fund 6G. Key to the future of 6G will be the choice to use the Network of Networks to embrace other access networks and support these with its Intelligence.

So, what is the future of 6G? Evolution or revolution? It will be a hybrid of both, with the realisation that it will include Wi-Fi, other IoT standards and a fixed network. This book has focused as much on what not to do as on what it could do. It has tried to give a broad perspective on the roles of all telecommunication and computer systems in delivering the broad digitalisation of industry and society, which is inevitable. The only question is the timing. The technologies we described in the book will have a life on their own and come into play regardless of their use in 6G.

References

[1] European Commission. "Europe Scales Up 6G Research Investments and Selects 35 New Projects Worth €250 Million." [Online]. 2022. Available from: https://digital-strategy.ec.europa.eu/en/news/europe-scales-6g-research-investments-and-selects-35-new-projects-worth-eu250-million.

[2] Wikipedia. "Fifth Generation Computer Systems." Edited May 2024 [Online]. Available from: https://en.wikipedia.org/wiki/Fifth_Generation_Computer_Systems.

[3] Gross A. "Beyond 5G: What Would 6G Look Like – and Do We Need It?" *Financial Times* [Online]. 2022. Available from: https://www.ft.com/content/74699854-3590-4245-a052-487bdda33a07.

[4] Wikipedia. "History of Nokia." Edited June 2024 [Online]. Available from: https://en.wikipedia.org/wiki/History_of_Nokia#cite_note-125.
[5] 3GPP. "The 5G Standard Release 18." [Online]. 2024. Available from: https://www.3gpp.org/specifications-technologies/releases/release-18.
[6] RCR Wireless News. "Private 5G will 'not reach potential' – Even as Spending Jumps 800% to $9bn by 2028." [Online]. 2024 [cited 23 January 2024]. Available from: https://www.rcrwireless.com/20240123/private-5g/private-5g-will-not-reach-potential-even-as-spending-jumps-800-to-9bn-by-2028.
[7] Mehdawi M, Fanan A, and Ammar M. "Spectrum Occupancy Survey in HULL-UK for Cognitive Radio Applications: Measurement & Analysis." *International Journal of Scientific & Technology Research*. 2013; 2(4): 231–236.
[8] Ericsson. "Ericsson Mobility Report November 2023." [Online]. 2023 [cited 23 November 2023]. Available from: https://www.ericsson.com/en/reports-and-papers/mobility-report/reports/November-2023.
[9] "The PLMN Problem for Private LTE/5G." [Online]. 2021. Available from: https://nickvsnetworking.com/the-plmn-problem-for-private-lte-5g/.
[10] Group Dell'Oro. "Mobile Core Market Stagnant Due to Lack of 5G Standalone Deployments, According to Dell'Oro Group." [Online]. 2024 [cited 23 January 2024]. Available from: https://www.delloro.com/news/mobile-core-market-stagnant-due-to-lack-of-5g-standalone-deployments/.
[11] UCTel. "5 Private 5G Use Cases for Commercial & Industrial Environments. " [Online]. 2024 [cited 4 March 2024]. Available from: https://www.uctel.co.uk/blog/wi-fi-vs-5g-the-difference-private-5g-brings-to-industrial-and-commercial-environments.
[12] Future Market Insights. "Privacy Enhancing Technology Market Outlook (2023 to 2033)." [Online]. 2023. Available from: https://www.futuremarketinsights.com/reports/privacy-enhancing-technology-market.
[13] Jobs S. [Online]. No date cited. Available from: https://www.azquotes.com/quote/1059248.
[14] Blanche PA. "Holography, and the Future of 3D Display." [Online]. 2021. Available from: https://www.light-am.com/article/doi/10.37188/lam.2021.028.
[15] RCR Wireless News. "Total IoT Connections to Surge, but Cellular IoT to Lose Share with 2G/3G Switch-off." [Online]. 2023 [cited 23 June 2023]. Available from: https://www.rcrwireless.com/20230622/internet-of-things-4/total-iot-connections-to-surge-but-cellular-iot-to-lose-share-with-2g-3g-switch-off.
[16] Ericsson. "Ericsson Technology Review. Extended Reality." [Online]. 2021 [cited 24 August 2021]. Available from: https://www.ericsson.com/4a492d/assets/local/reports-papers/ericsson-technology-review/docs/2021/xr-and-5g-extended-reality-at-scale-with-time-critical-communication.pdf.
[17] Wikipedia. "Apple Vision Pro." [Online]. 2024. Available from: https://en.wikipedia.org/wiki/Apple_Vision_Pro.

[18] Mason A. "The Apple Vision Pro's Mixed Reality Headset Makes Good Wi-Fi Connectivity More Important than Ever." [Online]. 2023 [cited 23 June 2023]. Available from: https://www.analysysmason.com/research/content/articles/apple-vision-pro-launch-rdvs0-rdmb0/.

[19] Uusitalo MA, Rugeland P, Boldi M, et al. "6G Vision, Value, Use Cases and Technologies from European 6G Flagship Project Hexa-X." *IEEE Access*. 2021; 9: 160004–160020, doi:10.1109/ACCESS.2021.3130030.

[20] Nokia. "6G AI-Native Air Interface." [Online]. 2023. Available from: https://www.bell-labs.com/research-innovation/what-is-6g/6g-technologies/ai-native-air-interface#gref.

[21] Gajdzik B, Grabowska S, and Saniuk S. "A Theoretical Framework for Industry 4.0 and Its Implementation with Selected Practical Schedules." *Energies*. 2021; 14(4): 940, https://doi.org/10.3390/en14040940.

[22] EU. "A Flagship for B5G/6G Vision and Intelligent Fabric of Technology Enablers Connecting Human, Physical, and Digital Worlds." [Online]. 2024 [cited 24 February 2024]. Available from: https://cordis.europa.eu/project/id/101015956.

Glossary

3GPP	3rd Generation Partnership Project
5G NR	5th Generation New Radio
5G-A	5G Advanced
5G-AKA	5G Authentication and Key Agreement
AAA	Authentication, Authorisation, and Accounting
ACK	Acknowledgement
ADC	Analogue-to-Digital Converter
ADSL	Asymmetric Digital Subscriber Line
AF	Application Function
AGV	Autonomous Guided Vehicle
AI	Artificial Intelligence
AMF	Access and Mobility Management Function
AMPDU	Aggregated MAC Protocol Data Units
AN	Autonomic Networking
ANDSF	Access Network Discovery and Selection Function
ANN	Artificial Neural Network
ANQP	Access Network Query Protocol
ANX	Automotive Network Exchange
AP	Access Point
API	Application Programming Interface
AQM	Active Queue Management
AR	Augmented Reality
ARIB	Association of Radio Industries and Businesses
ARPF	Authentication Credential Repository Function
ARPU	Average Revenue Per User
ASIC	Application-Specific Integrated Circuit
ATSSS	Access Traffic Steering, Switching and Splitting
AUSF	Authentication Server Function
AV	Augmented Video
AWS	Amazon Web Services
BBF	Broadband Forum

BBU	Base-Band Unit
BER	Bit Error Rate
BLE	Bluetooth Low Energy
BNG	Broadband Network Gateway
BT	British Telecommunications plc
CA	Carrier Aggregation
CA	Certificate Authority
CAGR	Compound Annual Growth Rate
CAPEX	CAPital EXpenditure
CBF	Coordinated Beam Forming
CBRS	Citizens Broadband Radio Service (FCC Part 96 service for 3550–3700 MHz band)
CBSD	Citizens Broadband Radio Service Device (eNodeB or base station in cellular context)
CCTV	Closed Circuit TV
CDN	Content Delivery Network
CEN	Comité Européen de Normalisation
CFN	Compute First Networking
CI/CD	Continuous Integration/Continuous Development
CMOS	Complementary Metal–Oxide–Semiconductor
CMTS	Cable Modem Termination System
COIN	Computing in the Network
COINRG	(IETF) Computing in the Network Research Group
CPRI	Common Public Radio Interface
CPU	Central Processing Unit
C-RAN	Centralised RAN
CSMA	Carrier-Sense Multiple Access
CSMA/CA	Carrier-Sense Multiple Access with Collision Avoidance
CSMA/CD	Carrier-Sense Multiple Access with Collision Detection
CSP	Communications Service Provider
CU	Centralised Unit
C-V2X	Cellular Vehicle to Everything
DAC	Digital to Analogue Converter
DC	Dual Connectivity
DHCP	Dynamic Host Configuration Protocol
DL	Down Link
DMTF	Distributed Management Task Force
DNS	Domain Name System
DOCSIS	Data Over Cable Service Interface Specification

DOT	(USA) Department of Transport
DSL	Digital Subscriber Loop
DSLAM	Digital Subscriber Line Access Multiplexer
DT	Digital Twin
DTaaS	Digital Twin as a Service
DTT	Digital Terrestrial TV
DU	Distributed Unit
EAP	Extensible Authentication Protocol
EAP-5G	Extensible Authentication Protocol for 5G
eBPF	Enhanced Berkeley Packet Filter
EC-GSM	Extended Coverage GSM IoT
ECN	Explicit Congestion Notification
eCPRI	Enhanced Common Public Radio Interface
EHT	Extremely High Throughput
eICIC	Enhanced InterCell Interference Coordination
EIRP	Equivalent Isotropic Radiated Power
eMBB	Enhanced Mobile Broadband
eMTC	Enhanced Machine-Type Communication
ES	End Station
eSIM	Embedded Subscriber Identity Module
ETSI	European Telecommunications Standards Institute
EU	European Union
eUICC	Embedded Universal Integrated Circuit Card
FCC	(USA) Federal Communications Commission
FD	Full Duplex
FDD	Frequency Division Duplex
FDMA	Frequency Division Multiple Access
FEC	Forward Error Correction
FL	Federated Learning
FLOPS	Floating Point Operations Per Second
F_{max}	Frequency where the power gain falls to unity, while f_t is the "transition frequency"
FMC	Fixed Mobile Convergence
FMCA	Fixed Mobile Convergence Alliance
FMIF	Fixed Mobile Interworking Function
FN-RG	Fixed-Network Residential Gateway
FPGA	Field Programmable Gate Array
FSPL	Free Space Path Loss

FSS	Fixed Satellite Service
f_t	Transition frequency – highest switching frequency for transistor
FTTC	Fibre to the Curb
FTTH	Fibre to the Home
FTTX	Fibre to the Home/Premise/Cabinet
FWA	Fixed Wireless Access
GaAS	Gallium Arsenide
GaN	Gallium Nitride
GAA	General Authorized Access license in CBRS band
GAN	Generative Adversarial Networks
GDPR	General Data Protection Regulation
GEO	Geostationary (satellites)
gNB	gNodeB (5G)
GPS	Global Positioning System
GPU	Graphics Processing Unit
GRE	Generic Routing Encapsulation
GSM	Global System for Mobile Communications (ETSI)
GSMA	GSM Association
GSMA OPG	GSMA Operator Platform Group
GTP	Generic Tunnelling Protocol
HAP	High-Altitude Platform
HARQ	Hybrid Automatic Repeat Request
HBT	Heterojunction Transistors
HD	High Definition
HD	Half Duplex
HE	Homomorphic Encryption
HEMT	High-Electron Mobility Transistors
HEVC	High-Efficiency Video Coding
HGV	Heavy Goods Vehicle
High Band	>6 GHz
HITL	Human in the Loop
HTTP	Hypertext Transfer Protocol
ICT	Information Communications Technology
IEC	International Electrotechnical Commission
IEEE	Institute of Electrical and Electronic Engineers
IETF	Internet Engineering Task Force
IF	Intermediate Frequency

iFFT	Inverse Fast Fourier Transform
IIoT	Industrial Internet of Things
IMS	IP Multi-media System
IMT	International Mobile Telecommunications
InP	Indium Phosphide
IoT	Internet of Things
IPoE	IP over Ethernet
ISG	(ETSI) Industry Specification Group
ISM	Industrial, Scientific and Medical
ISO	International Standards Organisation
ISP	Internet Service Provider
IT	Information Technology
ITU	International Telecommunications Union
ITU-T	International Telecommunications Union – Telecommunications
JCAS	Joint Communications and Sensing
JOTS NHIB	Joint Operator Technical Specifications for Neutral Host In-Building
JVM	Java Virtual Machines
KPI	Key Performance Indicator
L4S	Low Latency, Low Loss, and Scalable Throughput
LAA	Licensed-Assisted Access
LAN	Local Area Network
LBO	Local Break Out
LEO	Low Earth Orbit
LiDAR	Light Detection and Ranging
LISP	Locator/ID Separation Protocol
LLM	Large Language Model
LNA	Low Noise Amplifier
LNB	Low Noise Block
LoRaWAN	Long-Range Wide Area Network
LOS	Line of Sight
Low Band	Up to 1 GHz
LPWAN	Low Power Wide Area Network
LSA	Licensed Shared Access
LTE	Long-Term Evolution
LTE-M	Long-Term Evolution Machine Type Communication
LTE-NB	Long-Term Evolution – Narrow Band
MAC	Medium Access Control Layer

MANO	Management and Network Orchestration
MAPL	Maximum Allowable Path Loss
MARL	Multi-Agent Reinforcement Learning
MBP	Measurement-based Protection
MCC	Mobile Country Code
MEC	Multi-access Edge Computing
MEC	Mobile Edge Computing
MEO	Medium Earth Orbit
METIS	Mobile and Wireless Communications Enablers for Twenty-twenty (2020) Information Society
Mid Band	1–6 GHz
MIMO	Multiple Input and Multiple Output
MIT	Massachusetts Institute of Technology
ML	Machine Learning
MLO	Multi-Link Operation
mMTC	Massive Machine Type Communications
MNC	Mobile Network Code
MNC	Multi-National Corporation
MNO	Mobile Network Operator
MOCN	Multi-Operator Core Network
MORAN	Multi-Operator Radio Access Network
MPLS	Multi-Protocol Label Switching
MPTCP	Multipath Transmission Control Protocol
mTLS	Mutual Transport Layer Security
MUSIM	Multiple SIM
MVNO	Mobile Virtual Network Operator
N3IWF	Non-3GPP Interworking Function
N5CW	Not 5G Capable over WLAN
NACK	Negative Acknowledgement
NAS	Non-Access Stratum
NB-IoT	Narrowband IoT
NEF	Network Exposure Function
NF	Noise Figure
NF	Network Function
NFC	Near Field Communications
NFV	Network Functions Virtualisation
NGMN	Next Generation Mobile Networks Alliance
NHS	(UK) National Health Service

NICT	Japanese National Institute of Information and Communication Technologies
NIST	(USA) National Institute of Standards and Technology
NLOS	Non-Line of Sight
NPN	Non-Public Network
NR	(5G) New Radio
NRF	Network Repository Function
NR-U	New Radio Unlicensed
NSA	Non-Stand-Alone (core)
NSSF	Network Slice Selection Function
NTN	Non-Terrestrial Network
NTN-IoT	Non-Terrestrial Network Internet of Things
NTT	Nippon Telegraph and Telephone
OCP	Open Compute Project
ODM	Original Design Manufacturer
OEM	Original Equipment Manufacturer
OFCOM	(UK) Office of Communications
OFDM	Orthogonal Frequency Division Multiplexing
OFDMA	Orthogonal Frequency Division Multiple Access
OLT	Optical Line Terminator
ONF	Open Networking Foundation
OPEX	Operational Expenditure
OPS	Optical Packet Switching
OSI	Open Systems Interconnection
OSS	Operational Support System
OT	Operational Technology
OTT	Over-the-Top
PAL	Priority Access License in CBRS Band
PCF	Policy Control Function
PDCP	Packet Data Convergence Protocol
PET	Privacy-Enhancing Technology
PETS	Privacy-Enhancing Technologies
PHY	Physical Layer
PIN	Personal Identification Number
PKI	Public Key Infrastructure
PLC	Programmable Logic Controller
PLMN	Public Land Mobile Network
PMF	Performance Management Function

PNI-NPN	Public Network Integrated Non-Public Networks
PON	Passive Optical Network
PPPoE	Point to Point Protocol over Ethernet
Psat	Peak Output Power P_s
PSTN	Public Switched Telephone Network
QAM	Quadrature Amplitude Modulation
QoS	Quality of Service
QPSK	Quadrature Phase Shift Keying
QUIC	Quick UDP Internet Connections
RACH	Random Access Channel
RADAR	radio Detection and Ranging
RADIUS	Remote Authentication Dial-in Service
RADsec	Secure Radius
RAN	Radio Access Network
RBAC	Role based Access Control
REST	Representation State Transfer
RF	Radio Frequency
RFID	Radio Frequency Identification
RIC	Radio Intelligence Controller
RL	Reinforcement Learning
RSRP	Reference Signal Received Power
RSSI	Received Signal Strength Indication (in dBm)
RTCP	RTP Control Protocol
RTP	Real-time Transport Protocol
RTT	Round Trip Time
RU	Resource Unit
SA	Stand-Alone (core)
SA-Core	Stand-Alone Core
SAS	Spectrum Access System
SBA	Service-based Architecture
SCP	Service Communications Proxy
SDR	Software-defined Receiver
SDN	Software-defined Network
SDO	Standards Developing Organisation
SD-RAN	Software-defined Radio Access Network
SDWAN	Software-defined Wide Area Network
SEAF	SEcurity Anchor Function
SEPP	Security Edge Protection Proxy

SGI	Short Guard Interval
SIM	Subscriber Identity Module
SINR	Signal to Interference and Noise Ratio
SIP	Session Initiation Protocol
SLA	Service Level Agreement
SME	Small-Medium Enterprise
SMF	Session Management Function
SMPCR	Secure Multi-Party Computation
SMS	Short Message Service
SNR	Signal-to-Noise Ratio (in dB)
SNPN	Standalone Non-Public Network
SoC	System on (a) Chip
SSID	Service Set IDentifier
STA	Station (in WLANs)
SUCI	Subscription Concealed Identifier
SUPI	Subscription Permanent Identifier
TCP	Transmission Control Protocol
TDD	Time Division Duplex
TDMA	Time Division Multiple Access
TEE	Trusted Execution Environment
TLS	Transport Layer Security
TMT	Telecom, Media and Technology
TN	Terrestrial Network
TNGF	Trusted Non-3GPP Gateway Function
TVWS	TV White Space
TWIF	Trusted WLAN Interworking Function
UAV	Uncrewed Aerial Vehicle
UDM	Unified Data Management
UE	User Equipment
UHR	Ultra-High Reliability
UL	Up Link
UPF	User Plane Function
URLLC	Ultra-Reliable and Low Latency Communication
URSP	User Equipment Route Selection Policy
USIM	Universal Subscriber Identity Module
V2V	Vehicle to Vehicle
V2X	Vehicle to Infrastructure and Vehicles
Vo5G	Voice over 5G

VoIP	Voice over IP
VoLTE	Voice over LTE
VoNR	Voice over New Radio
VoWi-FI	Voice over Wi-Fi
VPN	Virtual Private Network
VR	Virtual Reality
WAG	Wi-Fi Access Gateway
W-AGF	Wired Access Gateway Function
WAIF	Who Am I Function
WAN	Wide Area Network
WBA	Wireless Broadband Alliance
WiMAX	Worldwide Interoperability for Microwave Access
WLAN	Wireless Local Area Network
WPA	Wi-Fi Protected Access
WSPR	Weak Signal Propagation Reporting
XAI	Explainable Artificial Intelligence
XR	Augmented or Virtual Reality
XR	eXtended Reality
ZTA	Zero Trust Architecture

Index

www.ingramcontent.com/pod-product-compliance
Lightning Source LLC
LaVergne TN
LVHW011537200926
848484LV00001B/277